# Lecture Notes in Networks and Systems

Volume 535

AF803428

The series "Lecture Notes in Networks and Systems" publishes the latest developments in Networks and Systems—quickly, informally and with high quality. Original research reported in proceedings and post-proceedings represents the core of LNNS.

Volumes published in LNNS embrace all aspects and subfields of, as well as new challenges in, Networks and Systems.

The series contains proceedings and edited volumes in systems and networks, spanning the areas of Cyber-Physical Systems, Autonomous Systems, Sensor Networks, Control Systems, Energy Systems, Automotive Systems, Biological Systems, Vehicular Networking and Connected Vehicles, Aerospace Systems, Automation, Manufacturing, Smart Grids, Nonlinear Systems, Power Systems, Robotics, Social Systems, Economic Systems and other. Of particular value to both the contributors and the readership are the short publication timeframe and the world-wide distribution and exposure which enable both a wide and rapid dissemination of research output.

The series covers the theory, applications, and perspectives on the state of the art and future developments relevant to systems and networks, decision making, control, complex processes and related areas, as embedded in the fields of interdisciplinary and applied sciences, engineering, computer science, physics, economics, social, and life sciences, as well as the paradigms and methodologies behind them.

Indexed by SCOPUS, INSPEC, WTI Frankfurt eG, zbMATH, SCImago.

All books published in the series are submitted for consideration in Web of Science.

For proposals from Asia please contact Aninda Bose (aninda.bose@springer.com).

Samarjeet Borah · Tapan K. Gandhi ·
Vincenzo Piuri
Editors

# Advanced Computational and Communication Paradigms

## Proceedings of ICACCP 2023

Springer

*Editors*
Samarjeet Borah
Department of Computer Applications
Sikkim Manipal Institute of Technology
Rangpo, Sikkim, India

Tapan K. Gandhi
Department of Electrical Engineering
Indian Institutes of Technology Delhi
Delhi, India

Vincenzo Piuri
Department of Computer Science
University of Milan
Milan, Italy

ISSN 2367-3370 ISSN 2367-3389 (electronic)
Lecture Notes in Networks and Systems
ISBN 978-981-99-4283-1 ISBN 978-981-99-4284-8 (eBook)
https://doi.org/10.1007/978-981-99-4284-8



This Springer imprint is published by the registered company Springer Nature Singapore Pte Ltd.
The registered company address is: 152 Beach Road, #21-01/04 Gateway East, Singapore 189721, Singapore

# Preface

Computational and communication paradigms play crucial roles in the field of computer science, shaping the way systems are designed and developed to solve complex problems. These paradigms provide frameworks and approaches that guide the organization and flow of computational processes and the exchange of information. Computational paradigms define the models and methods used to structure and perform computations. They determine how problems are formulated, algorithms are designed, and computations are performed. Some commonly accepted computational paradigms include—imperative paradigm, which focuses on explicit instructions that specify how a computation should be performed, while the declarative paradigm focuses on what should be computed rather than how it is computed. The object-oriented paradigm, on the other hand, organizes computations based on objects that encapsulate both data and methods for processing that data, and parallel and distributed paradigms focus on using multiple processing units or computers to solve problems more efficiently.

Communication paradigms define how information is exchanged and shared between components, processes, or systems. They specify protocols, interfaces, and mechanisms for data transmission and interaction. Some important communication paradigms are:—synchronous communication, in which the sender and receiver must be available at the same time. In turn, asynchronous communication allows independent processing and message transmission between sender and receiver. The publish-subscribe paradigm involves the decoupled exchange of messages between publishers and subscribers. Remote procedure call paradigm, which allows a client program to call procedures or methods on a remote server as if they were local.

It is observed that these paradigms are not mutually exclusive. In many cases, more than one paradigm is combined to address different aspects of computational and communication challenges.

In view of the importance of the research domain, the fourth International Conference on Advanced Computational and Communication Paradigms has been organized by the Department of Computer Science and Engineering, Sikkim Manipal Institute of Technology during 16–18 February 2023. The conference has witnessed responses from the researchers from diverse geographical locations. The papers included in this volume are categorized into four different sections based on the domain of applications such as—Soft Computing and Machine Learning, Biomedical Informatics, Image and Video Processing, and Communication Techniques. This is a rough categorization based on the contents.

The first section—soft computing and machine learning contains nineteen research works. The research works include studies on optimization algorithms, predictive models, neural networks, deep and reinforcement learning approaches, application of association rule mining algorithms, and natural language processing.

The second section of this volume includes eight works in total on biomedical informatics. Authors have used various machine learning and other computational methods for breast cancer detection, gait recognition, Parkinson's disease detection, COVID-19 lung patches segmentation, leukaemia detection, classification of diabetic retinopathy, tumour segmentation from CT images, and medicinal plants classification.

A total of eleven research works has been reported from the domain of image and video processing. These research works include real-time video stabilization framework for consumer camera, fine-grained recognition of twins, IoT based landslide monitoring system, change detection in land use and land cover, shadow removal in outdoor video sequences, segmentation of lymphatic filariasis parasite images, fish classification from underwater images, facial images and EEG signal, image cryptography, CT image super-resolution, and co-registration of the digitized sensor information and the brain MRI scan.

Finally, the communication techniques section includes six papers on LiDAR sensor, proportional integral derivative controlled phase-locked loop for communication systems, performance analysis of routing protocols in wireless sensor networks, design, and implementation of third and fourth order phase lock loop, IoT-based smart fisheries and optimal frequency control in a microgrid.

This volume mainly contains the ongoing work and research results of various universities, research laboratories, and institutions from different geographical areas. The editors hope that the results will ignite the fellow researchers and lead to the development of sought-after and thriving products. Papers included in the volume demonstrate the use of various advanced computational and communication paradigms to resolve important and real-life problems. The editors of this volume express their sincere gratitude to all the authors, reviewers, and especially to Springer for being this volume possible.

Rangpo, India — Samarjeet Borah
Delhi, India — Tapan K. Gandhi
Milan, Italy — Vincenzo Piuri

# Contents

**Soft Computing and Machine Learning**

**Performance Analysis of Clustering Using Modified Grey Wolf Optimization** ........ 3
B. M. Ahamed Shafeeq and Zahid Ahmed Ansari

**Designing of Advanced Machine Learning-Based Predictive Model for Maternal Mortality** ........ 15
Gaidinlung Kamei and Indrajit Banerjee

**Multi-criteria-Based Page Ranking Using Metaheuristic Swarm Optimization** ........ 31
Santosh Kumar and Sanjai Mohan Verma

***k*-NN-SSc: An Effective Similarity Score for *k*-NN Classifier** ........ 43
Robindro Singh Khumukcham, Linthoingambi Takhellambam, Boby Clinton Urikhimbam, Ranjan Yambem, and Nazrul Hoque

**Password Management System Using Blockchain** ........ 55
Rebecca Shruti Soren, Sandeep Gurung, and Dipendra Gurung

**Deep Reinforcement Learning Algorithms for D2D Communications: A Comparative Analysis** ........ 75
Devarani Devi Ningombam

**AI-Enabled IDS for Identifying Malicious Traffics in Cloud Systems** ... 85
Soham Biswas, Chandan Kumar, Suman Som, Md. Sarfaraj Alam Ansari, and Mahesh Chandra Govil

**NITSDN: Development of SDN Dataset for ML-Based Intrusion Detection System** ........ 99
Bipal Khanal, Chandan Kumar, and Md. Sarfaraj Alam Ansari

**Association Rule Mining in Distributed Environment: A Survey** ....... 113
Bhaswati Baishya, Shafiul Alom Ahmed, and Bhabesh Nath

**Performance Analysis of the AdaGrad Family of Algorithms** .......... 121
Abhishek Bagepalli and Sanjay Singh

**Time-Related Natural Language Query Handling to Extract Autism Spectrum Disorder Information from Temporal Database** ...... 133
Prasenjit Mukherjee, Chandrima Chakraborty, Nandan Banerji, Kailash Pati Mandal, and Baisakhi Chakraborty

**Emotion-Based Music Recommendation** ............................ 147
Akshata Kumble, Sriveda Medatati, and Ashwini Bhatt

**A Comprehensive Study on Natural Language Processing, It's Techniques and Advancements in Nepali Language** .................. 157
Sital Sharma, Kalpana Sharma, and Biswaraj Sen

**Quality Evaluation of Enterprise Architecture Frameworks** ........... 177
Priyanka Chakraborty and Anirban Sarkar

**Natural Language Query Processing System to Extract Autism Spectrum Disorder Information from Database** ..................... 187
Prasenjit Mukherjee, Nandan Banerji, Kailash Pati Mandal, Manish Godse, and Baisakhi Chakraborty

**Spiking Neural Network in Computer Vision: Techniques, Tools and Trends** ................................................ 201
Rohit Agarwal, Palash Ghosal, Narayan Murmu, and Debashis Nandi

**A Quantitative Study on the FaceNet System** ....................... 211
Rajesh Gopakumar, Karunakar A. Kotegar, and M. Vishal Anand

**Resampling Strategies for Mitigating Class Imbalance of ASD Dataset on the Performance of Machine Learning Classifiers** .......... 225
Rahul Kumar Gupta and Kamlesh Dutta

**Comparative Analysis of Minimax Algorithm with Alpha–Beta Pruning Optimization for Chess Engine** ............................ 237
Harshit Choudhary, Aniket Inamdar, Swapnil Kharade, Anunay Maheshwari, and Preeti Kale

**Biomedical Informatics**

**A Modified Microwave Based System Design for Early-Stage Breast Cancer Detection** ........................................ 251
Moutusi Mondal, Palash Ghosal, Amish Kumar, and Debashis Nandi

**Research Review and Literature Perception Towards Medicinal Plants Classification Using Deep Learning Techniques** ................ 259
Himanshu Kumar Diwedi, Anuradha Misra, Amod Kumar Tiwari, and Atif Mahmood

**Features Fusion-Based Gait Recognition with Covariate Conditions** .... 273
Margaret Kathing, Rishang Kumar Brahma, and Sarat Saharia

**Computational Modelling of Glucocerebrosidase Signalling Pathways in Parkinson's Disease** .... 281
Hemalatha Sasidharakurup, Kasi Viswanadh, Divya M. Sasidharan, Anu Sasidharan, Arushi Tiwari, Devi Krishna, Giovanni Naldi, Egidio D'Angelo, and Shyam Diwakar

**COVID-19 Lung Patch Segmentation Using COVSeg-NET** .... 291
Vivek Noel Soren, Sakshi Ahuja, B. K. Panigrahi, and Tapan K. Gandhi

**Research Study on Leukemia Detection Using Deep Learning Techniques** .... 301
Rohan Raj Maram, Sasibhushana Rao Pappu, Aravapalli Rama Satish, and Balajee Maram

**Automated Classification of Diabetic Retinopathy Using Deep Learning Architecture** .... 309
Noel D'Souza, P. C. Siddalingaswamy, and N. V. Subba Reddy

**Deep Learning-Based Tumor Segmentation from CT Images** .... 323
Nitya Jitani, Anup Basumatary, and Rosy Sarmah

**Image and Video Processing**

**DIVS: A Real-Time Video Stabilization Framework for Consumer Camera** .... 337
Varanasi L. V. S. K. B. Kasyap, Sumathi, Abhijit Adhikari, and V. S. Bhagavan

**Saliency-Driven 'Evidence CNN' for Fine-Grained Recognition of Twins** .... 351
Razia de Loyola Furtado e Sardinha

**IoT-Based Landslide Monitoring System Using LiDAR Sensors: A Review** .... 363
Bishal Gupta, Kalpana Sharma, Tawal Koirala, and Sital Sharma

**Change Detection in LULC for Durgapur: An Efficient GIS-Based Framework** .... 381
Kazi Amrin Kabir and Parag Kumar Guha Thakurta

**Morphological Operation-Based Background Subtraction Method for Shadow Removal in Outdoor Video Sequences** .... 391
Tannistha Pal

**Pre-processing Pipelines for Effective Segmentation of Lymphatic Filariasis Parasite Images** .... 403
Priyanka Kumar and Kanojia Sindhuben Babulal

**Fish Classification Using CNN and Logistic Regression from Underwater Images** ..... 415
Pratick Gupta, Pratima Sarkar, Bijoyeta Roy, and Shivam Kumar

**A Short Survey of Elucidating the Emotion Recognition Methodologies Using Facial Images and EEG Signals** ..... 425
Dilsheen Kaur, Anuradha Misra, and O. P. Vyas

**Novel Image Cryptography Method to Improve Security in Cloud Computing** ..... 439
Prathamesh Vijay Lahande and Parag Ravikant Kaveri

**CT Image Super-Resolution: A Perspective on Tissue-Specific Performance Analysis** ..... 449
Bakul Gohel, Nisarg Doshi, and Krishna Veer Singh

**Co-registration of the Digitised Sensor Information and the Brain MRI Scan for Electrical Source Imaging: Fiducial-Based Alignment Versus ICP-Based Approach** ..... 457
Bakul Gohel

**Communication Techniques**

**Relative Analysis of OCDMA-FSO and FSO Under Normal, Haze and Fog Conditions** ..... 471
Himali Sarangal, Butta Singh, Manjit Singh, Vinit Grewal, and Satveer Kour

**An Analytical Model for a Proportional–Integral–Derivative Controlled Phase-Locked Loop for Communication Systems** ..... 481
Geetamoni Konwar and Tulshi Bezboruah

**Performance Study and Comparative Analysis of Routing Protocols in Wireless Sensor Networks** ..... 491
Mandeep Kumar and Jahid Ali

**Design and Implementation of Third and Fourth Order Phase Lock Loop Using Simulink** ..... 505
Sneha Singh, Kashmiri Deka, and Geetamoni Konwar

**Survey on IoT Based Smart Fisheries in Context with Himalayan Region of Sikkim** ..... 515
Pritika Adhikari and Dhruba Ningombam

**Optimal Frequency Control in a Microgrid Under Wind Power and Load Uncertainties** ..... 531
Rachita R. Sarangi, Asit Mohanty, Prakash K. Ray, Ajit K. Barisal, and Suvendu M. Baral

**Author Index** ..... 545

# Editors and Contributors

## About the Editors

**Dr. Samarjeet Borah** is currently working as Professor and Head in the Department of Computer Applications, SMIT, Sikkim Manipal University (SMU), Sikkim, India. Dr. Borah has carried out various funded projects from AICTE (GoI), DST-CSRI (GoI), etc. He has published around 90 research papers in various journals and conferences. He has organized various workshops and conferences in national and international levels. Dr. Borah is involved with various book volumes and journals of repute from Springer, IEEE, Inderscience, and IGI Global as Editor or Guest Editor. He is Editor-in-Chief of the book and proceedings series—Research Notes on Computing and Communication Sciences, Apple Academic Press, USA. Dr. Borah has served as Keynote as well as Invited Speakers in several international conferences. His areas of research are data mining, information security, data science, and machine learning.

**Dr. Tapan K. Gandhi** is currently working as Associate Professor in the Department of Electrical Engineering, IIT Delhi, and Adjunct Faculty in the school of AI and data science, IIT Jodhpur. He is also Research Affiliate to MIT, USA. He received his Ph.D. fellowship from (MIT, USA) and obtained his Ph.D. in Biomedical Engineering from IIT Delhi. He has published papers in top ranking journals like Nature, PNAS, Current Biology, PloS Biology, IEEE Transactions. He has more than 150 publications in international journals and conference proceedings. He is PI and Co-PI of many funded projects from industry, international organizations as well as GoI organizations. He is serving as Expert Member in various research committees in GoI and UNESCO. He has also received Platinum Jubilee Young Scientist award 2015 by National Academy of Science in India and GYTI 2019 Award. Presently, he is Chairperson of Project Prakash Charitable Trust that helps in Restoration of Vision and providing education to visually impaired in India.

**Dr. Vincenzo Piuri** is Full Professor at the University of Milan, Italy (since 2000), where he was also Department Chair (2007–2012). He was Associate Professor at Politecnico di Milano, Italy (1992–2000), visiting professor at the University of Texas at Austin, USA (summers 1996–1999), and visiting researcher at George Mason University, USA (summers 2012–2017). He founded a start-up company, Sensure srl, in the area of intelligent systems for industrial applications (leading it from 2007 until 2010) and was active in industrial research projects with several companies. He received his M.S. and Ph.D. in Computer Engineering from Politecnico di Milano, Italy. His main research and industrial application interests are intelligent systems, computational intelligence, pattern analysis and recognition, machine learning, signal and image processing, biometrics, intelligent measurement systems, industrial applications, distributed processing systems, Internet of things, cloud computing, fault tolerance, application-specific digital processing architectures, and arithmetic architectures.

## Contributors

**Abhijit Adhikari** School of Computer Science and Engineering, VIT-AP University, Inavolu, India

**Pritika Adhikari** Sikkim Manipal Institute of Technology, Majhitar, India

**Rohit Agarwal** National Institute of Technology, Durgapur, India

**Shafiul Alom Ahmed** Department of Computer Science and Engineering, Tezpur University, Assam, India;
School of Computing Science and Engineering, VIT-Bhopal University, Sehore, Bhopal, India

**Sakshi Ahuja** Electrical Engineering Department, IITD, New Delhi, India

**Jahid Ali** Professor, Computer Applications, Sri Sai Iqbal College of Management and Information Technology, Pathankot, Punjab, India

**Md. Sarfaraj Alam Ansari** Department of Computer Science and Engineering, NIT Sikkim, Sikkim, India

**Zahid Ahmed Ansari** EES, University Polytechnic, Aligarh Muslim University, Aligarh, UP, India

**Kanojia Sindhuben Babulal** Department of Computer Science and Engineering, Central University of Jharkhand, Ranchi, India

**Abhishek Bagepalli** Department of Information and Communication Technology, Manipal Institute of Technology, Manipal Academy of Higher Education, Manipal, India

**Bhaswati Baishya** Department of Computer Science and Engineering, Tezpur University, Assam, India

**Indrajit Banerjee** Department of Information Technology, Indian Institute of Engineering Science and Technology Shibpur, Howrah, West Bengal, India

**Nandan Banerji** National Institute of Technology, Durgapur, India

**Suvendu M. Baral** Department of Electrical Engineering, Odisha University of Technology and Research, Bhubaneswar, India

**Ajit K. Barisal** Department of Electrical Engineering, Odisha University of Technology and Research, Bhubaneswar, India

**Anup Basumatary** Department of Computer Science and Engineering, Tezpur University, Assam, India

**Tulshi Bezboruah** Department of Electronics and Communication Technology, Gauhati University, Guwahati, Assam, India

**V. S. Bhagavan** Department of Engineering Mathematics, Koneru Lakshmaiah Education Foundation, Vaddeswaram, India

**Ashwini Bhatt** Department of Electronics and Communication, PES University, Bangalore, India

**Soham Biswas** Department of Computer Science and Engineering NIT Sikkim, Sikkim, India

**Rishang Kumar Brahma** Computer Science and Engineering, Tezpur University, Tezpur, Assam, India

**Baisakhi Chakraborty** National Institute of Technology, Durgapur, India

**Chandrima Chakraborty** Sasken Technologies Limited, Bengaluru, India

**Priyanka Chakraborty** Department of Computer Science and Engineering, National Institute of Technology, Durgapur, West Bengal, India

**Harshit Choudhary** Vishwanath Karad MIT World Peace University, Pune, India

**Razia de Loyola Furtado e Sardinha** Department of Information Technology, Padre Conceicao College of Engineering, Verna, Goa, India

**Kashmiri Deka** Department of Electronics and Communication Technology, Gauhati University, Guwahati, India

**Shyam Diwakar** Amrita Mind Brain Center, Amrita Vishwa Vidyapeetham, Kollam, India;
Department of Electronics and Communication Engineering, School of Engineering, Amrita Vishwa Vidyapeetham, Kollam, India

**Himanshu Kumar Diwedi** Computer Science and Engineering, Amity School of Engineering and Technology, Amity University, Lucknow, India

**Nisarg Doshi** Dhirubhai Ambani Institute of Information and Communication Technology (DA-IICT), Gandhinagar, India

**Kamlesh Dutta** Department of Computer Science & Engineeing, National Institute of Technology Hamirpur, Hamirpur, HP, India

**Egidio D'Angelo** Department of Brain and Behavioral Sciences, University of Pavia, Pavia, Italy;
Brain Connectivity Center, IRCCS Mondino Foundation, Pavia, Italy

**Noel D'Souza** Language Technologies Institute, Carnegie Mellon University, Pittsburgh, USA

**Tapan K. Gandhi** Electrical Engineering Department, IITD, New Delhi, India

**Palash Ghosal** Sikkim Manipal Institute of Technology, Majitar, Sikkim, India;
National Institute of Technology, Durgapur, India

**Manish Godse** bizAmica Software, Pune, India

**Bakul Gohel** Dhirubhai Ambani Institute of Information and Communication Technology (DA-IICT), Gandhinagar, India

**Rajesh Gopakumar** Computer Science and Engineering, Manipal Institute of Technology, Manipal Academy of Higher Education, Manipal, Karnataka, India

**Mahesh Chandra Govil** Department of Computer Science and Engineering NIT Sikkim, Sikkim, India

**Vinit Grewal** Department of Engineering and Technology, Guru Nanak Dev University, Regional Campus Jalandhar, India

**Bishal Gupta** Department of Computer Science and Engineering, Sikkim Manipal Institute of Technology, Sikkim Manipal University, Majhitar, Sikkim, India

**Pratick Gupta** Computer Science and Engineering, Techno International New Town, New Town, India

**Rahul Kumar Gupta** School Of Technology, Woxsen University, Hydrabad, India;
Department of Computer Science & Engineeing, National Institute of Technology Hamirpur, Hamirpur, HP, India

**Dipendra Gurung** Sikkim Manipal Institute of Technology, Sikkim, India

**Sandeep Gurung** Sikkim Manipal Institute of Technology, Sikkim, India

**Nazrul Hoque** Department of Computer Science, Manipur University, Imphal, Manipur, India

**Aniket Inamdar** Vishwanath Karad MIT World Peace University, Pune, India

**Nitya Jitani** Department of Computer Science and Engineering, Tezpur University, Assam, India

**Kazi Amrin Kabir** Department of Computer Science and Engineering, NIT Durgapur, Durgapur, West Bengal, India

**Preeti Kale** Vishwanath Karad MIT World Peace University, Pune, India

**Gaidinlung Kamei** Department of Information Technology, Indian Institute of Engineering Science and Technology Shibpur, Howrah, West Bengal, India

**Varanasi L. V. S. K. B. Kasyap** School of Computer Science and Engineering, VIT-AP University, Inavolu, India

**Margaret Kathing** Computer Science and Engineering, Tezpur University, Tezpur, Assam, India

**Dilsheen Kaur** Computer Science and Engineering, Amity School of Engineering and Technology, Amity University Lucknow Campus, India

**Parag Ravikant Kaveri** Symbiosis International (Deemed University), Symbiosis Institute of Computer Studies and Research, Pune, India

**Bipal Khanal** Department of Computer Science and Engineering, NIT Sikkim, Sikkim, India

**Swapnil Kharade** Vishwanath Karad MIT World Peace University, Pune, India

**Robindro Singh Khumukcham** Department of Computer Science, Manipur University, Imphal, Manipur, India

**Tawal Koirala** Department of Computer Science and Engineering, Sikkim Manipal Institute of Technology, Sikkim Manipal University, Majhitar, Sikkim, India

**Geetamoni Konwar** Department of Electronics and Communication Technology, Gauhati University, Guwahati, Assam, India

**Karunakar A. Kotegar** Data Science and Computer Applications, Manipal Institute of Technology, Manipal Academy of Higher Education, Manipal, Karnataka, India

**Satveer Kour** Department of CET, Guru Nanak Dev University, Amritsar, India

**Devi Krishna** School of Biotechnology, Amrita Vishwa Vidyapeetham, Kollam, India

**Amish Kumar** Zreyas Technology Pvt. Ltd., Kolkata, India

**Chandan Kumar** Department of Computer Science and Engineering, NIT Sikkim, Sikkim, India

**Mandeep Kumar** Research Scholar, Computer Science and Engineering, I.K. Gujral Punjab Technical University, Kapurthala, India

**Priyanka Kumar** Department of Computer Science and Engineering, Central University of Jharkhand, Ranchi, India

**Santosh Kumar** Department of Computer Science and Engineering, Galgotias University, Greater Noida, Uttar Pradesh, India

**Shivam Kumar** Computer Science and Engineering, Techno International New Town, New Town, India

**Akshata Kumble** Department of Electronics and Communication, PES University, Bangalore, India

**Prathamesh Vijay Lahande** Symbiosis International (Deemed University), Symbiosis Institute of Computer Studies and Research, Pune, India

**Anunay Maheshwari** Vishwanath Karad MIT World Peace University, Pune, India

**Atif Mahmood** Computer Science and Engineering, Pranveer Singh Institute of Technology, Kanpur, India

**Kailash Pati Mandal** National Institute of Technology, Durgapur, India

**Balajee Maram** Apex Institute of Technology—Computer Science and Engineering, Chandigarh University, Mohali, India

**Rohan Raj Maram** Chitkara University, Baddi, HP, India

**Sriveda Medatati** Department of Electronics and Communication, PES University, Bangalore, India

**Anuradha Misra** Computer Science and Engineering, Amity School of Engineering and Technology, Amity University, Lucknow, India

**Asit Mohanty** Department of Electrical Engineering, Odisha University of Technology and Research, Bhubaneswar, India

**Moutusi Mondal** National Institute of Technology, Durgapur, India

**Prasenjit Mukherjee** Manipur International University, Imphal, Manipur, India; Vodafone Intelligent Solutions, Pune, India

**Narayan Murmu** National Institute of Technology, Durgapur, India

**Giovanni Naldi** Department of Environmental Science and Policy, Università degli Studi di Milano, Milano, Italy

**Debashis Nandi** National Institute of Technology, Durgapur, India

**Bhabesh Nath** Department of Computer Science and Engineering, Tezpur University, Assam, India

**Devarani Devi Ningombam** Department of Computer Science and Engineering, National Institute of Technology Patna (NIT Patna), Bihar, India

**Dhruba Ningombam** Sikkim Manipal Institute of Technology, Majhitar, India

**Tannistha Pal** Department of Computer Science and Engineering, National Institute of Technology, Agartala, Barjala, Jirania, India

**B. K. Panigrahi** Electrical Engineering Department, IITD, New Delhi, India

**Sasibhushana Rao Pappu** CSE Department, GITAM(Deemed to Be University), GITAM School of Technology, Visakhapatnam, India

**Prakash K. Ray** Department of Electrical Engineering, Odisha University of Technology and Research, Bhubaneswar, India

**N. V. Subba Reddy** Department of Computer Science and Engineering, Manipal Institute of Technology, Manipal Academy of Higher Education, Manipal, Karnataka, India

**Bijoyeta Roy** Sikkim Manipal Institute of Technology, Pakyong, India

**Sarat Saharia** Computer Science and Engineering, Tezpur University, Tezpur, Assam, India

**Himali Sarangal** Department of Engineering and Technology, Guru Nanak Dev University, Regional Campus Jalandhar, India

**Rachita R. Sarangi** Department of Electrical Engineering, Odisha University of Technology and Research, Bhubaneswar, India

**Anirban Sarkar** Department of Computer Science and Engineering, National Institute of Technology, Durgapur, West Bengal, India

**Pratima Sarkar** Computer Science and Engineering, Techno International New Town, New Town, India

**Rosy Sarmah** Department of Computer Science and Engineering, Tezpur University, Assam, India

**Hemalatha Sasidharakurup** Amrita Mind Brain Center, Amrita Vishwa Vidyapeetham, Kollam, India;
School of Biotechnology, Amrita Vishwa Vidyapeetham, Kollam, India

**Anu Sasidharan** School of Biotechnology, Amrita Vishwa Vidyapeetham, Kollam, India

**Divya M. Sasidharan** School of Biotechnology, Amrita Vishwa Vidyapeetham, Kollam, India

**Aravapalli Rama Satish** School of Computer Science and Engineering, VIT-AP University, Vijayawada, Andhra Pradesh, India

**Biswaraj Sen** Sikkim Manipal Institute of Technology, Sikkim Manipal University, Sikkim, India

**B. M. Ahamed Shafeeq** Department of Computer Science and Engineering, Manipal Institute of Technology, Manipal Academy of Higher Education, Manipal, India

**Kalpana Sharma** Department of Computer Science and Engineering, Sikkim Manipal Institute of Technology, Sikkim Manipal University, Majhitar, Sikkim, India

**Sital Sharma** Department of Computer Science and Engineering, Sikkim Manipal Institute of Technology, Sikkim Manipal University, Majhitar, Sikkim, India

**P. C. Siddalingaswamy** Department of Computer Science and Engineering, Manipal Institute of Technology, Manipal Academy of Higher Education, Manipal, Karnataka, India

**Butta Singh** Department of Engineering and Technology, Guru Nanak Dev University, Regional Campus Jalandhar, India

**Krishna Veer Singh** Dhirubhai Ambani Institute of Information and Communication Technology (DA-IICT), Gandhinagar, India

**Manjit Singh** Department of Engineering and Technology, Guru Nanak Dev University, Regional Campus Jalandhar, India

**Sanjay Singh** Department of Information and Communication Technology, Manipal Institute of Technology, Manipal Academy of Higher Education, Manipal, India

**Sneha Singh** Department of Electronics and Communication Technology, Gauhati University, Guwahati, India

**Suman Som** Department of Computer Science and Engineering NIT Sikkim, Sikkim, India

**Rebecca Shruti Soren** Sikkim Manipal Institute of Technology, Sikkim, India

**Vivek Noel Soren** Electrical Engineering Department, IITD, New Delhi, India

**Sumathi** School of Computer Science and Engineering, VIT-AP University, Inavolu, India

**Linthoingambi Takhellambam** Department of Computer Science, Manipur University, Imphal, Manipur, India

**Parag Kumar Guha Thakurta** Department of Computer Science and Engineering, NIT Durgapur, Durgapur, West Bengal, India

**Amod Kumar Tiwari** Computer Science and Engineering, Rajkiya Engineering College, Sonbhadra, India

**Arushi Tiwari** School of Biotechnology, Amrita Vishwa Vidyapeetham, Kollam, India

**Boby Clinton Urikhimbam** Department of Computer Science, Manipur University, Imphal, Manipur, India

**Sanjai Mohan Verma** ABES Engineering College, Ghaziabad, Uttar Pradesh, India

**M. Vishal Anand** Computer Science and Engineering, Manipal Institute of Technology, Manipal Academy of Higher Education, Manipal, Karnataka, India

**Kasi Viswanadh** School of Biotechnology, Amrita Vishwa Vidyapeetham, Kollam, India

**O. P. Vyas** Information Technology, Indian Institute of Information Technology Allahabad, Prayagraj, India

**Ranjan Yambem** Department of Computer Science, Manipur University, Imphal, Manipur, India

# Soft Computing and Machine Learning

# Performance Analysis of Clustering Using Modified Grey Wolf Optimization

**B. M. Ahamed Shafeeq and Zahid Ahmed Ansari**

**Abstract** Data clustering is the widely used technique in academia and industry to analyse large volumes of data with unknown patterns. Data clustering approaches that draw inspiration from biology are increasingly widely used. In this research, we propose a parallelized automated data clustering using a modified Grey Wolf Optimization technique based on the hunting style of the grey wolf which involves Tracking, chasing and approaching the prey. It will find the optimal solution from the generated 'N' solutions. However, the nature of massive data available in the repositories is unknown. So, it is a tedious task to guess the right number of clusters for the massive data. By repeating the procedure with clusters K = 2 to N, the suggested technique determines the best number of clusters. The ideal number of clusters are detected based on the best values of Silhouette index, Davies-Bouldin index and Calinski-Harabasz index. This research aims to propose a more efficient Intelligent clustering framework. The suggested approach operates in both the scenarios, i.e. with a predetermined number of clusters and an uncertain number of clusters. The user can either fix the number of clusters or let the system identify the optimal number of clusters. The proposed method parallelizes and automates cluster analysis in the most effective manner for determining the best clusters and forming natural clusters.

**Keywords** Clustering · Grey-Wolf optimization · Parallelization · Quality metrics

---

B. M. A. Shafeeq (✉)
Department of Computer Science and Engineering, Manipal Institute of Technology, Manipal Academy of Higher Education, Manipal, India
e-mail: ahamed.shafeeq@manipal.edu

Z. A. Ansari
EES, University Polytechnic, Aligarh Muslim University, Aligarh, UP 202002, India
e-mail: Zansari.ubp@amu.ac.in



S. Borah et al. (eds.), *Advanced Computational and Communication Paradigms*,
Lecture Notes in Networks and Systems 535,
https://doi.org/10.1007/978-981-99-4284-8_1

# 1 Introduction

Clustering is a fundamental problem that emerges regularly in many disciplines, including data exploration, knowledge discovery and pattern categorization [1]. Traditional data clustering approaches, such as partition cluster analysis or hierarchical cluster analysis, use their full capacity in dealing with a rise in data size [2, 3]. These algorithms are limited either by the need to store all of the data in memory or by their computational time complexity [4]. Bio-inspired computing has recently emerged as a more feasible optimization tool for data clustering. Bio-inspired computing is becoming increasingly popular for a wide range of domains and applications.

Bio-inspired approaches are those that are influenced by biological systems. The term 'bio-inspired optimization' refers to a class of methodologies that are inspired by biological processes. Bio-inspired techniques' versatility has tempted academics to apply them to a wide range of applications. The optimization process seeks the best solution to a problem. It is critical to select the appropriate algorithm for a certain application. There are several algorithms developed by the researchers to simulate the behaviour of animals or insect groups by imposing methods that may be used for a variety of applications. Grey Wolf Optimizer is a novel meta-heuristic optimization tool with high efficiency [5]. It works effectively for composite functions while avoiding local minima and has high exploitation skills for unimodal problems and better exploration skills for multimodal ones.

The Grey Wolf Optimization-based Algorithm (GWOC) is a recent new meta-heuristic optimization technique [6]. The main objective of the technique is to mimic grey wolves' natural behaviour of searching for prey systematically. One of the most current research areas has been their social hierarchy, hunt down, surrounding and attacking target features. The exploration capacity of the grey wolf algorithm overcomes the problem of local optima. In order to evaluate the quality of the clusters, the method was assessed with several data sets. This research paper is ordered into different modules. The background and related work of K-means method is discussed in detail in Sect. 2. The methodology followed is elaborated in Sect. 3. Section 4 contains the demonstration of empirical analysis results. Section 5 concludes the research work carried.

## 2 Related Work

Knowledge discovery and data mining rely heavily on data cluster analysis. Cluster analysis tries to categorise data groups by discovering similarities and contrasts between data points. The process can be completed under supervision, semi-supervised or unsupervised [7]. The strategies suggested are independent of data set size and make them ideal for huge data sets. The experimental results show that the suggested technique surpasses the best cluster number estimation algorithms while using far less computer resources [8]. An automatic hyper-parameter determination approach for k-means clustering that can handle large datasets and a wide range of applications [4]. As described, a hierarchical K-means technique improves clustering performance in huge data situations. One of the main problems with K-Means clustering is locating a locally optimum solution; the recommended method resolves this problem [9].

S. Mirjalili et al. suggested Grey Wolf Optimizer (GWO) [5], a novel memetic motivated by grey wolves. The GWO approach is based on grey wolves' inherent leadership structure and hunting mechanism. Grey wolves are said to emulate four forms of leadership behaviour: alpha, beta, delta and omega. The algorithm's performance is tested and compared against Particle Swarm Optimization PSO, Gravitational Search Algorithm, Differential Evolution, Evolutionary Programming and Evolution Strategy utilising a range of benchmark data sets. When compared against these well-known meta-heuristics, the findings suggest that the GWO algorithm may produce extremely competitive outcomes. The suggested technique can tackle complex issues with unknown search spaces, as proven by typical engineering design considerations and a real-world application. Kumar et al. [6] introduced the Grey wolf clustering technique, which is based on Grey wolf prey hunting behaviour. Their social structure, tracking, surrounding and attacking prey traits are among the most recent research fields. The grey wolf algorithm's capacity to explore the search space solves the problem of local optima. Grey wolf clustering is a concept that has been offered. Tripathi et al. [10] presented an enhanced grey wolf optimizer as a novel form of the grey wolf optimizer. To improve its capacity to find prey, the grey wolf's hunting method is combined with binomial crossover and lévy flying steps. Data is piling up in the repositories at breakneck speed. Effective analytical techniques are required for making proper decisions based on data. In today's context, improvements in standard procedures are required to understand massive data sets. The extended grey wolf optimizer, which is based on Map Reduce, is introduced as a helpful clustering tool for large data sets. Several benchmark data set from UCI are used to validate the proposed strategy. Experiments have shown that the proposed technique is a reliable and powerful replacement.

Aljarah et al. combined the Grey wolf optimizer (GWO) and the well-known Tabu Search (TS) algorithms [11]. Grey wolves' hunting behaviour and social cooperation inspired GWO, a typical meta-heuristic algorithm. The common issue of local optima in clustering can be resolved in this approach. This technique identifies the best search agents and computes the best search agent by taking the average of the first three best search agents. The theoretical fusion of GWO and TS is validated against well-known memetic algorithms on various clustering datasets to check how well it performs. According to considerable testing and research, the suggested GWOTS outperforms GWO and can be utilised for clustering applications. The Grey Wolf Optimizer (GWO) technique for segmenting satellite images was suggested by Kapoor et al. [12]. The initial GWO has been altered to function as a clustering technique. Two satellite photos were used to evaluate this method. GWO is a unique memetic optimization strategy that has gained a lot of traction in the optimization area. It excels at exploitation for unimodal problems and excels at composite functions while avoiding local minima. It also strengthens multimodal problem-solving abilities. It is quick to compute and frequently outperforms in terms of the Davies-Bouldin index, average inter-cluster distance and average intra-cluster distance.

## 3 Materials and Methods

The suggested hybrid K-Means Grey Wolf Optimization-based Algorithm (HKGWOC) is divided into two stages. The K-Means method is used in the first step to identify the ideal number of clusters, and the GWOC is used in the second stage to locate the final clusters based on the optimal number of clusters identified in the first stage. The main objective of the technique is to mimic grey wolves' natural behaviour of searching for prey systematically. One of the most current research areas has been their social hierarchy, hunt down, surrounding and attacking target features. The exploration capacity of the grey wolf algorithm overcomes the problem of local optima. The suggested algorithm replaces the search agent with lower fitness by a new search agent. The proposed algorithm was compared with various data sets.

Algorithm: Hybrid K-Means and Grey Wolf Optimization Based Clustering (DHKGWOC)
Input : Data set D, N-Maximum number of clusters, Number of Search agents
Output : Best search agent with optimal k clusters
1) Repeat for different values of k =5 to N
2) Randomly choose k cluster centres for each search agent.
3) Repeat
4) Compute the distance between each data point and k cluster centres. Identify the cluster centre to which the item is the most similar.
5) Modify the cluster average value, that is, compute the average value of the items for each cluster.
6) Find the variance using Eq. 3.1 of each cluster and identify the cluster with highest variance.
7) If current maximum variance > previous maximum variance then Replace highest variance cluster with randomly selected new cluster.
8) Calculate the sum of squared error using Eq. 3.2.
9) Until E is smaller than the threshold.
10) Determine the best k from all the solutions
11) Create a population of n potential solutions, each with an arbitrarily selected optimal k cluster centres obtained from K-Means clustering.
11) Calculate the fitness of the clusters in every potential solution in the assigned node.
12) Determine the top three best search agents in the master node based on fitness.
13) Repeat
15) Loop -for all search agents
16) Calculate control coefficients A and C
17) Modify the location of the present search agent (potential solution)
18) Update the cluster centres attached to each search agent based on modified locations
19) End Loop
20) Modify the coefficient a
21) Calculate the fitness of all probable solutions
22) Modify the value of three best solutions based on fitness.
23) Identify the search agent with lowest fitness values and replace it with a new search agent
24) Until maximum number of iterations
25) Identify the best solution with optimal k clusters

$$S^2 = \frac{\sum_{x \in c_j} \left| x - \overline{x}_j \right|}{n - 1} \quad (1)$$

$S^2$ = Variance; $\overline{x_j}$ = mean of the jth cluster; n = number of data points in the cluster;

$$E = \sum_{i=1}^{k} \sum_{x \in c_i} |x - \overline{x}_i|^2 \tag{2}$$

$E$ is the summation of the squared error denoting all points in the dataset; $x$ is the point in a specific cluster $C_i$ and $\overline{x_i}$ is the mean of cluster $C_i$.

$$\mathrm{A} = 2 * \mathrm{a} * \mathrm{r}_1 - \mathrm{a} \tag{3}$$

$$\mathrm{C} = 2 * \mathrm{r}_2 \tag{4}$$

where a decrease gradually from 2 to 0 on every phase of iteration and $\mathrm{r}_1$ and $\mathrm{r}_2$ are arbitrary values in the range [0, 1].

$$\mathrm{D}_\alpha = \|\mathrm{C}_1 \cdot \mathrm{Y}_\alpha - \mathrm{Y}\|, \mathrm{D}_\beta = \left\|\mathrm{C}_2 \cdot \mathrm{Y}_\beta - \mathrm{Y}\right\|, \ \mathrm{D}_\delta = \|\mathrm{C}_3 \cdot \mathrm{Y}_\delta - \mathrm{Y}\| \tag{5}$$

$$\mathrm{Y}_1 = \mathrm{Y}_\alpha - \mathrm{A}_1 \cdot \mathrm{D}_\alpha, \ \mathrm{Y}_2 = \mathrm{Y}_\beta - \mathrm{A}_2 \cdot \mathrm{D}_\beta, \ \mathrm{Y}_3 = \mathrm{Y}_\delta - \mathrm{A}_3 \cdot \mathrm{D}_\delta \tag{6}$$

$$\mathrm{Y}(\mathrm{t}+1) = \frac{\mathrm{Y}_1 + \mathrm{Y}_2 + \mathrm{Y}_3}{3} \tag{7}$$

- $\mathrm{D}_\alpha$ denotes the space between the location of α wolf ($\mathrm{Y}_\alpha$) and a grey wolf (Y).
- t is the current iteration.
- A and C are control coefficients.

**Table 1** Glossary of terms

| Abbreviation | Meaning |
|---|---|
| DKMEANS | Dynamic K-Means clustering |
| PDKMEANS | Parallel Dynamic K-Means clustering |
| FFA | Firefly Optimization Algorithm |
| WOC | Whale Optimization-based Clustering |
| DFFA | Dynamic Firefly Optimization Algorithm |
| DWOC | Dynamic Whale Optimization-based Clustering |
| DHKGWOC | Dynamic Hybrid K-Means Grey-wolf Optimization-based Clustering |
| DHKWOC | Dynamic Hybrid K-Means Whale Optimization-based Clustering |
| DHKFFA | Dynamic Hybrid K-Means Firefly Optimization Algorithm |
| PDFFA | Parallel Dynamic Firefly Optimization Algorithm |
| PDWOC | Parallel Dynamic Whale Optimization-based Clustering |
| PDHKWOC | Parallel Dynamic Hybrid K-Means Whale Optimization-based Clustering |
| PDHKGWOC | Parallel Dynamic Hybrid K-Means Grey-wolf Optimization Clustering |
| PDHKFFA | Parallel Dynamic Hybrid K-Means Firefly Optimization Algorithm |

- Y(t) is the best search agent at iteration t.
- $Y_\alpha$, $Y_\beta$, $Y_\delta$—The best 3 candidate solutions.

## 4 Results and Discussions

The five algorithms, namely, ACO, ABC, ESA, FFA and GWOC, are implemented on MATLAB. Its performance is analysed for census data with silhouette, Davies-Bouldin and Calinski-Harabasz. Table 2 records the measured quality values of all the algorithms for the different data sets. Figure 1 shows the performance of the algorithms for the LPG usage data, crime data and Census data. The various quality metrics are used to validate the clusters formed using different clustering techniques. The quality metrics used are Silhouette index, Calinski-Harabasz Index, Davies-Bouldin Index, Dunn's Index, C-Index and Sum-of-squared error. The best clustering techniques are identified using majority voting method. The GWOC method beat the other algorithms, as seen in Figure. The performance FFA is very close to GWOC. The algorithms ACO and ABC produce low-quality clusters.

Census Dataset: The data collected is from the census 2011 repository (https://census.gov.in). The census data repository constitutes more than six lakh records and 130+ features from thirty-three states, including union territories.

LPG Usage Data Set: This data set contains the information of LPG Usage per households in the districts of India and data set includes six hundred and thirty-four districts' LPG Usage/household information. The data is available in http://census.gov.in.

Crime Data Set: The crime dataset is collected from US Crime data repository. For the experiment analysis, the data is collected from the site https://catalog.data.gov/. This data constitutes 5283 records with ten features (different crimes) of Washington state, USA.

The performance evaluation of SIX clustering techniques is discussed in this section. The six clustering techniques considered for the analysis are K-Means,

**Table 2** Performance analysis of k-means, FFA, GWOC and WOC-based clustering

| Data set | Quality criteria | ACO | ABC | ESA | FFA | GWOC |
|---|---|---|---|---|---|---|
| LPG-data set (k = 3) | Silhouette coefficient | 0.67 | 0.66 | 0.69 | 0.73 | 0.79 |
| | Davies-Bouldin index | 0.78 | 0.79 | 0.73 | 0.78 | 0.65 |
| | Calinski-Harabasz index | 458 | 481 | 463 | 565 | 578 |
| Crime data (k = 5) | Silhouette coefficient | 0.69 | 0.68 | 0.65 | 0.72 | 0.78 |
| | Davies-Bouldin index | 0.93 | 1.14 | 1.13 | 0.85 | 0.74 |
| | Calinski-Harabasz index | 849 | 825 | 837 | 928 | 1041 |
| Census data (k = 5) | Silhouette coefficient | 0.74 | 0.7 | 0.73 | 0.73 | 0.84 |
| | Davies-Bouldin index | 1.04 | 1.08 | 0.96 | 0.80 | 0.39 |
| | Calinski-Harabasz index | 367 | 382 | 399 | 1126 | 1195 |

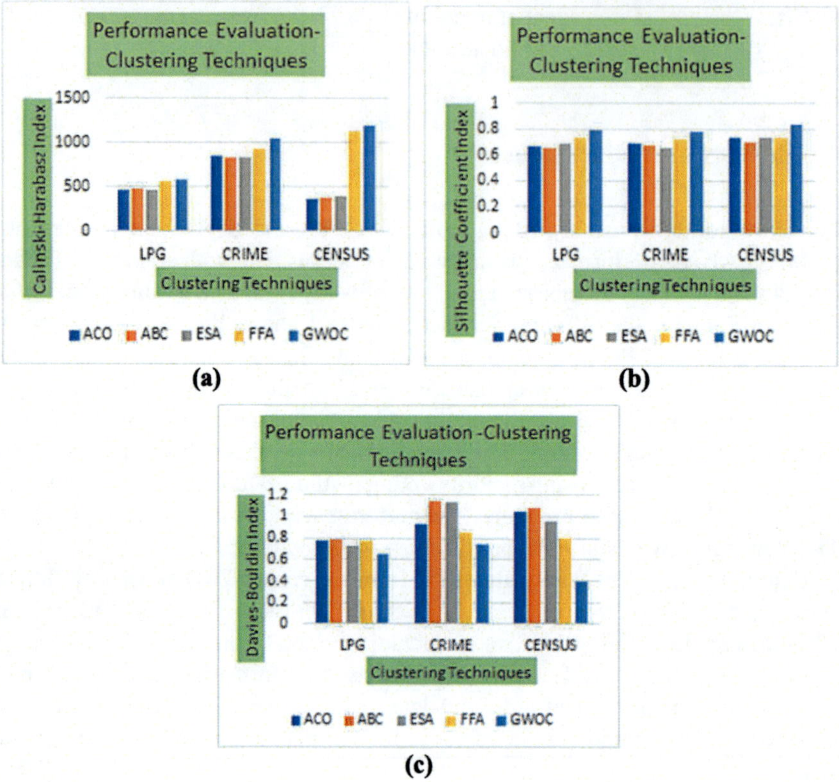


**Fig. 1** Performance evaluation **a** Calinski-Harabasz index **b** Silhouette index **c** Davies-Bouldin index

Dynamic Firefly optimization-based clustering, Dynamic Grey Wolf optimization-based clustering, Hybrid K-Means and Grey wolf optimization-based clustering, Hybrid K-Means and Firefly optimization-based clustering, and hybrid K-Means and Whale optimization-based clustering. The empirical analysis results are recorded in Table 3. The census data set is used for the analysis of proposed modified algorithms. Figure 2 shows the graph plot of analysis of clustering techniques using the Silhouette index, DB Index, C-Index and Dunn's Index. The findings of the analysis from Fig. 2 are listed as follows.

(i) The values of Silhouette index and Davies-Bouldin index suggest that the best quality cluster is produced by DHKGWOC technique.
(ii) Dunn's Index shows the highest value for DHKGWOC and the least value for DHKFFA.

**Table 3** Comparative analysis of KMEANS, DHKGWOC, DHKFFA and DHKWOC clustering algorithms

| Clustering techniques | Silhouette index | Calinski-Harabasz index | Davies-Bouldin index | Dunn's index | C-index | SSE |
|---|---|---|---|---|---|---|
| DKMEANS | 0.802 | 840 | 0.651 | 0.784 | 0.823 | 120 |
| DFFA | 0.745 | 1025 | 0.728 | 0.713 | 0.736 | 144 |
| DGWOC | 0.814 | 1157 | 0.569 | 0.823 | 0.732 | 184 |
| DHKGWOC | 0.856 | 1365 | 0.522 | 0.898 | 0.828 | 198 |
| DHKFFA | 0.784 | 1041 | 0.761 | 0.725 | 0.732 | 129 |
| DHKWOC | 0.796 | 1137 | 0.728 | 0.713 | 0.756 | 148 |

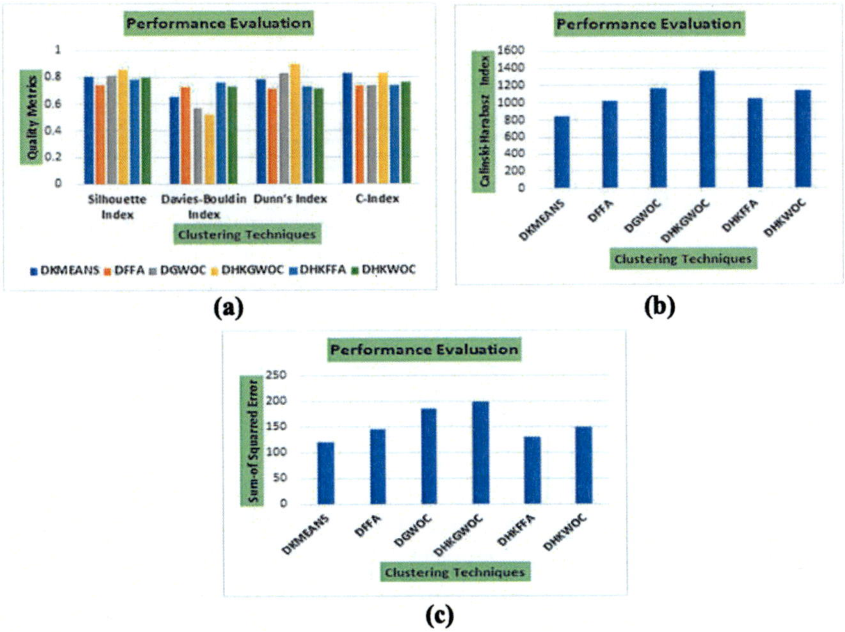


**Fig. 2** Clustering techniques versus quality indices **a** Silhouette, Davies-Bouldin, Dunn's and C-index **b** Calinski-Harabasz index **c** Sum-of-squared error

(iii) C-Index shows that the DHKGWOC produces better quality clusters than other techniques and the C-index value of the KMEANS is very close to DHKGWOC.

The execution time of five clustering techniques for census data with six lakhs thirty thousand records are recorded in Table 4. The PDKMEANS technique outperforms all her clustering techniques. The next nearest best performer is the PDHKWOC technique, and the PDHKFFA technique shows the least performance.

The graph plot of Execution Time in seconds as a function of various data sizes is shown in Fig. 3. The algorithms' time performance is tested using census data with

**Table 4** Time performance of clustering techniques

| #Cores/techniques | PDKMEANS | PDFFA | PDHKGWOC | PDHKFFA | PDHKWOC |
|---|---|---|---|---|---|
| Single core | 773.5 | 1547 | 1076 | 1863 | 1476 |
| Two core | 559 | 1008 | 684 | 1164 | 933 |
| Four core | 427 | 685 | 443 | 769 | 641 |
| Six core | 182 | 488 | 325 | 512 | 468 |
| Eight core | 92.32 | 322 | 198 | 341 | 292 |

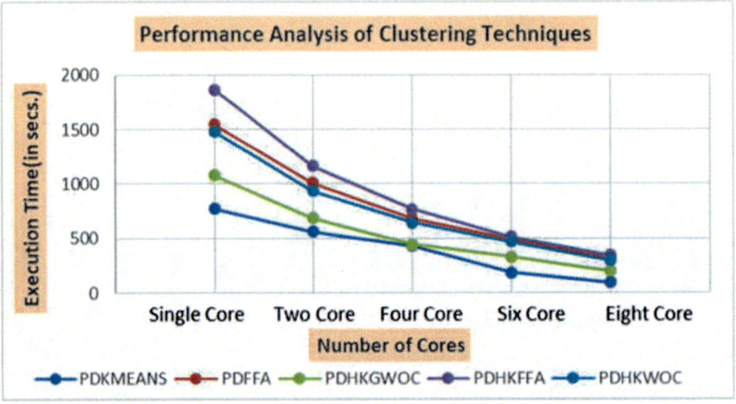


**Fig. 3** Performance analysis of execution time of the clustering techniques (single, two, four, six and eight core processors)

six lakhs thirty thousand records. According to the results, the PDHKFFA clustering algorithm is the slowest, while parallel Dynamic K-Means (PDKMEANS) with 8-Core is the fastest. The execution time of PDKMEANS is closer to the parallel dynamic Grey Wolf-based optimization clustering (PDHKGWOC).

## 5 Conclusion

Soft computing has shown significant progress in solving numerous complex optimization problems in knowledge discovery. In recent years many soft computing-based clustering techniques have been introduced. A few state-of-the-art algorithms that were recently introduced are compared for the quality of clusters created. These algorithms have shown better performance over traditional approaches. The technique overcomes the most common and complex problem of getting stuck in local minima. Some of the areas to focus on while using soft computing techniques are parameter tuning, scalability and intelligent dynamic algorithms. The ideal number of clusters is determined using the suggested updated clustering approaches. The empirical study shows that Grey-wolf optimization-based clustering delivers higher-quality solutions than other strategies. For huge data, the findings reveal that parallel dynamic K-Means clustering is the quickest, while parallel dynamic hybrid K-Means and Firefly optimization-based clustering is the slowest. The execution speed of parallel dynamic Grey Wolf-based optimization clustering (PDHKGWOC) is extremely close to that of parallel dynamic K-Means clustering (PDKMEANS). Among all clusters investigated in the experiment, the PDHKGWOC delivers the highest quality clusters. The best option is PDHKGWOC, although there is a trade-off in execution speed.

## References


1. Li W (2008) Modified K-means clustering algorithm. In: IEEE computer society congress on image and signal processing, pp 618–621
2. Cohen-Addad V, Kanade V, Mallmann-Trenn F, Mathieu C (2019) Hierarchical clustering: objective functions and algorithms. J ACM (JACM) 66(4):1–42
3. Yu T, Zhao W, Liu P, Janjic V, Yan X, Wang S, Fu H, Yang G, Thomson J (2019) Large-scale automatic k-means clustering for heterogeneous many-core supercomputer. IEEE Trans Parallel Distrib Syst 31(5):997–1008
4. Bae J, Helldin T, Riveiro M, Nowaczyk S, Bouguelia MR, Falkman G (2020) Interactive clustering: a comprehensive review. ACM Comput Surv (CSUR) 53(1):1–39
5. Mirjalili S, Mirjalili SM, Lewis A (2014) Grey Wolf optimizer. Adv Eng Softw 69:46–61
6. Kumar V, Chhabra JK, Kumar D (2016) Grey Wolf algorithm-based clustering technique. J Intell Syst 26(1)
7. Shafeeq A (2012) Dynamic clustering of data with modified k-means algorithm. In: Proceedings of the 2012 conference on information and computer networks, pp 221–225
8. Gupta A, Datta S, Das S (2018) Fast automatic estimation of the number of clusters from the minimum inter-center distance for k-means clustering. Pattern Recogn Lett 116:72–79
9. Xu T-S, Chiang H-D, Liu G-Y, Tan C-W (2015) Hierarchical k-means method for clustering large-scale advanced metering infrastructure data. IEEE Trans Power Deliv 32(2):609–616
10. Tripathi AK, Sharma K, Bala M (2018) A Novel Clustering method using enhanced Grey Wolf optimizer and mapreduce. Big Data Res 14:93–100
11. Ibrahim A, Majdi M, Heidari AA (2020) Clustering analysis using a novel locality-informed Grey Wolf-inspired clustering approach. Knowl Inf Syst 62:507–539
12. Kapoor S, Zeya I, Singhal C, Nanda SJ (2017) A grey wolf optimizer based automatic clustering algorithm for satellite image segmentation. Procedia Comput Sci 115:415–422

# Designing of Advanced Machine Learning-Based Predictive Model for Maternal Mortality

**Gaidinlung Kamei and Indrajit Banerjee**

**Abstract** Maternal mortality is a vital indicator in determining the development of a country. To reduce maternal death, the government takes many healthcare schemes and measures. The incorporation of machine learning in public health contributes to significant improvement. A predictive machine learning model with some provided inputs foretelling maternal death will help the planning to be improved. This study conducts to design an advanced machine learning-based predictive model that predicts the probable maternal death given a few selected attributes to estimate the effectiveness and improvement of health schemes. Health Management Information System (HMIS) datasets from 2018 to 2020 from the Open Government Data (OGD) platform are used to train the proposed advanced machine learning-based predictive model. The proposed predictive model shows its significant performance in the test results.

**Keywords** Data preprocessing · Machine learning · Random forest · Healthcare · Maternal mortality

## 1 Introduction

Maternal mortality refers to the death of women while giving birth or during pregnancy. Maternal health is considered one of the essential aspects of a government concerning socio-economic development. Curving maternal mortality is still a challenge in a developing country like India. The well-being of mothers is to be improved and hence reduce maternal mortality. The maternal mortality ratio is the number of deaths per 100,000 live births due to childbirth/pregnancy and is a commonly used metric for maternal mortality. At the same time, maternal mortality rate stands for the number of females who die between 15 and 49 years of age.

G. Kamei (✉) · I. Banerjee
Department of Information Technology, Indian Institute of Engineering Science and Technology Shibpur, Howrah, West Bengal 711103, India
e-mail: 2020itp003.gaidinlung@students.iiests.ac.in

 
S. Borah et al. (eds.), *Advanced Computational and Communication Paradigms*, Lecture Notes in Networks and Systems 535,
https://doi.org/10.1007/978-981-99-4284-8_2

An extensive study of worldwide maternal mortality between 2003 and 2009 suggests that hemorrhage, hypertensive disorders, and sepsis were the primary causes of maternal deaths worldwide [1]. Most of these cases are preventable [2], yet it is still a dire challenge in developing countries [3–5]. Few special measures to reduce maternal mortality include improved health policies and programs and adequate funding [1]. Most importantly, proper data maintenance at the regional and global levels is essential for extensive systematic analytical study [1].

India has a declining maternal mortality rate trend from 556 (in 1990) to 122 (in 2017). Across the globe, the rate is also declining from 385 (in 1990) to 211 (in 2017) [6]. The goal is to keep maternal mortality as minimum as possible. Strategies are employed to reduce maternal mortality worldwide [5, 7] as well as in India [6]. Sustainable development goals include a plan for reducing maternal mortality.

This study focuses on designing a new predictive model based on advanced machine learning techniques. The model will be trained and tested with the available Health Management Information System data that records each sub-district of Indian states and union territories.

Section 1 introduces maternal health in India and how machine learning and data analytics can be implemented to solve related issues. An extensive literature review is discussed in Sect. 2. The proposed predictive model data acquisition, preprocessing for training and testing in machine learning models are discussed in detail in Sect. 3. The performance of the model is discussed in Sect. 4 and concluded in Sect. 5.

## 2 Literature Review

The abundance of data in data stores of various forms, and the development of new algorithms which can be accessible to low-cost computation systems, give rise to the application of machine learning in various fields of study [8]. There is a recent trend where machine learning has become a significant tool in medical science, healthcare, and epidemiology [9].

Although statistics and machine learning have fuzzy boundaries, there exist major differences. On the other hand, machine learning works on the population to fit the model rather than samples and hence makes better modeling with much data [10]. The requirement of domain knowledge of the subject to model with statistics makes it difficult for computer scientists, data scientists, data analysts, and statisticians. In contrast, the empirical capabilities of the machine learning model itself suffice.

Few selected research studies have been conducted closely related to maternal mortality in recent years and are considered part of the literature survey. In Ghana, maternal mortality is forecasted for 2011 to 2015 using ARIMA data from 2000 to 2010, with a predicted maternal mortality ratio of 959.0, while the actual maternal mortality recorded by hospitals is 996.7 [11]. As a part of prenatal care, the model predicts possible threats to the fetus during the gestation period using records related to maternal history in the Philippines [12]. A similar model is used in Bangladesh [13]. Indian districts having maternal mortality ratio extremities are predicted [14].

Notable studies have been conducted on maternal health in various countries or at least the datasets. There is a need for an exploratory study of maternal mortality from a machine learning and data analytics perspective to improve maternal health in India. It is highly motivated by advancements in practical machine learning techniques and the abundant availability of public health datasets from the relevant source of the Government of India.

## 3 Proposed Predictive Model

### 3.1 *Overview of the Proposed System*

This paper shows a predictive model for probable maternal death fitting the Health Management Information System (HMIS) data. It aims to reduce maternal death during the prenatal period till postpartum recovery. The availability of HMIS data for public access makes it possible to design the predictive machine learning model using a machine learning algorithm. Figure 1 shows the proposed system in the flow diagram.

### 3.2 *Data Acquisition*

Machine-readable HMIS data of 26 states and 3 union territories for the three consecutive years (March 2017–March 2020) in comma-separated values (CSV) file format are collected from a publicly accessible government-owned shareable data web portal called Open Government Data (OGD) Platform India which the National Informatics Centre hosts under the government of India. Total 986 datasets are collected. Each dataset represents the data related to a district.

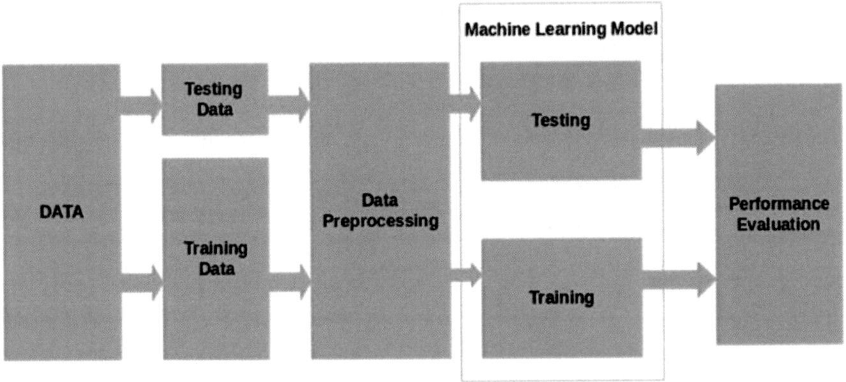


**Fig. 1** Proposed system

### 3.3 Data Preprocessing

The datasets collected consist of missing values, inconsistent datatype, and fragments of data into datasets. Preprocessing is required, i.e., to integrate the datasets into a single dataset, clean to remove noise, and fill in the missing values. Only significant features are retained, and the rest are discarded. The values other than the target feature are normalized. The data preprocessing steps are given in Fig. 2. The frequency distribution of maternal mortality is shown in Fig. 3.

The 986 datasets are collected to be merged into a single dataset. The integrated dataset contains data of the same features for the year 2017–2018, 2018–2019, and 2019–2020 and is aligned, and the dataset of 17,430 rows and 398 columns is obtained. The rows and columns containing all or above 70% of the rows and columns size as the value NA or missing values are discarded. The remaining missing values are filled using weighted k-nearest neighbor imputation [15]. Few instances are detected as outliers using the box plot visualization and are capped to the 90th percentile. The random oversampling method balances the skewed data distribution (Figs. 3 and 4) [16].

**Algorithm 1** : CART Regressor for feature selection

---

**Input**: Data $D$ with $p$ features and N observation
**Output**: Selected features

---

1. The data $D$ of p features and N observations are taken.
2. The target ranges $c_{min}$ to $c_{max}$.
3. $R_M$is the region of partition.
4. For regions 1 to $M$:
   a. The two regions $R_M(j,s\ ) = \{D|D_j \geq s\ \}$ and $R_{M+1}(j,s\,) = \{D|D_j \leq s\}$, $c_M$ and $c_{M+1}$ the response of the respective regions, given $j$ features such that $j \subseteq p$ split from the split point $s$ is decided based on the value that solves $min_{j,s}[min_{c_M} \sum_{x_i \in R_M(j,s)} (y_i - c_M)^2 +$ $min_{c_{M+1}} \sum_{x_i \in R_{M+1}(j,s)} (y_i - c_{M+1})^2]$ $\hat{c}_M = average(y_i|x_i \in R_M(j,s))$ and $\hat{c}_{M+1} = average(y_i|x_i \in R_{M+1}(j,s))$
5. Thus the tree is created.
6. Features involved in the trees are considered for feature selection.

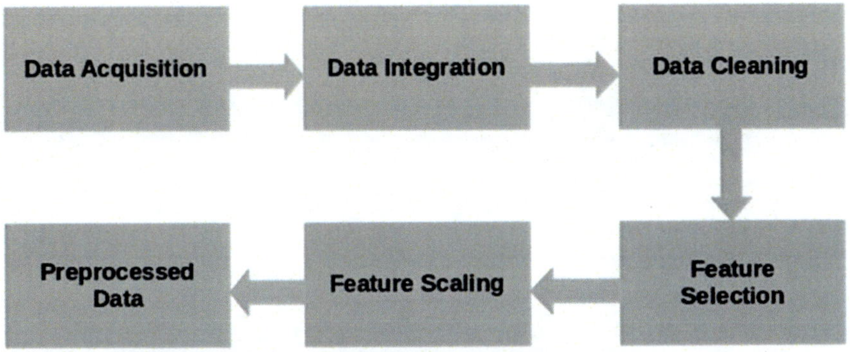


**Fig. 2** Data preprocessing

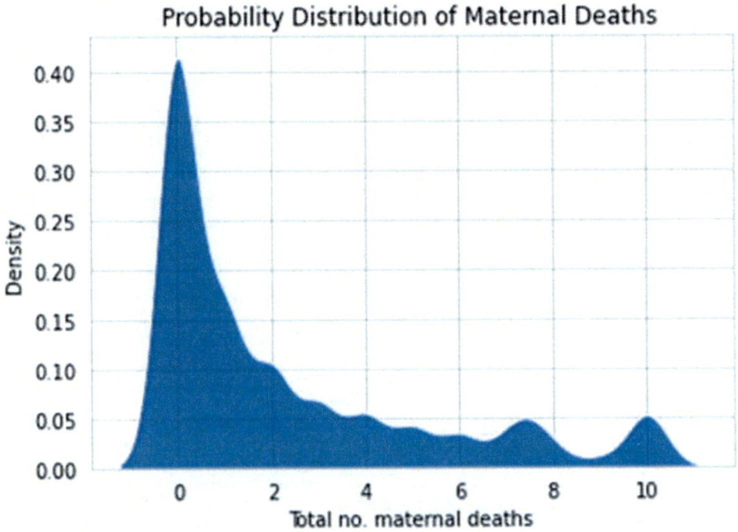


**Fig. 3** Probability distribution of maternal mortality

Using an intrinsic tree-based feature selection algorithm based on a highly randomized Extra Tree Classifier (Algorithm 1) for the target "Total No. of Maternal Death," 398 columns are reduced to 15 features (Table 1). The inclusion of features that are irrelevant and redundant to the target reduces the predictive accuracy, this phenomenon is also known as the curse of dimensionality. The feature values other than the target values are normalized using Min–Max Scaler, this procedure is also called feature scaling. The values range from 0 to 1. The dataset can be considered ready for designing a machine learning model.

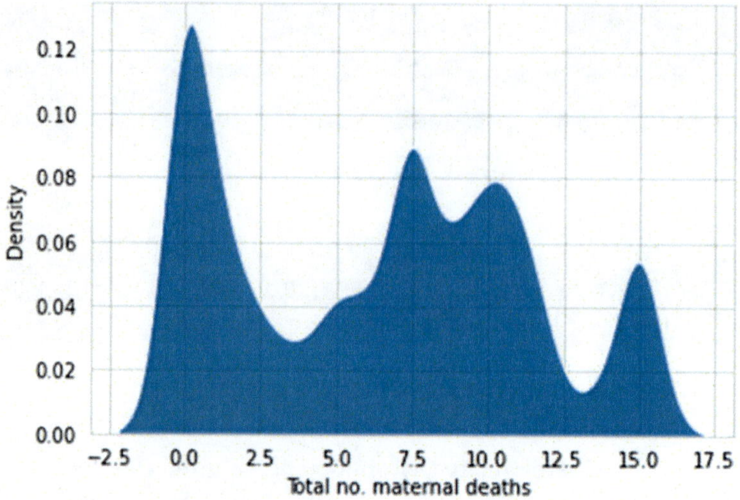


**Fig. 4** Probability distribution of Maternal Mortality after oversampling

**Table 1** Selected features

| Sl. no | Selected features |
|---|---|
| 1 | Number of pregnant women registered within first trimester |
| 2 | TT2 given to pregnant women (numbers) |
| 3 | TT booster given to pregnant women (numbers) |
| 4 | Number of pregnant women having Hb level <11 (tested cases) |
| 5 | Number of home deliveries attended by non-SBA trained (trained TB/Dai) |
| 6 | Deliveries conducted at public institutions |
| 7 | Number of women discharged under 48 h of delivery in public facilities |
| 8 | Women received 1st postpartum check-up between 48 h and 14 days |
| 9 | Total number of reported stillbirths |
| 10 | Number of home deliveries |
| 11 | Number of newborns weighed at birth |
| 12 | Number of newborns having weight less than 2.5 kg |
| 13 | Number of newborns breastfed within 1 h |
| 14 | Number of PW given one Albendazole tablet after 1st trimester |
| 15 | Number of pregnant women tested for Syphilis |

## 3.4 Proposed Model

To design the proposed model, it considers the machine learning algorithms (Table 2), and the best-performing model is anticipated for designing the predictive model. The linear, Lasso, and ridge regressions are defined by the general linear regression Eq. (1), whereas Bernoulli random forest, decision tree, and KNN follow polynomial regression shown in Eq. (2). The performances of each model are evaluated using K-Fold cross-validation on the preprocessed data. Ten random subsets called folds are taken from the dataset, ninefolds are used for training, and the rest for testing [17].

$$\hat{y} = h_\theta(x) = \theta.x \tag{1}$$

where $\theta$ is a model parameter with bias $\theta_0$ is included, $\hat{y}$ is a predicted value, and $h_\theta(.)$ is a regression function with model parameter $\theta$.

$$\hat{y} = f(X, \beta) + \varepsilon \tag{2}$$

where X is the instance's feature vector, β is the vector of model parameters, *f(.)* = a known regression function, and ε = error.

To find the optimal parameter $\theta$ that gives the best cost function mean squared error, a normal equation Eqs. (3) and (4) is commonly used.

$$\hat{\theta} = \left(X^T X\right)^{-1} X^T y \tag{3}$$

$$\hat{\theta} = X^+ y \tag{4}$$

where $X$ is the instance, $X^T$ is the transpose of $X$, $y$ is the vector of target values, and $X^+$ is the pseudoinverse of $X$.

**Table 2** Performances measurement of the models

| Model | RMSE |
|---|---|
| Linear | 2.936432 |
| Ridge regression | 2.935068 |
| Lasso regression | 3.106856 |
| CART regressor | 1.738600 |
| Bernoulli random forest | 0.799740 |
| Random forest regressor | 0.897561 |

The performances of all the possible permutations are evaluated with mean absolute error, root mean squared error, and explained variance score, which are specified in Eqs. (5), (6), and (7) respectively. We will use the root mean squared error and mean absolute error for model performance evaluation and explained variance for precision.

$$MAE = \frac{1}{n}\sum_{i=0}(y_i - \hat{y_i}) \tag{5}$$

$$RMSE = \frac{1}{n}\sum_{i=0}^{n}(y_i - \hat{y_i})^2 \tag{6}$$

$$Explained\,Variance = 1 - \frac{var(y - \hat{y})}{var(y)} \tag{7}$$

## 3.5 Model Design

The model based on Bernoulli random forest regressor gives the best performance (Table 1), and it is considered for the predictive model design [18]. The proposed predictive model based on Bernoulli random forest regressor is fed with the preprocessed data. The dataset is divided into two sets, training data and test data in a ratio of 4:1. The proposed model is trained using the training data. And, the performance measurement is carried out using the test data.

### 3.5.1 Training Dataset Partitions

The structured part is used for constructing trees and forests. Selecting the best features and splitting points in each node is based on the data in structured parts. The estimation part is used for prediction, i.e., averaging predicted targets from leave nodes.

### 3.5.2 Decision Tree Construction

In the proposed predictive model, the decision tree construction (Algorithm 2) is based on the Bernoulli distribution. One candidate feature is selected in probability *p* or *1−p*. The splitting point selection is based on impurity optimization. The nodes are further split into structured and estimation parts recursively.

**Algorithm 2** : Decision Tree Construction

---

**Input**: Structured part, Estimation part, parameter $k_n$
**Output**: Decision Tree

---

7. while stop condition is FALSE do
    a. Assume a node of the tree is D. And D consists of the structured part $D_S$ and $D_E$.
    b. Select a candidate features subset. Choose one candidate features in probability $p_1$ or $\sqrt{D}$ candidate features in probability $1 - p_1$.
    c. Select the splitting point s using structured part $D_S$. For each candidate feature, randomly sample a point to split with probability $p_2$or $1 - p_2$. The splitting point is selected by optimizing given the impurity criterion.
    d. The splitting feature and point is obtained with the equation $I(s) = MSE(D^s) - MSE(D^{ls}) - MSE(D^{rs})$. The structured part is divided into two child nodes, $D^l$ and $D^l$. They are further divided into, ($D^{ls}$and $D^{le}$) and ($D^{rs}$and $D^{re}$).
    e. if the number of estimation part $D^{le}$and $D^{re}$ are larger than $k_n$ then go to 1.
    f. else stop condition is True
8. end while
9. return a decision tree.

### 3.5.3 Forest Construction

M decision trees are constructed and prediction (Algorithm 3) is based on averaging prediction from individual decision trees.

**Algorithm 3** : Forest Construction algorithm

---

**Input**: Training data $D_n$, and the number of trees M $\in N^+$, parameter $k_n$
**Output**: Prediction for a query point x

---

1 **for** i = 1 to i = M do
   1.1. Partition the training data into Structured and Estimation.
   1.2 Decision Tree for BRF is created using Structured and Estimation parts (Algorithm 1).
   1.3. Predict for x with individual trees,
   $$\hat{y} = \frac{1}{N(A^E(x))} \sum_{(x,y) \in A^E} Y$$
2 **end for**
3 Predict for x using Bernoulli Random Forest by averaging trees, $\underline{\hat{y}} = \frac{1}{M} \sum_{i=1}^{M} \hat{y}_i$
4 **return** predicted value for x

---

The learning performance of the proposed predictive model is consistent for even for the larger variance and size of the dataset. The proposed predictive model can be applied for both classification and regression with consistency. It is proven with the lemmas given below.

For regression, the dataset $D_n$ is taken into an $f$ estimator. If the risk function $R(f)$ satisfies the following condition:

$$R(f) = E\left[\left(f(X, C, D_n) - f(X)\right)^2\right] \to 0, n \to 0$$

We check for the consistency of each individual tree in the proposed predictive model. Two basic lemmas and a corollary for proving the consistency of the proposed model are given below.

**Lemma 1** *The consistency of each* $\{f\}$ *predictor ensures the averaging estimator* $\underline{f^{(N)}}$ *for N predictors with randomness* $C$.

**Lemma 2** *For (X, Y) distribution, estimators {f} are consistent if*

$$\lim_{n \to 0} E[(f(X, C, D_n) - f(X))^2 | I] = 0$$

**Corollary 1** In a partition of the training set, if the leaf predictors are fitted and $E\left[Y^2\right] < \infty$, then the consistency of the tree construction is ensured when $n \to +\infty$, the leaf node cell is approximated 0, but at the same time, the leaf node has an infinite estimation part.

***Proof*** If $n$ tends to 0, the instance size of the leaf nodes $K\left(N^E(X)\right) \rightarrow \infty$ is trivial because $K\left(N^E(X)\right) \geq k_n$.

Diameter of the leaf nodes, $dia(N^E(X))$ approximates 0 in probability. $size(r)$ is the size of the $r^{th}$ feature of $N(X)$ nodes. The expectation of $size(r)$ approximates 0, for $r = \{1, 2, 3, \ldots, D\}$ $size_{max}(a)$ is the greatest size in the child nodes of features. Random sampling in [0, 1] with probability $p_2$ or by impurity criterion optimization with $1 - p_2$ to obtain the splitting point.

$$E[size_{max}(a)] \leq (1 - p_2) + p_2 E[max(U, 1 - U)] = (1 - p_2) + \frac{3}{4}p_2 = 1 - \frac{1}{4}p_2$$

$$E[Size_c(a)] = P(C_1)E[size(a)|]$$

where $U$ is a random sample from $Uniform[0, 1]$ that gives a sample with $p_2$. A candidate feature $F_1$ is selected with $p_1$ and $\sqrt{D}$ candidate features $F_2$.

$$\begin{aligned} E[size_{child}(a)] &= P(F_1)E\left[size_{child}(a)\middle|F_1\right] + P\left(\underline{F_1}\right)E\left[size_{child}(a)\middle|\underline{F_1}\right] \\ &\leq p_1 P(F_1)E[size_{child}(a)|F_1] + (1 - p) \\ &= p_1 P(F_2|F_1)E[size_{child}(a)|F_1, F_2] + P(\underline{F_2}|F_1)E[size_{child}(a)|F_1, \underline{F_2}] \\ &\quad + (1 - p) \leq p_1\left(\frac{1}{D}E[size_{max}(a)] + 1 + \frac{1}{D}\right) + (1 - p_1) \leq 1 - \frac{p_1 p_2}{4D} \end{aligned}$$

The height of the tree is *h;* for *h* iterations,

$$E[size(a)] \leq \left(1 - \frac{p_1 p_2}{4D}\right)^h \tag{8}$$

The consistency for the proposed predictive model is shown in Eq. (10) for $n \rightarrow \infty$.

## 4 Result and Discussion

The implementation of data preprocessing, training model, cross-validation, and testing are performed in a personal computer system with system memory of 8 GB RAM, storage of 512 SSD, AMD Ryzen 7 processor, and AMD Renior integrated graphics. The system configuration is sufficient for designing a mid-size machine learning model. We use the environment Anaconda 36.16.5, which includes NumPy and SciPy for basic mathematical and statistical operation, Scikit-learn for machine learning model designing, Pandas for data analysis and preprocessing, and Seaborn for visualization.

The primary objective of the proposed predictive model is to predict the probable maternal mortality using the supervised data model. The performance of the proposed model on the test set is given in Tables 3 and 4 tenfold cross-validation score of the model on the training data. Figure 5 shows the learning curve.

**Table. 3** Performance of Bernoulli random forest on the test data

| Performance metrics | RMSE values |
|---|---|
| Mean absolute error | 0.1393912937743190 |
| Root mean squared error | 0.6034566810909524 |
| Mean squared error | 0.3641599659533074 |

**Table. 4** Cross-validation score on the training data with Bernoulli random forest

| Metrics | RMSE values |
|---|---|
| K-Fold cross-validation score $K = 10$ | 0.67837271<br>0.60089165<br>0.64645753<br>0.61822632<br>0.65254681<br>0.67818794<br>0.64397005<br>0.60580572<br>0.71082762<br>0.64735344 |
| Mean score | 0.64826397 |
| Standard deviation | 0.03272027 |

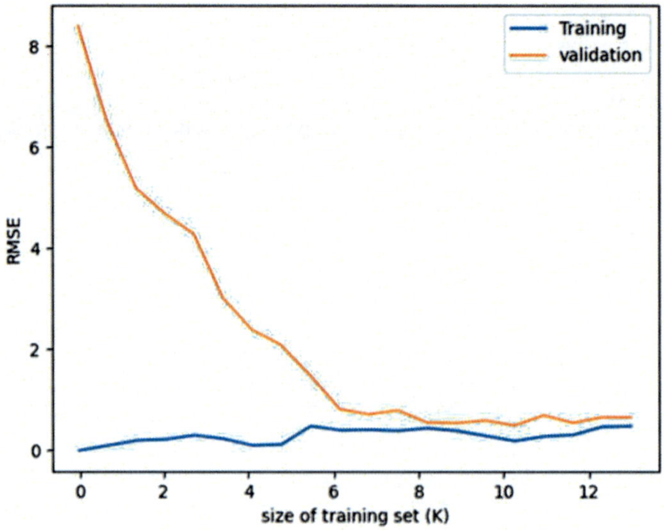


**Fig. 5** Learning curve

With the increasing size of the data points, training RMSE increases and validation RMSE decreases and stabilizes. The difference in training and validation error is considerably large, and the training error in RMSE is smaller than that of validation. This depicts overfitting in the model.

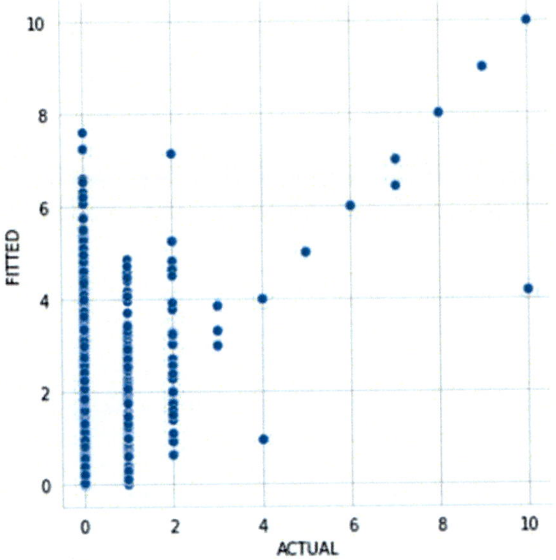


**Fig. 6** Fitted versus actual

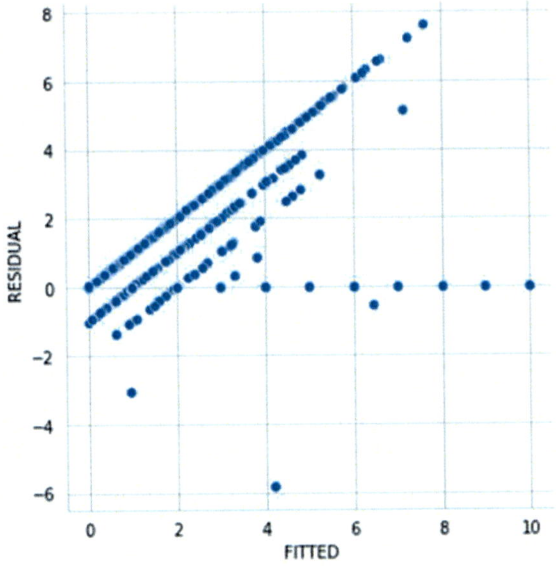


**Fig. 7** Residual versus fitted

The model depicts better accuracy for the data points with "Total number of maternal deaths" values farther away from zero (Figs. 6 and 7). However, errors are prevalent for the values near zero. The target "Total number of maternal death" values range from 0 to 10, and the prediction on the test set with an RMSE value of 0.6482 is acceptable. The difference in RMSE of validation and test scores is 0.03 (approx.).

Despite being the best cross-validation score among the models, hyperparameter tuning for the proposed predictive regressor model further enhances its performance. Figure 4. shows some scale of overfitting, and it can further be improved to some extent with the implementation of hyperparameter tuning with a greater range of parameters. Hyperparameter tuning usually takes high computation power to perform in a shorter time. For this very reason, it is not always feasible for hyperparameter tuning. Additional datasets of the preceding and succeeding years can also help curve the skewness in the data and the overall performance of the model.

## 5 Conclusion

In a study, planning and strategical analysis for the improvement of maternal health and reduction of maternal mortality in India, advancement in technology, namely algorithms and computation power, and the abundant readily availability of data related to public health makes it possible for the field of machine learning to contribute to solving problems in public health. In this paper, the prediction model predicts maternal mortality based on the operation, conduct, and management that affects maternal deaths. The proposed maternal mortality predictive model gives promising results compared to its counterparts. Planning healthcare schemes and measures can be improved using this predictive model to estimate effectiveness. The proposed model can further be improved by including more data points and higher computation power facilitating exhaustive hyperparameter tuning for optimal learning and predictive performance.

## References

1. Say L, Chou D, Gemmill A, Tunçalp Ö, Moller AB, Daniels JD et al (2014) Global causes of maternal death: A WHO systematic analysis. Lancet Global Health 2(6):e323–e333
2. Oona MRC, Graham WJ (2006) Strategies for reducing maternal mortality: getting on with what works. Lancet 368(9543):1284–1299, ISSN 0140–6736. https://doi.org/10.1016/S0140-6736(06)69381-1.
3. Reyes Espinoza IS (2020) Obstetric Hemorrhage, its role in maternal morbidity and mortality, and the importance of its diagnosis, prevention and timely management. Mexican J Med Res ICA 8(15):37–44. https://doi.org/10.29057/mjmr.v8i15.3906
4. Atallah F, Goffman D (2020) Improving healthcare responses to obstetric hemorrhage: Strategies to mitigate risk. In: Risk Management and Healthcare Policy, vol 13, pp 34–43
5. Emmons AB (1929) Maternal mortality and its prevention. Am J Obstetr Gynecol 17(5):738–740. https://doi.org/10.1016/s0002-9378(29)90991-1

6. Ministry of Health and Family Welfare. Annual Report 2019–20. Health report. Govt. of India
7. Schuitemaker NW, Gravenhorst JB, Van Geijn HP, Dekker GA, Van Dongen PW (1991) Maternal mortality and its prevention. Europ J Obstetr Gynecol Reprod Biol 42(Suppl: S31-S35). PMID: 1809606.
8. Jordan M, Mitchell T (2015) Machine learning: trends, perspectives, and prospects. Science 349:6245. https://doi.org/10.1126/science.aaa8415
9. Wiens J, Shenoy ES (2018) Machine learning for healthcare: on the verge of a major shift in healthcare epidemiology. Clin Infect Dis 66(1):149–1531. https://doi.org/10.1093/cid/cix731
10. Bzdok D, Altman N, Krzywinski M (2018) Statistics versus machine learning. Nat Methods 18(4). https://doi.org/10.1038/nmeth.4642
11. Sarpong SA (2013) Modeling and forecasting maternal mortality; an application of ARIMA models. Int J Appl Sci Technol 3(1)
12. Bautista JM, Quiwa QAI, Reyes RSJ (2020) Machine learning analysis for remote prenatal care. In: 2020 IEEE region 10 conference (tencon), pp. 397–402. https://doi.org/10.1109/TENCON50793.2020.9293890.
13. Bin Alam MS, Patwary MJA, Hassan M (2021) Birth mode prediction using bagging ensemble classifier: a case study of Bangladesh. In: 2021 International conference on information and communication technology for sustainable development (ICICT4SD), pp 95–99. https://doi.org/10.1109/ICICT4SD50815.2021.9396909
14. Kumar R, Sharma AK, Barik S, Kumar V (2021) A nested stacking ensemble model for predicting districts with high and low maternal mortality ratio (MMR) in India. Int J Inf Technol 13:433–446
15. Troyanskaya O, Cantor M, Sherlock G, Brown P, Hastie T, Tibshirani R, Botstein D, Altman RB (2001) Missing value estimation methods for DNA microarrays. Bioinformatics 17(6):520–525
16. Menardi G, Torelli N (2014) Training and assessing classification rules with imbalanced data. Data Min Knowl Disc 28(1):92–122
17. Ojala M, Garriga GC (2010) Permutation tests for studying classifier performance. J Mach Learn Res 11(6)
18. Wang Y, Xia ST, Tang Q, Wu J, Zhu X (2017) A novel consistent random forest framework: Bernoulli random forests. IEEE Trans Neural Netw Learn Syst 29(8):3510–3523

# Multi-criteria-Based Page Ranking Using Metaheuristic Swarm Optimization

**Santosh Kumar and Sanjai Mohan Verma**

**Abstract** The size of data on the internet has grown enormously and also growing continuously. Finding any information for a user in response to query has become a challenging task. In this, search engines are playing a vital role. Yet, these are unable to a retrieve the relevant information. These search engines use some limited criteria to fetch the results in response to the user's query after ranking and indexing the pages. In this paper, a multi-criteria-based approach has been proposed to rank web pages. These criteria include keywords, user access count, unique visitors, stay time, hubs, and authority values. The proposed approach uses particle swarm optimization to find the optimal results. And, this algorithm has been compared with some existing algorithms based on iterative improvement and genetic algorithm. It has been observed that the proposed approach is able to find the most relevant top-ranked pages.

**Keywords** Web mining · Web usage mining · Web content mining · Web structure mining · Metaheuristic · Particle swarm optimization

## 1 Introduction

These days the internet is growing in extremely fast pace and has become an integral part of every person's life. Web mining is the data mining technique to extract the useful information from the web based on its usage, content, and structure of the web pages. The internet content is growing day by day and as it grows, the search for getting the right information from the web is getting challenging and has become a hot topic for researchers. Using web mining techniques, the web

S. Kumar (✉)
Department of Computer Science and Engineering, Galgotias University, Greater Noida, Uttar Pradesh 201310, India
e-mail: santoshg25@gmail.com

S. M. Verma
ABES Engineering College, Ghaziabad, Uttar Pradesh 201009, India

 
S. Borah et al. (eds.), *Advanced Computational and Communication Paradigms*, Lecture Notes in Networks and Systems 535,
https://doi.org/10.1007/978-981-99-4284-8_3

can be made more beneficial and the efficiency of interaction with the web can be increased. There is much research going on for web mining techniques [1] based on the metaheuristic swarm-based stochastic methods such as Particle Swarm Optimization (PSO), ant colony optimization algorithms, etc. Metaheuristics can be defined as a high-level algorithm framework independent of the problems which provided the guidelines or methods to develop heuristic optimization algorithms. Meta means beyond high level and heuristic means to search. Meta-heuristics can be divided into genetic/evolutionary, nature-inspired, physics/chemistry-based, human-based algorithms, etc. Metaheuristic algorithms try to find the best (feasible) solution from the possible solutions of an optimization problem. There have been many researches on the usage of PSO with some customization to improve the efficiency of web mining, e.g., web search, clustering, and structure of the websites. Some of the papers as mentioned in the next section have been studied for reference. Most of the papers have proposed to use PSO with some hybrid or customization to improve the solution of the problems which may be termed as NP-hard problems. Association rules mining ARM is a data mining technique used to find the relationship between interrelated data. GA and PSO being population-based stochastic search methods which can be used for finding the association rules. In Agarwal and Nanavati [2] hybrid algorithm GA-PSO is proposed to find important and correct mined rules. These days, people use the internet a lot and give their opinion on any book, topic, or movie if it is good or bad. Basari et al. [3] employed the support vector machine (SVM) supervised learning method to identify data by categorizing it into two groups and determining the opinion. Hybrid particle swarm optimization technique has been used to solve this dual optimization technique to enhance the solution. The resources are widely dispersed due to the internet's rapid growth and lack of a unified management and structure, which decreases the effectiveness of using the internet. In Liangtu and Xiaoming [4], Vector Space Model (VSM) and improved PSO with reverse thinking particles (PSORTP) are used to improve web text processing. Web log mining based on the intrinsic information from server logs [5] is a crucial component of online use mining to increase the functionality of web servers. The improved PSO technique is suggested for phase-wise process improvement (EHPSO). To enhance web clustering using the web pattern, a new clustering model called RVPSO-K [6] based on PSO and K-means method is developed. There has been some level of success seen by most of the authors in their papers with the customized/hybrid algorithms they have proposed. In order to improve the performance of web clustering utilizing web mining techniques, this study uses swarm-based optimization algorithm, PSO, in web mining.

Due to the large size of data on the internet, it has been challenging to show users the most relevant sites based on the search criteria. Sometimes, there are many irrelevant sites displayed to the user. In this paper, an approach to find the optimal solution with the most relevant sites using PSO by using the appropriate fitness function and the suitable variables to find the better solution.

### 1.1 Problem Statement

Finding optimal p page out of a large number(n) of pages is an NP-complete problem, the value nCp is very large and computationally not feasible. So, it is a combinatorial optimization problem. So, urls have been given a unique ID which is an integer value as shown in Table 1. Combination of size p we will represent a candidate solution as shown in Table 2. The algorithm PSO has been used to find the more relevant fee page based on the multi-criteria.

$$Possible\ Candidate\ Solutions\ i.e.\ Size\ of\ Search\ Space = C_p^n = \frac{n!}{p! \times (n-p)!} \quad (1)$$

where $n$ is possible URLs and is large;

$t$ is top-ranked '$t$' web documents.

## 2 Related Work

The approaches discussed in the research papers that were mentioned as part of the current research paper's literary review are briefly described in this section. In these papers, various variations of the particle swarm-based optimization techniques are used with some hybrid approaches to get better results.

In Chen and Zhang [7], the authors proposed an approach to organize the structure of a website by looking into web access, structure, and web contents. The link clicks are used for constructing web user utility. The purpose is to get a better web structure using the web utility according to the criteria using the frequencies and time division of the web links. The hybrid method GA/PSO is tested on the real website and it resulted with web user benefits by links. In Rajdeepa and Sumathi [8], using PSO algorithm, the authors proposed a fresh strategy called PSOHITS. Topic drift results

**Table 1** Mapping of urls with ids

| Webpages | WebP-1 | WebP-2 | WebP-3 | … | WebP-n |
|---|---|---|---|---|---|
| Ids | 1 | 2 | 3 | … | n |

**Table 2** Candidate solution

| Sample one candidate solution of top-5 web pages | | | | |
|---|---|---|---|---|
| Unique ids representing URLs generated randomly, random (1, n) | | | | |
| 40 | 22 | 54 | 325 | 21 |

from the Hyperlink-Induced Topic Search (HITS) algorithm's equal weighting of links inside web pages. Thus, method looks for the relevant pages to predefined topics using PSO algorithm and increases the crawling options. The topic relevancy of the web page is determined using the hyperlink metadata. This approach seeks pertinent subjects and increases the precision of data collection. To obtain accurate and deterministic mined rules, writers in Agarwal and Nanavati [2] suggested using association rule mining with a hybrid genetic algorithm and particle swarm optimization (GA-PSO algorithm). The key benefit of this suggested method is that the exploration and exploitation tasks are balanced by a mix of multi-objective GA and multi-objective PSO. Authors have evaluated this method on the Bakery data set to show that this model converges four times faster than min-objective hybridization. In Basari et al. [3], authors talked about the text mining from online disclosure of sources like Twitter about the opinion of any movie as good or bad based on the text messages. Opinion mining is the use of computational linguistics, natural language mining, and other techniques to determine if a movie is good or poor based on message opinion. According to the authors, utilizing hybrid PSO to increase the accuracy level of parameter selection has increased it from 71.87 to 77%. In Zhan [9], the authors emphasized the use of the PSO technique to aid in the clustering of web data to enhance the website's structure. Because there is so much information available, resources are widely dispersed online and lack a uniform management and structure. The two biggest hurdles in data mining are extracting web text feature and clustering. Web text data clustering uses such nature-based evolutionary methods, and the PSO algorithm will increase processing efficiency. In Ghorpade-Aher and Bagdiya [10], the authors discussed the use of the web in social media and the use of big data for extracting the summary of content. It is a painful task to extract the essence of social media, here the mining is re-formulated into multi-objective optimization (MOD) for an extractive summary. A Gravitational Search Algorithm (GSA) is used to perfect extracting the concise summary and mixing it with the PSO has produced better results which outperformed other methods. In Mosa [11], the authors proposed the use of PSO for the classification of documents over the internet which are growing exponentially day by day. In this paper, a PSO-based algorithm for document classification is presented. The authors compared their proposed algorithm with other conventional document classification algorithms and saw much better performance. In Wang et al. [12], the authors talked about the new web session clustering algorithm that uses PSO algorithm. The authors proposed a swarm intelligence-based PSO—clustering algorithm for the clustering of web user sessions. Results have shown better results than the benchmark k-means clustering algorithm for clustering web user sessions. In Liu et al. [13], the authors talked about the benefits of the particle swarm optimization algorithm (PSO) but it suffers from early convergence problems. In this paper, the authors proposed a modified PSO initialized by OLHD (Optimal Latin Hypercube Design) which improves the efficiency of the basic version. The extended algorithm OLCPSO (Optimal Latin hypercube and Classification and regression improving PSO) (Optimal Latin hypercube and Classification and regression improving PSO) successfully improves the efficiency of basic PSO. In Liangtu and Xiaoming [4], the authors talked about the growth of the internet at an outstanding rate and therefore the quantity of data on the net is overwhelming. Web text feature extraction is considered the main downside in text mining. There is a tendency to use Vector Space Model (VSM) because of the description of internet text and gives a completely unique feature extraction algorithmic rule that relies on the improved particle swarm optimization with reverse thinking particles (PSORTP). This algorithmic rule can greatly improve the potency of internet texts. In Panda et al. [14], the authors proposed some abstract theories to extract data through completely different web mining techniques like cluster, FIS, ANN, LGP, etc. We tend to additionally target many aspects of applications of internet mining in E-commerce & Business Intelligence. Finally, we tend to mentioned Swarm Intelligence (SI) techniques that are supported by distributive self-organized systems like Ant Colony Optimization (ACO), Stochastic Diffusion Search (SDS), and PSO briefly during this survey that is most popular due to its broad uses and ease. In Kundra et al. [5], the authors proposed an enhanced method of web log mining and online direction prophecy to boost the performance of log mining. In Kothari et al. [15], the authors presented a literary review of the papers on PSO algorithm which is inspired by the natural flock of birds. It describes the journey of PSO from its invention and its application in different real problems. They have presented the comparative table of implementation of PSO and reviewed its success. This paper gives inspiration to the readers to join the PSO world. In Chen and Zhang [6], the authors proposed a new clustering method RVPSO-K which is based on combining K-means and PSO algorithms as per the type of clustering of web usage patterns and the problems of PSO. Experiments on the web log data after data preprocessing have shown that RVPSO-K is better than the K-means based on PSO algorithm for stability, precision, and convergence speed. Additionally, as a proposal of another threshold, the agility of recommended web was enhanced to meet the specific requirements. In Grosan et al. [16], the authors presented the theory of nature-based biological motivation and swarm intelligence and its usage in participle swarm optimization and ant colony optimization algorithms. The basic terms are described and matched with some of the past and current processes using swarm intelligence techniques. In Kumar and Kumar [17], the authors proposed a novel approach using web content, structure, and usage data for prioritizing web documents. They have used many areas such as adaptive web development, web personalization, search engine optimization, recommender systems, etc. The proposed approach has produced better page ranking of the web documents in comparison to other approaches like WDPGA, WDPSA, etc. In Chaudhary and Gupta [18], the authors introduced a method to prioritize the pages based on web content and web usage data. The proposed method is based on GA to supply quality results of user query. Prioritization of websites is a type of NP-complete problem. The method used the parameters from web content and web usage mining and produced better results than the existing approach. In Gupta and Singh [19], authors proposed an approach to web mining using GA which in turn results in searching better results from a simple search on the web instead of getting irrelevant information. The novel approach uses the fitness function to find the goodness of the solution. It considers the web access based on the frequency, unique users, and duration and giving various costs to parameters to find the better results. It also takes account of the Hubs (Links to other sites) and Authorities (Other pages that are referring to any page). In Sarzaeim et al. [20] and Zou et al. [21], the authors described that Teaching Learning-Based Optimization (TLBO) has been successfully applied on several optimization problems like web page ranking [22].

### 2.1 *Observations from the Literature*

After reviewing the research papers while literary reviews that have been mentioned in the above-related work section, it is observed that raking in suitable sites has not been researched much and can be taken into account for utilizing the metaheuristic swarm-based optimization techniques to get better results.

## 3 Proposed PSO-Based Approach

### 3.1 *Particle Swarm Optimization (PSO)*

This is based on the nature-inspired social behavior of bird flocks flying in the sky to search for food where they do not know the actual location of the food but know the distance of the food. The group of birds flock is known as swarm and each bird is considered as a particle. The PSO is a stochastic algorithm which is based on the population for solving NP-hard problems to get the best solution, although it does not guarantee the best solution but tries to get the best solution in the minimum time. In this algorithm, all the birds follow the bird which is closest to the food particle. In this algorithm, parameters are adjusted appropriately to find the best solution (Fig. 1).

## 4 Proposed Approach

Below is the approach to utilize the Genetic Algorithm and PSO techniques to find the optimal solution to get the most efficient set of websites from all the websites based on the provided text while searching on the web. Following are the steps that will be followed to implement the algorithm:

1. For each generation pBest of each particle's fitness value is compared with its previous pBest and if better then pBest will be updated with the better value.
2. For each particle, do
   a. $v_p = \text{compute\_velocity}\ (x_p, \text{pBest}, \text{gBest})$

      $$v_{i+1} = w \times v_i + c1 \times r1(\text{pBest} - x_i) + c2 \times r2(\text{gBest} - x_i)$$
   b. $\text{update\_position}\ (x_p, v_p)$

Algorithm: pseudo code of the PSO

```
For each particle i
        Initialize particle present[i]
        Initialize particle velocity V[i]
        Initialize particle position P[i]
        Initialize particle best pBest[i]
End For
Initialize inertia weight w
Initialize cognitive constant c1
Initialize social constant c2
Do
For each particle i
    Calculate fitness value
    If fitness value >pBest[i]
        Set particle present value as new pBest[i]
End For
Choose the particle with best fitness values of all
the particles as gBest
For each particle i
   Calculate particle velocity V[i] = w × V[i]
   + c1 ×rand(0,1) × (pBest[i] – present[i])
   + c2 ×rand(0,1) × (gBest – present[i])
   Update particle position P[i] = P[i] + V[i]
End For
While maximum iteration is exhausted or min error
is attained
```

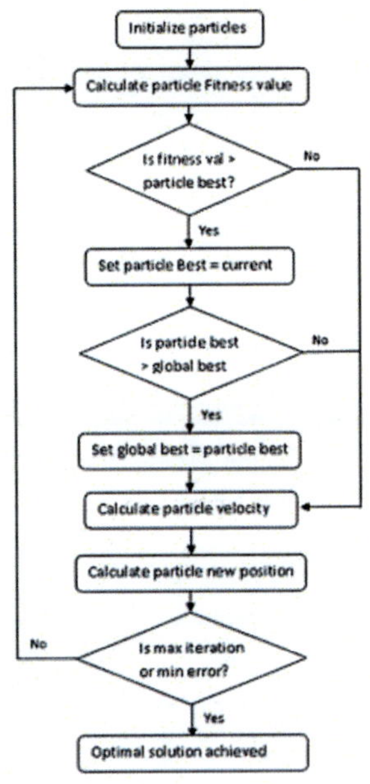


**Fig. 1** PSO algorithm

$$x_{i+1} = x_i + v_{i+1}$$

End For

3. Final aim is to update the position $x_p$ $(x_1, x_2, , \ldots.)$, consider chromosome (e.g., P1) and find five neighbors of P1 say PN1, PN2,.. PN5
4. Calculate X1 values by combining P1 with its neighbors to find new generation (e.g., P1 & PN1, P1 & PN2, …, P1 & PN5).
5. The value which is near to X1, updates P1 with that neighbor. Thus, we find the better solution after multiple iterations.

## *4.1 Existing Page Ranking Algorithm Comparison*

Table 3 gives a comparative study of existing PageRank algorithms and the proposed one. The limitation of the existing PageRank algorithms is that it done is using the multi-criteria which are used in the proposed algorithm.

**Table 3** Comparison of PageRank algorithms

| Algorithm | Criteria selected | Complexity |
|---|---|---|
| PageRank [23] | Inlinks and outlinks | O log(N) |
| Weighted page rank [24] | Inlinks and outlinks | Less then O Log(N) |
| SEO [25] | Page speed, page views, and page layout | – |
| Optimized weighted PageRank [26] | Blogs, inlinks, and out-links | Less then O Log(N) |
| ITS [27] | Web usage log | Log(N) |
| Web content algorithm [28] | Web content data | – |
| MCIR [29] | Inlinks, outlinks, keywords, page speed, domain age, content quality, and ignoring bad links for trusted web page results | O log(N) |
| WPPSO | Keywords, unique visitors, access count, hubs, authority, and time stayed | – |

## 5 Result and Discussion

The algorithms have been implemented using Java on the Windows platform. Table 4 shows the comparative analysis of cost of selecting optimal p pages by the three algorithms based on iterative improvement, genetic algorithm swarm-based optimization algorithm. It can be seen from the table that the algorithm WPPPSO gives better quality p pages.

**Table 4** Comparative analysis of cost of selecting optimal p pages

| Population_size = 100 | | | |
|---|---|---|---|
| Optimal-P pages | WPPII | WPPGA | WPPPSO |
| | Neighbors = 50 | Selection = BTS, Crossover = 70%, Mutation = 5% | C1 = 1, C2 = 1 and W = 1 |
| 5 | 2190 | 2315 | 2489 |
| 6 | 2623 | 2748 | 2957 |
| 7 | 3268 | 3393 | 3446 |
| 8 | 3509 | 3634 | 3922 |
| 9 | 4178 | 4303 | 4466 |
| 10 | 4470 | 4595 | 4851 |

The algorithm, WPPII has been run for a number of neighbors equal to 50 for a different number of pages. The algorithm WPPGA has been run for different combinations of crossover and mutation rates (60, 70, 80, 90) and (0.05, 0.10). In all the cases, it has seen from Fig. 2 that the proposed WPPPSO algorithm is able to select better quality p pages in comparison to other algorithms.

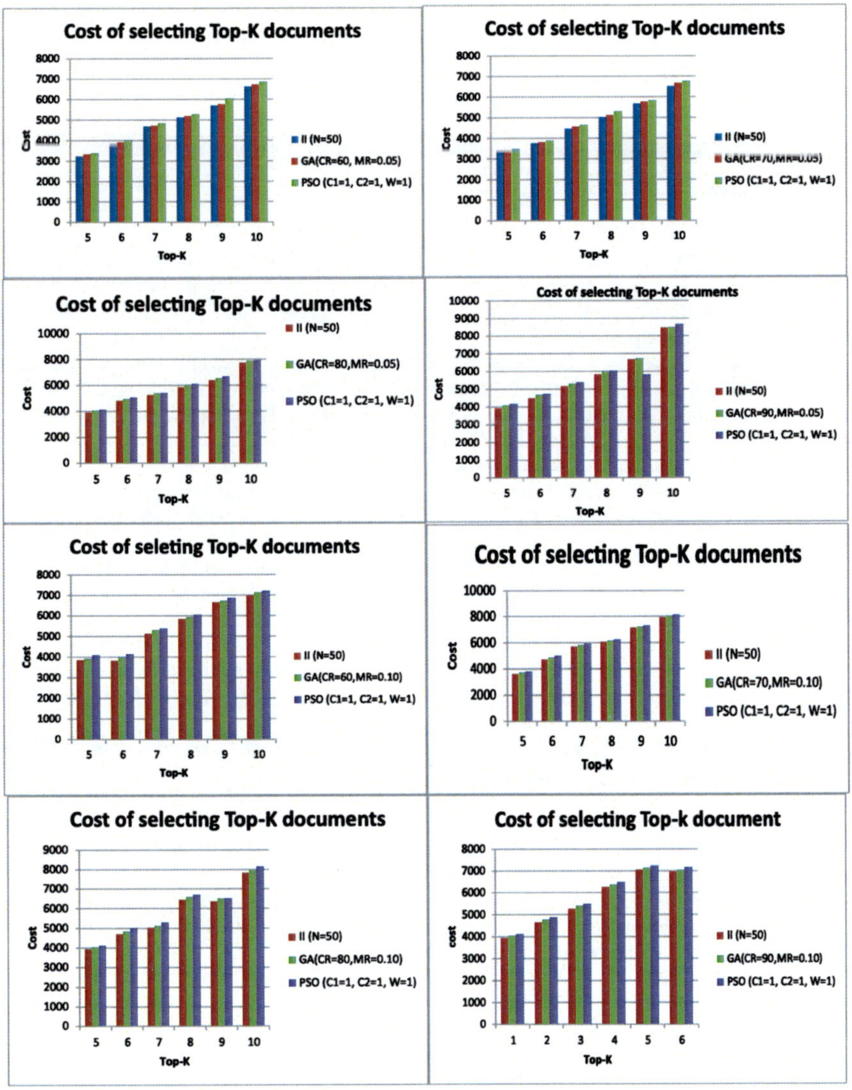


**Fig. 2** PSO-based selection of top-p (top-K) pages versus cost

## 6 Conclusion

In this paper, multi-criteria-based algorithm using PSO has been proposed to find optimal p pages from a huge source of information. The novelty of the algorithm is the multi-criteria reads exploit the web content, web usage, and web structure data to rank the web page. Experimentally, it has been shown that the proposed algorithm for selecting top-p web pages is performing better in comparison to the legacy-based algorithms based on evolutionary approach in terms of the cost of selection of top-ranked pages.

## References

1. Kumar R, Kumar S (2021) A study on web mining classification and approaches. In: International conference on recent trends in communication and electronics recent trends in communication and electronics, 1st edn. CRC Press, pp 72–75. https://doi.org/10.1201/9781003193838-20. (First Published: 2021)
2. Agarwal A, Nanavati N (2016) Association rule mining using hybrid GA-PSO for multi-objective optimization. In: 2016 IEEE international conference on computational intelligence and computing research (ICCIC), pp 1–7. https://doi.org/10.1109/ICCIC.2016.7919571.n
3. Basari ASH, Hussin B, Ananta IG, Zeniarja J (2013) Opinion mining of movie review using hybrid method of support vector machine and particle swarm optimization. Procedia Eng 53:453–462. ISSN 1877-7058. https://doi.org/10.1016/j.proeng.2013.02.059
4. Liangtu S, Xiaoming Z (2007) Web text feature extraction with particle swarm optimization summary
5. Kundra K, Kaur U, Singh D (2015) Efficient web log mining and navigational prediction with EHPSO and scaled markov model. Smart Innov Syst Technol 33:529–543. https://doi.org/10.1007/978-81-322-2202-6_48
6. Chen J, Zhang H (2007) Research on application of clustering algorithm based on PSO for the web usage pattern. In: 2007 International conference on wireless communications, networking and mobile computing, WiCOM 2007, pp 3705–3708. https://doi.org/10.1109/WICOM.2007.916
7. Rajdeepa B, Sumathi P (2014) Web structure mining for users based on a hybrid GA/PSO approach. J Theor Appl Inf Technol 70:573–578
8. Zhan HQ (2010) A improved topics search algorithm based on PSO strategy for web mining. KEM.https://doi.org/10.4028/www.scientific.net/kem.439-440.1481
9. Ghorpade-Aher J, Bagdiya R (2014) A review on clustering web data using PSO. Int J Comput Appl 108:31–36
10. Mosa MA (2020) A novel hybrid particle swarm optimization and gravitational search algorithm for multi-objective optimization of text mining. Appl Soft Comput 90:106189. ISSN 1568-4946. https://doi.org/10.1016/j.asoc.2020.106189
11. Wang Z, Zhang Q, Zhang D (2007) A PSO-based web document classification algorithm, pp 659–664. https://doi.org/10.1109/SNPD.2007.72
12. Alam S, Dobbie G, Riddle P (2008) Particle swarm optimization based clustering of web usage data, pp 451–454. https://doi.org/10.1109/WIIAT.2008.292
13. Liu Z, Zhu P, Chen W, Yang R-J (2015) Improved particle swarm optimization algorithm using design of experiment and data mining techniques. Struct Multidiscip Optim 52. https://doi.org/10.1007/s00158-015-1271-7
14. Panda AK, Dehuri SN, Patra MR, Mitra A (2011) A survey on swarm and evolutionary algorithms for web mining applications, vol7077, pp 9–16. https://doi.org/10.1007/978-3-642-27242-4_2

15. Kothari V, AnuradhaJ, Shah S, Mittal P (2012) A survey on particle swarm optimization in feature selection. https://doi.org/10.1007/978-3-642-29216-3_22
16. Grosan C, Abraham A, Chis M (2007) Swarm intelligence in data mining. https://doi.org/10.1007/978-3-540-34956-3_1
17. Kumar S, Kumar R (2021) WDPMA: an MA-based model for web documents prioritization. Int J Inf Technol Web Eng 16:1–24. https://doi.org/10.4018/IJITWE.2021040101
18. Chaudhary K, Gupta SK (2014) Prioritizing web links based on web usage and content data. In: 2014 International conference on issues and challenges in intelligent computing techniques (ICICT), pp 546–551. https://doi.org/10.1109/ICICICT.2014.6781340
19. Gupta SK, Singh D, Doegar A (2016) Web documents prioritization using genetic algorithm. In: 2016 3rd International conference on computing for sustainable global development (INDIACom), pp 3042–3047
20. Sarzaeim P, Bozorg-Haddad O, Chu X (2018) Teaching-learning-based optimization (TLBO) algorithm. https://doi.org/10.1007/978-981-10-5221-7_6
21. Zou F, Chen D, Xu Q (2018) A survey of teaching-learning-based optimization. Neurocomputing335. https://doi.org/10.1016/j.neucom.2018.06.076
22. Kumar S, Anand C (2020) Comparative study of web page ranking algorithms. Int J Adv Sci Technol 29(5s):322–331
23. Lawrence P, Sergey B, Rajeev M, Terry W (1998) The PageRank citation ranking: bringing order to the web
24. Xing W, Ghorbani A (2004) Weighted PageRank algorithm. In: Proceedings second annual conference on communication networks and services research, pp 305–314
25. Egri G, Bayrak C (2014) The role of search engine optimization on keeping the user on the site. Procedia Comput 36:335–342
26. Fujimura K, Inoue T, Sugisaki M (2005) The Eigen Rumor algorithm for ranking blogs. In: 2nd Annual workshop on the weblogging ecosystem: aggregation, analysis and dynamics
27. Alhaidari F, Alwarthan S, Alamoudi A (2020) User preference based weighted page ranking algorithm. In: 3rd International conference on computer applications & information security (ICCAIS), pp 1–6
28. Hao Z, Qiumei P, Hong Z, Zhihao S (2015) An improved pagerank algorithm based on web content. In: 14th International symposium on distributed computing and applications for business engineering and science (DCABES), pp 284–287
29. Attia M, Abdel-Fattah MA, Khedr AE (2022) A proposed multi criteria indexing and ranking model for documents and web pages on large scale data. J King Saud Univ-Comput Inf Sci 34(10):8702–8715

# $k$-NN-SSc: An Effective Similarity Score for $k$-NN Classifier

**Robindro Singh Khumukcham, Linthoingambi Takhellambam, Boby Clinton Urikhimbam, Ranjan Yambem, and Nazrul Hoque**

**Abstract** This paper proposes a new similarity score known as $k$-NN-SSc that is incorporated into the traditional $k$-NN classifier. The proposed similarity score computes the similarity between two instances, and the score is used to find the $k$-nearest neighbors of an unknown instance for the $k$-NN classifier. The effectiveness of the proposed $k$-NN-SSc similarity score measure is compared with Euclidean, Minkowski, Manhattan, Chebyshev, and Cosine distance measures. From the experimental results, we observed that the proposed measure performs better as compared to other competing measures in terms of Accuracy, F1 Score, and Matthew's Correlation Coefficient (MCC) on 20 UCI repository datasets.

**Keywords** $k$-NN · Classification · Precision · Euclidean · Supervised learning

## 1 Introduction

In machine learning (ML), $k$-Nearest Neighbors ($k$-NN) is a widely used supervised learning classifier. The $k$-NN classifier is widely used because of its effective performance in the prediction with simple implementation logic. This classifier computes the class label of a test instance based on its $k$-nearest neighbors in the training set. The label of the test instance is the label that occurred highest in the $k$-nearest neighbors according to the majority voting rule. The method does not have any training phase and is known as a non-parametric classifier. The $k$-NN classifier has been used in diverse applications of ML for the last four decades. A significant number of modified $k$-NN classifiers are developed for future data prediction in the field

R. S. Khumukcham (✉) · L. Takhellambam · B. C. Urikhimbam · R. Yambem · N. Hoque
Department of Computer Science, Manipur University, Canchipur, Imphal 795003, Manipur, India
e-mail: rbkh@manipuruniv.ac.in

B. C. Urikhimbam
e-mail: boby.phd.csc@manipuruniv.ac.in

R. Yambem
e-mail: ranjan.phd.csc@manipuruniv.ac.in

 
S. Borah et al. (eds.), *Advanced Computational and Communication Paradigms*, Lecture Notes in Networks and Systems 535,
https://doi.org/10.1007/978-981-99-4284-8_4

of pattern recognition and data science. Although it is very simple to understand or implement, it also has some limitations. Especially the complexity of the $k$-NN classifier increases when the number of training instances increases. The value of $k$ and the proximity measure used to find the nearest neighbors in the $k$-NN classifier can significantly affect the classifier's performance. The proximity measure may be a distance or a similarity measure. It is very challenging to decide the value of $k$ as well as an effective distance measure in a real-life scenario. Many researchers developed methods for picking or determining the value of $k$ based on some heuristics or employing statistical measures. In addition to the value of $k$, the other important issue is selecting an appropriate distance measure.

A good distance measure helps in improving the performance of classification. It allows the algorithms to recognize similarities between testing and training instances by computing the distance or similarity. If the distance is zero, then the instances are equivalent; otherwise, they are different. The distance function can differ across different distance measures. Since the $k$-NN is a distance-based classifier, the classifier's performance depends on the distance measure. The mainly used distance measures in the $k$-NN classifier are Euclidean, Manhattan, Chebyshev, Minkowski, Cosine, Jaccard, etc. In this paper, we proposed a new similarity_score measure known as $k$-NN-SSc to compute the similarity between two instances. Instead of computing the distance between two instances, we compute the similarity between two instances using the new similarity measure. The measure computes a similarity_score which resembles similarity or closeness between the two instances. Based on this similarity_score, we can find the $k$-nearest neighbors for the $k$-NN classifier. Moreover, the proposed similarity_score known as $k$-NN-SSc aims to perform efficiently on a big dataset with many features.

The $k$-NN is one of the simplest and most widely used classifiers in ML, but the performance of the classifier depends on a distance measure used to find the nearest neighbors. Although people used a large number of proximity measures in the $k$-NN classifier, in some high-dimensional datasets, most of the measures yield poor performance. So it motivates us to develop an effective similarity score computation method for finding the nearest neighbors.

Let us consider a dataset having n number of instances. The dataset is divided into $d_{train}$ and $d_{test}$. To classify a test instance $O_j$ first apply an effective distance or similarity measure that can pick up the $k$ number of nearest neighbors based on which the $k$-NN classifier predicts the class label for the test instance $O_j$.
The main contributions of the paper are listed here.

1. An effective similarity score measure is introduced to find the nearest neighbors in the $k$-NN classifier.
2. The similarity score measure is incorporated into the $k$-NN classifier to predict the class label of an unknown instance.

The novelty of the proposed method is the development of a new similarity measure that can compute the similarity score between any two real-valued data instances. The proposed measure does not work as a generic similarity measure but is designed only to fit the traditional $k$-NN classifier. Unlike Pearson, Spearman, and Kendall correlation measures which are used to find a correlation between two instances in the interval of $-1$ to 1 or 0 to 1, the proposed measure computes similarity in terms of real value, which may be either +ve or −ve real value.

## 2 Related Work

A significant number of research works using $k$-NN classifiers are available in the field of data science and Machine Learning (ML). An optimal subset of features can help the $k$-NN in predicting the unseen objects with high prediction accuracy [26, 27]. In the paper [1], a novel $k$-NN classifier is introduced to overcome the dependency of the $k$-NN algorithm on $k$ value and to make classification faster. The experimental results show that the new method gives better classification accuracy than the standard $k$-NN. In [2], a modified version of the $k$-NN algorithm is proposed by introducing an affinity function to calculate the distance between a test instance and a training instance. The classification accuracy of the proposed method from the experimental results is found to surpass some familiar algorithms. Hassanat, Ahmad Basheer [3] presented a similarity measure to be employed in the $k$-NN algorithm that is invariant to large differences in dimensions. The result shows that the proposed metric outperformed existing standard $k$-NN distance measures. Ougiaroglou et al. [4] developed an adaptive $k$-NN classifier using dynamic $k$ and utilizes three different heuristics for searching the $k$ value. The authors illustrated the applicability of the proposed method by accomplishing better results than standard methods. An improved $k$-NN classifier named D$k$-NN, with dynamic $k$ value, is developed by Zhong et al. [5] by considering an interval of $k$ as $[k_{\min}, k_{\max}]$ and then evaluates an optimal value of $k$ based on the distribution of instances in a particular class. Soucy and Mineau [7] designed a simple non-weighted features $k$-NN algorithm for text categorization and making the $k$-NN algorithm be applicable to the contexts where both the volume of documents and the size of the vocabulary are high.

Moutafis et al. [8] introduced a new method for categorizing the distance learning methods into categories that can be used by the researchers to keep their research works in the right direction and compare to the appropriate methods. Prasatha et al. [9] have conducted a survey on the effects of distance measures and found that the performance of the $k$-NN classifier depends significantly on the distance measures. Kahraman [10] came up with a new and powerful similarity measurement method named fuzzy distance metric to enhance the performance of the $k$-NN algorithm by combining the proposed method with existing weighted methods and the $k$-NN algorithm. The results from the experimental analysis show that the proposed method has better classification accuracy than other classification algorithms. A novel approach of $k$-NN named two-level nearest neighbor algorithm (TLNN) is proposed in [11] to minimize the mean absolute error of the misclassification rate in $k$-NN. The proposed method learns the prior knowledge from the training data and minimizes the dependence on the statistical methods. Bilge et al. [12] came up with a new algorithm of $k$-NN using the Lorentzian Distance Metric, which gives a zero distance for far points. The proposed method is tested by comparing with some classification methods and gives better results than others. Wang et al. [13] recommended a unified optimization aspect of distance metric learning for unsupervised, supervised, or semi-supervised learning algorithms to be fitted into a common framework to give future directions for distance metric learning.

Hu et al. [14] illustrated the effect of distance function in the accuracy of $k$-NN classifier in the medical datasets. The authors had shown that the $k$-NN classifier with Chi-square distance function could give better results in medical datasets. Suarez et al. [15] presented a theoretical background and foundations of distance metric learning and demonstrated the capabilities of distance metric in standard classification. Todeschini et al. [16] proposed two classification methods, $N$-Nearest Neighbors ($N3$) and Binned Nearest Neighbors (BNN), which are inspired by the principles of the $k$-NN algorithm. The working principles of these methods are based on object pairwise similarities. Geng et al. [17] introduced a new density-based clustering method, RECOME, to exploit a novel density measure to detect core objects with various densities and generate atom clusters. The authors also came up with a merging strategy based on the $k$-NN graph to refine the atom clusters. It has been observed in the experimental evaluations that the proposed method can discover clusters of different shapes, densities, and scales with parameters chosen in stable ranges. Gerhana et al. [18] presented a study on the implementation of the $k$-NN algorithm in the Case-Based Reasoning model. Kumar et al. [19] initiated a similarity-preserving function named Sequence and Set Similarity Measure ($S^3M$) to capture the order of occurrence of items in sequences and the constituent items of sequences. The authors demonstrated the usefulness of the proposed measure for classification and regression problems.

Jiang et al. [20] presented an enhanced $k$-NN algorithm for text categorization by combining a constrained one-pass clustering algorithm and $k$-NN text categorization. The experimental analysis results show that the proposed method outperforms the state-of-the-art classifiers in text categorization problems. Zhang et al. [21] proposed a new method $k$Tree to learn non-identical optimal $k$ values for every new sample by involving a training stage in the $k$-NN classification. The proposed $k$Tree method has a higher classification accuracy, in contrast to the traditional $k$-NN methods. A novel mechanism to combine heterogeneous data with $k$-NN framework for gene function prediction is presented in [22]. The method includes a regression-based weighting technique and a probabilistic voting scheme. It has been demonstrated that the proposed method considerably outperforms the traditional $k$-NN for gene function prediction from heterogeneous data. Wang et al. [23] introduced a weight-based $k$-NN recommendation algorithm named IR-IUF++ to update the limitations of the Pearson correlation coefficient in Collaborate Filtering. The experimental analysis shows that the proposed method performed better than the traditional similarity algorithm. Heui Seok Lim [24] proposed a new method by tuning the parameters of the $k$-NN algorithm to improve the performance of text classification problems. It has been demonstrated from this study that the performance of the $k$-NN algorithm can be enhanced by tuning the parameters rather than developing new learning methods. Yean et al. [25] conducted a study on the effect of utilizing the different distance metrics in the $k$-NN classification algorithm on the emotional EEG signal of stroke patients and normal people. The experimental result showed that the performance of the $k$-NN classification algorithm was hugely affected by the distance metric used in the classifier.

## 3 Proposed Method

In this paper, we propose an effective measure to compute the similarity between two numeric objects having $n$ number of features. Although we found a significant number of similarity computation measures in the literature, there is no measure that can ensure high classification accuracy using $k$-NN on all types of data considered from various applications. The proposed $k$-NN-SSc computes the absolute distance between a pair of features as shown in Eq. 1 and finds the number of exactly matching features of the two corresponding objects or instances. Similarly, it computes a number of features that do not match the corresponding objects. The method computes the similarity count and penalty count for the matching and non-matching features, respectively, as given in Eqs. 2 and 3. For non-matching features, a penalty value is computed by taking the sum of the absolute difference between every pair of features as given in Eq. 4. The similarity grade represents the number of matching features out of the total features, computed as shown in Eq. 5. Penalty grade is computed for non-matching features as given in Eq. 6. Finally, the method computes similarity-score value considering similarity count ($\alpha$), similarity grade ($\delta$) and penalty grade ($\theta$) either using Eqs. 7 or 8. The $k$-NN algorithm employs distance measures to calculate the distance between the new instance and all the instances in the training dataset. The smallest distances are selected by choosing an appropriate value of $k$, and a prediction of the new instance is made.

The steps of the $k$-NN classifier with the new similarity measure $k$-NN-SSc are described in Algorithm 1.

The proposed modified $k$-NN classifier uses a new similarity score $k$-NN-SSc. The following definitions help us discuss the proposed $k$-NN in a theoretical manner. For a test object $O_i$ that has $n$ number of features (excluding the class label information), $k$-NN computes the nearest neighbors of $O_i$ using the following definitions.

1. Absolute distance between $O_j$ and $O_i$ over feature $f_k$ as defined in Eq. 1.

$$dist(O_i, O_j) = \|O_i(f_k) - O_j(f_k)\| \tag{1}$$

2. Similarity count: Similarity Count represented as $\alpha$ is the number of features for which the distance between two objects $O_i$ and $O_j$ over feature $f_k$ is 0, i.e., $dist_{f_k}(O_i, O_j) = 0$, where $1 \leq k \leq n$.

$$dist_{f_k}(O_i, O_j) = 0 \tag{2}$$

**Algorithm 1** $k$-NN Classifier with New Similarity Measure $k$-NN-SSc

**Require:** Dataset $D$ with $N$ instances and $d$ dimensions of the dataset.
**Ensure:** Prediction of the class label for each test instance.

```
for each test object O_i ∈ d_test do
    for each training object O_j ∈ d_train do
        Compute dist(O_i, O_j) = ||O_i(f_k) − O_j(f_k)||
    end for
    Find similarity count α between objects O_i and O_j using Eq. 2.
    Find penalty count β between objects O_i and O_j using Eq. 3.
    if α > β then
        Find Similarity Score using Eq. 7.
    else
        Find Similarity Score using Eq. 8.
    end if
    Find the k-nearest neighbors for each test object O_i ∈ d_test from d_train.
    Decide the label of O_i from the maximum occurrence of labels in k instances.
end for
```

3. Penalty count: Penalty Count represented as $\beta$ is the number of features for which the distance between two objects $O_i$ and $O_j$ over feature $f_k$ is not 0, i.e., $dist_{f_k}(O_i, O_j) \neq 0$, where $1 \leq k \leq n$.

$$dist_{f_k}(O_i, O_j) \neq 0 \tag{3}$$

4. Penalty Value: Then the penalty value, denoted by $\omega$, is computed by summing up all the distances for which the distance is non-zero between the instances of the training and the test object.

$$\omega = \sum_{f_k} dist_{f_k}(O_i, O_j) \quad \text{if} \quad dist_{f_k}(O_i, O_j) \neq 0 \tag{4}$$

5. Similarity Grade represented as $\delta$ is computed using Eq. 2

$$\delta(o_i, o_j) = \frac{\alpha(o_i, o_j)}{n} \tag{5}$$

6. Penalty Grade represented as $\theta$ is computed as shown in the Eq. 3.

$$\theta(O_i, O_j) = \frac{\omega(O_i, O_j)}{n} \tag{6}$$

$$similarity\_score = \alpha(O_i, O_j) + (\delta(O_i, O_j) - \theta(O_i, O_j)), \quad \text{when } (\alpha > \beta), \tag{7}$$

$$similarity\_score = \alpha(O_i, O_j) - (\delta(O_i, O_j) + \theta(O_i, O_j)), \quad \text{otherwise.} \tag{8}$$

**Table 1** Training instances

| Obj | $F_1$ | $F_2$ | $F_3$ | $F_4$ | $F_5$ | $F_6$ |
|---|---|---|---|---|---|---|
| $O_1$ | 89 | 22 | 48 | 18 | 77 | 29 |
| $O_2$ | 89 | 22 | 89 | 17 | 78 | 10 |
| $O_3$ | 88 | 55 | 48 | 18 | 11 | 14 |
| $O_4$ | 89 | 22 | 41 | 18 | 77 | 30 |
| $O_5$ | 89 | 56 | 23 | 34 | 77 | 14 |
| $O_6$ | 76 | 22 | 48 | 18 | 77 | 20 |
| $O_7$ | 87 | 48 | 48 | 18 | 77 | 10 |

**Table 2** Computed similarity scores

| $F_1$ | $F_2$ | $F_3$ | $F_4$ | $F_5$ | $F_6$ | Similarity score |
|---|---|---|---|---|---|---|
| 89 | 22 | 48 | 18 | 77 | 29 | 7 |
| 89 | 22 | 89 | 17 | 78 | 10 | −8.66 |
| 88 | 55 | 48 | 18 | 11 | 14 | −17.5 |
| 89 | 22 | 41 | 18 | 77 | 30 | 3.66 |
| 89 | 56 | 23 | 34 | 77 | 14 | −13.33 |
| 76 | 22 | 48 | 18 | 77 | 20 | 1 |
| 87 | 48 | 48 | 18 | 77 | 10 | −4.33 |

### *3.1 Working Example*

Let us consider a dataset with seven instances, each with six features, say $F_1$ to $F_6$. Let us consider one testing instance $O_j = \{89, 22, 48, 18, 77, 29\}$ and seven training instances $O_1$ to $O_7$. The training instances are shown in Table 1.

The computed similarity score is shown in Table 2.

If we consider the value of k as three, then objects $O_1$, $O_4$, and $O_6$ will be similar objects having high similarity scores as 7, 3.66, and 1, respectively. So these three objects will be considered as the nearest objects for the k-NN classifier.

## 4 Experimental Results

We compared our method with some existing methods on 20 datasets downloaded from Kaggle and the UCI machine learning repository. The performance of the proposed similarity measure is evaluated in terms of accuracy. In our experiment, we used fivefold cross-validation and computed the average accuracy. The performance of the $k$-NN classifier incorporating our proposed measure is compared with other measures as shown in Table 3.

**Table 3** Performance of the proposed similarity measure

| Dataset | Euclidean | Minkowski | Manhattan | Chebyshev | k-NN-SSc | Cosine |
|---|---|---|---|---|---|---|
| Wine | 44.5 | 44.5 | 46.1 | 46.4 | 48.8 | 45.8 |
| Titanic | 69.7 | 69.7 | 72.64 | 68.69 | 78.57 | 72.7 |
| Iris | 96.66 | 96.66 | 96.66 | 96 | 95.33 | 96 |
| Ionosphere | 82.34 | 82.34 | 88.61 | 87.16 | 86.6 | 82.9 |
| Bank-note authentication | 100 | 100 | 99.92 | 100 | 100 | 99.56 |
| Water potability | 56.4 | 56.4 | 56.77 | 55.7 | 56.87 | 56.47 |
| Fish | 44.54 | 44.54 | 41.74 | 49.625 | 49 | 65.37 |
| Drug | 73 | 73 | 78 | 67 | 81.5 | 66 |
| Breast cancer | 88.58 | 88.58 | 88.05 | 88.58 | 88.58 | 86.81 |
| Mushroom | 94.61 | 94.61 | 94.82 | 92.15 | 95.68 | 94.56 |
| Yeast | 54.37 | 54.37 | 55.68 | 54.84 | 55.72 | 44.87 |
| Monk's 1 | 56.36 | 56.36 | 59.07 | 56.36 | 84.37 | 67.90 |
| Monk's 2 | 59.55 | 59.55 | 62.04 | 60.05 | 68.68 | 68.18 |
| Sonar | 59.30 | 59.30 | 63.19 | 50.52 | 58.38 | 58.95 |
| Liver | 64.04 | 64.04 | 65.21 | 66.08 | 63.76 | 65.47 |
| Heart failure | 64.19 | 64.19 | 64.19 | 62.83 | 75.22 | 64.52 |
| Parkinson's disease | 67.17 | 67.17 | 72.82 | 62.56 | 77.57 | 42.56 |
| Diabetes | 73.44 | 73.44 | 73.44 | 73.48 | 73.82 | 67.84 |
| Labor | 80.75 | 80.75 | 84.09 | 66.96 | 85.33 | 92.87 |
| Balance scale | 73.60 | 73.60 | 72.80 | 76.17 | 89.27 | 65.92 |

## 4.1 Performance Analysis Measures

The performance of the proposed method $k$-NN-SSc is evaluated using Accuracy which is defined next. Definitions of the performance measures are given below.

1. $Accuracy = \frac{TP+TN}{TP+FP+FN+TN}$.

## 4.2 Result Analysis

As shown in Table 3, the proposed $k$-NN-SSc measure yields better results on most of the datasets. From the experimental results, we observed that the proposed measure outperformed on 12 datasets as compared to other measures. Moreover, it shows very similar results to others on many datasets. In terms of accuracy, the proposed method gives the best result on Wine, Titanic, Bank Note Authentication, Water Potability,

Drug, Breast Cancer, Mushroom, Labor, Monk's 2, Yeast, Monk's 1, Heart failure, Parkinson's Disease, Diabetes, and Balance scale datasets.

### *4.3 Discussion*

From the experimental analysis, we observed that the proposed similarity score used in the $k$-NN classifier yields better performance during the classification of data. The measure is found effective to the traditional $k$-NN classifier used for classification. We evaluate the measure on different types of data, such as high-dimensional gene expression, disease diagnosis, and UCI ML data. The proposed similarity measure's accuracy is the highest compared to other methods on Wine, Titanic, Mushroom, Monk's 2, Monk's 1, Yeast, Drug, Water Potability, Parkinson's Disease, Diabetes, Heart failure, and Balance scale datasets. However, the proposed method $k$-NN-SSc shows poor performance on the Iris dataset. On the Fish dataset, $k$-NN-SSc gives better results on all the compared method except Cosine similarity. The experimental analysis reveals that the proposed similarity score measure $k$-NN-SSc used in the $k$-NN classifier can effectively classify instances of various application datasets. Thus, we can conclude that the problem of low efficiency of the classical $k$-NN algorithm in dealing with large features is handled by the effective proposed measure $k$-NN-SSc.

## 5 Conclusion and Future Work

In this paper, we have presented a novel solution for dealing with the shortcomings of $k$-NN by developing a new similarity score of the classical $k$-NN algorithm. The experimental results clearly indicate that the proposed method has outperformed some well-known variants of the $k$-NN algorithm. A properly chosen value of $k$ can potentially improve the classification results. Thus, in the future, we will try to improve our classification accuracy by choosing a suitable value of $k$. In this paper, we have experimented only with numerical data. So, another direction of future research will be to extend the current approach for the classification of categorical data.

## References


1. Guo G et al (2003) KNN model-based approach in classification. In: OTM confederated international conferences on the move to meaningful internet systems. Springer, Berlin, Heidelberg
2. Bhattacharya G, Ghosh K, Chowdhury AS (2012) An affinity-based new local distance function and similarity measure for kNN algorithm. Pattern Recogn Lett 33(3):356–363
3. Dimensionality Invariant Similarity Measure, Ahmad Basheer Hassanat. arXiv:1409.0923 [cs.LG]

4. Ougiaroglou S, Nanopoulos A, Papadopoulos AN, Manolopoulos Y, WelzerDruzovec T (2007) Adaptive k-nearest-neighbor classification using a dynamic number of nearest neighbors. In: East European conference on advances in databases and information systems. Springer, pp 66–82
5. Zhong XF, Guo SZ, Gao L, Shan H, Zheng JH (2017) An improved k-NN classification with dynamic k. In: Proceedings of the 9th international conference on machine learning and computing, pp 211–216
6. https://machinelearningmastery.com/distance-measures-for-machine-learning/
7. Soucy P, Mineau GW (2001) A simple KNN algorithm for text categorization. In: Proceedings of the IEEE international conference on data mining, pp 647–648. https://doi.org/10.1109/ICDM.2001.989592
8. Moutafis P, Leng M, Kakadiaris IA (2017) An overview and empirical comparison of distance metric learning methods. IEEE Trans Cybern 47(3):612–625. https://doi.org/10.1109/TCYB.2016.2521767
9. Prasatha VS et al (2017) Effects of distance measure choice on KNN classifier performance-a review. arXiv:1708.04321
10. Kahraman HT (2016) A novel and powerful hybrid classifier method: development and testing of heuristic k-nn algorithm with fuzzy distance metric. Data Knowl Eng 103:44–59
11. Gao Y et al (2012) A novel two-level nearest neighbor classification algorithm using an adaptive distance metric. Knowl Based Syst 26:103–110
12. Bilge HŞ, Kerimbekov Y, Uğurlu HH (2015) A new classification method by using Lorentzian distance metric. In: International symposium on innovations in intelligent systems and applications (INISTA), pp 1–6. https://doi.org/10.1109/INISTA.2015.7276764
13. Wang F, Sun J (2015) Survey on distance metric learning and dimensionality reduction in data mining. Data Min Knowl Disc 29(2):534–564
14. Hu L-Y et al (2016) The distance function effect on k-nearest neighbor classification for medical datasets. SpringerPlus 5(1):1–9
15. Suárez JL, García S, Herrera F (2021) A tutorial on distance metric learning: mathematical foundations, algorithms, experimental analysis, prospects and challenges. Neurocomputing 425: 300–322
16. Todeschini R et al (2015) N3 and BNN: two new similarity based classification methods in comparison with other classifiers. J Chem Inf Model 55(11):2365–2374
17. Geng Y et al (2018) RECOME: a new density-based clustering algorithm using relative KNN kernel density. Inf Sci 436:13–30
18. Gerhana YA et al (2017) The implementation of K-nearest neighbor algorithm in case-based reasoning model for forming automatic answer identity and searching answer similarity of algorithm case. In: 2017 5th international conference on cyber and IT service management (CITSM). IEEE
19. Kumar P, Raju BS, Radha Krishna P (2010) A new similarity metric for sequential data. Int J Data Warehous Min (IJDWM) 6(4):16–32
20. Jiang S et al (2012) An improved K-nearest-neighbor algorithm for text categorization. Exp Syst Appl 39(1):1503–1509
21. Zhang S et al (2017) Efficient kNN classification with different numbers of nearest neighbors. IEEE Trans Neural Netw Learn Syst 29(5):1774–1785
22. Yao Z, Ruzzo WL (2006) A regression-based K nearest neighbor algorithm for gene function prediction from heterogeneous data. BMC Bioinf 7(1). BioMed Central
23. Wang B, Liao Q, Zhang C (2013) Weight-based KNN recommender system. In: 2013 5th international conference on intelligent human-machine systems and cybernetics, vol 2. IEEE, 2013
24. Lim HS (2004) Improving kNN-based text classification with well-estimated parameters. In: International conference on neural information processing. Springer, Berlin, Heidelberg
25. Yean CW et al (2018) Analysis of the distance metrics of KNN classifier for EEG signal in stroke patients. In: 2018 international conference on computational approach in smart systems design and applications (ICASSDA). IEEE

26. Shekhar S, Hoque N, Bhattacharyya DK (2022) PKNN-MIFS: a parallel KNN classifier over an optimal subset of features. In: Intelligent systems with applications, vol 14. Elsevier
27. Hoque N, Bhattacharyya DK, Kalita JK (2014) MIFS-ND: a mutual information-based feature selection method. Expert Syst Appl 41(14):6371–6385

# Password Management System Using Blockchain

**Rebecca Shruti Soren, Sandeep Gurung, and Dipendra Gurung**

**Abstract** Password management is the process of safeguarding and maintaining passwords from creation to expiration by adhering to a set of sustainable standards. As the IT landscape evolves, passwords serve as the first line of security for sensitive data, but they are also a prominent target for hackers and may cause significant issues when mishandled or hacked. Due to a rise in businesses falling victim to cyber-attacks because of inadequate password security, secure administration of privileged account credentials has gained importance in recent years. However, consumers have no real assurance that their passwords will remain secure, as loopholes and defective coding make it possible for even security solutions to be compromised. However, blockchain technology, with its immutability, resistance to hacking, and entirely decentralized nature, promises far greater security. Combining security solutions like password managers with blockchain considerably increases the likelihood that stored data is secure. This article aims to demonstrate how, by utilizing blockchain, the security of password management applications may be improved since user credentials are no longer kept on centralized servers.

**Keywords** Blockchain · Password storage · Password manager · Credential storage · Ethereum · Security

R. S. Soren (✉) · S. Gurung · D. Gurung
Sikkim Manipal Institute of Technology, Sikkim, India
e-mail: sorenrebeccashruti@gmail.com

S. Gurung
e-mail: sandeep.gu@smit.smu.edu.in

D. Gurung
e-mail: dipendra.gu@smit.smu.edu.in



S. Borah et al. (eds.), *Advanced Computational and Communication Paradigms*, Lecture Notes in Networks and Systems 535,
https://doi.org/10.1007/978-981-99-4284-8_5

# 1 Introduction

A password is a group of characters that is used to verify a user's identity during the authentication process. They are intended to be known only to the user and provide access to a device, application, or website. A password manager facilitates the retrieval of complicated passwords and may store them in an encrypted database. With the growing number of services requiring authentication, password managers are becoming more important than ever. Password managers provide enhanced security and simplicity when using passwords to access Internet services. A recent study [1] was able to quantify that password managers do influence password security; however, it was also noted that the underlying mechanisms of the password manager determine whether this influence is beneficial or exacerbates existing problems.

The best password manager is LastPass, which is also Editors' Choice for PC Magazine and CNET. It is one of the most feature-rich password management programs out there. It incorporates features like bookmarklets, one-time passwords, and two-factor authentication [2] and is compatible with almost all popular browsers, mobile devices, and desktop platforms. Users of LastPass may access their saved credentials through the LastPass website, a bookmarklet, or a browser plugin. Users may exchange passwords with one other using LastPass, which also maintains the credential database encrypted on its servers [2]. However, having a single point of failure, LastPass is rendered vulnerable to malicious actors and various cyber-attacks. Blockchain is a shared, immutable ledger as unlike traditional databases. It makes network asset tracking and transaction recording simpler. The asset may be tangible and/or intangible. A blockchain database is a shared database that is continually coordinated.

The main goal of this article is to provide an extensive body of knowledge on the use of Blockchain technology to help the security of sensitive data. In addition, we supplement the topic with essential insights into the utilized methodologies, their merits, limitations, and feasibility to real-world scenarios by establishing a password management system on a private blockchain using open-source software.

# 2 Definitions

## *2.1 Blockchain*

Blockchain is a decentralized, immutable record. It makes it simpler to keep track of assets and record transactions in a network. Both tangible and intangible assets may be present. A blockchain database is a shared—and constantly coordinated—database.

It is extremely impossible to tamper with a record that has been added to the chain. The network performs continual checks, comparing the node's copy of the chain with those of the other nodes in the network (P2P), to guarantee that all copies

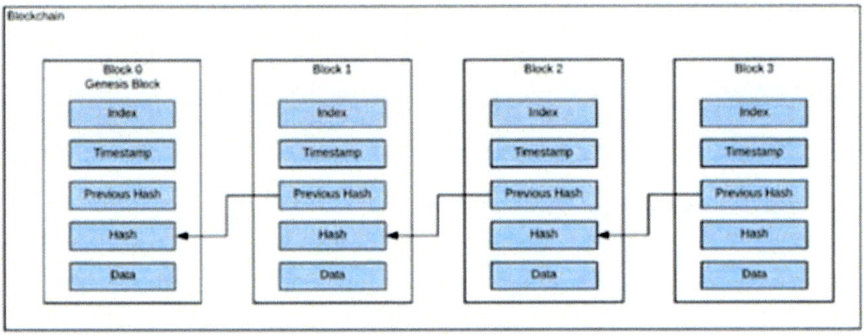


**Fig. 1** Blockchain

of the database are identical. It is "an open, distributed ledger that can efficiently and permanently record transactions between two parties" [3].

For each transaction, all parties have to agree to its terms and validate it. The transaction is encoded into a block of digital data and uniquely signed or identifiable once it has been verified. Each block is linked to the one before and after it. This results in an irreversible and unchanging chain. Blocks are chained together to prevent any block from being changed or placed between two existing blocks. To falsify a single record, all instances of the chain would have to be falsified or tampered with. That is virtually impossible as there exist millions of instances in the chain (Fig. 1).

## 2.2 *Transactions*

A blockchain is a distributed transactional database. Participating in the network enables access to the database's entries for all users. If a database modification is desired, a transaction must be established and accepted by all parties. While the transaction is being processed, no other transactions can modify the database. Furthermore, the sender always cryptographically signs a transaction. This makes it simple to restrict access to certain database updates.

## 2.3 *Ethereum*

Ethereum is a decentralized, global software platform that runs on a blockchain. Ethereum is made to be scalable, decentralized, secure, and programmable.

## 2.4 Ether

The cryptocurrency Ether is utilized on the Ethereum platform.

## 2.5 Ethereum Virtual Machine

The environment in which Ethereum smart contracts are used is called the Ethereum Virtual Machine, or EVM. Code running within the EVM has no access to the network, disk, or any other processes since it is completely isolated.

## 2.6 Accounts

There are different kinds of accounts in Ethereum that can access the same address space:

- **External accounts**: The address of an external account may be found using the public key.
- **Contract accounts**: At the time of the contract's creation, the address is established.

## 2.7 Storage

Each account has access to permanent memory storage. Storage is done using a key-value store. A contract is only permitted to read and write to its own storage.

## 2.8 Memory

For each message call, a contract obtains a fresh instance of the second memory region, known as memory. However, reads are only allowed to be 256 bits wide, whereas writes can be either 8 or 256 bits wide. Memory is linear and can be accessed at the byte level. With increasing size, memory gets more expensive as it scales quadratically.

## 2.9 *Stack*

All calculations on the EVM are done on the stack because it is a stack computer. It supports words with 256 bits and up to 1024 elements. The top end of the stack is the only part of the stack that may be accessed: either one of the top 16 elements can be transferred to the top of the stack, or the top element can be switched out for one of the 16 items below it. The outputs of all subsequent operations are added to the stack once the top two (or one, or more, depending on the method) stack components have been removed.

## 2.10 *Message Calls*

Like transactions, message calls also have a source, a target, a data payload, ether, gas, and return data. A top-level message call is a component of every transaction that might lead to additional message calls. A contract can decide how much of its unused gas to keep and how much to deliver with the inner message call. An error value is added to the stack if an out-of-gas exception (or any other exception) occurs in the inner function.

## 2.11 *Smart Contracts*

A smart contract is a software program or transaction protocol that carries out, manages, or records legally significant events in line with the terms of a contract or other agreement.

## 2.12 *State Variables*

State variables are values that are persisted in contract storage indefinitely.

## 2.13 *Functions*

A smart contract's functions are the executable pieces of code. Function calls can occur internally or externally, with varying degrees of visibility (private versus public) toward other contracts.

## 3 Related Works

Despite being widely available, password managers are not widely used, according to a new study paper [2]. The paper [2] reported on three studies that were conducted: a usability test and interview research; a study of password manager users and non-users that contrasts experts and non-experts in their usage (or non-use) of password managers; and a study on the user interfaces and functionality of three password managers. Lack of transparency and lack of trust are the key determinants of password manager use, according to the study [2].

Another recent study [4] found that smart contracts, which are executable programs that run on top of the blockchain to allow, execute, and enforce an agreement between untrusted parties without the assistance of a trusted third party, to be yet another promising aspect of the blockchain technology. The paper [4] highlighted the study topics on blockchain-based smart contracts that have been studied as well as the existing challenges that need to be resolved in future research. To do this, a thorough mapping study was conducted, and the article [4] focused on the importance and effectiveness of smart contracts in Blockchain, the relevance of which and their potential future prospects were mapped using data from previous research publications.

Another study [5] provided insight into the creation and implementation of the world's first decentralized replicated relational database with blockchain features, termed blockchain relational databases. Several parallels were found between the functionalities provided by blockchain systems and a replicated relational database.

## 4 Problems with Traditional Password Storage Systems

Based on our findings, the problems posed by using passwords and traditional password management systems can be summarized as follows:

- Easy to remember passwords are usually weak and can compromise the user's access to a service.
- Users forgetting their passwords and not being able to retrieve their accounts and losing out important information.
- Passwords are vulnerable to cyber-attacks such as brute force attack, offline cracking, and dictionary attacks.
- There is just one point of failure with traditional password managers. If the user forgets or loses his master password or other identifying information, he or she may lose access to all of their passwords at once. A single point of failure (SPOF) is a possible danger because a single flaw or malfunction causes the entire system to fail.
- Most password managers use central server systems and are prone to various cyber-attacks for the same reason. If a malicious user gains access to the server, they may choose to delete or modify entries of other users on the system.

- For a password manager service managed by a third party, the user would have to trust the third party to not reveal their plain passwords.
- The user would have to trust that appropriate encryption algorithms are used to protect the confidentiality of their passwords.

## 5 Solution Strategy

Without a single point of failure, systems that are built on the blockchain are much more resistant to attack. A blockchain is a ledger of transactions that is shared globally. By just connecting to the network, anyone may read the database entries. Using the attributes of a blockchain, we can ensure safe and secure retrieval of encrypted passwords from the password management system (Fig. 2).

The user's email ID and password will be saved as a key-value pair in a smart contract. Thereafter, the contract will be deployed on a private blockchain network. Users will be able to interact with the smart contract via an interactive interface.

### *5.1 User Operations*

The software solution should provide basic CRUD functionalities to the user.

1. **Add Password**: The user will be able to save their email ID and passwords in the contract. If an entry with an email ID existing in the contract is entered, the software will give error and no changes will be made.
2. **Retrieve Password**: The user will be able to retrieve their saved passwords by giving the corresponding email ID as input. It is mandatory that the entry exists in the contract, else software will throw error.
3. **Update Password**: The user will be able to update their saved passwords by giving the corresponding email ID as input. It is mandatory that the entry exists in the contract, else software will throw error.
4. **Delete Password**: The user will be able to delete their saved details by giving the corresponding email ID as input. It is mandatory that the entry exists in the contract, else software should throw error.

**Fig. 2** Using smart contract for storing passwords in the blockchain

### 5.2 *Non-functional Requirements*

1. **Performance**: The system must be interactive, with as little delays as possible. As a result, there are no instant delays in any system action-response. The operations are performed in less than 30 s.
2. **Safety**: The blockchain securely stores information and is tamper-proof.
3. **Reusability**: The system can be easily accessed on a different computer if the user knows the address of the deployed contract. The deployed contract will retain all saved information.

## 6 Implementation

### 6.1 *Architecture*

When a smart contract is compiled, an Application Binary Interface (ABI) and byte-code are created. The ABI generates a local copy of the smart contract. The web3 interface leverages the ABI to establish method names, arguments, data types, and event structure for smart contracts. The bytecode is stored on the Blockchain network and executed when we interact with the smart contract (Fig. 3).

Next.js is a framework that builds static webpages and enables server-side rendering for web applications. The browser-based application makes use of the web3

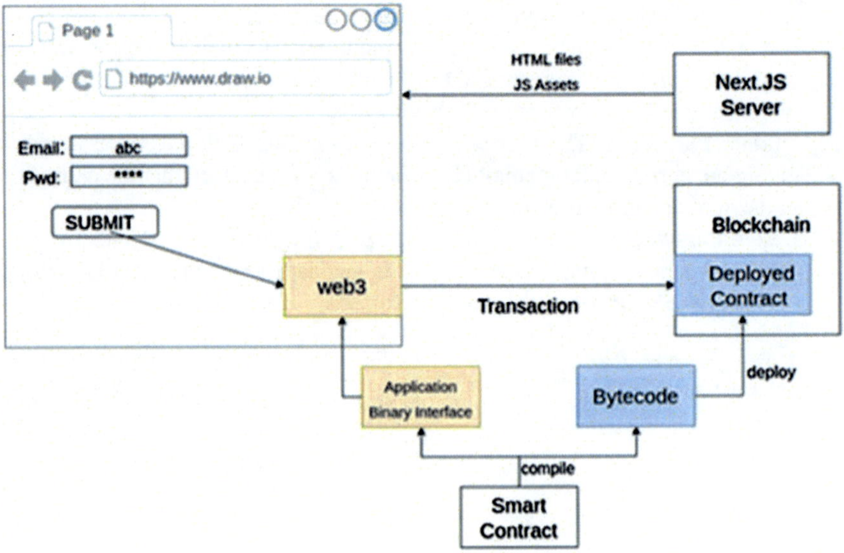


**Fig. 3** Architecture for password management system

library interface, which generates a transaction and transmits it to the blockchain network where the smart contract is deployed.

## 6.2 *Flowcharts*

**Deploying a private blockchain**

In order to construct a private blockchain, the genesis block, the initial block in a blockchain, must be configured. The chain data then needs to be initialized using the configuration of the genesis block. This establishes a private blockchain network (Fig. 4).

Geth's JavaScript console may be linked to this blockchain network to facilitate account creation, which is equivalent to adding a block to the blockchain, and mining of ethers.

**Deploy smart contract on the private blockchain**

A smart contract has to be compiled with Solidity, which produces bytecode, in order to be deployed on a blockchain network. Using Node.js and the web3 framework, this bytecode is published to the blockchain network (Fig. 5).

**Fig. 4** Block diagram for deploying a private blockchain

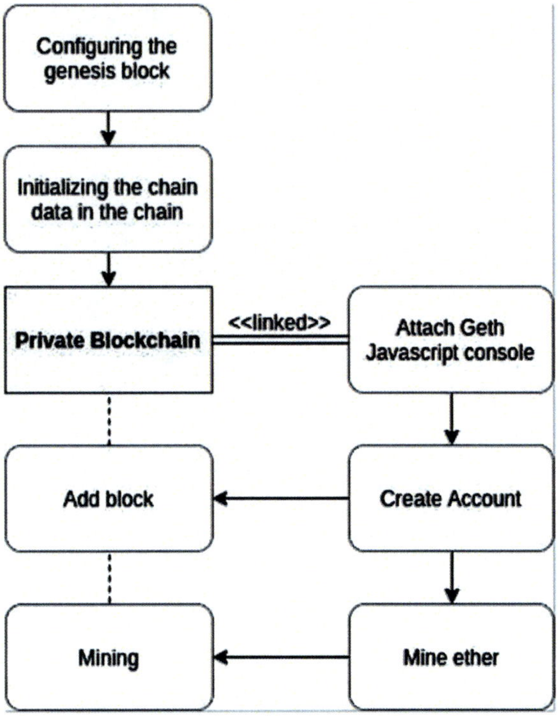

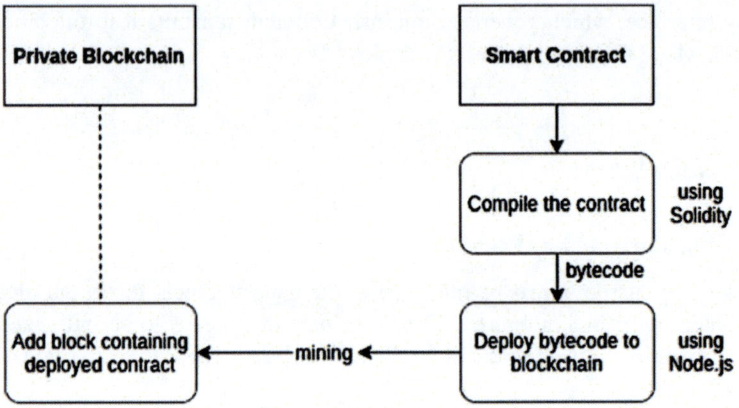


**Fig. 5** Block diagram illustrating the deployment of smart contracts on a private blockchain

To store contracts in a block and make them accessible to the whole blockchain network, mining of the blockchain is required.

**Develop Front end to facilitate interaction with smart contract**

An interface is deployed to simplify user interaction with the deployed smart contract.

To do this, an instance of the deployed contract is created by combining the Application Binary Interface and the deployed contract's address from the blockchain. This instance is then exported to the web interface, where it may be used to perform add, update, and delete operations on the smart contract. Each transaction will be recorded in a block on the blockchain (Fig. 6).

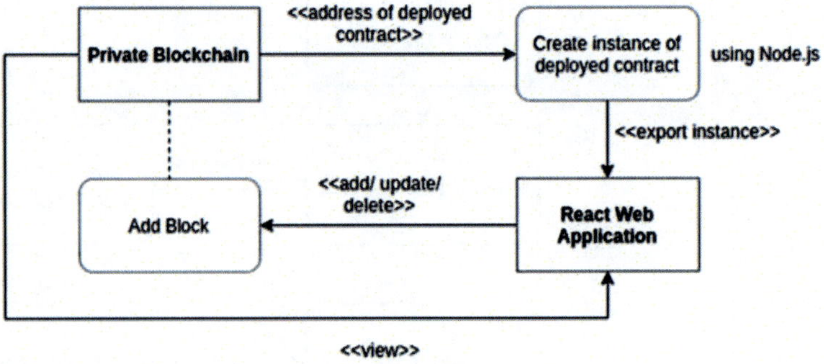


**Fig. 6** Block diagram for front-end interaction with smart contract

## 6.3 *Activity Diagram*

**Making changes to contract (Add/Update/Delete)**

If the user must do an add, update, or delete operation, they are required to provide their email address and/or password. If not, the flow is concluded (Fig. 7).

Once the email address and/or password are supplied, a transaction is generated and transmitted to the deployed contract. Mining occurs on the blockchain, which records the transaction in a block confirming that the contract has been changed. Therefore, the flow terminates.

**Retrieving data from contract**

The program will require the user's email address in order to view information stored on the contract. A transaction is sent to the contract in order to determine whether or not the email address exists in the smart contract. If not, the flow will terminate (Fig. 8).

If the email address is contained in the contract, the corresponding password is retrieved and displayed to the user.

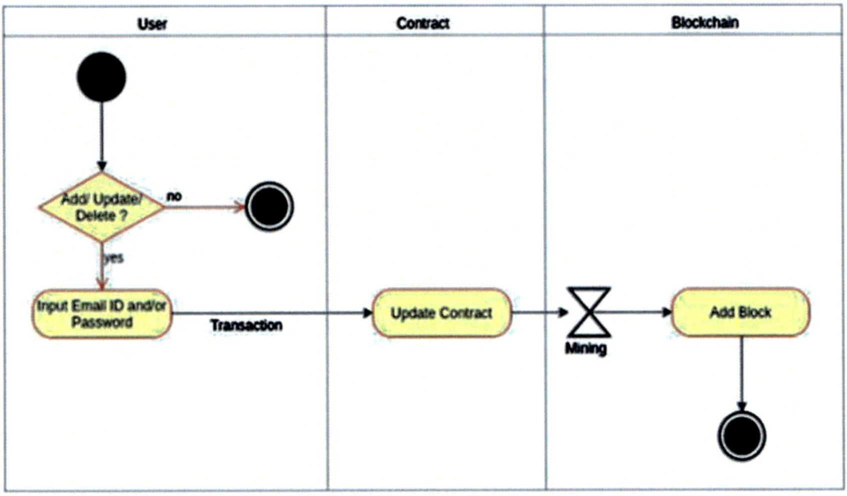


**Fig. 7** Activity diagram for making changes to contract

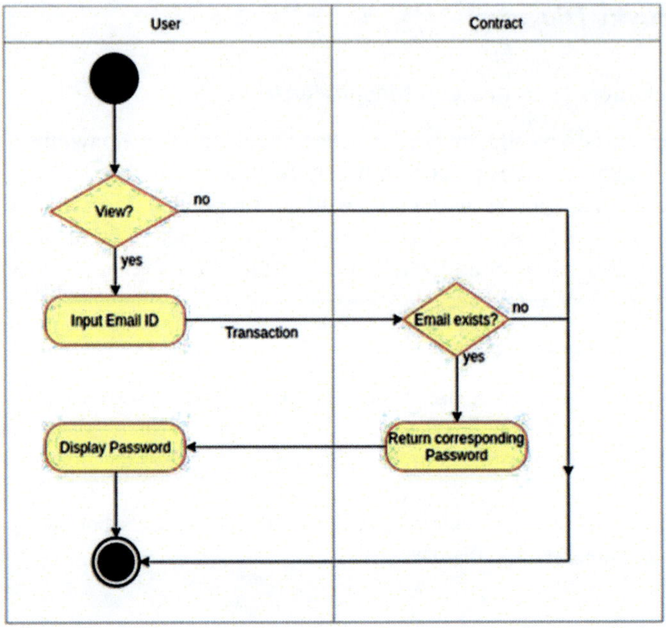


**Fig. 8** Activity diagram for retrieving data from contract

## 6.4 Algorithms

### Add Password

- User deploys the contract using their account credentials.
- The user enters the email ID and password value as string inputs to the function.
- The function verifies that the user is the password's manager (modifier).
- The email ID and password are stored in the contract.

Expected Input: Two strings—email ID and password.
Expected Output: Values are successfully stored.

### Retrieve Password

- User deploys the contract using their account credentials.
- The user enters the index of the email ID as input to the function.
- The function verifies that the user is the password's manager (modifier).
- The password value for the corresponding index stored in the contract is fetched and displayed.

Expected Input: An integer.
Expected Output: The password string in plain text.

**Update Password**

- User deploys the contract using their account credentials.
- The user enters the index of the email ID and password value as string inputs to the add function.
- The function verifies that the user is the password's manager (modifier).
- The password stored in the contract is updated.

Expected Input: Index and new value of saved password.
Expected Output: Password is updated.

**Delete Password**

- User deploys the contract using their account credentials.
- The user enters the index of the email ID as input to the function.
- The function verifies that the user is the password's manager (modifier).
- The email ID and password stored in the contract are deleted.

Expected Input: Index of the saved password.
Expected Output: Password is deleted.

# 7 Experimental Results

## *7.1 Software Specifications*

All software used in testing this implementation are as follows:

- **Solidity**: Solidity is a statistically typed, object-oriented, high-level programming language used to create smart contracts on various blockchain platforms, most notably Ethereum [6].
- **Geth**: Geth is an application that functions as a node for the Ethereum blockchain, allowing users to mine ether and create smart contracts.
- **React**: React is a JavaScript library for making user interfaces. It is commonly known as React.js or ReactJS.
- **Remix IDE**: Remix is a server-side component-free browser-based integrated development environment (IDE) with a built-in compiler and solidity runtime environment. It enables you to build solidity smart contracts right from the browser before debugging, deploying, and running them.

## 7.2 *The Private Blockchain Network*

**Configuration of the Genesis block**

The initial block in any blockchain is known as a genesis block. The genesis block in Geth is customized using a genesis.json file, the path to which Geth must be supplied at startup. When creating a genesis block, a few initial parameters for the private blockchain must be defined [6] (Fig. 9).

For the purposes of our experiment, the genesis block in Geth was configured keeping the following parameters in mind:

- **difficulty**: Controls the mining puzzle's difficulty. A lower number allows for faster mining.
- **gasLimit**: Sets a maximum number of smart contracts that can be executed.
- **alloc**: Allows for the pre-allocation of ether from the genesis block to one or more accounts.

**Deployment of the Private Blockchain**

The init command prepares the network using the genesis.json file configuration. The following output is produced on executing the command: geth init /genesis.json –datadir test (Fig. 10).

The datadir command line option can be used to indicate the directory in which the chain data will be stored. This folder will grow in size as data is added to the blockchain. If this folder is removed, the blockchain must be reinitialized.

**Initial Operations**

Geth responds to commands encoded as JSON objects in accordance with the JSON-RPC-API definition [6]. The Geth JavaScript console can be started using the command: geth attach ipc:/test/geth.ipc.

The following one-time procedures were executed through the console:

blockchain > {} genesis.json > ...

```
{
    "config": {
        "chainId": 2105,
        "homesteadBlock": 0,
        "eip150Block": 0,
        "eip155Block": 0,
        "eip158Block": 0
    },
    "difficulty": "0x0",
    "gasLimit": "3000000",
    "alloc": {}
}
```

**Fig. 9** Genesis block configuration

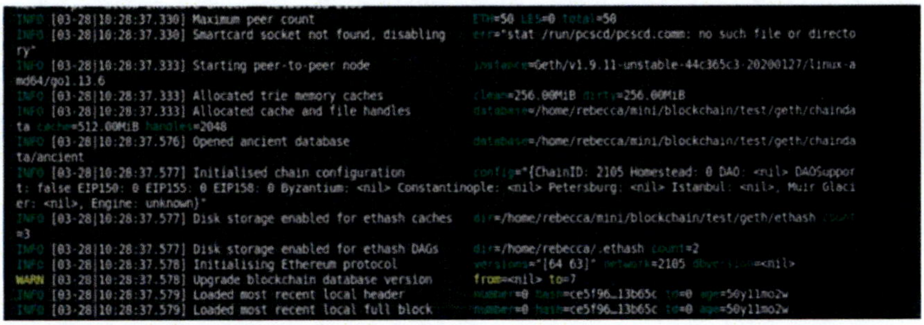

**Fig. 10** Console output-running the chain

**Fig. 11** Creation of new account

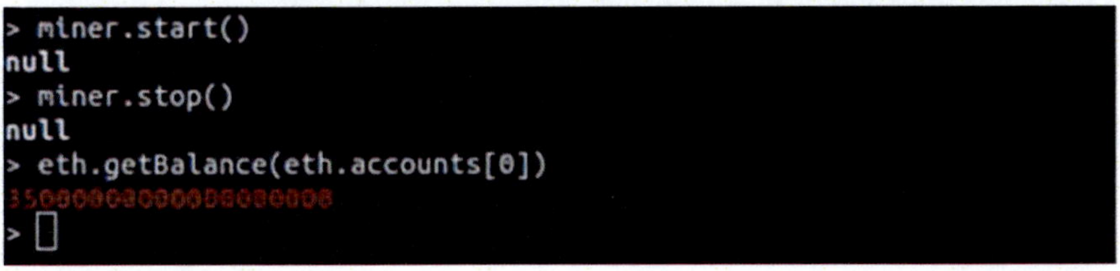


**Fig. 12** Mining ethers

***Creation of accounts*** A new account is created with a passphrase. Initially, the account has 0 ether (Fig. 11).

Command: personal.newAccount().

***Mining blocks*** Using the "eth.getBalance()" function, it is possible to determine how many ethers have been mined on the specified account. The returned value is in wei, the smallest unit of ether (Fig. 12).

Command: miner.start().

## 7.3 *The Smart Contract*

The smart contract contains the implementations of the algorithms mentioned in Sect. 6.

**Compilation of the smart contract**

The contract was then compiled using Remix IDE, which returns a bytecode upon successful compilation. This bytecode needs to be retained in a JSON file for deployment (Fig. 13).

**Deployment of the smart contract**

A group of libraries known as web3.js enable communication with a nearby or far-off Ethereum node through HTTP, IPC, and WebSocket. On Ethereum blockchain networks, interacting with smart contracts is made simple via the eth.Contract object.

Using the "deploy" and "send" functions provided by the library, the smart contract can be submitted to the blockchain network. Contracts once submitted need to be mined in order to be made available to the network.

## *7.4 The User Interface*

In order to enable user operations on the smart contract that was deployed in the preceding part, a front-end application was deployed using Node.js.

The functionalities tested in the application are as follows:

**Add Operation**

To add a pair of email ID and password, the user navigates to the /add URL and enters the information into the input fields. Upon clicking the submit button, the information is saved in the contract (Fig. 14).

The front-end interface sends a transaction with email ID and password as inputs to the addPassword() method of the smart contact via the web3 instance. Every smart contract deployed into the blockchain has its own personal database, known as smart contract storage. Thus, the email ID-password pair gets stored in the smart contract at a unique block index on the blockchain following the successful execution of the transaction.

**Retrieval Operation**

The user navigates to the /view URL in order to view the data. When the user hovers the mouse pointer over the hidden password field, the password is revealed (Fig. 15).

The front-end interface invokes the smart contract to access its state variable and read value of all key-value pairs from the user's copy of blockchain state.

**Update Operation**

To update a password, the user navigates to the /update URL and enters the new values into the corresponding input fields. Upon clicking the update button, the data in the contract is updated (Fig. 16).

The front-end application sends a transaction containing the index and new password to the updatePassword() function of the smart contract through the web3

Fig. 13 Remix IDE is used to test smart contract deployment

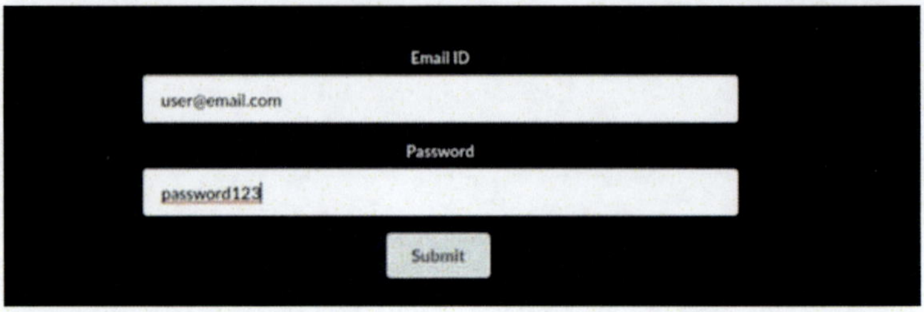


Fig. 14 Adding credentials

Fig. 15 Viewing the saved credentials

Fig. 16 Updating the credentials

instance. The new password is recorded in the smart contract at the unique block index on the blockchain following the successful execution of the transaction.

### Delete Operation

The front-end application instructs the smart contract to delete the key-value pair with the provided index number. Once the transaction is successfully executed, the block index points to NULL value in the blockchain.

# 8 Conclusion

The blockchain procedures are completely transparent to the end user since they are perfectly incorporated into the application. This makes the application as user-friendly as any conventional password manager. The user is no longer required to remember multiple passwords for different websites, nor is it required to entrust the corporation with their passwords and privacy.

Thereby, not only providing a secure and persistent way to store credentials, but also eliminating the need to trust third parties with sensitive information and they also only require a minimum setup cost.

# 9 Limitations and Future Scope

For the purposes of this system, the official Go implementation of the Ethereum protocol, Geth, was utilized. It currently uses a Proof of Work (PoW)-based consensus procedure, with every full node storing every piece of information. There is no one entity in charge of the network. One can operate a complete node and join the network at any time [7]. Therefore, this Password Management System runs on a fully decentralized blockchain system.

Like any other blockchain application, this application too is limited by the DCS trilemma: The DCS triangle states that we can have any two of the three key attributes of blockchain: Decentralization; Consensus; Scalability, but not all three at the same time [8].

Ethereum protocol embodies systems that are Decentralized and Consistent, but not planetary Scale. Because each node must keep all data, or what is known as a "replication factor of n" in databases, Ethereum has trouble scaling throughout the entire world [7].

Ethereum can approach planetary scale through *sharding*—where each node only preserves a limited subset of the data. However, this is not yet implemented in Ethereum protocol yet. Ethereum Foundation's Casper research shows promising results toward slowly shifting Ethereum toward being scalable [9].

**Acknowledgements** In appreciation, we extend our sincerest gratitude to Sikkim Manipal Institute of Technology, Sikkim, India.

# References

1. Lyastani SG, Schilling M, Fahl S, Backes M, Bugiel S (2018) Better managed than memorized? Studying the impact of managers on password strength and reuse. In: 27th USENIX security symposium, pp 203–220

2. Alodhyani F, Theodorakopoulos G, Reinecke P (2020) Password man-agers—It's all about trust and transparency. Future Internet 12(11):189. https://doi.org/10.3390/fi12110189
3. Pymnts.com. Blockchain, distributed ledger. https://www.pymnts.com/tag/blockchain/
4. Alharby M, Van Moorsel A (2017) Blockchain-based smart contracts: a systematic mapping study. arXiv preprint arXiv:1710.06372
5. Nathan S, Govindarajan C, Saraf A, Sethi M, Jayachandran P (2019) Blockchain meets database: design and implementation of a blockchain relational database. arXiv preprint arXiv:1903.01919
6. Solidity. Solidity 0.6.8 documentation. https://solidity.readthedocs.io/en/v0.6.8/
7. BigchainDB, Trent McConaghy. The DCS Triangle (2016). https://blog.bigchaindb.com/the-dcs-triangle-5ce0e9e0f1dc
8. Medium, Hoa Phan. Solving The Blockchain Trilemma (2018). https://medium.com/amate/solving-the-blockchain-trilemma-91fa9ca4e007
9. Buterin V, Griffith V (2017) Casper the friendly finality gadget. arXiv preprint arXiv:1710.09437. Aldhafferi N, Watson C, ASM S (2013) Personal information privacy settings of online social networks and their suitability for mobile internet devices. Int J Secur Privacy Trust Manag 2(2):1–17. https://doi.org/10.5121/ijsptm.2013.2201

# Deep Reinforcement Learning Algorithms for D2D Communications: A Comparative Analysis

Devarani Devi Ningombam

**Abstract** Device-to-device (D2D) communications, which permit direct communication among two mobile devices and are enabled by the widely used cellular network, may offer a viable answer to the issue of promoting trustworthy and efficient communication. In D2D communications, the efficient management of resources while minimizing interference is a common concern. This study considers the deep reinforcement learning (DRL) method to present new resource techniques for D2D communications in order to improve decision-making and control. We investigate the actor-critic (AC)-based DRL algorithm utilizing a centralized training framework that can meet the "quality of service (QoS)" requirements of cellular and D2D users while simultaneously optimizing the system's overall QoS. Comparing the performance of the proposed optimization strategy to that of previous research, simulation findings reveal a significant improvement in the system throughput and faster convergence.

**Keywords** Device-to-device communications · Deep reinforcement learning · Centralized training and decentralized execution · Actor critic · Quality of Service (QoS)

## 1 Introduction

"Device-to-device communication (D2D)" denotes to a set of technologies that enable devices or users to connect directly over a network without the use of an evolved NodeB (eNB) or other intermediary devices [1]. As a result, D2D communication becomes an indispensable tool for solving difficulties such as interference control and coverage. D2D communication, which forms the basis of cellular networks, increases system capacity and user data rates [2]. Despite this, average mobile and D2D users continue to endure severe intrusions due to bandwidth sharing.

D. D. Ningombam (✉)
Department of Computer Science and Engineering, National Institute of Technology Patna (NIT Patna), Bihar, India
e-mail: devarani.cs@nitp.ac.in

 
S. Borah et al. (eds.), *Advanced Computational and Communication Paradigms*,
Lecture Notes in Networks and Systems 535,
https://doi.org/10.1007/978-981-99-4284-8_6

Cellular networks face a variety of obstacles in order to meet user Quality-of-Service (QoS) expectations and limit user intrusion, including the simultaneous selection of modes enabled by D2D and the escalation of channel allocation [3].

Controlling the precision of the D2D communications that underpin the cellular network is a serious concern. The effectiveness and output of the system are a direct result of the management strategy. In the CTDE machine learning architecture, D2D protocol users must learn to choose the appropriate channel to maximize their global reward. The problem of data balance has been tackled by a number of different machine learning strategies, including "Random Forest, Naive Bayes, and Support Vector Machine," among others [4]. To acquire a better understanding of how cellular networks function, we compare various DRL algorithms for D2D communication. In this section, we describe the technical contribution of this work:

- DRL algorithms are used to come up with a way to share resources between traditional mobile users and D2D users.
- The efficiency of different DRL methods, such as "Deep Q-Networks (DQN), Deep Deterministic Policy Gradient (DDPG), and Quantile Regression Deep Q-Network (QR-DQN)," is compared with possible network speed and spectrum efficiency.

The structure of the paper is as follows: In Sect. 2, mathematical models and formulations for problems are offered. In Sect. 3, the learning framework and DRL algorithms are designed. In Sect. 4, numerical simulations are used to demonstrate the algorithm's validity. Section 5 is dedicated to the conclusion.

## 2 Related Work

The authors of [5] present a DRL-based joint optimization approach. The authors assessed the efficiency with which the Proximal Policy Optimization (PPO) method resolved a maximization game. To increase the sum-rate while simultaneously satisfying the QoS needs of both regular cell users and D2D clients, a distributed multi-agent deep Q-network architecture is proposed. [6] examines the challenge of solving the joint optimization problem of channel estimation and radio resource in D2D-enabled heterogeneous cellular networks using mm-wave and cellular bands in order to ensure that the QoS criteria of the users are met and to efficiently handle any interference between users. In order to reduce signaling costs, it is also described how teaching D2D users the optimal mode and channel while ignoring global information can be performed by partial information exchange. Using shallow and deep convolutional neural network (CNN) models, [7] suggests two preventative channel assignment methods: controlled and intelligent prediction. Controlled prediction figures out what the values of different actions will be and changes the policy to choose the best actions at each stage.

## 3 System Architecture and Formulation of Problems

Figure 1 illustrates the system architecture used in the paper. Consideration is given to an uplink cellular network that enables D2D communications, allowing D2D users to reuse uplink cellular resources. The evolved NodeB (eNB) is positioned in the center of a 500-m-diameter cell in which all users are evenly distributed. We specify the traditional mobile users' groups and D2D pairs as two sets, $TU = \{1, 2, \ldots, t\}$ and $DU = \{1, 2, \ldots, i\}$, respectively. Subsequently, the received SINRs for $p$th cellular user and $q$th D2D pair that reuse the resource of the $p$th cellular user are illustrated as follows (1)–(2):

$$SINR_{q,p} = \frac{P_q.d_q^{\omega(\varphi-1)}.|h_q|^2}{P_p.d_{pq}^{\omega(\varphi-1)}.|h_{pq}|^2 + \sum_{i=1}^{2} I_{c_i} + P_N} \tag{1}$$

and

$$\gamma_p = \frac{P_p.d_p^{\omega(\varphi-1)}.|h_p|^2}{P_q.d_q^{\omega(\varphi-1)}.|h_q|^2 + \sum_{i=1}^{2} I_{c_i} + P_N} \tag{2}$$

where the power through the $q$th D2D pair is represented by $P_q$, the power through the $p$th cellular user is represented by $P_p$, and the power through the neighboring co-channel users is represented by $P_N$. The distance between $p$th D2D pair and eNB is represented by $d_p$; the distance between $q$th cellular user and eNB is represented by $d_q$; the distance between $q$th cellular user and $p$th D2D pair is represented by $d_{pq}$; $h_q$, $h_p$, and $h_{pq}$ are the channel coefficients between the $q$th D2D pair and eNB, $p$th cellular user and eNB, $p$th cellular user and $q$th D2D pair, and target device and

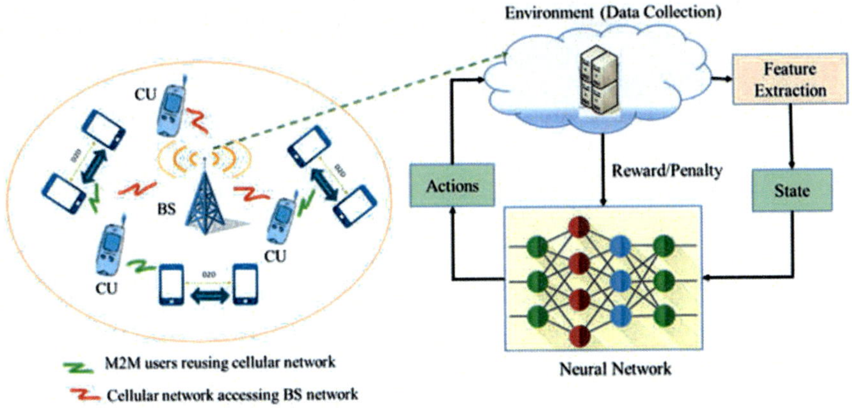


**Fig. 1** Architecture of DRL-based D2D communications underlaying cellular network

neighboring devices, respectively. The path loss exponent is represented by $\alpha$ and the power control factor is represented by $\varphi$, where $\varphi\epsilon[0, 1]$.

This study's primary objective is to improve frequency reuse by investigating D2D users while minimizing the likelihood of system interference. Frequency reuse (FR) is the practice of having several devices or users use the same channels and bandwidths in a communication system in an effort to increase the space in the system and the gradation effectiveness. In this situation, cellular users are the primary consumers of resources, whereas D2D users are the secondary consumers [8]. This indicates that the eNB has allocated all available resources to mobile users. Thus, each cellular user believed there was no cellular user incursion since they each used a distinct asset block for the uplink and downlink distribution. D2D users who are present in any portion of the cell region may only utilize the frequency band assigned for regular cellular users. The device's received signal is illustrated as

$$S(\tau) = s_o(\tau) + \sum_{Neigh=1}^{2} s_{Neigh}(\tau) + \vartheta(\tau) \tag{3}$$

where $s_o(\tau)$ is the signal received from the desired device of the cell, $s_{Neigh}(\tau)$ is the total of the signals that the target cell part of the cell has received from nearby devices, and $\vartheta(\tau)$ is the power spectral density in the noise $P_N$.

## 3.1 Partially Observable Markov Decision Process

In our research, we consider a fully cooperative multi-agent system with partially observable "Markov Decision Processes (PoMDPs)," which is given by a tuple$\langle \varnothing, \delta, \rho, \mathcal{P}, \theta, \mu, n, \alpha_0 \rangle$, where $\varnothing = \{1, 2, \ldots, i\}$ represent the set of $i$ agents, $\delta = \{\delta_1, \delta_2, \ldots, \delta_n\}$ represent the observation space of the agents, $\rho = \{\rho_1, \rho_2, \ldots, \rho_n\}$ represent the action space of the agent, $\theta$ is the global reward, $\mu\epsilon[0, 1]$ is the discount factor, and $\alpha_0$ is the initial state distribution. Every agent at each time step $q\epsilon\varnothing = \{1, 2, \ldots, i\}$ selects an action, $a_i\epsilon\delta$ and coordinated action,$\boldsymbol{a}\epsilon\delta$. Then the transition of state takes place which is represented by a transition function $\alpha(s\prime|s, \boldsymbol{a}) : \delta \times \rho \times \varnothing \rightarrow [0, 1]$. In the MARL setting, each agent seeks to optimize their own projected return, $J(\pi) = \mathbb{E}_{\delta,\rho}[r]$, where$r = \sum_{i=0}^{\infty} \gamma^i \varepsilon$.

## 3.2 Centralized Training Decentralized Execution

During the training phase, the centralized critic is taught how to update the decentralized rules using the "Centralized Training and Decentralized Execution (CTDE)" method, which is widely used in RL. Each agent's policy is updated to maximize $\mathbb{E}[r_{ext}]$ using the policy gradient, as shown below, when we consider $i$ independent parameterized policies $\pi_{\psi_t}$ for $t \epsilon \varnothing$.

$$\nabla_{\psi_t} J_{ext}(\psi_1, \psi_2, \ldots, \psi_n) = \mathbb{E}_{\delta,\rho}\left[\nabla_{\psi_i} \log \pi_{\psi_i}(\rho_i|\delta_i) A_\pi(\delta, \boldsymbol{a})\right] \tag{4}$$

where $A_\pi(\delta, \boldsymbol{a})$ stands for the centralized critics and $(\delta, \boldsymbol{a})$ is the combined state-action.

# 4 Proposed Algorithm

## 4.1 *Deep Q-Network (DQN)*

A reinforcement learning strategy incorporating Q-learning and deep neural networks, or DQN, is presented to allow RL to operate in complex and high-dimensional settings. The Q-value function is assumed by DQN using a neural network (NN) [9]. In the setting, the state is provided as the input, and the Q-value for every action that could be taken is created as the output.

| Algorithm 1. DQN |
|---|
| **Step 1**: Initialize memory and $Q$-value with random weights |
| **Step 2: for** $\tau$ =1 to T **do** |
| **Step 3:** Select action by using $\epsilon - \boldsymbol{greedy}$ |
| **Step 4**: Execute action and calculate reward $\boldsymbol{r}(\boldsymbol{\delta}, \boldsymbol{a})$ |
| **Step 5**: Calculate total reward ($Q$-value) as: $Q(\boldsymbol{\delta}, \boldsymbol{a}) = \boldsymbol{r}(\boldsymbol{\delta}, \boldsymbol{a}) + \upsilon \max_{\boldsymbol{a}} Q\left(\boldsymbol{\delta}', \boldsymbol{a}\right)$ |
| **Step 6:** Store the past experiences in a memory |
| **Step 7:** Determine the next action considering the maximum output of the $Q$-network |
| **Step 8:** Calculate the loss function as: $\boldsymbol{Loss} = \boldsymbol{MSE}\left(Q_{\boldsymbol{Predicted}}, Q_{\boldsymbol{Target}}\right)$ |
| **Step 9**: Update the $Q$-value as: $Q(\boldsymbol{\delta}_t, \boldsymbol{A}_t) \leftarrow Q\left(\boldsymbol{\delta}_t, \boldsymbol{\rho}_t\right) + \boldsymbol{\alpha}\left[\boldsymbol{R}_{t+1} + \boldsymbol{\gamma} \max_{\boldsymbol{a}} Q(\boldsymbol{\delta}_{t+1}, \boldsymbol{a}) - Q\left(\boldsymbol{\delta}_t, \boldsymbol{\rho}_t\right)\right]$ |
| **Step 10**: end **for** |

## 4.2 *Deep Deterministic Policy Gradient (DDPG)*

An RL method that associates Q-learning and policy gradients, deep deterministic policy gradient (DDPG) is used. As an actor-critic (AC) technique, DDPG uses two models: the actor and the critic. Rather than generating a dispersion of possibilities across acts, the actor learns from policies and outputs the precise action (continuous) based on the state as input [10]. A $Q$-value network, the critic, receives inputs like state and action and outputs a $Q$-value. In Algorithm 2, we could see the DDPG pseudocode described.

| Algorithm 2: DDPG |
|---|
| **Step 1:** Assign critic network and actor with different weights |
| **Step 2:** Assign target network and replay buffer |
| **Step 3: for** $\tau$ =1 to $T$ **do** |
| **Step 4:** Select a course of action depending on the prevailing policy and exploration noise |
| **Step 5:** Act and observe reward $\boldsymbol{r}(\boldsymbol{\delta}, \boldsymbol{a})$ and new state |
| **Step 6:** In the buffer memory, keep the prior experiences |
| **Step 7:** Sample minibatch transitions from reply buffer |
| **Step 8:** By reducing the loss function, update the critic |
| **Step 9:** Utilizing the sampled gradient, update the actor policy |
| **Step 10:** Update target networks |
| **Step 11:** end **for** |

## 4.3 *Quantile Regression Deep Q-Network (QR-DQN)*

One area where deep learning has not been fully explored is the uncertainty in estimations. The best estimate as defined by a loss function is prioritized by the majority of deep learning frameworks at the moment. When making decisions, the majority of people rely on quantiles rather than actual estimation uncertainty [11].

In lieu of a fixed value support, the Quantile Regression DQN (QR-DQN) takes a fixed probability support into account. QR-DQN which uses a parametric model to compute the quantiles of the target distribution assumes that the Cumulative Probability Function (CDF) has discrete support.

## 4.4 *Channel Model*

In the proposed approach, the eNB evaluates the channel characteristics between itself and cellular users in order to allocate available resources to cellular users. Likewise, the eNB determines the channel characteristics between itself and D2D pairings. This proposed technique employs the network model described in [3].

$$z_{TU} = 40.9 + 36.7log_{10}(d) + 26log_{10}\left(\frac{w_c}{5}\right) \tag{5}$$

$$z_{DU} = 49 + 40log_{10}\left(\frac{d}{1000}\right) + 30log_{10}\left(\frac{w_c}{1000}\right) \quad (6)$$

where $z_{TU}$ stands for the path loss model for conventional mobile users and $z_{DU}$ stands for the path loss for D2D users, and $d$ stands for distance bounded by the cellular users and D2D users in meters, while $w_c$ stands for the carrier frequency between the transmitter and receiver of the device in GHz.

## 5 Simulation Result and Discussion

Table 1 lists the simulation parameters and their corresponding values. We suppose that the cell has a 500-m radius and, depending on the scenario under consideration, has a variety of D2D pairs and conventional mobile users. We offer simulated results based on a comparative examination of several frequency reuse environments.

Our solution is compared to the method of random resource allocation, in which D2D pairs are able to utilize the resources of classical cellular users that are already accessible in the cell, regardless of their position within the cell.

Figure 2 depicts a comparison of the throughput performance of several DRL algorithms for D2D-enabled uplink cellular networks. In the figure, the efficacy of each DRL approach was evaluated using different D2D user counts for each cell. The study found that policy gradient-based learning is better than value-based, quantile-based, and random methods for allocating resources because it has a high throughput value. The main reason for this performance improvement is the coexisting scheme of cellular and D2D communication underlying cellular users, which provides more communication opportunities.

Figure 3 depicts a comparison of the spectral efficiency of various DRL algorithms for D2D communications in a cell with an arbitrary number of D2D users. The figure demonstrates that the DDPG method delivers the maximum spectrum efficiency in comparison to other algorithms. From the figure, we can see that the proposed system intelligently selects promising communication modes and allocates spectrum resources in order to maximize utilization while ensuring variable user QoS.

**Table 1** Simulation parameters and values

| Abbreviation | Meaning | Values |
|---|---|---|
| $P_{eNB}$ | Base station power | 42 dBm–48 dBm |
| $P_{DU}$ | D2D user power | 8 dBm–15 dBm |
| $P_N$ | Noise power | −174 dBm/Hz |
| $d_p$ | Distance between 2 D2D users | 10 m–50 m |
| $d_{eNB,p}$ | Distance between base station and D2D user | 10 m |
| $w_c$ | Carrier frequency | 2 GHz |

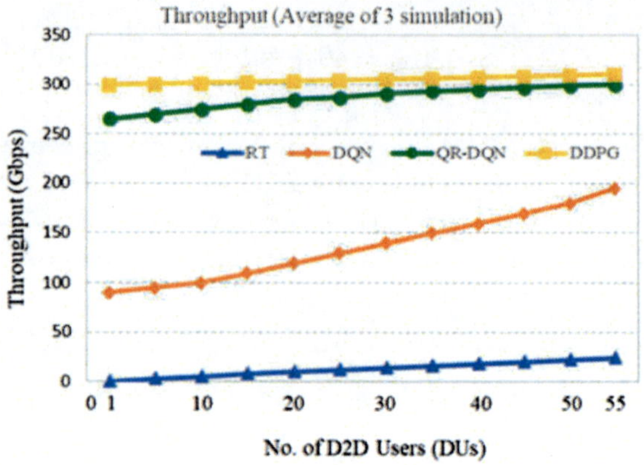


**Fig. 2** Throughput performance comparison for D2D communications using different DRL algorithms

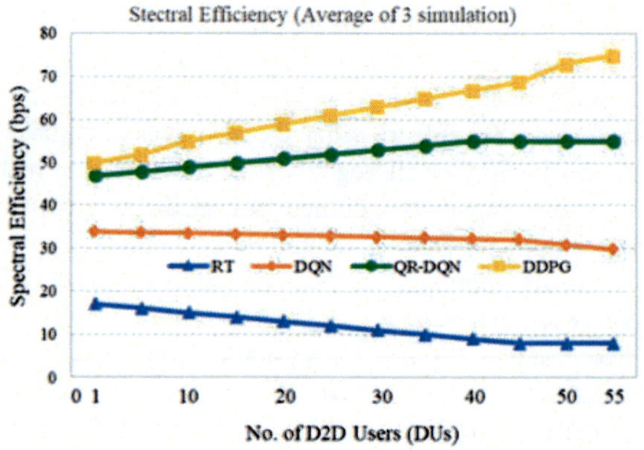


**Fig. 3** Spectral efficiency performance comparison for D2D communications using different DRL algorithms

## 6 Conclusions

This study introduces a deep reinforcement learning-based approach to resource management that enables D2D users to share uplink cellular resources. A new approach to improving the spectral efficiency and capacity of a system was presented in this research. In the proposed technique for resource distribution, the measure of distance between D2D users and cellular users is taken into consideration. The results of our simulations indicate that the proposed method offers superior spectrum utilization and throughput compared to the methods that are considered to be state of the art.

## References


1. Ningombam DD, Shin S (2019) Optimal resource management and binary power control in network-assisted D2D communications for higher frequency reuse factor. Sensors 19:251. https://doi.org/10.3390/s19020251
2. Li Z, Sanchez Moya F, Gabor F, Da Silva Jr, J, Koufos K (2016) Device-to-device (D2D) communications. In: Dohler M, Nakamura T, Osseiran A, Monserrat J, Marsch P (eds), 5G mobile and wireless communications technology (pp 107–136). Cambridge: Cambridge University Press. https://doi.org/10.1017/CBO9781316417744.006
3. Ningombam DD, Lee CG, Shin S (2019) Interference mitigation for multicast D2D communications underlay cellular networks. International conference on artificial intelligence in information and communication (ICAIIC), 2019, pp 1–4, https://doi.org/10.1109/ICAIIC.2019.8668982
4. Telu VS, Padimi V, Ningombam DD (2022) Optimizing predictions of brain stroke using machine learning. J Neutrosophic Fuzzy Syst 2(2):31–43
5. Nyuyen KK, Masaracchia A, Yin C, Nyuyen LD, Dobre OA, Duong TQ (2021) Deep reinforcement learning for intelligent reflecting surface-assisted D2D communications
6. Yu S, Lee JW (2022) Deep reinforcement learning based resource allocation for D2D communications underlay cellular networks. Sensors 22:9459. https://doi.org/10.3390/s22239459
7. Sakib S, Tazrin T, Fouda MM, Fadlullah ZM, Nasser N (2021) A deep learning method for predictive channel assignment in beyond 5G networks. IEEE Netw 35:266–272
8. Kim K (2022) Multi-agent deep q network to enhance the reinforcement learning for delayed reward system. Appl Sci 12:3520. https://doi.org/10.3390/app12073520
9. Jafari R, Javidi MM, Kuchaki M (2019) Using deep reinforcement learning approach for solving the multiple sequential alignment problem. Appl Sci 1(6):592
10. Che Y, Lin F, Liu J (2021) Deep reinforcement learning in M2M communication for resource scheduling. world conference on computing and communication technologies (WCCCT), 2021, pp 97–100, https://doi.org/10.1109/WCCCT52091.2021.00025
11. Xu X, Wang W, Shin Y, Zheng C (2021) Dynamic network quantile regression model

# AI-Enabled IDS for Identifying Malicious Traffics in Cloud Systems

**Soham Biswas, Chandan Kumar, Suman Som, Md. Sarfaraj Alam Ansari, and Mahesh Chandra Govil**

**Abstract** The widespread usage of the Internet has increased network security issues. Some of the serious threats are Brute force, Heartbleed, Botnet, Web attacks, DoS, DDoS, and Infiltration of the network. With the increased usage of online services and the development of new technologies like IoT, Cloud, Edge, and Mobile Computing, the detection of these threats is important and becoming paramount. This study presents an Intrusion Detection System (IDS) framework by proposing various hybrid models for the detection of malicious traffic. The models are developed using feature selection techniques, namely Recursive Feature Elimination, Chi-Square, Principal Component Analysis, and five Machine Learning (ML) techniques. In this experiment, the standard CSE-CIC-IDS 2018 dataset is used to examine the performance evaluation of the proposed approaches. In this paper, according to the comparison of the proposed hybrid approaches, Recursive Feature Elimination with Random Forest performs better than the other approaches with an accuracy of 99.23%. The proposed approach provides better accuracy as compared to earlier reported similar approaches in the literature.

**Keywords** Network security · IDS · Malicious traffic · Machine learning · Feature selection

S. Biswas · C. Kumar · S. Som · M. S. A. Ansari (✉) · M. C. Govil
Department of Computer Science and Engineering NIT Sikkim, Sikkim 737139, India
e-mail: sarfaraj@nitsikkim.ac.in

S. Biswas
e-mail: m210006@nitsikkim.ac.in

C. Kumar
e-mail: phcs220029@nitsikkim.ac.in

S. Som
e-mail: m210007@nitsikkim.ac.in

 
S. Borah et al. (eds.), *Advanced Computational and Communication Paradigms*,
Lecture Notes in Networks and Systems 535,
https://doi.org/10.1007/978-981-99-4284-8_7

## 1 Introduction

According to Cisco's annual report, over two-thirds of the world's population will have access to the Internet by 2023 and there will be 5.3 billion users worldwide, rising from 3.9 billion in 2018 [1]. Even though more people are using the Internet and benefiting from the services, however, users are also confronted by more serious security threats that are rapidly growing and changing dynamically [2]. The cloud computing infrastructure has numerous flaws, which attract intruders to explore and exploit various attacks. The latest attacks on cloud computing are Brute force, Heartbleed, Botnet, DoS, DDoS, Web attacks, and Infiltration of the network [3, 4].

Traditional network intrusion detection techniques are typically passive and unable to correctly assess a wide range of unknown intrusions. Therefore, to meet the cloud security requirements, it is essential to explore more precise and effective intelligent network intrusion detection techniques. An intrusion detection system (IDS) basically keeps track on network traffic in real time and can send alerts or initiate preventative measures when it detects unusual transmissions [5, 6]. The four basic IDS operations are event gathering, pre-processing, detection, and alert mechanism, where the detection is the main focus. Recently, machine learning (ML) and data mining techniques are playing an important role to address the issue of network intrusion detection by classifying the attacks [7, 8]. Machine learning has the ability to extract out decent representation from large amounts of data and produce considerably improved results [9]. For identifying the attacks, a high value of accuracy, precision, recall, and low False Alarm Rate (FAR) is required. To put the discussion in brief, the primary goal of intrusion detection is to raise the accuracy of machine learning methods in correctly identifying invasive behaviour [10].

In this study, we discuss the growing issue of data security in cloud-based networks. We have proposed the hybrid models using feature selection techniques and machine learning for identifying the intrusion effectively. The proposed approaches have been implemented on popular IDS dataset CICIDS 2018. The results are compared with other developed models in this study and the similar approaches developed for the same purpose. The contribution of our study consists of the following:

1. For the detection of malicious traffic in the cloud network, fifteen (15) different hybrid models are developed using the above three feature selection techniques and five machine learning methods: Decision Tree (C4.5), Naive Bayes Classifier, Logistic Regression, Random Forest, and Stochastic Gradient Descent.
2. Selection of the best features using the feature selection techniques such as Chi-Square, Recursive Feature Elimination, and Principal Component Analysis.
3. Performance analysis and evaluation of the best model among the proposed hybrid models for identifying malicious traffics using benchmark dataset CSE-CIC-IDS 2018.
4. The performance of the proposed models is also compared with existing approaches.

The remaining of the paper is organized as follows: the related works are presented in Sect. 2. Section 3 describes the proposed methodology for IDS. The results and discussion are stated in Sect. 4. Finally, the study is concluded at Sect. 5.

## 2 Related Works

In recent times, we have seen a range of research works on identifying malicious traffic [11, 12]. Many of researchers have used feature selection and machine learning methods in their work to develop an effective intrusion detection system. Keserwani et al. [8] have proposed an intrusion detection model named GWO-BES-DSAE to secure the virtual cloud network (VCN) and achieved an accuracy of 96.06% in the multi-class classification of the CICIDS 2017 dataset. In this hybrid model, Grey-Wolf Optimization (GWO) with Bald Eagle Search (BES) is used for feature selection and deep sparse auto-encoder (DSAE) is used for classification. Barbhuiya et al. [13] proposed a real-time anomaly detecting system with 90–95% accuracy. The authors got a low false-positive rate of 0–3% while detecting DDoS and crypto-mining attacks.

Alharbi and Alshubi [14] used graph-based ML approaches to detect malware and achieved 99% accuracy with the ExtraTrees classifier and Pearson's correlation features. Another method named BotChase developed using machine learning achieved up to 95% accuracy in graph-based bot detection [15]. Elsayed et al. [16] proposed a detailed study of various approaches on classical ML techniques for detecting attacks in software-defined networks and got the best accuracy of 81.5% using J48 decision tree.

Nguyen et al. [17] have proposed a deep learning approach and applied over three datasets along with ten machine learning approaches. In their work, the highest accuracy of 97.11% is achieved by the proposed method compared to other machine learning approaches. Syms et al. [18] proposed a hybrid intrusion detection system which achieved 98.6% accuracy on the UNSW NB15 dataset with two classes. They compared their proposed HCNN with RNN and showed that the proposed work outperformed.

In this study, CICIDS 2018 dataset is considered which includes five malicious traffics with cloud network traffic. The hybrid models are developed using three different feature selection (FS) algorithms and five machine learning (ML) techniques. Further, all the models are analysed on the CICIDS 2018 dataset and the comparative study is performed. The proposed methods are also compared with existing state-of-the-art algorithms. In the following section, we will discuss in detail the proposed methodology, framework, dataset, FS & ML techniques, and proposed algorithm.

## 3 Proposed Methodology

This section presents the proposed framework for the IDS. The framework consists of data collection, pre-processing, and FS- and ML-based hybrid models for training and testing. This section is concluded with a proposal of an efficient algorithm. A brief discussion is given below.

### 3.1 Framework

The proposed framework consists of five phases. The first phase consists of data collection. Data is collected from the cloud. Subsequently, the collected data is pre-processed to make it suitable for the ML algorithms. Further, since all the features available in the dataset might not be important, we have to carry out the feature selection task and select the best features. The fourth phase consists of splitting the dataset into testing and training parts. We opt for the ratio of 70:30 for splitting the dataset into training and testing. The model is trained by applying five different ML algorithms. Finally, the proposed system is validated by detecting the attacks. Figure 1 illustrates the proposed framework.

### 3.2 Dataset

The task of anomaly detection is one of the challenging jobs which could be reduced by selecting the most up-to-date dataset. This is why for our work a benchmark dataset CSE-CIC-IDS2018 [3] is used to train and validate the proposed framework by applying various ML algorithms for the identification of malicious traffic from the massive traffic flows in the cloud. Seven different attacks are labelled in this dataset: *Brute force*, *Heartbleed*, *Botnet*, *DoS*, *Web attacks*, *DDoS*, and *Infiltration* traffics. We choose the 5 most significant classes from this dataset and reduce the dataset volume before training the models. The traffic flows are, namely (1) Benign, (2) Botnet, (3) GoldenEye, (4) HOIC, and (5) Infiltration. The dataset has a total of 80 features [4]. The traffic flow volumes are shown in Table 1.

### 3.3 Pre-processing

Pre-processing is a crucial step in increasing the consistency of the data and the accuracy of any proposed solution. Pre-processing of data includes the cleansing and normalization of data. The cleaning or retrieval of data is an important phase as

**Table 1** Dataset

| No. | Network traffic name | Volumes |
|---|---|---|
| 1 | Benign network traffic | 287 |
| 2 | Botnet network traffic | 315 |
| 3 | GoldenEye network traffic | 262 |
| 4 | HOIC network traffic | 406 |
| 5 | Infiltration network traffic | 37 |

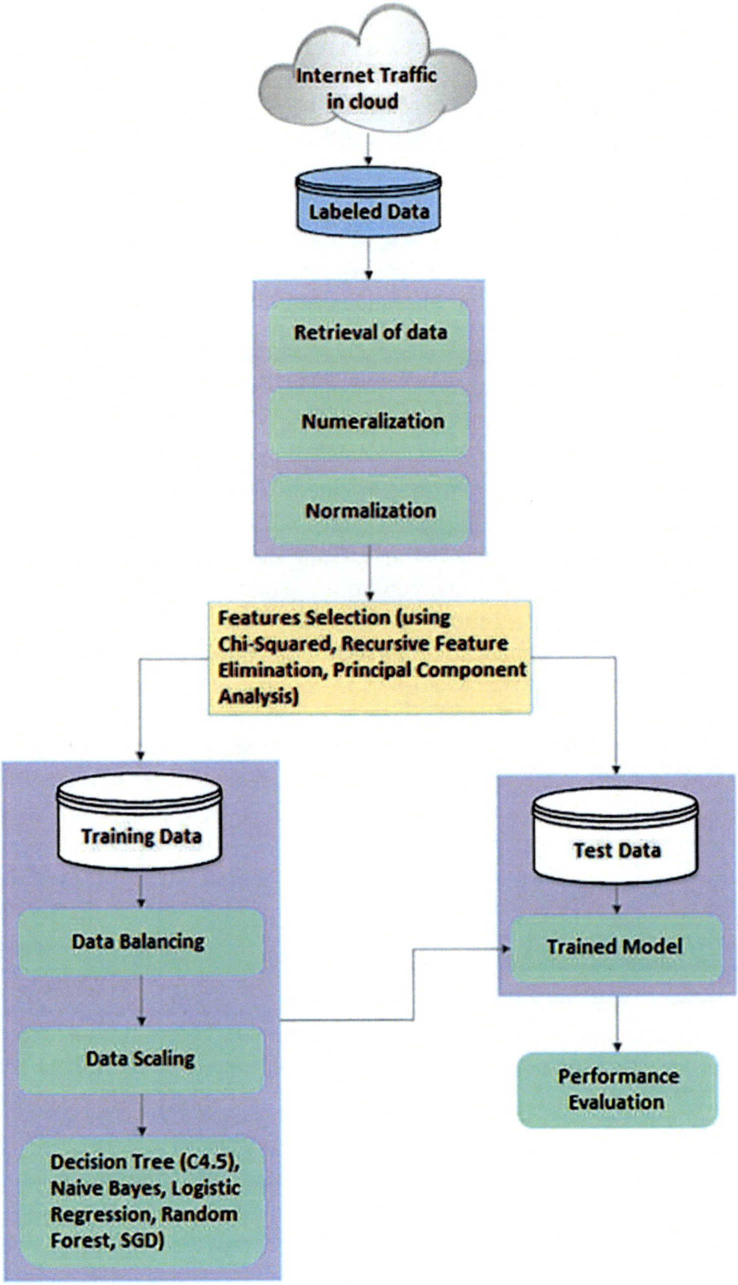


**Fig. 1** Framework

duplication and noise are eliminated. In addition, data obtained can be incomplete, incorrect, or null. All incoherence is eliminated in data cleaning. This task is carried out in the Python environment. Encoding or Numeralization typically, in most dataset columns, include several forms of labels, which are categorical or string types, in comprehensible human terms. But on numerical data, the system works better. Many encoding methods are given to translate words into numerical types. Here, a sequence number is assigned to each nominal feature. The difference between the minimum and maximum values of attributes is significantly large. It makes the estimation of variance from the central tendency of the distribution large and may lead to erroneous results. This need normalization of features and Z-score transformation is used to resolve this issue.

### 3.4 Feature Selection

Feature selection helps in building the model with less computing resources and limited computational time. So, before training the model, three feature selection techniques are used to reduce the redundant features and select the best out of 80 feature vectors. By reducing the horizontal space feature vector within the dataset, it becomes possible to construct the model using limited computing resources and less computational time. We have used *Chi-Squared* technique using *SelectKBest*, *Recursive Feature Elimination (RFE)* methods, and *Principal Component Analysis (PCA)* for feature selection. We have selected the important features by using the aforementioned three different feature selection techniques. Out of total 80 features from PCA, 20 best features are selected. In Table 2, 20 best features that are selected using those different feature selection techniques are presented.

### 3.5 Classification Techniques

For model building, we implemented five different ML algorithms and selected the best one based on the obtained accuracy. Five different ML algorithms are Naive Bayes (NB), Decision Tree (DT), Random Forest (RF), Logistic Regression (LR), and Stochastic Gradient Descent (SGD).

### 3.6 Proposed Model

Using the methods discussed above, we proposed total 15 different models of IDS for identifying malicious and legitimate traffics in cloud. The proposed models' details are shown in Table 3.

**Table 2** Selected features

| Feature selection methods | Selected features |
|---|---|
| Chi-Squared | Flow Duration, Flow IAT Std, Flow IAT Mean, Flow IAT Max, Fwd IAT Tot, Flow IAT Min, Fwd IAT Mean, Fwd IAT Max, Fwd IAT Std, Fwd IAT Min, Bwd IAT Mean, Bwd IAT Tot, Bwd IAT Std, Bwd IAT Min, Bwd IAT Max, Active Mean, Idle Mean, Active Max, Idle Max, Idle Min |
| RFE | Flow Duration, Dst Port, Tot Fwd Pkts, Tot Len Fwd Pkts, Tot Bwd Pkts, Tot Len Bwd Pkts, Fwd Pkt Len Mean, Fwd Pkt Len Max, Fwd Pkt Len Std, Flow Byts/s, Bwd Pkt Len Min, Fwd IAT Min, Pkt Len Mean, Fwd Header Len, Pkt Len Std, SYN Flag Cnt, Pkt Len Var, Pkt Size Avg, Init Bwd Win Byts, Init Fwd Win Byts |
| PCA | Dst Port, Tot Len Bwd Pkts, Tot Len Fwd Pkts, Fwd Pkt Len Max, Fwd Pkt Len Mean, Fwd Pkt Len Min, Bwd Pkt Len Min, Fwd IAT Min, Flow Byts/s, Bwd Pkts/s, Pkt Len Var, Pkt Len Std, PSH Flag Cnt, RST Flag Cnt, URG Flag Cnt, Pkt Size Avg, ECE Flag Cnt, Subflow Bwd Byts, Init Bwd Win Byts, Init Fwd Win Byts |

**Table 3** Details of the proposed models

| Model name | Feature selection technique | Classification technique |
|---|---|---|
| $\chi^2$_DT_IDS | Chi-squared | DT(C4.5) |
| $\chi^2$_NB_IDS | Chi-squared | Naive Bayes |
| $\chi^2$_LR_IDS | Chi-squared | Logistic regression |
| $\chi^2$_RF_IDS | Chi-squared | Random forest |
| $\chi^2$_SGD_IDS | Chi-squared | SGD |
| RFE_DT_IDS | Recursive feature elimination | DT(C4.5) |
| RFE_NB_IDS | Recursive feature elimination | Naive Bayes |
| RFE_LR_IDS | Recursive feature elimination | Logistic regression |
| RFE_RF_IDS | Recursive feature elimination | Random forest |
| RFE_SGD_IDS | Recursive feature elimination | SGD |
| PCA_DT_IDS | Principal component analysis | DT(C4.5) |
| PCA_NB_IDS | Principal component analysis | Naive Bayes |
| PCA_LR_IDS | Principal component analysis | Logistic regression |
| PCA_RF_IDS | Principal component analysis | Random forest |
| PCA_SGD_IDS | Principal component analysis | SGD |

### 3.7 *Proposed Algorithm*

We have proposed an algorithm for IDS to get the best output over using the proposed 15 different models. This algorithm runs for three different feature selection methods along with five different ML techniques. The proposed Algorithm 1 takes the list of features (*l*) from the dataset (*D*) and fits it to the ML model (*S*). Initially, feature selection is done, which produces the best selected features (K). For feature selection, '*n*' the number of different feature selection methods is applied. Based on the selected features by each feature selection method, '*m*' number of classification algorithms is applied. The total of '*M*' number of models is trained for the classification task. Finally, performance '*R*' is measured against each trained model.

**Algorithm 1:** Best Model by the IDS

```
Input: list with 80 features (l)
Output: Identify malicious traffics
D ← ReadDataset(l)
S ← Model()
for i = 1 to n do
    ▷ n is the number of feature selection methods
    S ← FeaturesSelection(D,K)
    for i = 1 to m do
        ▷ m is the number of classification techniques
        S ← TrainingModels(M,S)
        R ← StoredScore
    end
end
S ← EvaluatedModel(S,R)
IdentifyMaliciousTraffics(S,D)
return S
```

## 4 Experimental Results

In this section, the experimental environment for the simulation of the model is presented. The performance of the proposed models for IDS to identify the malicious traffics in cloud computing is also analysed below.

**Table 4** System configurations

| Hardware and software | Characteristics |
|---|---|
| Memory | 16 Gb DDR4 3600 MHz |
| Processor | i7-10700K @ 5.1 GHz |
| Operating system | Windows 10 Pro 64 bits |
| Programming language | Python 3.6 |
| Library | Scikit-learn |

## *4.1 Experimental Environment*

The simulation of the proposed framework was conducted in a system environment comprising an Intel Core-i7-10700K CPU, 16 GB of DDR4 RAM operating at a clock speed of 3600 MHz. Additionally, various software and libraries were used, and it is presented in Table 4. Python is used to accomplish the simulation and modelling.

## *4.2 Evaluation Metrics*

To measure the performance of the proposed models' behaviours, some standard metrics are available. We have opted the following metrics for the evaluation of our proposed model:

$$\text{Recall} = \frac{T^+}{T^+ + F^-} \tag{1}$$

$$\text{Precision} = \frac{T^+}{T^+ + F^+} \tag{2}$$

$$\text{F1-score} = \frac{2 \times \text{Precision} \times \text{Recall}}{\text{Precision} + \text{Recall}} \tag{3}$$

$$\text{Accuracy} = \frac{T^+ + T^-}{T^+ + T^- + F^+ + F^-} \tag{4}$$

Here, we benchmark those five ML techniques and evaluate their performance in detecting malicious traffic, which is described in Table 5. From Fig. 2, it can be clearly observed that when DT is combined with PCA, it attains an accuracy of 98.98% and the misclassification of 0.01%. From Fig. 3, it is shown that NB has significantly better performance with RFE than other two. This model achieved 86.51% accuracy with 13% misclassification. In Fig. 4, the LR has better accuracy with PCA and it

**Table 5** Evaluation of benchmarking ML methods

| Proposed models | Precision (%) | Recall (%) | F1-score (%) | Accuracy (%) |
|---|---|---|---|---|
| $\chi^2$_DT_IDS | 0.95 | 0.97 | 0.97 | 97.70 |
| $\chi^2$_NB_IDS | 0.69 | 0.67 | 0.60 | 68.95 |
| $\chi^2$_LR_IDS | 0.78 | 0.75 | 0.74 | 80.15 |
| $\chi^2$_RF_IDS | 0.94 | 0.97 | 0.95 | 97.45 |
| $\chi^2$_SGD_IDS | 0.46 | 0.52 | 0.47 | 65.13 |
| RFE_DT_IDS | 0.92 | 0.96 | 0.94 | 97.45 |
| RFE_NB_IDS | 0.81 | 0.77 | 0.76 | 86.51 |
| RFE_LR_IDS | 0.73 | 0.76 | 0.74 | 82.18 |
| RFE_RF_IDS | 0.97 | 0.98 | 0.98 | 99.23 |
| RFE_SGD_IDS | 0.87 | 0.84 | 0.85 | 94.14 |
| PCA_DT_IDS | 0.96 | 0.97 | 0.97 | 98.98 |
| PCA_NB_IDS | 0.78 | 0.76 | 0.72 | 83.20 |
| PCA_LR_IDS | 0.84 | 0.89 | 0.85 | 90.33 |
| PCA_RF_IDS | 0.97 | 0.98 | 0.98 | 98.72 |
| PCA_SGD_IDS | 0.85 | 0.86 | 0.85 | 93.12 |

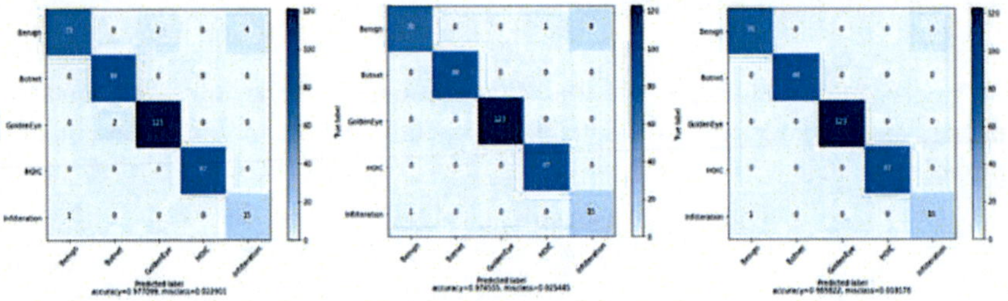

(a) $\chi^2$_DT_IDS (b) RFE_DT_IDS (c) PCA_DT_IDS

**Fig. 2** Confusion matrix of various DT_IDS models

achieved 90.33% of accuracy. In Fig. 5, the RF achieved better accuracy of 99.23% using RFE. Also, we can see in Fig. 6 that the SGD has better accuracy with RFE and its accuracy is 97.45%.

### *4.3 Comparative Study*

We have compared our proposed model with [13, 17, 18]. From Table 6, it can be seen that our proposed model outperforms and surpasses the results of those works. We got better accuracy as the model iteratively runs to select the best feature and against each set of best selected feature again runs iteratively to apply various classification

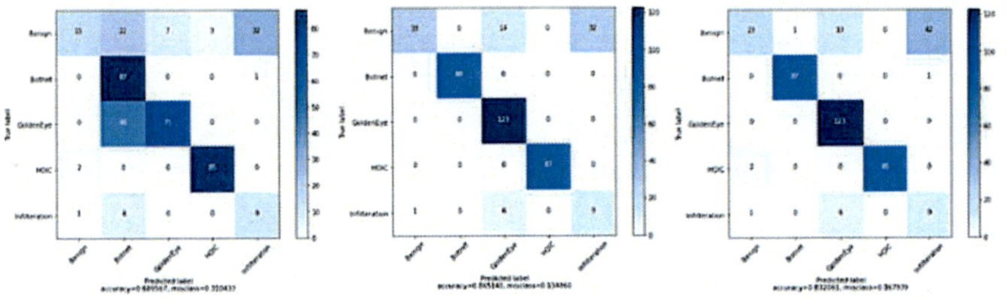

(a) $\chi^2$_NB _IDS (b) RFE _NB _IDS (c) PCA _NB _IDS

**Fig. 3** Confusion matrix of various NB_IDS models

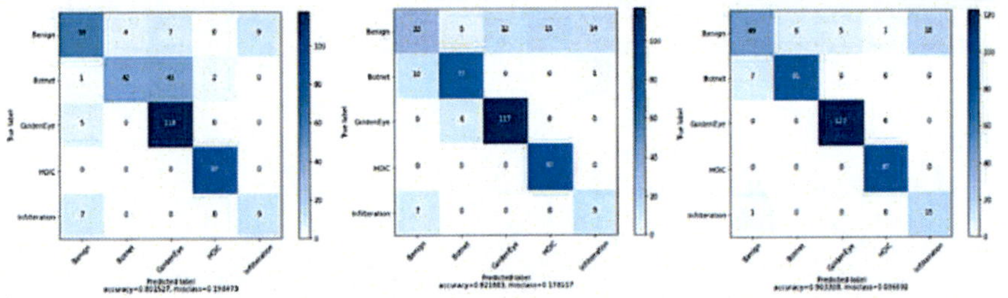

(a) $\chi^2$_LR _IDS (b) RFE _LR _IDS (c) PCA _LR _IDS

**Fig. 4** Confusion matrix of various LR_IDS models

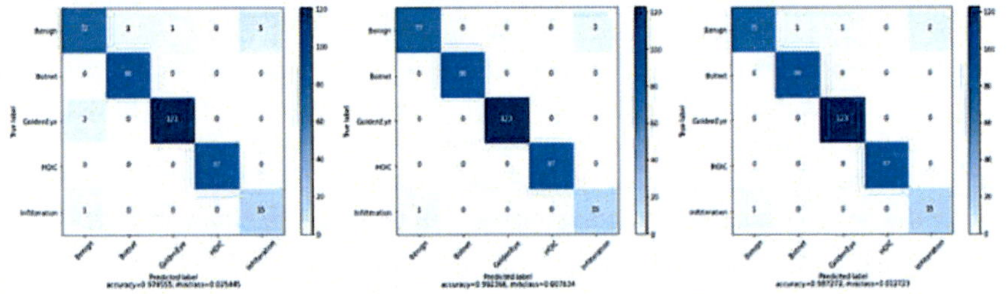

(a) $\chi^2$_RF _IDS (b) RFE _RF _IDS (c) PCA _RF _IDS

**Fig. 5** Confusion matrix of various RF_IDS models

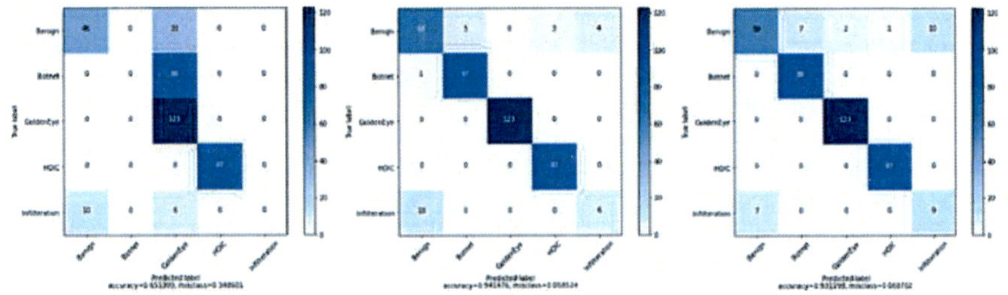

(a) $\chi^2$_SGD _IDS (b) RFE _SGD _IDS (c) PCA _SGD _IDS

**Fig. 6** Confusion matrix of various SGD_IDS models

**Table 6** Comparative study

| References | Year | Accuracy (%) |
|---|---|---|
| [13] | 2018 | 95.00 |
| [17] | 2018 | 97.11 |
| [18] | 2020 | 98.60 |
| [8] | 2021 | 96.06 |
| Proposed model | 2022 | 99.23 |

algorithm. Finally, it selects the best classification model and calculates the accuracy. The performance of the proposed model (RFE_RF_IDS) is better than [13, 17, 18] in terms of accuracy and the same is shown in Table 6.

## 5 Conclusion and Future Work

Nowadays, maximum number of services are offered over the Internet which is made available by cloud. It is very crucial to secure the cloud services which could be smoothly available to each user. Therefore, strengthening cloud security is very important. To address the same, hybrid models are proposed for IDS of cloud computing. The performance of the models are analysed and the best optimized model is obtained for identifying malicious traffics in cloud. The RFE_RF_IDS model has outperformed to other models and achieved the highest accuracy of 99.23% on CSE-CIC-IDS2018 dataset. The model became exclusive due to the inclusion of Recursive Features Elimination and Random Forest. Since the uses of the Internet are increasing exponentially, a heuristic-based IDS approach in the near future is necessary to get better performance for detecting intrusion in cloud computing networks.

## References

1. Executive Perspectives Cisco Report (2022) Cisco Annual Internet Report–Cisco Annual Internet Report (2018–2023) White Paper. [online] Cisco. https://www.cisco.com/c/en/us/solutions/collateral/executive-perspectives/annual-internet-report/white-paper-c11-741490.html. Accessed 4 September 2022
2. Selvakumar K, Karuppiah M, Sairamesh L, Hafizul Islam SK, Hassan MM, Fortino G, Raymond Choo K-K (2019) Intelligent temporal classification and fuzzy rough set-based feature selection algorithm for intrusion detection system in WSNs. Inf Sci 497:77–90
3. Sharafaldin I, Lashkari AH, Ghorbani AA (2018) Toward generating a new intrusion detection dataset and intrusion traffic characterization. ICISSp 1:108–116
4. Unb.ca. 2022. IDS 2018 | Datasets | Research | Canadian Institute for Cybersecurity | UNB. https://www.unb.ca/cic/datasets/ids-2018.html. Accessed 4 September 2022

5. Samrin R, Vasumathi D (2017) Review on anomaly based network intrusion detection system. In: 2017 international conference on electrical, electronics, communication, computer, and optimization techniques (ICEECCOT). IEEE, pp 141–147
6. Ansari Md SA, Chattopadhayay A, Das S. A kernel level VFs logger for building efficient file system intrusion detection system. In: 2010 second international conference on computer and network technology. IEEE, pp 273–279
7. Elshafie HM, Mahmoud TM, Ali AA (2019) Improving the performance of the snort intrusion detection using clonal selection. In: 2019 international conference on innovative trends in computer engineering (ITCE). IEEE, pp 104–110
8. Keserwani PK, Govil MC, Pilli ES, Govil P (2021) An optimal NIDS for VCN using feature selection and deep learning technique: IDS for VCN. Int J Digit Crime Forensics (IJDCF) 13(6):1–25
9. Ansari Md SA, Pal K, Govil MC, Govil P, Srivastava A (2021) Ensemble machine learning for P2P traffic identification. Int J Comput Digit Syst
10. Yin C, Zhu Y, Fei J, He X (2017) A deep learning approach for intrusion detection using recurrent neural networks. IEEE Access 5:21954–21961
11. Nisioti A, Mylonas A, Yoo PD, Katos V (2018) From intrusion detection to attacker attribution: a comprehensive survey of unsupervised methods. IEEE Commun Surv Tutor 20(4):3369–3388
12. Ansari Md SA, Govil MC, Pal K, Samaddar SG (2020) Tools and techniques in optimization of network resources. In: 2020 3rd international conference on emerging technologies in computer engineering: machine learning and internet of things (ICETCE). IEEE, pp 1–7
13. Barbhuiya S, Papazachos Z, Kilpatrick P, Nikolopoulos DS (2018) RADS: real-time anomaly detection system for cloud data centres. arXiv:1811.04481
14. Alharbi A, Alsubhi K (2021) Botnet detection approach using graph-based machine learning. IEEE Access 9:99166–99180
15. Abou Daya A, Salahuddin MA, Limam N, Boutaba R (2020) BotChase: graph-based bot detection using machine learning. IEEE Trans Netw Serv Manage 17(1):15–29
16. Elsayed MS, Le-Khac NA, Dev S, Jurcut AD (2020) Detecting abnormal traffic in large-scale networks. In: 2020 international symposium on networks, computers and communications (ISNCC). IEEE, pp 1–7
17. Nguyen KK et al (2018) Cyberattack detection in mobile cloud computing: a deep learning approach. In: 2018 IEEE wireless communications and networking conference (WCNC). IEEE
18. Smys S, Basar A, Wang H (2020) Hybrid intrusion detection system for internet of things (IoT). J ISMAC 2(04):190–199

# NITSDN: Development of SDN Dataset for ML-Based Intrusion Detection System

**Bipal Khanal, Chandan Kumar, and Md. Sarfaraj Alam Ansari**

**Abstract** Decision for the routing of traffic and packet forwarding is done by the embedded layers at the traditional distributed networks. Due to the embedded layer, management of the network is complex in traditional distributed networks. The solution to this complexity lies in the use of a new networking approach known as Software Defined Networks (SDN). SDN separates the data and control planes, and due to this separation, it has become widely adopted for various networking applications. Besides its benefit there are some security flaws as if the SDN controller gets hampered then the complete network starts malfunctioning. To enhance the security and stability of SDN, an intrusion detection system (IDS) is employed. To build efficient IDS the SDN-specific feature is required, for this we have created a NITSDN dataset. It contains the network traffic flow of the SDN environment. The evaluation metrics of the IDS build using the NITSDN dataset attain an accuracy of 99.48% using the Decision Tree machine learning classifiers.

**Keywords** Data collection · Software defined networks · Feature extraction · Mininet · Intrusion detection system

## 1 Introduction

In traditional distributed networks, decision-making and traffic forwarding are embedded together, referred to as the control and data plane. Both the control and data plane are embedded within the networking devices (e.g., Routers or switches). On each device, architectural configuration for routing and switching of the traffic is

B. Khanal · C. Kumar · Md. S. A. Ansari (✉)
Department of Computer Science and Engineering, NIT Sikkim, Sikkim 737139, India
e-mail: sarfaraj@nitsikkim.ac.in

B. Khanal
e-mail: m210012@nitsikkim.ac.in

C. Kumar
e-mail: phcs220029@nitsikkim.ac.in


S. Borah et al. (eds.), *Advanced Computational and Communication Paradigms*,
Lecture Notes in Networks and Systems 535,
https://doi.org/10.1007/978-981-99-4284-8_8

carried out independently. The traditional distributed networks have an operational overhead of management in terms of maintenance costs. This problem gets rectified by introducing the new networking paradigm which is referred to as Software-defined networks (SDN). The major benefit of SDN is that it decouples the control plane and data plane which makes the network more manageable and flexible. Thus, in SDN the whole system is controlled from a centralized device referred to as a controller. Central management of the networking environment using an SDN controller is one of the prominent reasons that SDN is adopted widely.

Although SDN has various advantages, it has some security flaws which need to rectify. For instance, any security breach within traditional distributed networks results in meager damage for only a small portion of the network, a whistle if any attack is mounted against the SDN controller then it might lead to critical consequences over the entire network. In a situation like this, if attackers thrive in bringing down the controller, then it might expose the network to severe hampers. The aforementioned security challenges get rectified by implementing the Intrusion Detection System (IDS). IDS has the capability to detect the attack within the system and generates an alarm to inform the administrator. IDSs are classified into two types: signature-based and anomaly-based IDS. To design signature-based IDS there is a need for a knowledge base of the malicious and normal signatures from the traffic flow of the SDN environment. Most of the research for designing the IDS is carried out using the conventional networks dataset. However, SDN is prone to novel threats that are distinct from those present in traditional networks. So, to design an efficient IDS for the SDN environment there is a need for the network traffic dataset of the SDN environment. Since there is only one public SDN dataset available which has only seven attack categories. In a real networking scenario, there is a requirement for a higher number of attack categories. Developing an SDN dataset that has a larger number of attack categories is one of the prime research domains.

## 2 Related Works

Various dataset is available to build the IDS. In this section, the existing literature on the various publicly available datasets is discussed. The datasets are widely used by ML algorithms to develop an ML-based IDS. We also present some existing study that uses the ML algorithm to build an IDS.

Divekar et al. [1] set the experimental setup to develop a dataset referred as KDD'99 using the traditional network architecture. The KDD'99 dataset is widely used in various research study to develop IDS. It was taken from DARPA packet traces. There are 41 traffic features in total, and they are divided into three categories.: traffic features, basic features, and content features. KDD'99 dataset contains four attack categories, Remote to Local (R2L), User to Root (U2R), (Denial of Service (DoS), and probe attacks besides the normal data. KDD'99 dataset has a problem of redundant records, where about 78% of the training data and 75% of the testing data are duplicated records.

The author of [2] created the NSL-KDD dataset, which is a variation of the KDD'99 dataset. The aim was to address certain issues present in the KDD'99 dataset, such as repeated records. In the two subsets of the dataset, the training and testing sets, the testing set has a higher frequency of attacks, with 17 additional attack types that are not included in the training set. The authors of [2] develop the NSL-KDD dataset. It is an extension of the KDD'99 dataset. It was generated to reduce some of the problems of the KDD'99 dataset such as duplicate records. In the two subsets, the training and testing set the distribution of attacks is higher in the testing set the training set where 17 additional attacks are not represented in the training set. Both NSL-KDD and KDD'99 are not captured using the real network and attack trends as they were generated two decades ago. Koyoto Dataset 2006+ [3] was developed by collecting the real network traffic of honeypot servers at Koyoto University. Koyoto Dataset contains 24 statistical features. The concern with this dataset is that the dataset contains the traffic of conventional networks. The Canadian Institute of Cybersecurity developed the dataset for intrusion detection which is referred to as CICIDS 2017 [5]. The CICIDS-2017 dataset consists of 80 flow-based features. This dataset contains a broad range of attack scenarios, i.e., it has a total of 13 different types of attack The existing literature related to the IDS dataset discussed so far is collected using traditional networks. Until the literature survey carried out for this study, there is only one publicly available dataset which is collected using the SND environment. The InSDN [7] dataset is the only known publicly available dataset that was generated from a virtual SDN environment that covers a variety of attack scenarios. The dataset is divided into three sets on the basis of network traffic type and victim machine. The first group consists solely of normal network traffic, while the second group contains malicious traffic aimed at the Mealsplotable-2 server and the third group features malicious traffic targeting the OVS machine. The InSDN dataset has a total of 343,939 instances, with 68,424 instances being normal traffic and 275,515 instances being malicious. The significant disadvantage of this dataset is the stark imbalance between normal and malicious traffic instances.

Varghese et al. [8] proposed a method for detecting and mitigating DDoS attacks in an SDN environment using the Data Plane Development Kit (DPDK). The framework consists of two primary modules: the DPDK-based detection module and the control plane mitigation module. The detection process employs a single statistical feature and is based on an anomaly detection algorithm. The algorithm was tested on the CICDoS2017 and CICDDoS2019 datasets and achieved accuracy rates of 88.33% and 96.59%, respectively. A multi-class classification model that utilizes stacked autoencoders and a softmax classifier was proposed by Choobdar et al. [9]. The model was trained and evaluated using the NSL-KDD and CICID2017 datasets, achieving average accuracy rates of 98.50% and 94.88%, respectively, for multi-class classification. Similarly, in [15] an SDN-based real-time network intrusion detection system utilizing an ensemble of autoencoders was employed and it achieved a prediction accuracy of 97%. The model was trained and tested using various open datasets and its real-time performance was simulated in a Mininet emulator. Saritha et al. [12] proposed an IDS for SDN using a multi-stage optimized deep forest classifier.

They implemented a modified deep forest classifier along with various prepossessing methodologies like normalization and oversampling and tested the model on the NSL-KDD dataset which yielded 95.32% accuracy in the binary classification.

# 3 Proposed Methodology

This section presents the testbed setup and topology description of the SDN for the collection of network traffic. The proposed methodology consists of data capture, data conversion, and feature extraction, pre-processing, performance evaluation by applying an ML classifier [13]. Figure 1 presents the graphical abstract of the proposed framework.

## *3.1 Data Collection*

Network traffic data is collected in the SDN environment. This section presents the mechanism applied for the data collection in an SDN environment.

### 3.1.1 Testbed Setup

Table 1 presents the configuration of the testbed setup of the system used for the data collection. Mininet is a network emulation tool used vastly for performing network simulations because it uses lightweight virtualization for emulation of the network components (end hosts, switches, network links, routers) on a single Linux kernel. OpenvSwitch is a virtual switch that uses programmatic extensions to provide network automation. For the controller of our SDN network, the Ryu is used which is an SDN framework based on components. It constituents of software elements with well-defined APIs which helps the developers to develop and implement new network administration and control applications efficiently. Wireshark is a free and open-source network analyzer that is widely used to capture and analyze network traffic through a network interface.

### 3.1.2 Topology Description

A tree topology presented in Fig. 2 is used for the simulation. It consists of four switches (s1, s2, s3, s4) and nine hosts (h1–h9) where host (h1, h2, and h3) are connected to switch (s2, h4, h5) and host (h6) are connected to switch (s3) and host (h7, h8) and host (h9) are connected to switch (s4). In the topology host h8 is chosen as the victim node from which the data is captured packet tracer tool (Wireshark). Rest other hosts are used to creating normal traffic as well as malicious traffic.

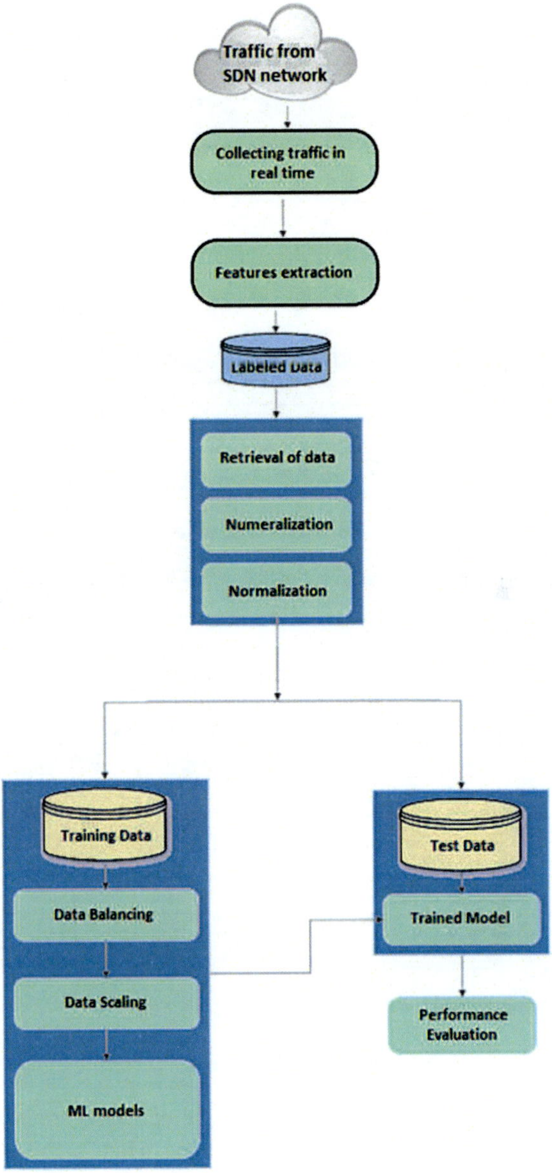


**Fig. 1** Graphical abstract of the proposed framework

**Table 1** System configuration used for data collection

| System resource | Configuration |
|---|---|
| Mininet | Ver2.2.2 |
| OpenFlow | Ver1.3 |
| OpenvSwitch | Ver2.13.8 |
| Ryu | Ver4.34 |
| Wireshark | Ver3.7.6 |

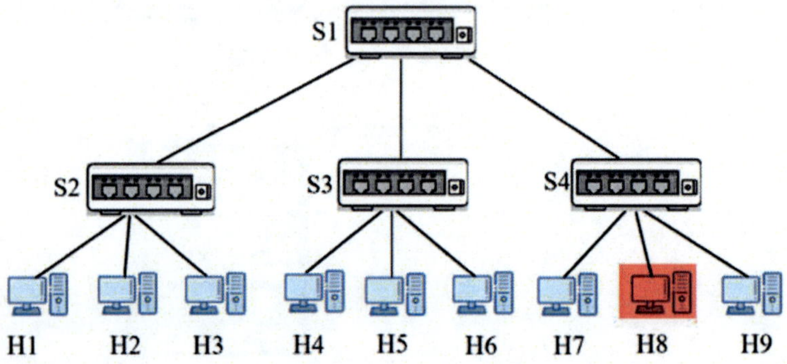


**Fig. 2** Tree topology

**Table 2** Instance of data available in NITSDN dataset

| Data label | Instance |
|---|---|
| Normal | 79068 |
| DoS | 41823 |
| DDoS | 351055 |
| PortScan | 224566 |
| BOTNET | 3677 |

### 3.1.3 Data Generation

After setting up the testbed all the necessary tools are installed. The topology was created in Mininet which is presented in Fig. 2. The necessary setup done in Mininet is remote Ryu controller, OpenvSwitch, and OpenFlow protocol. Now Wireshark starts running and selects the interface of the victim host (h8) and starts capturing packets. Normal as well as malicious traffic is generated from all other hosts available in the network architecture. To generate malicious packets primarily attacks like DDoS, SYN flood attacks, and IP spoofing attacks from multiple hosts in the network are employed. Table 2 presents the instances of data captured for each type of attack and normal traffic.

## 3.2 Feature Extraction

Network traffic packet captured using Wireshark exports the data into pcap file. For the feature extraction, the pcap file is used. In this study, CICFlowMeter [21] is a network traffic analyzer tool used for feature extraction using the pcap file. CICFlowMeter exports a CSV file consisting of more than 75 network traffic analysis features like the number of packets, length of packets, flow duration, number of bytes, etc. The feature extracted using CICFlowMeter from the network traffic is presented in Table 3.

**Table 3** List of features extracted

| Sl. no | Features | Sl. no | Features | Sl. no | Features |
|---|---|---|---|---|---|
| 1 | SrcIP | 29 | PacketLength Std | 57 | Down/Up_Ratio |
| 2 | SrcPort | 30 | FlowIAT Min | 58 | Idle_Min |
| 3 | DstIP | 31 | BwdInit WinBytes | 59 | Idle_Max |
| 4 | DstPort | 32 | SubflowFwd Bytes | 60 | Idle_Std |
| 5 | Fwd_Packet/Bulk_Avg | 33 | Protocol | 61 | Idle_Mean |
| 6 | TimeStamp | 34 | FwdIAT Max | 62 | Active_Min |
| 7 | FlowDuration | 35 | FwdIAT Min | 63 | URG_Flag Count |
| 8 | TotalFwd Packet | 36 | BwdIAT Total | 64 | CWR_Flag Count |
| 9 | TotalBwd packets | 37 | BwdIAT Mean | 65 | Flow_ID |
| 10 | TotalLength of FwdPacket | 38 | BwdIAT Std | 66 | Active_Mean |
| 11 | TotalLength of BwdPacket | 39 | BwdIAT Max | 67 | ECE_Flag Count |
| 12 | FwdPacket LengthMax | 40 | BwdIAT Min | 68 | Active_Max |
| 13 | FwdPacket LengthMin | 41 | FwdPSH Flags | 69 | Bwd_Packets/s |
| 14 | FwdPacket LengthMean | 42 | BwdPSH Flags | 70 | Active_Std |
| 15 | FwdPacket LengthStd | 43 | Fwd_URG Flags | 71 | ACK_Flag Count |
| 16 | BwdPacket LengthMax | 44 | BwdURG Flags | 72 | PSH_Flag Count |
| 17 | BwdPacket LengthMin | 45 | FwdHeader Length | 73 | RST_Flag Count |
| 18 | BwdPacket LengthMean | 46 | BwdHeader Length | 74 | SYN_Flag Count |
| 19 | BwdPacket LengthStd | 47 | FwdPackets/s | 75 | FIN_Flag Count |
| 20 | BwdPacket/BulkAvg | 48 | FwdBytes/BulkAvg | 76 | FlowIAT Max |
| 21 | FwdSegment SizeAvg | 49 | AveragePacket Size | 77 | Flow_Bytes/s |
| 22 | BwdSegment SizeAvg | 50 | PacketLength Min | 78 | Flow_Packets/s |
| 23 | FWDInit WinBytes | 51 | PacketLength Max | 79 | FwdIAT Total |
| 24 | Subflow BwdBytes | 52 | PacketLength Mean | 80 | FlowIAT Mean |
| 25 | Subflow BwdPackets | 53 | BwdBulk RateAvg | 81 | FlowIAT Std |
| 26 | PacketLength Variance | 54 | SubflowFwd Packets | 82 | FwdIAT Std |
| 27 | BwdBytes/BulkAvg | 55 | FwdAct DataPkts | 83 | FwdIAT Mean |
| 28 | FwdBulk RateAvg | 56 | FwdSeg SizeMin | 84 | Label |

### 3.3 Network Traffic Classification

The features extracted from the flow of the network traffic are used to build the ML classification model which helps in distinguishing normal and malicious traffic. The classification model is built using the Decision Tree [17], Naive Bayes [18], Logistic Regression [19], and Multi-Layer Perceptron classifier [20] algorithm. Experimental simulation of the aforementioned classifier is implemented using the *sklearn* library.

## 4 Result and Discussion

This section presents a detailed discussion of the performance assessment of the NITSDN dataset. Five ML classifiers are applied over the NITSDN dataset and performance is calculated. In the underneath section, the performance of the ML classifiers is presented.

### 4.1 Data Generation for Attack Packets

To generate a comprehensive dataset various malicious packets are generated using a variety of tools available. Table 4 shows different attack classes, types of attack, and different tools used along with the IP addresses of the victim and attacker machine.

**Table 4** Attack classes for dataset generation

| Attack class | Attack type | Tools used | Attacker device | Victim host |
|---|---|---|---|---|
| DoS | TCP_SYN flood<br>UDP_flood<br>HTTP_flood<br>TCP flood | HULK<br>Torshammer<br>Nping<br>Hping3 | Host h1<br>IP: 10.0.0.1 | Host h8<br>IP: 10.0.0.8 |
| DDoS | TCP_SYN_flood<br>UDP_flood<br>ICMP_flood | HULK<br>Torshammer<br>Nping<br>Hping3 | Host h1, h2<br>h1 IP: 10.0.0.1<br>h2 IP:10.0.0.2 | Host h8<br>IP: 10.0.0.8 |
| Probe | Port scan<br>Version scan | Nmap | Host h1<br>h1 IP: 10.0.0.1 | Host h8<br>IP: 10.0.0.8 |
| R2L | Password guessing attack | Burp suite<br>Hydra | Host h1<br>h1 IP: 10.0.0.1 | Host h8<br>IP: 10.0.0.8 |
| Malware | Botnet attack | ARES | Host h1<br>h1 IP: 10.0.0.1 | Host h8<br>IP: 10.0.0.8 |

**Table 5** Performance of ML classifier metrics

| Attack | Metrics | DT | NB | LR | MLP | QDA |
|---|---|---|---|---|---|---|
| Normal | Precision | 0.92 | 0.99 | 0.99 | 1.00 | 1.00 |
| | Recall | 1.00 | 1.00 | 0.86 | 1.00 | 1.00 |
| | F1-score | 0.96 | 1.00 | 0.92 | 1.00 | 1.00 |
| DoS | Precision | 1.00 | 1.00 | 0.97 | 1.00 | 0.50 |
| | Recall | 1.00 | 0.79 | 0.99 | 1.00 | 1.00 |
| | F1-score | 0.88 | 0.97 | 0.88 | 1.00 | 1.00 |
| DDoS | Precision | 1.00 | 0.92 | 0.98 | 1.00 | 1.00 |
| | Recall | 1.00 | 0.68 | 0.80 | 0.97 | 1.00 |
| | F1-score | 1.00 | 0.78 | 0.88 | 0.98 | 1.00 |
| PortScan | Precision | 1.00 | 0.82 | 0.89 | 0.98 | 1.00 |
| | Recall | 1.00 | 1.00 | 0.97 | 1.00 | 0.78 |
| | F1-score | 1.00 | 0.90 | 0.93 | 0.99 | 0.87 |
| BOTNET | Precision | 0. 00 | 0.60 | 0.26 | 1.00 | 1.00 |
| | Recall | 0.00 | 0.83 | 0.93 | 0.96 | 0.99 |
| | F1-score | 0.00 | 0.70 | 0.40 | 0.98 | 1.00 |

## *4.2 Performance of ML Classifiers for Multi-class Classification*

Table 5 displays the performance of the ML classifier on the NITSDN dataset. Higher evaluation metrics are obtained for the DoS, DDoS, and PortScan types of attacks while BOTNET gives poor performance metrics. However, the overall accuracy of the DT, NB, LR, MLP, and QDA classifiers are 99.48%, 86.98%, 91.18%, 94.88%, and 99.84% respectively.

### 4.2.1 Comparison with the Existing Dataset

The performance metrics of the ML classifier applied to the NITSDN dataset are compared with the performance metrics of the ML classifier applied to the InSDN dataset through a binary classification analysis. Table 6 presents the comparison of the results.

**Table 6** Comparision with InSDN

| Dataset | Metrics | DT | NB | MLP |
|---|---|---|---|---|
| InSDN [7] | Precision | 0.9940 | 0.9519 | 0.9888 |
| | Recall | 0.9961 | 0.9804 | 0.9978 |
| | F1-score | 0.9951 | 0.9660 | 0.9935 |
| NITSDN (proposed) | Precision | 1.00 | 0.9514 | 0.9999 |
| | Recall | 1.00 | 0.9486 | 0.9999 |
| | F1-score | 1.00 | 0.9418 | 0.9999 |

### 4.2.2 Confusion Metrics

The performance of ML classifiers on the NITSDN dataset is also calculated in terms of the confusion metrics. Confusion metrics (CM) present a glimpse of the correct and false predictions made by the classifier. Figure 3 illustrates the confusion metrics obtained for the multi-class classification of the various attack labels available in the NITSDN dataset.

## 5 Conclusion and Future Work

Nowadays, the nature of network traffic is dynamic, due to which the networking devices such as routers and switches have to be smart. The programmability capacity of the SDN allows the networking device to make it efficient and intelligent in decision-making. At the same time, various security measures hamper its services. Multiple systems have been employed to make it robust in terms of security, and IDS is one of them. To build an efficient IDS there is a need for an SDN-specific feature that can be gathered from the traffic of the SDN environment. We have captured the SDN traffic flow and created a dataset named NITSDN which is presented in this paper. However, the attacks are not limited to what is presented in this study. In the future, we will create a more complex topology of the SDN environment and try to obtain the dataset having a larger number of attack categories for which we will perform statistical analysis on the dataset [22]. Also, in this study, the simulation is done using the virtual emulator environment but in the near future we will try to use the real SDN environment.

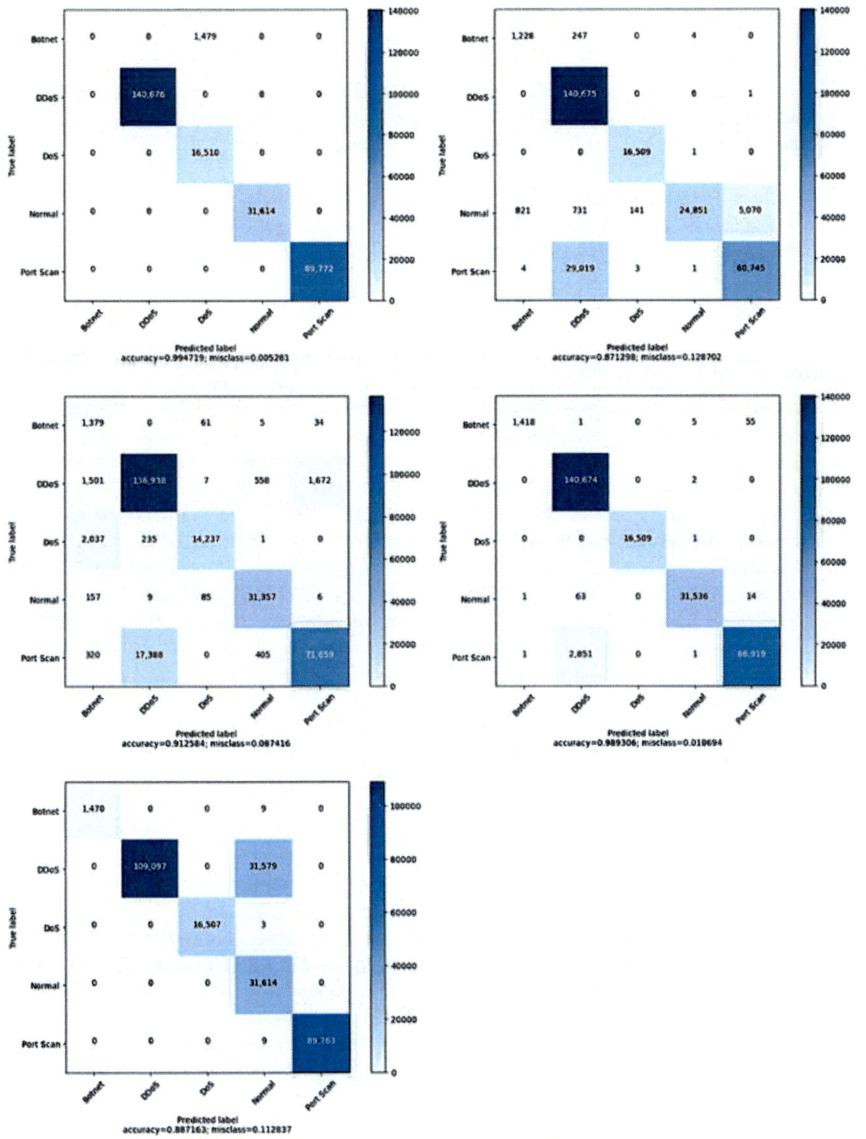


**Fig. 3** Confusion metrics obtained using decision tree (top left), Naive Bayes (top right), logistic regression (middle left), multi-layer perceptron (middle right), and quadratic discriminant analysis (bottom left)

## References


1. Divekar MP, Savla V, Mishra R, Shirole M (2018) Benchmarking datasets for anomaly-based network intrusion detection: KDD CUP 99 alternatives. In: Proceedings on 2018 IEEE 3rd international conference on computing, communication and security (ICCCS), Oct. 2018, pp 1–8
2. Tavallaee M, Bagheri E, Lu W, Ghorbani AA (2009) A detailed analysis of the KDD CUP 99 data set. In: Proceedings of the 2009 IEEE symposium on computational intelligence in security and defense applications, July 2009, pp 1–6
3. Song J, Takakura H, Okabe Y (2006) Description of Kyoto University benchmark data. http://www.takakura.com/Kyoto_data/BenchmarkData-Description-v5.pdf. Accessed 21 November 2022
4. Shiravi HS, Tavallaee M, Ghorbani AA (2012) Toward developing a systematic approach to generate benchmark datasets for intrusion detection. Comput Secur 31(3):357–374
5. Sharafaldin AHL, Ghorbani AA (2018) Toward generating a new intrusion detection dataset and intrusion traffic characterization. In: Proceedings of the ICISSP, pp 108–116
6. Canadian Institute of Cybersecurity (2018) CSE-CIC-IDS2018. https://www.unb.ca/cic/datasets/ids2018.html. Accessed 10 February 2020
7. Elsayed MS, Le-Khac N-A, Jurcut AD (2020) InSDN: a novel SDN intrusion dataset. IEEE Access 8:165263–165284. https://doi.org/10.1109/ACCESS.2020.3022633
8. Varghese JE, Muniyal B (2021) An efficient IDS framework for DDoS attacks in SDN environment. IEEE Access 9:69680–69699. https://doi.org/10.1109/ACCESS.2021.3078065
9. Choobdar P, Naderan M, Naderan M (2022) Detection and multi-class classification of intrusion in software defined networks using stacked auto-encoders and CICIDS2017 dataset. Wirel Pers Commun 123:437–471. https://doi.org/10.1007/s11277-021-09139-y
10. Sharafaldin I, Lashkari AH, Ghorbani AA (2018) Toward generating a new intrusion detection dataset and intrusion traffic characterization. ICISSp 1:108–116
11. Unb.ca (2022) IDS 2018 | Datasets | Research | Canadian Institute for Cybersecurity | UNB. https://www.unb.ca/cic/datasets/ids-2018.html. Accessed 4 September 2022
12. Saritha Reddy A, Ramasubba Reddy B, Suresh Babu A (2022) An improved intrusion detection system for SDN using multi-stage optimized deep forest classifier. Int J Comput Sci Netw Secur 22(4):374–386. https://doi.org/10.22937/IJCSNS.2022.22.4.44
13. Kumar P, Babulal KS (2022) Hematological image analysis for segmentation and characterization of erythrocytes using FC-TriSDR. Multimed Tools Appl 1–26
14. Badotra S, Panda SN (2021) SNORT based early DDoS detection system using Opendaylight and open networking operating system in software defined networking. Cluster Comput 24:501–513. https://doi.org/10.1007/s10586-020-03133-y
15. Yang L, Song Y, Gao S, Hu A, Xiao B (2022) Griffin: real-time network intrusion detection system via ensemble of autoencoder in SDN. IEEE Trans Netw Serv Manage 19(3):2269–2281. https://doi.org/10.1109/TNSM.2022.3175710
16. Radhi M, Mohammed A (2022) An efficient deep learning approach for network intrusion detection system on software defined network. Int J Netw Secur Appl 14:1–14. https://doi.org/10.5121/ijnsa.2022.14401
17. Quinlan JR (1986) Induction of decision trees. Mach Learn 1(1):81–106
18. Vikramkumar VB, Trilochan (2014) Bayes and Naive Bayes classifier. https://arxiv.org/abs/1404.0933
19. Wright RE (1995) Logistic regression. Washington, DC, US, pp 217–244
20. Suter BW (1990) The multilayer perceptron as an approximation to a Bayes optimal discriminant function. IEEE Trans Neural Netw 1(4):291

21. Lashkari AH, Draper-Gil G, Mamun MSI, Ghorbani AA (2017) Characterization of Tor traffic using time based features. In: Proceeding of the 3rd international conference on information system security and privacy, SCITEPRESS, Porto, Portugal
22. Ansari MS, Pal K, Govil P, Govil MC, Awasthi LK (2022) A statistical analysis of Sampark dataset for peer-to-peer traffic and selfish-peer identification. Multimed Tools Appl. https://doi.org/10.1007/s11042-022-13556-x

# Association Rule Mining in Distributed Environment: A Survey

**Bhaswati Baishya, Shafiul Alom Ahmed, and Bhabesh Nath**

**Abstract** Due to the increasing popularity of digital mediums and easy access to Internet technology, society has become dependent on digital resources. This generates a huge amount of data and identifying relationships or patterns among these data has become a predominant field of research. Traditional rule-mining techniques are primarily centralized, which makes these techniques ineffective when dealing with distributed databases. Distributed Association Rule Mining (D-ARM) is being developed for finding association rules from distributed databases stored in different locations. We have explored various D-ARM algorithms based on FP-Growth and Apriori, and we also examined the advantages and limitations of these algorithms.

**Keywords** Association rule mining · Distributed association rule mining · Apriori · FP-growth · Frequent itemset mining

## 1 Introduction

Data mining is the process of extracting relevant and useful information from large databases. The primary concern of data mining is to uncover hidden patterns and associations among large databases [1, 2]. ARM explores meaningful correlations between itemsets in a transactional dataset. Market basket analysis [3], fraud detection [4], medical diagnosis [5], weather forecasting [6], and bio-informatics [7] are some of the widely used application areas of ARM.

B. Baishya · S. A. Ahmed (✉) · B. Nath (✉)
Department of Computer Science and Engineering, Tezpur University, Assam, India
e-mail: tezu.shafiul@gmail.com

B. Nath
e-mail: bnath@tezu.ernet.in

S. A. Ahmed
School of Computing Science and Engineering, VIT-Bhopal University, Sehore, Bhopal, India



S. Borah et al. (eds.), *Advanced Computational and Communication Paradigms*, Lecture Notes in Networks and Systems 535,
https://doi.org/10.1007/978-981-99-4284-8_9

Users have become reliant on digital resources as a result of widespread access to Internet technologies. As a result, the data is increasing very rapidly every second. As a result, a tremendous amount of transactions are produced, and the database size is growing rapidly. Since these huge amounts of data are not physically possible to store in a centralized data storage, the database is primarily distributed in various locations. Distributed association rule mining is used on geographically dispersed databases to discover the associations between the items in each distributed database. Therefore, traditional association rule-mining models are inadequate while handling geographically located databases. As a result, this leads to the emergence of distributed association rule mining [8, 9]. This article discusses some existing D-ARM algorithms and their types by considering their advantages and limitations.

## 2 Association Rule Mining (ARM)

With massive data being generated rapidly, identifying relationships or patterns among data has become an exciting field of research. ARM follows a support and confidence threshold framework. To find associations among the items, we have to follow two steps [1, 2, 10]:

- Find all the frequent itemsets: Frequent itemsets are those whose support count is greater than the minimum support value defined by the user.
- Association rule generation: Strong association rules are those that satisfy the minimum confidence value and are created from frequent itemsets found in the previous step.

The algorithm for finding frequent patterns can be categorized into two categories:

- Frequent pattern with candidate generation: In this category, the algorithms find frequent itemsets level wise and generate a large number of candidate itemsets. Thus needs more memory space to store the itemsets. Apriori is the most well-known algorithm in this category [11].
- Frequent pattern without candidate generation: The algorithms without candidate generation employ a divide-and-conquer strategy instead of repeatedly scanning the database. The algorithms that fall under this category generate a lesser number of candidate itemsets compared to with candidate generation algorithms and also reduce computational costs. FP-Growth is the most used algorithm without generating the candidate itemsets [12].

### 2.1 *Preliminaries*

A few metrics are used to extract frequent patterns or itemsets from a transactional database. Assume that D be a transactional database, with items I = $[i_1],[i_2],\ldots,[i_m]$, and set of all transactions T = $[t_1],[t_2],\ldots,[t_n]$ [2].

**Definition 1** (*Support Count* ($\sigma$)) Support count denotes the total number of appearances of an item or itemset in a transaction [2].

**Definition 2** (*Support (Supp)*) If P is an item of a transaction, then the Supp of P is represented as the total number of occurrences of P in the transaction divided by the number of transactions present in the database. Mathematically [2]

$$Supp(P) = \frac{\sigma(P)}{N} \tag{1}$$

**Definition 3** (*Frequent Patterns (or) Itemsets*) To find the frequent itemsets, users set a support threshold value. Those items satisfy the user-given supp value are chosen to be frequent itemsets. An item P is frequent if [2]:

$$Supp(P) \geq minsupp \tag{2}$$

**Definition 4** (*Confidence (Conf)*) Conf of two items or itemsets P and Q is defined as how frequently items with P also contain items with Q [2].

$$Conf(P \rightarrow Q) = \frac{\sigma(P \cup Q)}{\sigma(P)} \tag{3}$$

**Definition 5** (*Association Rule*) The rules which satisfy the two evaluation metrics supp and conf are selected as association rules of the transaction [2].

**Definition 6** (*Strong Association Rules*) The rules which satisfy both user-given minimum supp value and conf value are considered as strong association rules [2].

# 3 Association Rule Mining in Distributed Environment

Distributed association rule mining find rules from various geographically distributed databases. It can be referred to "finding frequent patterns from massive databases which are distributed among different locations." Due to the rapid development of digital media and easy access to Internet technology users are generating a large amount of data every second. Therefore, a huge number of transactions are generated every day; hence the database sizes are increasing exponentially. Analyzing such data is essential because raw data may not be directly helpful. Still, proper handling or processing can extract hidden patterns among data, information, or knowledge from such massive databases. Due to the distributed nature of the database and the centralized nature of the standard algorithms, they cannot handle such massive databases as these algorithms suffer from some computational problems. A large database is split up into smaller databases to different computational processing nodes that require their own processing resources to make the mining techniques scalable. Computational processing nodes are computer systems that are physically separated but communicate with each other through messages passing. This whole process of finding association rules between large database spread into different locations need to communicate externally, thus resulting in higher communication costs [8, 13, 14].

# 4 Literature Survey

The traditional algorithms are suitable for centralized systems. Applying those algorithms to distributed database becomes challenging when the data size increases rapidly. The main issues with centralized systems are memory issues, low processing ability, and bandwidth limitations. This paper attempts to depict a timeline of existing association rule-mining algorithms in a distributed environment taking Apriori [11], and FP-Growth [12] as parent algorithms.

## *4.1 Apriori-Based Algorithms Over Distributed Environment*

In comparison to the conventional Apriori technique, the Fast Distribution Mining (FDM) [9] algorithm was proposed to create lesser number of candidate itemsets. The authors have proposed local and global pruning methods to prune out the itemsets.

Distributed Mining of Association Rules (DMA) [15] was proposed to mine association rules over the distributed database. This algorithm follows some steps (i) candidate sets generation, (ii) local partition scanning, (iii) local pruning, (iv) support count exchange, and (v) broadcast mining results. DMA also reduces the candidate itemsets and increases efficiency.

Distributed Frequent Itemset Mining Algorithm (DFIMA) [16] reduces the number of candidate itemsets and used spark [17] environment to implement the algorithm. As this algorithm uses Apriori as the parent algorithm, it scans the database multiple times, and this issue has become the bottleneck of the algorithm. The performance study shows that this algorithm works better with a higher support value.

Another Apriori-based parallel FIM algorithm Yet Another Frequent Itemset Mining (YAFIM) [18] was proposed. This algorithm gave better performance over different real-life datasets. Performance study shows that YAFIM is scalable in terms of computational speed.

To discover frequent patterns over distributed databases using Spark as the implementation platform, Adaptive-Miner [19] was proposed. Frequent singleton generation and frequent itemsets creation are the two phases of this approach.

In Table 1, we have summarized the advantages and limitations of the mentioned Apriori-based algorithms computed in a distributed environment.

**Table 1** Advantage and limitations of Apriori-based distributed mining algorithms

| Algorithm name | Advantages | Limitations |
|---|---|---|
| FDM [9] | (i) Reduces the number of candidate itemset<br>(ii) Does not exchange data tuples between processors, only exchanges the counts | (i) Performs poorly due to uneven distribution of database<br>(ii) Takes more computational cost |
| DMA [15] | (i) Generates a few number of candidate itemset | (i) Differences between the database partition sizes are not fixed among the computational nodes |
| DFIMA [16] | (i) Efficient and scalable | (i) Scans the database multiple times |
| YAFIM [18] | (i) Time complexity is less | (i) Multiple scans of the database |
| Adaptive Miner [19] | (i) Scalable in terms of execution time | (i) Generates a huge number of candidate sets |

## 4.2 *FP-Growth-Based Algorithms Over Distributed Environment*

H-mine [20] is a memory-based hyperstructure frequent itemset mining algorithm to generate frequent items from large databases. The authors have also proposed a hyper-linked data structure, H-Struct. Performance evaluation shows that the algorithm is scalable and also has minimal space complexity.

Parallel Frequent Pattern (PFP) [21] is proposed to mine web data for query recommendation. PFP parallelizes the FP-Growth algorithm on distributed computing nodes. It partitions the large database into smaller databases and distributes them to the different processing nodes. It is a Hadoop map-Reduce [22] approach.

Parallel Randomized Algorithm for approximate association rules (PARMA) [23] was proposed to show experimentally that PARMA performs well than the previously existing algorithms implemented on Hadoop map-reduce. The algorithm is compared with DistCount [24] and PFP [21]. This first randomized map-reduce technique operates almost linearly to find frequent patterns.

Xiujin S. et al. have proposed Distributed FP-Growth Algorithm (DFPS) [25]. This algorithm partitions the database into smaller parts and distributes it into different processing nodes. Each node builds its own conditional FP-Tree. The authors have implemented the speed performance and scalability of the proposed algorithm.

Distributed Frequent Itemset Mining (DFIM) [26] is proposed to solve the issue of space and time complexity of the existing algorithms. The authors have proposed two frameworks for the DFIM algorithm: local global and redistribution mining. These two frameworks' performances were compared, and it was discovered that redistribution mining performed better.

**Table 2** Advantage and limitations of FP-Growth-based distributed mining algorithms

| Algorithm name | Advantages | Limitations |
|---|---|---|
| H-mine [20] | (i) Scalable in terms of database partitioning | (i) Memory issue |
| PFP [21] | (i) Reduces computational time. (ii) Scalable than other alternative approaches | (i) Mines only top k-patterns |
| PARMA [23] | (i) Runs faster than previously introduced FIM algorithms for the same platform<br>(ii) Shows higher accuracy | (i) Memory cost is higher |
| DFPS [25] | (i) Scalable in large databases<br>(ii) Speed performance is high | (i) High communication load |
| DFIM [26] | (i) Scalable and computational cost is minimum | (i) Memory cost is higher |
| SSPFP [27] | (i) Effective and feasible in real-time streaming data | (i) Suffers from communication loss between nodes<br>(ii) Merging FP-Tree is very time-consuming |
| DFP [24] | (i) Provides fast and scalable service (ii) Overcomes the issue of insufficient memory and multiple database scans | (i) Efficiency can be improved |
| EBPFP [28] | (i) Enhances the load-balancing strategy | (i) Execution time is more and bandwidth limitation is another issue |

To solve the issues of traditional mining algorithms, Longtao et al. proposed a parallel FP-Growth approach named as SSPFP [27]. This algorithm gives a better speed performance and efficiency in different clusters.

Peng et al. proposed an algorithm to perform in a distributed computing environment and to give fast and scalable service over a big database named DFP [24]. This algorithm gives better mining and computing efficiency.

Essam et al. introduced Enhanced Balanced Parallel Frequent Pattern (EBPFP) [28]. This algorithm is proposed for load balancing of the databases into various processing nodes. This algorithm enhances the load-balancing strategy and reduces the time complexity.

In Table 2, we have discussed the advantages and limitations of the mentioned FP-Growth-based algorithms computed in a distributed environment.

## 5 Discussion

In this literature review, a few methods for mining frequent patterns from large distributed databases are identified and examined. In Tables 1 and 2, we have discussed the advantages and limitations of few existing D-ARM algorithms. In the literature, it has been found that the algorithms which find frequent itemsets without candidate generation work faster. The Apriori-based algorithms scan the database multiple times to find frequent itemsets and also generate a huge number of candidate itemsets. As a result, it requires more memory space and execution time. FP-Growth algorithms scan the database only twice and store the frequent itemsets in a prefix tree. The prefix tree is mainly known as FP-Tree. The FP-Tree is a main memory-resident data structure. When the dataset size becomes huge, the FP-Tree may grow beyond the computer's physical memory.

## 6 Conclusion

Finding association rules from distributed databases spread over numerous sites is a significant challenge for distributed association rule mining. This paper mainly gives the idea about several algorithms proposed in a distributed environment to find frequent itemsets. The researchers have been focusing on distributed association rule mining to improve the algorithm's efficiency, scalability, and rule-mining process. It is observed that the space and time requirements are significant issues with most of the existing algorithms. Future algorithms and techniques ought to take scalability and efficiency into account while developing distributed association rule-mining algorithms.

## References


1. Han J, Pei J, Tong H (2022) Data mining: concepts and techniques. Morgan Kaufmann
2. Tan PN, Steinbach M, Kumar V (2016) Introduction to data mining. PearsonEducation India
3. Kaur M, Kang S (2016) Market Basket Analysis: identify the changing trends of market data using association rule mining. Procedia Comput Sci 85:78–85
4. Tripathi D, Nigam B, Edla DR (2017) A novel web fraud detection technique using association rule mining. Procedia Comput Sci 115:274–281
5. Ahmed SA, Nath B (2021) Identification of adverse disease agents and risk analysis using frequent pattern mining. Inf Sci 576:609–641
6. Guo Z, Dai X, Lin H (2004) Application of association rule in disaster weather forecasting. Geograph Inf Sci 10(1):68–72
7. Creighton C, Hanash S (2003) Mining gene expression databases for association rules. Bioinformatics 19(1):79–86
8. Sawant V, Shah K (2014) A survey of distributed association rule mining algorithms. J Emerg Trends Comput Inf Sci 5(5):391–398
9. Cheung DW, Han J, Ng VT, Fu AW, Fu Y (1996) A fast distributed algorithm for mining association rules. In: Fourth international conference on parallel and distributed information systems, pp 31–42. IEEE
10. Agrawal R, Imieliński T, Swami A (1993) Mining association rules between sets of items in large databases. In: Proceedings of the 1993 ACM SIGMOD international conference on management of data, pp 207–216
11. Agrawal R, Srikant R (1994) Fast algorithms for mining association rules. In: Proceedings 20th international conference very large data bases, VLDB (Vol 1215, pp 487–499)

12. Han J, Pei J, Yin Y (2000) Mining frequent patterns without candidate generation. ACM SIGMOD Rec 29(2):1–12
13. Zaki MJ (2000) Parallel and distributed data mining: an introduction. Lect Notes Comput Sci 1759:1–23
14. Paul S (2011) Parallel and distributed data mining. In: new fundamental technologies in data mining. IntechOpen
15. Cheung DW, Ng VT, Fu AW, Fu Y (1996) Efficient mining of association rules in distributed databases. IEEE Trans Knowl Data Eng 8(6):911–922
16. Zhang F, Liu M, Gui F, Shen W, Shami A, Ma Y (2015) A distributed frequent itemset mining algorithm using spark for big data analytics. Clust Comput 18:1493–1501
17. Salloum S, Dautov R, Chen X, Peng PX, Huang JZ (2016) Big data analytics on Apache Spark. Int J Data Sci Analyt 1:145–164
18. Qiu H, Gu R, Yuan C, Huang Y (2014) Yafim: a parallel frequent itemset mining algorithm with spark. In: 2014 IEEE international parallel distributed processing symposium workshops, pp 1664–1671. IEEE
19. Rathee S, Kashyap A (2018) Adaptive-miner: an efficient distributed association rule mining algorithm on Spark. J Big Data 5:1–17
20. Pei J, Han J, Lu H, Nishio S, Tang S, Yang D (2001) H-mine: hyper-structure mining of frequent patterns in large databases. In proceedings 2001 IEEE international conference on data mining, pp 441–448. IEEE
21. Li H, Wang Y, Zhang D, Zhang M, Chang EY (2008) Pfp: parallel fp-growth for query recommendation. In: Proceedings of the 2008 ACM conference on Recommender systems, pp 107–114
22. Dean J, Ghemawat S (2008) MapReduce: simplified data processing on large clusters. Commun ACM 51(1):107–113
23. Riondato M, DeBrabant JA, Fonseca R, Upfal E (2012) PARMA: a parallel randomized algorithm for approximate association rules mining in MapReduce. In: Proceedings of the 21st ACM international conference on Information and knowledge management, pp 85–94
24. Huang PY, Cheng WS, Chen JC, Chung WY, Chen YL, Lin KW (2021) A distributed method for fast mining frequent patterns from big data. IEEE Access 9:135144–135159
25. Shi X, Chen S, Yang H (2017) DFPS: distributed fp-growth algorithm based on spark. In: 2017 IEEE 2nd advanced information technology, electronic and automation control conference (IAEAC), pp 1725–1731. IEEE
26. Xiao W, Hu J (2021) Paradigm and performance analysis of distributed frequent itemset mining algorithms based on Mapreduce. Microprocess Microsyst 82:103817
27. Liu L, Wen J, Zheng Z, Su H (2021) An improved approach for mining association rules in parallel using Spark Streaming. Int J Circuit Theory Appl 49(4):1028–1039
28. Essam A, Abdel-Fattah MA, Abdelhamid L (2021) Towards enhancing the performance of parallel FP-growth on spark. IEEE Access 10:286–296

# Performance Analysis of the AdaGrad Family of Algorithms

**Abhishek Bagepalli and Sanjay Singh**

**Abstract** Optimization algorithms in Machine Learning play a vital role in proper convergence during function evaluation to get better results, thereby forming one of the most important parts of Machine Learning. The AdaGrad family of algorithms, particularly is noteworthy for several reasons. Their performance on certain types of data distributions is yet to be studied for robust adaptation. In this paper, we looked into the performance of these algorithms when pitted against different types of data distributions, the effect of cosine annealing on different hyperparameters, and how one can improve upon them. We found that cosine annealing improves the overall error rate average of the AdaGrad family by a factor of 2.5. These results would be vastly helpful in noise modeling and removal techniques, primarily in image and signal processing.

**Keywords** Optimization · AdaGrad · Cosine annealing · Convergence · Machine learning

## 1 Introduction

Modern architectures in deep learning have revolutionized several application areas, from computer vision to speech recognition. Optimization methods have played an enormous role in the success of such models and have taken a significant place in artificial intelligence.

The gradient descent method (and its variants), which is by far the most popular of them, has several problems in the form of severe dependence on learning rates, causing issues in some instances, hence calling for better methods. It lays the foundation of an algorithm that adaptively scales the gradient value following the data or in correspondence with time. It begs the question, 'why not do both?'. In this paper,

A. Bagepalli · S. Singh (✉)
Department of Information and Communication Technology, Manipal Institute of Technology, Manipal Academy of Higher Education, Manipal 576104, India
e-mail: sanjay.singh@manipal.edu



S. Borah et al. (eds.), *Advanced Computational and Communication Paradigms*, Lecture Notes in Networks and Systems 535,
https://doi.org/10.1007/978-981-99-4284-8_10

we ponder this question and perform experiments to determine if the simultaneous adaption of the learning rate, along with the annealing of hyperparameters, improves the performance of such algorithms. Existing algorithms that adaptively scale learning rates include the AdaGrad family of algorithms, which are noteworthy for their performance [2]. Coming to the chronological update of hyperparameters, Cosine Annealing [8] exists, reducing the learning rate's value over time.

The data distributions we considered are by the Central Limit Theorem [7], which focuses on commonly occurring data distributions in nature. The algorithms we considered are AdaGrad [9], RMSProp(root mean square propagation), AdaDelta, and AMSGrad [12].

Our approach included the creation of specific data distributions, monitoring the performance of said algorithms, and critical comparison among them. We also looked into the workings of each algorithm in order to better their performance and came up with interesting and helpful conclusions. This paper's main contributions are:

- This paper proves that simultaneous updation of hyperparameters in different optimization algorithms for both time and the data itself results in improving the performance of all considered algorithms, namely AdaGrad, RMSProp, AdaDelta, and AMSGrad.

This paper is organized as follows. Section 2 describes the related work. Details of the datasets used and experiments are given in Sects. 3 and 4. The results are presented in Sect. 7 and are discussed in Sect. 7. Section 8 is where we wrap up our paper and discuss future work.

## 2 Related Work

Issues fueled the need for better optimization algorithms in the performance of the gradient descent algorithm (and other related algorithms). The unstable convergence of gradient descent is excellently summarized in [1]. Several solutions were suggested in the paper, but no concrete details were provided. A much more comprehensive study on the derivation and performance of better optimization algorithms was done by [4], where the AdaGrad family of algorithms is introduced. The features of the AdaGrad algorithms are further summarized in [2], where the central focus is on AdaGrad's performance in the presence of saddle points. Finally, suggestions were made by [10], where the emphasis was placed on endowing algorithms with long-term memory of gradients. The use of the decreasing nature of the cosine function is summarized in [3], while the use of a cosine annealing scheduler was made in [11] To test the previously stated hypothesis, our paper uses the contributions made in optimization methods, data distributions, and learning rate schedules to effectively determine the simultaneous usage of hyperparameter updates in coherence with both time and data distribution.

## 3 Data Creation and Collection

This section describes the dataset creation process and the rationale for choosing those data distributions. The Central Limit Theorem [7] states that when a large sample size has a finite variance, the samples will be normally distributed. Hence, it seemed logical to consider datasets involving the Gaussian distribution, which is the most commonly occurring one. We considered data in two dimensions, which called for multivariate Gaussian distributed data. The distribution of Gaussian Data in two dimensions involves information about the mean values ($\mu$) of the data in each dimension (which, in our case, is a vector of dimension $2 \times 1$) and a variance–covariance matrix ($\Sigma$) (dimension $2 \times 2$) which has variances ($\sigma^2$) along its main diagonal and covariances ($\sigma(x, y)$) along its counter diagonal. It gives rise to three cases of data distributions, each with varying values in the variance–covariance matrix as follows:

(i) $\Sigma$ is arbitrary
(ii) $\sigma^2 = 2\sigma(x, y)$
(iii) $\Sigma = \sigma^2 I (\sigma(x, y) = 0)$ (where I is an identity matrix of dimension $2 \times 2$).

Out of these three cases, we have considered the first two cases. In addition to the above data, we considered two cases of mixed distributions [5]. The reason for such a consideration is that most real-life random variables are not derived from a single distribution but a mixture of distributions. True to its name, a mixed distribution contains a mixture of two or more distributions. Before delving into the cases considered, let us look into another type of data distribution—the Exponential Distribution [6]. We have a mixed distribution as, "Exponential + Gaussian ($\mu_1, \Sigma_1$) + Gaussian ($\mu_2, \Sigma_2$)," where $\mu_1 = [0,\ 0]^T$, $\mu_2 = [3,\ 3]^T$, $\Sigma_1 = \begin{bmatrix} 0.1 & 0 \\ 0 & 0.1 \end{bmatrix}$), and $\Sigma_2 = \begin{bmatrix} 5 & 5 \\ 5 & 5 \end{bmatrix}$.

The creation of each dataset considered in our study was meticulously designed, keeping in mind the challenges posed by each distribution. All datasets were created using the "`numpy.random`" library in Python. To ensure uniformity across cases involving different parameter updates, we ensured that all distributions shared a common seed node. A total of five hundred samples were considered for each case of the data distribution, and each of those samples was considered a starting point for each algorithm.

For each of the five hundred data points in each data distribution, four algorithms (as described below) were run for five hundred iterations each, providing ample metrics to measure the error rates of each algorithm.

Hence, for each point in data distribution, we collected the error data at each of the five hundred iterations for all four algorithms. A circular loss function was considered for our study to ensure a definite minimum.

# 4 Methodology

This section describes the methodology used in this paper.

## 4.1 *AdaGrad Family of Algorithms*

The AdaGrad family has several members, of which we decided to monitor the performance of four prominent members:

(i) **AdaGrad** or the adaptive gradients algorithm is a wonderful solution to the rigidity of learning rates in Gradient Descent algorithms. It adaptively scales the learning rate for each dimension. Let us look into the working of AdaGrad. We consider the sum of squared partial derivatives and use it as a scaling factor for the learning rate. The scaling is done by dividing the current learning rate by the square root of the sum of squared partial derivatives until that iteration. A small value is added to the denominator to avoid division by zero. With the new step size in hand, the parameter can be updated, and the next iteration can be performed, and thus the algorithm continues. AdaGrad has one hyperparameter, $\alpha$, which denotes the initial step size or learning rate. The set of equations for AdaGrad is as follows:

$$\alpha_{t+1} = \frac{\alpha_t}{\sqrt{\epsilon + S_{t+1}}} \tag{1}$$

where $S_t$ denotes the sum of squared partial derivatives at time $t$. The parameter of AdaGrad is updated as per the following equation:

$$\theta_{t+1} = \theta_t - \alpha_{t+1}\nabla f(\theta_t) \tag{2}$$

(ii) **RMSProp**: A problem with AdaGrad is that the search can become too slow if the denominator becomes too large, resulting in minimal learning rates for each dimension or parameter. It could cause the algorithm to stop before the intended end is reached. RMSProp or Root Mean Square Propagation follows a central idea of storing the changing average of the squared gradients for each parameter and then dividing the gradient by the square root of the mean square. To prevent issues like the ones posed by AdaGrad, RMSProp used an exponentially decaying average to not give as much importance to gradient information from the extreme past, enabling it to converge rapidly the moment it located a convex bowl.
The process of updating, while similar to that of AdaGrad, replaces the square root of the sum of squared gradients with a root mean square (RMS) measure of the sum of squared gradients up to that iteration. Once the new step size is calculated, the parameter is updated similarly as followed in AdaGrad. We note that RMSProp has two hyperparameters, $\eta$ denoting initial step size or

learning rate, and $\rho$ denoting a measure of momentum. The required relations for RMSProp are as follows:

$$S_{t+1} = \rho S_t + (1 - \rho)(\nabla f(\theta_t))^2 \tag{3}$$

and the updated step size is given by

$$\eta_{t+1} = \frac{\eta_t}{\epsilon + RMS(S_{t+1})}. \tag{4}$$

The final parameter is updated as per the following equation

$$\theta_{t+1} = \theta_t - \eta_{t+1} \nabla f(\theta_t) \tag{5}$$

(iii) **AdaDelta** is a more robust Adagrad extension that adapts learning rates based on a moving window of gradient updates rather than accumulating all previous gradients. AdaDelta continues to learn in this manner even after numerous updates have been completed. It is designed to boost the optimization process by typing to reduce the number of iterations. It involves the calculation of a step size for each parameter of the objective function for every iteration. The calculation of the learning rate is similar to that of AdaGrad, and the sum of squared partial derivatives is similar to that of RMSProp. However, there is a minor change in the form of a decaying moving average of the squared change to the parameter, $\delta$.
The parameter $\delta$ is further updated in successive iterations using the hyperparameter, $\rho$, like in RMSProp. The updated delta value calculates the learning rate in the next iteration. Following this, the change to the parameter is calculated, and finally, the parameter is updated using the latest value of the change. AdaDelta has one hyperparameter, $\rho$-a measure of momentum.
AdaDelta does not require manual setting and tuning of an initial learning rate like the other algorithms. The updation in the learning rate for AdaDelta is given by

$$\gamma_{t+1} = \frac{\epsilon + \sqrt{\delta_t}}{\epsilon + \sqrt{S_t}} \tag{6}$$

where $S_t$ denotes the decaying moving average of the squared partial derivative. The intermediate change to the parameter in the current iteration is given by

$$\Delta_{t+1} = \gamma_{t+1} \nabla f(\theta_t). \tag{7}$$

The updated decaying moving average is given by

$$\delta_{t+1} = \delta_t + (1 - \rho)\Delta_{t+1}^2. \tag{8}$$

Finally, the relation for parameter updation is given by

$$\theta_{t+1} = \theta_t - \Delta_{t+1}. \tag{9}$$

(iv) **AMSGrad**: It is a stochastic optimization method designed to address a convergence problem with Gradient Descent and Adam-based optimizers [10]. AMSGrad updates the parameters using the maximum of previously squared gradients rather than the exponential average. It maintains a first and second-moment vector and a maximum value of the second-moment vector for each parameter undergoing updation as part of the search process. Initially, it is initialized to zero, and each value undergoes updation as they progress through the respective iteration. The first-moment vector employs a hyperparameter, $\beta_1$, and is updated using the objective function gradient and $\beta_1$.
The second-moment vector undergoes a similar updation but uses the square of the gradient and a new hyperparameter, $\beta_2$. The maximum value of the second-moment vector is computed, and the parameter is updated using the first-moment vector, the maximum of the second-moment vector, the initial step size, or the learning rate. AMSGrad has three hyperparameters: $\zeta$-initial step size or learning rate, $\beta_1$-decay factor for the first-moment vector, and $\beta_2$-decay factor for the second-moment vector. The first-moment vector is given by

$$m(t) = \beta_1(t)m(t-1) + (1-\beta_1(t))\nabla f(\theta(t-1)). \tag{10}$$

The second-moment vector is given by

$$v(t) = \beta_2(t)v(t-1) + (1-\beta_2(t))(\nabla f(\theta(t-1)))^2. \tag{11}$$

The maximum of the second-moment parameter vector is given by

$$\hat{v}(t) = \max\{\hat{v}(t-1), v(t)\} \tag{12}$$

and the final updated parameter is given by

$$\theta_{t+1} = \theta_t - \frac{\zeta_t m(t)}{\sqrt{\hat{v}(t)}}. \tag{13}$$

## 5 Learning Schedule

We have used cosine annealing [8] to study the behavior of various optimization algorithms discussed in Sect. 4. Cosine Annealing is a type of learning rate schedule that has the effect of starting with a high learning rate and rapidly decreasing to a low value before rapidly increasing again. It is done in coherence with the fact that as an algorithm approaches a particular minimum, it needs to take smaller steps as the scope narrows rapidly. The cosine function, a decreasing function, offers a unique opportunity to tune a parameter accordingly. The resetting of the learning rate acts as a simulated restart of the learning process, and the use of good weights as the restart's starting point is referred to as a "warm restart," as opposed to a "cold restart," which may use a new set of small random numbers as a starting point.

Cosine Annealing requires specific values as input for the update process. Such an update enabled a highly successful implementation of previously mentioned algorithms, which needed to perform satisfactorily on datasets created with certain distributions in mind. The learning schedule as per the cosine annealing [8] is given by

$$\eta = \eta_{\min} + \frac{1}{2}\left(\eta_{\max} - \eta_{\min}\right)\left(1 + \cos\left(\frac{T_{\text{cur}}}{T}\pi\right)\right) \tag{14}$$

where $\eta_{\min}$ and $\eta_{\max}$ are ranges for learning rate, and $T_{\text{cur}}$ accounts for how many iterations have been performed since the last restart, and $T$ is the total number of iterations.

## 6 Evaluation Metrics

As mentioned in the data section, we collected error data in Euclidean distance from the origin, which is the minimum of our circular loss function. For each of the five hundred points in every distribution, Euclidean error data were collected for five hundred iterations for each of AdaGrad, RMSProp, AdaDelta, and AMSGrad. The error data for each algorithm was stored in a NumPy array considering ease of access and management. Equipped with the above data, we computed the average error of each algorithm for every single one of the five hundred data points.

Next, we equipped each of the four algorithms with cosine annealing for each hyperparameter (as stated above) and again computed the error norms for all data points. We derived meaningful conclusions from the data and plots based on the observed results and careful comparative analysis. All plots were created using "matplotlib" in Python.

## 7 Results and Discussion

This section presents the findings from the experiment. We evaluate the performances of AdaGrad, RMSProp, AdaDelta, and AMSGrad with and without the incorporation of cosine annealing for all involved hyperparameters. The main idea behind introducing cosine annealing is to improve algorithm performance by reducing the number of iterations before convergence to a minimum.

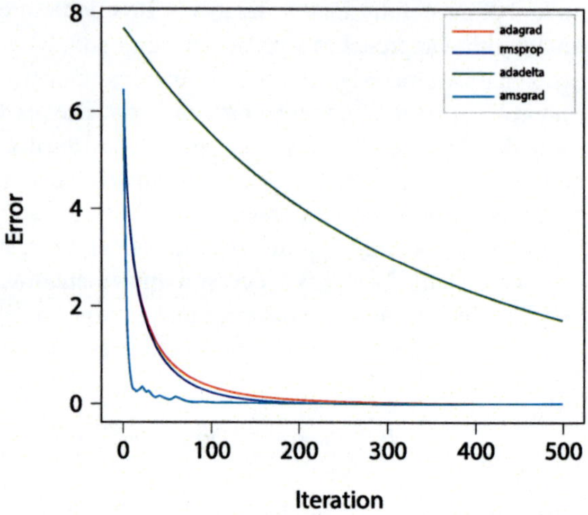


**Fig. 1** Plot of average sum of squared error versus iteration for case-1 of multivariate Gaussian data

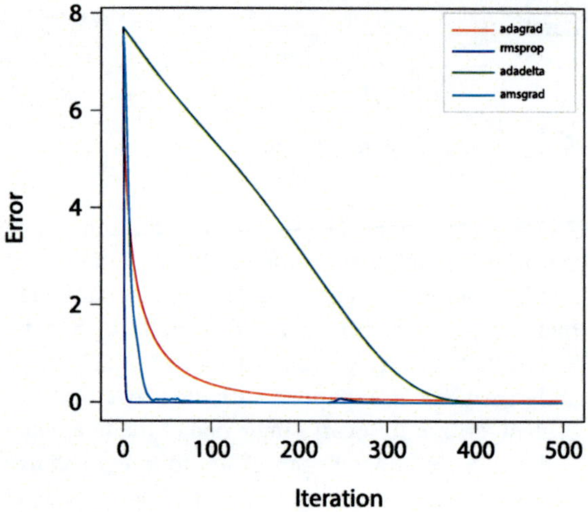


**Fig. 2** Plot of average sum of squared error versus iteration for case-1 of multivariate Gaussian data with cosine annealing

Now we look at the respective performances of the four algorithms on the following data distributions:

1. Multivariate Gaussian Case-1 is included in the Figs. 1 and 2.
2. Multivariate Gaussian Case-2 is included in the Figs. 3 and 4.
3. Mixed Case-2 is included in the Figs. 5 and 6.

Owing to the limited availability of GPU resources, we limited the data size to five hundred samples per data distribution and five hundred iterations per data point. Observing the algorithm's performance over much larger datasets that would strain the GPU would give a better idea of their feasibility.

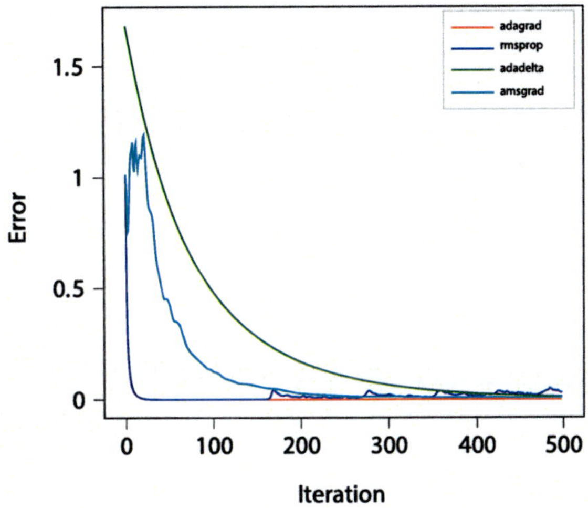


**Fig. 3** Plot of average sum of squared error versus iteration for case-2 of multivariate Gaussian data

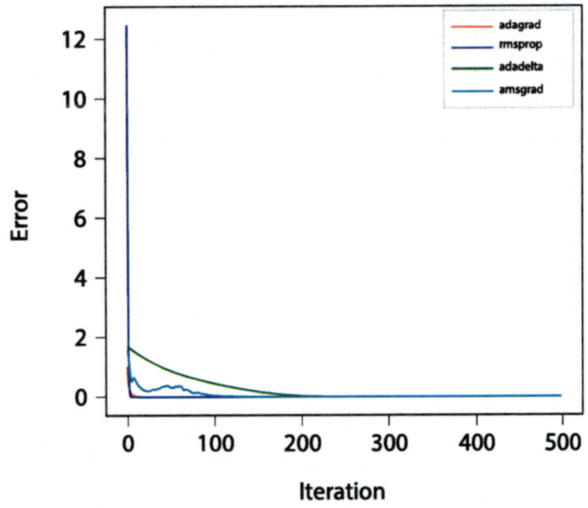


**Fig. 4** Plot of average sum of squared error versus iteration for case-2 of multivariate Gaussian data with cosine annealing

The limited number of mixed distributions considered is also a potential improvement where multiple other distributions can also be factored in to observing and deciding on the usage of the algorithms. Domain-specific knowledge would aid in creating mixed distribution data, which would be immensely helpful in areas involving noise modeling for image and signal processing tasks.

The study of recurrent data distribution types in domain-specific areas is another area with scope for improvement. It would make conclusions derived from those results tailormade for the respective domain.

As seen in case-1 of the multivariate Gaussian data distribution, before cosine annealing is performed, AMSGrad outperforms its counterparts, rapidly converging to a shallow error rate. While the subsequent best algorithms, i.e., AdaGrad and

**Fig. 5** Plot of average sum of squared error versus iteration for case-2 of mixed data

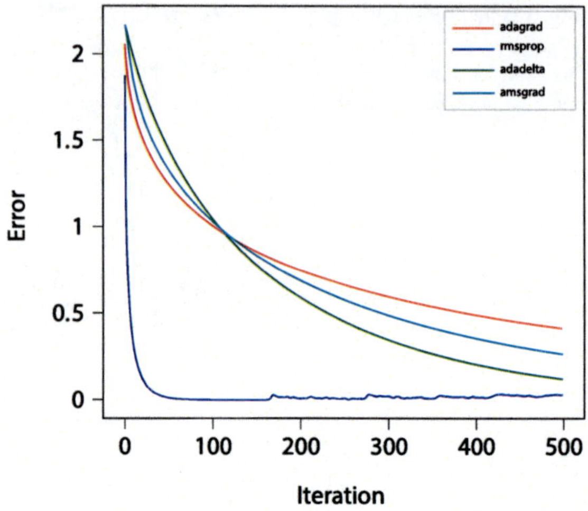


**Fig. 6** Plot of average sum of squared error versus iteration for case-1 of mixed data with cosine annealing

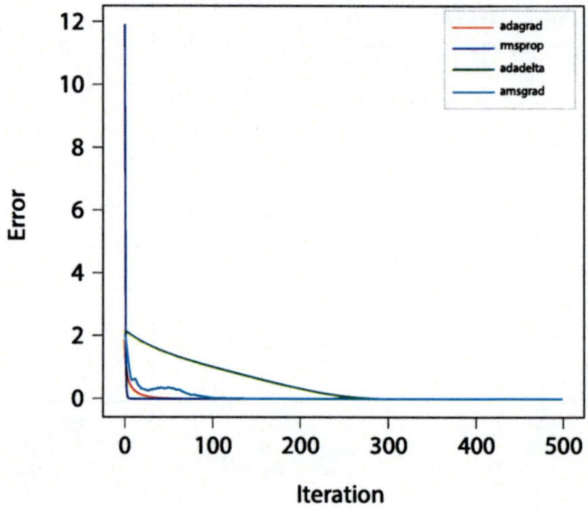


RMSProp, take about a hundred iterations to converge, AMSGrad converges almost instantly. The performance of AdaDelta, however, is inferior.

The same case is rerun, but now with cosine annealing for all hyperparameters, giving rise to some incredible observations. RMSProp outperforms all other algorithms, converging almost instantly. AMSGrad holds its ground, with a slightly better performance than before. AdaGrad's performance is almost unchanged, while a good improvement is seen in the case of AdaDelta, which now converges accurately, albeit taking about 400 iterations to do so. In case-2 of the multivariate Gaussian data distribution, RMSProp and AdaGrad outperform their counterparts before cosine annealing. However, a slight issue is observed in the case of RMSProp, as there is a very slight increase in its error rate, even after initial convergence. AMSGrad takes a wayward path before finally converging, with the error rate initially increasing before falling toward the end toward a decent convergence value. AdaDelta, while monotonically decreasing, takes about 400 iterations to converge properly to the minimum.

The same case is rerun but with cosine annealing for all hyperparameters; the observations are noteworthy for the following reasons. There is a marked improvement in the performance of all algorithms, with RMSProp and AdaGrad still converging faster than the rest. The initial error rate for AdaGrad, however, is much smaller when compared to RMSProp. AMSGrad, while having a near-perfect convergence, initially dabbles uncertainly, while AdaDelta converges monotonically before reaching the minimum around iteration 200.

The inferences that can be drawn from the above results are that RMSProp and AdaGrad respond excellently to cosine annealing. However, RMSProp needs to be adjusted to stop further movement after the minimum is found. Due to its not settling in local minima, however, RMSProp's biggest strength becomes its biggest weakness, causing the error rate to fluctuate perennially. Therefore, AdaGrad is the best choice for all three cases of multivariate Gaussian data when cosine annealing is involved. When cosine annealing is not involved, RMSProp is still the best bet for case-2, while AMSGrad is a better choice for case-1.

It proves our hypothesis of simultaneous updation of hyperparameters according to data and time. We look at two cases with a mixture of data distributions to further verify the hypothesis.

In case of the mixed distributions, we consider a mixture of exponential and two Gaussian data mixtures, as mentioned in the data section. We now look at the performance of different algorithms before cosine annealing is incorporated. Only RMSProp reaches the minimum, while all other algorithms fail to do so. However, the convergence rate is prolonged for RMSProp, with the best minimum reached around iteration 23.

On incorporating cosine annealing, however, A marked improvement is observed in all algorithms, with RMSProp being the fastest. However, there is a deviation from the best minimum to a slightly increased error rate. AdaGrad, AMSGrad, and AdaDelta converge to the same minimum, although AdaDelta is considerably slower, reaching the minimum around iteration 250.

Once again, based on the marked improvement of the algorithms' performance, the hypothesis that simultaneous updation of hyperparameters concerning data and time stands correct.

## 8 Conclusion

The AdaGrad family of algorithms is an integral part of the optimization process, thus forming a vital part of Machine Learning and Deep Learning. However, the performance of each algorithm varies from one data distribution to the other. In this paper, we have investigated the performance of the AdaGrad family of algorithms on different data distributions, which are representative of data distributions most commonly seen in real life. We have observed that with conventional learning schedules, AMSGrad converges almost instantly, while others do not give a satisfactory performance in the first case of multivariate Gaussian data. The second case saw RMSProp and AdaGrad outperforming their counterparts. On incorporating cosine annealing, a marked improvement was observed in the performance of all algorithms in all cases of data distribution that were considered, thereby confirming our initial hypothesis that the simultaneous usage of hyperparameter updates in coherence with both time and the data distribution would lead to an improvement in algorithm performance.

## References

1. Ahn K, Zhang J, Sra S (2022) Understanding the unstable convergence of gradient descent. In: Chaudhuri K, Jegelka S, Song L, Szepesvári C, Niu G, Sabato S (eds) International conference on machine learning, ICML 2022, 17–23 July 2022, Baltimore, Maryland, USA. Proceedings of machine learning research, vol 162. PMLR, pp 247–257. https://proceedings.mlr.press/v162/ahn22a.html
2. Antonakopoulos K, Mertikopoulos P, Piliouras G, Wang X (2022) AdaGrad avoids saddle points. In: Chaudhuri K, Jegelka S, Song L, Szepesvari C, Niu G, Sabato S (eds) Proceedings of the 39th international conference on machine learning. Proceedings of machine learning research, vol 162. PMLR (17–23 Jul 2022), pp 731–771. https://proceedings.mlr.press/v162/antonakopoulos22a.html
3. Cazenave T, Sentuc J, Videau M (2022) Cosine annealing, mixnet and swish activation for computer go. In: Browne C, Kishimoto A, Schaeffer J (eds) Advances in computer games. Springer International Publishing, Cham, pp 53–60
4. Duchi JC, Hazan E, Singer Y (2011) Adaptive subgradient methods for online learning and stochastic optimization. J Mach Learn Res 12:2121–2159. https://doi.org/10.5555/1953048.2021068
5. Ghojogh B, Ghojogh A, Crowley M, Karray F (2019) Fitting a mixture distribution to data: tutorial. arXiv:1901.06708
6. Gupta A, Zeng WB, Wu Y (2010) Probability and statistical models: foundations for problems in reliability and financial mathematics. In: Exponential distribution. Birkhäuser, pp 23–43. https://doi.org/10.1007/978-0-8176-4987-6_2
7. Kwak S, Kim J (2017) Central limit theorem: the cornerstone of modern statistics. Korean J Anesthesiol 70:144. https://doi.org/10.4097/kjae.2017.70.2.144
8. Loshchilov I, Hutter F (2016) SGDR: stochastic gradient descent with warm restarts. arXiv:1608.03983
9. Lydia A, Francis S (2019) Adagrad - an optimizer for stochastic gradient descent. Int J Inf Comput Sci 6:566–568
10. Reddi SJ, Kale S, Kumar S (2018) On the convergence of Adam and beyond. In: 6th international conference on learning representations, ICLR 2018, Vancouver, BC, Canada, April 30–May 3, 2018, Conference track proceedings. OpenReview.net. https://openreview.net/forum?id=ryQu7f-RZ
11. Wang Z, Hou F, Qiu Y, Ma Z, Singh S, Wang R (2022) Cyclic augment: speech data random augmentation with cosine annealing scheduler for automatic speech recognition. In: Proceedings of the Interspeech 2022, pp 3859–3863. https://doi.org/10.21437/Interspeech.2022-526
12. Zaheer R, Shaziya H (2019) A study of the optimization algorithms in deep learning. In: 2019 third international conference on inventive systems and control (ICISC), pp 536–539. https://doi.org/10.1109/ICISC44355.2019.9036442

# Time-Related Natural Language Query Handling to Extract Autism Spectrum Disorder Information from Temporal Database

**Prasenjit Mukherjee, Chandrima Chakraborty, Nandan Banerji, Kailash Pati Mandal, and Baisakhi Chakraborty**

**Abstract** A temporal database may contain time-related data and handling such kind of database is a challenging task. It is difficult to extract data from a temporal database using natural language (NL) queries. Many NL queries to database interfaces have been developed but these interfaces are not suitable to handle time-related NL queries. The proposed system can handle time-related NL queries in the healthcare domain. A temporal database has been created where various Autism Spectrum Disorders (ASD) real-life problem statements have been stored with time. The proposed system is able to extract data from a temporal database using time-related NL queries with aggregation. Time tick has been assigned for handling temporal queries whereas aggregation methods have been used for handling computational type queries. Auxiliary verbs are used as an indication of whether the NL query is a temporal query. The proposed system has the capacity to produce structured query language (SQL) from temporal and computational types of queries which are frequently occurring in NL. This system will help the parents of an autistic child to enhance their knowledge about autism after getting many real-life autism problems according to time.

**Keywords** Natural language queries · Autism spectrum · Temporal query · Structured query language · Healthcare

P. Mukherjee (✉)
Manipur International University, Imphal, Manipur, India
e-mail: prasen.mscit09@gmail.com

Vodafone Intelligent Solutions, Pune, India

C. Chakraborty
Sasken Technologies Limited, Bengaluru, India

N. Banerji · K. P. Mandal · B. Chakraborty
National Institute of Technology, Durgapur, India

 
S. Borah et al. (eds.), *Advanced Computational and Communication Paradigms*, Lecture Notes in Networks and Systems 535,
https://doi.org/10.1007/978-981-99-4284-8_11

## 1 Introduction

Autism spectrum disorder detection at an early age in a child is a challenging task. The growth of Artificial Intelligence (AI) in the last decade is very high and today many systems have been developed using AI that can detect ASD at an early age in a child. But the detection is not enough because the maximum parents do not have good knowledge about ASD and ASD-related problems. Parents should know about the various ASD problems according to the age of a child because it is mandatory to understand the ASD problem and according to the specific problem, the treatment can be started which will be result oriented. There is an urgent need for widespread early detection of autism spectrum disorder (ASD), as effective early therapies can ameliorate the symptoms of ASD. In this work, the authors suggest an automated behavior-based screening strategy to detect high-risk ASD in infants between the ages of 8 and 24 months. To induce a baby's spontaneous social behavior through face-to-face interaction, the still-face paradigm (SFP) was used. These indicators were utilized to create an automated model for HR-ASD detection as in [1]. The severe neurological illness known as autism spectrum disorder (ASD) has an effect on every aspect of the brain and has an adverse effect on a person's physical, mental, emotional, and social well-being. Social interaction and communication with others are challenging for them. Parents, family, or friends are always needed to provide support and act as mentors for them. Sadly, autism has no known cure, although early identification can aid in effective therapy. Autism disorder can be identified by looking at a person's behavioral patterns. Therefore, there is a need for quick, affordable ASD screening to assist people in determining if they ought to get a medical diagnosis and get help as in [2]. Many NL queries to database interfaces have been developed to extract information from healthcare databases. An NL database interface (NLDBI) is an identical interactive interface for novice users because it is very simple for interaction in NL and retrieves more correct information from the database than a generic search as in [3]. Supervised learning, unsupervised learning, and deep learning are advanced technologies in artificial intelligence and are highly acceptable for structured data. But these technologies are not efficient for unstructured data like processing NL as in [4]. NL interfaces accept NL text or NL queries. Time-related queries or temporal queries aim to extract knowledge about data related to time instances or time ticks. The NL interface for the temporal database is another challenging task in NLP. Sometimes, the requirement of data is time-dependent. The temporal database is used for time-related data.

The proposed system works on handling time-related NL queries and generates responses from the temporal database. The time ticks may be categorized as past, recent past, and present. The proposed system generates the semantic table for each NL query. The SQL will be generated from this semantic table where an auxiliary verb will be used for SQL that is related to the temporal query sentence. The aggregation like count, sum, total sum, max, min, etc., are also handled by the proposed system. ASD-related problems have been stored in a database with time and the proposed system has the ability to capture ASD problems using the time-related NL queries.

## 2 Related Works

Many AI-based systems have been developed that are useful today in ASD detection. ASD detection-type systems and some other NLP-based systems have been discussed in this section. Impulses from neuroimaging are fractal in nature has been shown to influence estimates of functional connectivity in the brain. P-value analysis was used to uncover a number of important linkages, and they had strong ties to autism. Feed a machine learning (ML) classifier to classify subtypes against normal controls. Both multiclass and binary categorization fared better than traditional correlation-based Pearson classification approaches for accuracy, sensitivity, and specificity. This work exemplifies the enormous possibilities of the non-oscillatory connection approach for diagnosing autism and other neurological disorders as in [5]. Growing empirical evidence suggests that practicing mindfulness might help people with autism spectrum disorder control their emotions in a better manner than those who do not practice mindfulness. It is possible to develop mindfulness using meditation techniques. Given the emphasis on self-report measures, evaluating the contribution of mindfulness in enhancing emotion control is difficult. Thirty-five children with ASD underwent an EEG-based investigation at the time of the touch-the-ground conceptual framework. In particular, we created a characteristic and channel selection, together with a classification algorithm predictor method that distinguishes between participants' resting states before and after a mindfulness meditation activity. With an average accuracy, sensitivity, and specificity of 80.76%, 78.24%, and 82.14%, respectively, a classifier utilizing a nonlinear combination of chosen EEG variables from chosen channel locations differentiated pre-MF and post-MF resting states. Finally, we confirmed that the pre-MF and post-MF separations are caused by the MF prime rather than linear-temporal drift as in [6]. Around 1% of the world's population is affected by its symptoms, which often appear during the formative phases, or during the first two years following birth. ASD is mostly brought on by genetic or environmental causes. Currently, the only procedures utilized to diagnose ASD are clinically based standardized testing. Machine learning approaches are being utilized to supplement traditional methods in order to increase the accuracy and turnaround time for diagnosis. On the basis of the results from applying models like support vector machines, random forest classifier, naive Bayes, logistic regression, and KNN to their dataset, built prediction models. In order to speed up the diagnostic process, the major goal of the article is to ascertain if the kid is predisposed to ASD in its early stages as in [7]. AAC is a term that refers to a group of techniques or tactics that together make up any device, such as a speech-generating device, program, or procedure that improves a person's capacity for communication. This is because other studies only concentrate on low- or high-tech AAC, or interventions to offer insights into how ASD children respond to a particular approach. A total of 60 research publications were reviewed using the PRISMA methodology, indexed by the Springer, Science Direct, Scopus, ACM, and IEEE databases, and published in the AAC magazine. This comprehensive overview gives a summary of the paradigm change in AAC modalities from PECS to applications based on artificial intelligence, machine learning, and augmented reality as in [8]. Substring formation is very useful in computer programs. The palindrome problem is a well-known problem in a computer program, where S is a string. S = A1A2 where A1 and A2 are substrings. The reverse of substring A2 is to be matched exactly with A1 and this is the simple algorithm of palindrome as in [9]. The database is playing an important role as a major source of information in the computer field. Two new systems have been identified by the researchers that are NL interface to the database (NLDBI) and keyword-based interface to the database (KBIDB). The authors have proposed a system that is an NL and a keyword-based interface to the database which processes the syntactically correct and incorrect NL input query. The proposed system has been applied to the agriculture survey database that managed 53% of the incorrect query as in [10]. Extraction of meaningful data from the repository requires a good knowledge of the database language. Structure query language (SQL) is one of the important and popular database languages in the database field. Users may not be able to write complex type SQL for needful data extraction from the database. Authors have introduced an interface that is NL-based and a user may query in their own language. The proposed system is a database intelligent querying system (DBIQS) that is an automated, fast and reliable, and simple type system to extract valuable data from a database using NL processing as in [11]. Desheng Wu and Yuhui Guan stated that the retrieval algorithm of artificial intelligence does not consider the range of content from several sources of information which is the cause of inadequate recovery performance and poor evaluation correctness. They have proposed an information processing strategy for artificial intelligence based on correlation calculations as in [12]. Yoruba [13] is an English language translator that has been developed by John Oladosu and Justice Emuoyibofarhe. The proposed model is based on Doctor–Patient Mobile Chat Application. Authors have conducted the assessments on medical practitioners and patients. The proposed model has been developed by the Java language as in [13].

## 3 Architecture

The proposed system will accept time-related NL queries in English. The NL query will be validated first. A part of speech (POS) table will be used to match each word from the sentence. If a single word does not match with the part of speech (POS) table then the sentence will be validated as true and the query sentence will go to the next step. In the next step, the query sentence will be tokenized and POS Tagger will parse the token as determiner, pronoun, preposition, conjunction, wh-word, and be-verb. Nouns, verbs, adjectives, and adverbs are not present in the POS Table. After parsing, tokens will be detected as determiner, pronoun, preposition, conjunction, wh-word, and be-verb. These identified tokens will be removed from the token list. If an auxiliary verb like is, are, was, were, have, and had is detected then that will be removed from the token list but it will be stored in another variable called an auxiliary verb. The auxiliary verb has been used in time-related SQL query generation (Fig. 1).

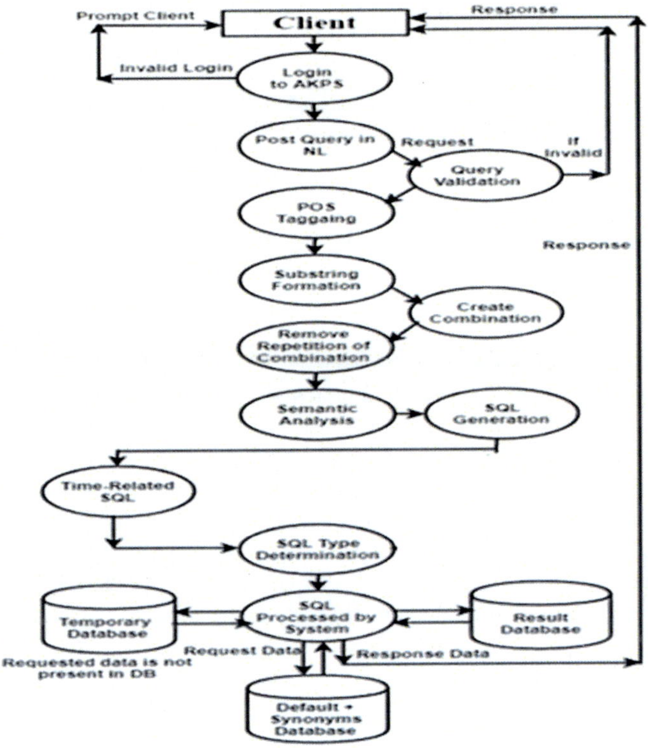


**Fig. 1** Architecture of the proposed system

Unidentified tokens will be termed unique tokens. The substring will be formed on unique tokens and each substring will be sent to create combinations as in [14]. After creating combinations, each combination will be checked for repetition, and if any repetition occurs in combinations then the repetitive combination will be removed. Now, each combination will go for three types of searching procedures like entity search, attribute search, and value search. First, each combination will go for entity search where this search procedure matches the selected combination in the entity table where database table names are given as entity names with synonyms. The primary key, foreign key, and candidate key are also initialized. If the selected combination is matched with any entity name or with any synonym then the corresponding row will be selected and to be inserted into the semantic table except synonym words. If the entity search does not return any entity name with the corresponding row from the entity table then the selected combination will go for the attribute search in the attribute table where this table contains the entity name, attribute name synonym of the attributes, primary key, foreign key, and candidate key of the tables of the temporal database. If the selected combination is matched with the attribute name or synonym of the attribute then the corresponding row will be selected except the synonym and will be inserted in the semantic table. If there is no match with any

**Table 1** Time durations of different time ticks

| Sl. no | Time tick | Time duration |
|---|---|---|
| 1 | Present Time (current year) | 2019–2022 |
| 2 | Recent-Past Time (Last decade) | 2015–2018 |
| 3 | Past Time (Past of last decade) | 2010–2014 |

value in the attribute table, then the selected combination will go for the value search procedure where the selected combination will be searched in the whole database. If any value is matched, then the corresponding attribute name will be selected and it will go for the attribute search from where the matched attribute name and selected row value except synonym will be fetched and will be inserted into the semantic table. The semantic table will be prepared using these three search techniques. The SQL will be prepared using the semantic table. The SQL may be generated as time-related with aggregation and time-related without aggregation. This classification of SQL generation has been done using the Auxiliary verb. Auxiliary verbs "am, is, are" have been used as present time tick. Auxiliary verbs "was, were, had" have been used as past time ticks, whereas "has, have" have been used as recent-past time ticks. The Time duration of different time ticks has been given in Table 1.

The Response will be generated from created classified SQL by the proposed system.

## *3.1 Algorithm of Proposed System*

The steps of the algorithm have been discussed below:

1. **NLP Query reading & Validation**

The proposed NLPBQRS will first read the NL Query which is given by the user after a valid login in the system. Now, the system will tokenize the query to generate individual words or tokens. POS will be used to validate the list of tokens of the query.

2. **POS Tagging**

The POS tagger of this proposed system contains a list of WH words, auxiliary verbs, determiners, pronouns, prepositions, and conjunctions except nouns, adjectives, verbs, and adverbs. These words are stored in various string arrays like Wh[], Aux [], Determiners [], Conjunction [], Preposition [], Pronoun [], and BVerb []. If a word matches with any word from the POS list, then move the matched word termed as a known word from the token array. If the word is not matched with any word from the POS list, then this word will be termed as an unknown word. The unknown word will not be removed from its position in the token array. An example has been given below. A token array of the query sentence has been given in Table 2 and after POS tagging, unique tokens have been generated according to Table 3.

NL query as example 1: *"what is the total number of autism problem in current time"* (Table 4).

3. **substring Formation**

The substring will be formed using the unique tokens in this step that has been given in Table 3.

4. **Create Combination**

The combination will be formed from each substring. According to the above example, combinations have been given here—*"total"*, *"number"*, *"total number"*, *"autism"*, *"problem"*, *"autism problem"*, *"current"*, *"time"*, and *"current time"*.

**Table 2** Token Array for example 1

| 0 | 1 | 2 | 3 | 4 | 5 | 6 | 7 | 8 | 9 | 10 |
|---|---|---|---|---|---|---|---|---|---|---|
| What | Is | The | Total | Number | Of | Autism | Problem | In | Current | Time |

**Table 3** Unique tokens of example 1

| 0 | 1 | 2 | 3 | 4 | 5 | 6 | 7 | 8 | 9 | 10 |
|---|---|---|---|---|---|---|---|---|---|---|
| | | | Total | Number | | Autism | Problem | | Current | time |
| | | | Unique words | | | Unique words | | | Unique words | |

**Table 4** Substring array

| 0 | 1 | 2 | 3 | 4 |
|---|---|---|---|---|
| Total number | Autism problem | Current time | Null | Null |

*Pseudo Code:*

*// Pseudo Code for combination Creation method*

```
FOR x=0 to x<subs3.Length z=0;
IF(subs3[x]  !=  null)  THEN  words2  =
subs3[x].Split();  FOR  y=0  to  y<words2.Length
1=words2[y].CompareTo(" "); IF(x<0) THEN
Do nothing; ELSE
Copywords2[z] = words2[y]; z is incremented
by 1 END IF
END FOR
CALL GetCombination2 (copywords2, z) Method GetCombination2 (string[] list1,
int a) Method Int x, y, z, vcount1;
Semantic sm3 = new Semantic(); String calc1 = null;
Double count1 = Math.Pow(2, a); FOR x=0 to x<=count1
String str1 = Convert.ToString(x, 2).PadLeft(a, '0'); FOR y = 0 to y<str1.Length
IF (str1[y] == '1') THEN
calc1 = calc1 + list1[y] + " "; END IF
END FOR
Sm3.InsertCombination1(calc1); Method.
Calc1 = null; END FOR
END GetCombination2
.......
```

5. **Remove Repetition:**

If any repetition occurs after completion of the combination step then repeated or duplicate combination will be removed.

6. **Semantic representation**

The system captures every variety of combinations from the combination table. The system will match each combination with each word in the specifiers array, computational adjective array, and computational noun array. If a word is matched, then the matching word will be inserted into the semantic table. If the combination is not matched with any word of these arrays, then three searching methods will be applied as per the matching requirements which are discussed above. Table 5 represents the semantic table of the above example.

7. **SQL Type Determination and SQL Generation:**

The proposed NLPBQRS creates the Structure Query Language (SQL) from the semantic table (Table 5). The semantic table contains the specifiers, Comp_Adj,

**Table 5** Semantic Table Representation

| SID | Specifier | Comp_ Noun | Comp_ Adj | Entity | Attributes | Auxiliary Verb | Be Verb | PK | FK | CK | Value |
|---|---|---|---|---|---|---|---|---|---|---|---|
| 1 | Total | Number | Null | Autism | Autism _Problem | Is | Null | AID | Null | Null | Null |

Comp_Noun, Entity name, Attribute name, auxiliary verb, be verb, Primary Key, Foreign Key, Candidate Key, and Value. Auxiliary verb has been used to generate time-related query. But the proposed system will generate time-related query with aggregation due to the presence of auxiliary verb "is". "is" will select the present time tick according to Table 1. SQL generation has been classified as time-related SQL with aggregation termed as temporal computational SQL and time-related SQL without aggregation termed as temporal non-computational SQL as in Fig. 2.

The existence of the auxiliary verbs in an NL query determines the SQL generation type. If the auxiliary verb exists, then temporal SQL will be generated. If an auxiliary verb is not present in the NL sentence then normal SQL will be generated. Auxiliary verbs "am, is, are" have been used as present time tick. Auxiliary verbs "was, were, had" have been used as past time ticks, whereas "has, have" have been used as recent-past time ticks.

The pseudo code:

*The basic algorithm of the classification of SQL at run time will be-*
*a) Read Semantic Table.*
*b) SQL Creation Condition as per diagram in Figure 5.*

```
    IF Auxiliary Verb == "Present Time" THEN
    IF (Specifiers == TRUE && Comp_Noun == TRUE && Comp_Adj == TRUE) || (
Comp_Noun == TRUE && Comp_Adj
    == TRUE) THEN
    // Temporal Computational SQL in Present Time frame END IF



    ELSE
    IF Auxiliary Verb == "Recent Past" THEN
    IF (Specifiers == TRUE && Comp_Noun == TRUE && Comp_Adj == TRUE) || (
Comp_Noun == TRUE && Comp_Adj
    == TRUE) THEN
    // Temporal Computational SQL in Recent Time frame END IF
    ELSE
    IF Auxiliary Verb == "Past Time" THEN
    IF (Specifiers == TRUE && Comp_Noun == TRUE && Comp_Adj == TRUE) || (
Comp_Noun == TRUE && Comp_Adj
    == TRUE) THEN
    // Temporal Computational SQL in Past Time frame END IF
    ELSE
    IF (Specifiers == TRUE && Comp_Noun == TRUE && Comp_Adj == TRUE) || (
Comp_Noun == TRUE && Comp_Adj
    == TRUE) THEN
    // Conventional Computational SQL END IF
    ELSE
    // Conventional SQL
```

**Temporal Computational SQL (Temporal Query Detection)**

The temporal SQL with aggregation methods may be generated as per the requirement of the NL query. The Temporal query depends on the time tick that has been used in Date_From and Date_To attributes in the Autism table. Time tick refers to time duration that has been stated as present time, past time, and recent past. If an NL query contains words like "sum, average, maximum, total, etc.", then these words are referred to as computational words. The words are categorized as specifiers, computational adjectives, and computational nouns. Specifiers, computational adjectives, and computational noun lists have been used in the entity search method and attribute search method to detect the computational words from NL queries. The computational words will be inserted into the semantic table once detected and SQL will be generated with a time tick and aggregation method termed as temporal computational SQL. The SQL will be generated according to the above example:

**SQL Query:** *"SELECT COUNT(Autism_Problem) FROM Autism WHERE Date_From BETWEEN 2019 AND 2022."*

The output has been given in Fig. 3 where the number of autism problems has been given 11. Count () aggregation method will be utilized due to computational noun and computational adjective detection.

If the query is—"what was the total number of autism problem," the SQL will be—*"SELECT COUNT(Autism_Problem) FROM Autism WHERE Date_From BETWEEN 2010 AND 2014"* (Fig. 3).

**Temporal Non-Computational SQL (Temporal Query Detection)**

*what was the autism problem"* is a good example of temporal non-computational SQL generation where "was" will be selected as a time tick but this query is not computational because "sum", "number", "total number", "maximum" or "minimum" types words are not present in the query. According to the time tick, the SQL will be—*"SELECT Autism_Problem FROM Autism WHERE Date_From BETWEEN 2010 AND 2014.*

## 3.2 Database of the Proposed System

A database has been implemented where autism is a table which is containing columns that are described below.

i. AID—AID column is a unique id of each autism problem statement.
ii. Date_From—this column contains starting year of different time ticks.
iii. Date_To—this column contains the ending year of different time ticks.
iv. Autism_Problem—this column contains autism problem statements that have been collected from the internet.

Table 6 has been given which is representing the autism table. Other tables are also developed like Doctors, Employees, and MRecords. The autism table has been highlighted here due to the maximum temporal queries that will be applied to the autism table.

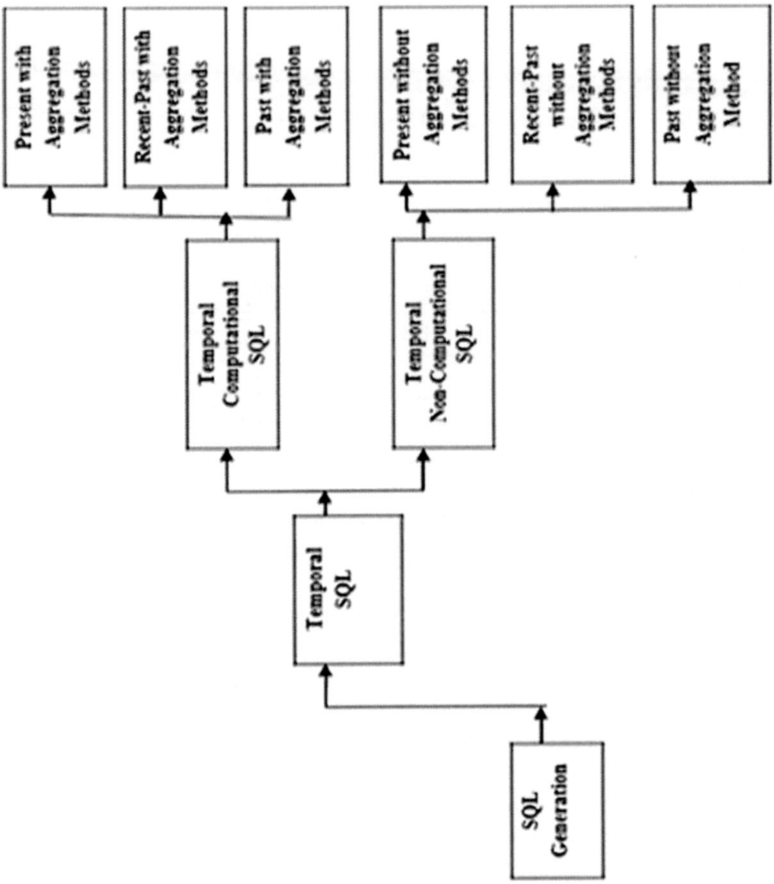


**Fig. 2** Diagram of the classification of the SQL Generation

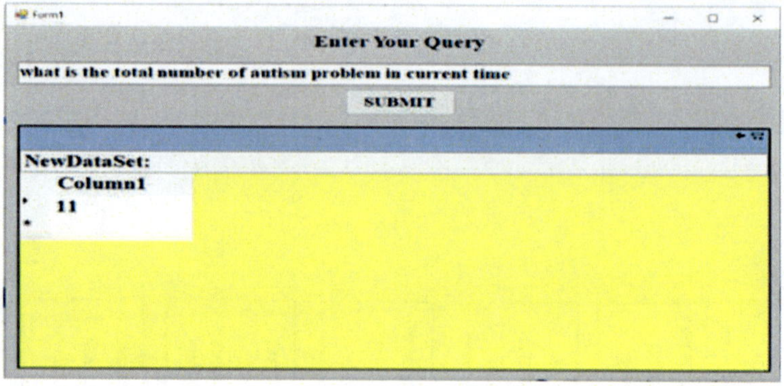


Fig. 3 Output of the Example 1 query in natural language

Table 6 Autism Table in the proposed database

| AID | Date_ To | Date_ From | Autism_Problem |
|---|---|---|---|
| 1 | 2010 | 2014 | I'm confused with my 3-year-old son. All he does is to mumble only with no proper words but he goes to speech therapy every month. Does speech therapy help such patients? |
| 2 | 2015 | 2018 | I can't show him how to do potty during the day while his dad is at work. I've realized he needs examples and sounds. |
| 3 | 2015 | 2018 | Hi! Please advise on potty training for a 4-year-old girl with ASD. |
| 4 | 2019 | 2022 | I need help for my son who is 2 years old and he has meltdowns but, today after every 5 min, he is having a meltdown. Can anyone help me with some tips on what to do at this point? |

## 4 Limitation and Future Work

The proposed system is flexible to handle time-related NL queries with aggregation and time-related queries without aggregation. But the system has some limitations to generate transitive-type SQL from NL queries. If two tables are needed for joining where both tables are not in a direct relationship but in an indirect relationship, then the tables can be joined in a transitive manner. Tense has been used to detect temporal query and its time tick from NL query. But perfect continuous tense has not been used as the proposed system is not been able to handle "for", "since", "till", "before", etc., time-specifying words. A single NL query may ask for two or more answers which mean a single NL query is containing two or more questions. The limitations can be removed after the enhancement of this proposed system which will be addressed and resolved in future work.

# 5 Conclusion

The proposed system generates SQL from NL queries for autism data extraction from the database. This system is able to handle conventional as well as temporal-type NL queries. The proposed system uses an auxiliary verb that acts as a time tick for querying data within a specific time duration. NL queries most of the time expect aggregated output. To identify the keywords that specify the aggregation, the system identifies the computational specifiers, adjectives, and nouns that are used to detect the right aggregation method at the time of SQL generation.

# References

1. Tang C, Zheng W, Zong Y, Qiu N, Lu C, Zhang X, Ke X, Guan C (2020) Automatic identification of high-risk autism spectrum disorder: a feasibility study using video and audio data under the still-face paradigm. IEEE Trans Neural Syst Rehabil Eng 28(11):2401–2410
2. Bhuvaneshwari R, Bavan PS, Harika PL, Mathubaala N (2022) Detection of autism spectrum disorder using machine learning. Int J Eng Res Technol (IJERT) 11(7):533–536
3. Fulford K, Olmsted A (2017) Mobile natural language database interface for accessing relational data. International conference on information society (i-Society), pp 86– 87
4. Kulkarni R, Kulkarni H, Balar K, Krishna P (2018) Cognitive natural language search using calibrated mesh. IEEE 17th international conference on cognitive informatics & cognitive computing, pp 174–178
5. Sadiq A, Al-Hiyali MI, Yahya N, Tang TB, Khan DM (2022) non-oscillatory connectivity approach for classification of autism spectrum disorder subtypes using resting-state fMRI. IEEE Access, 14049–14061
6. Susam BT, Riek NT, Beck K, Eldeeb S, Hudac CM, Gable PA, Conner C, Akcakaya M, White S, Mazefsky C (2022) Quantitative EEG changes in youth with ASD following brief mindfulness meditation exercise. IEEE transactions on neural systems and rehabilitation engineering, 2395–2405
7. Vakadkar K, Purkayastha D, Krishnan D (2021) Detection of autism spectrum disorder in children using machine learning techniques. SN Comput Sci 2(5):1–9
8. Farzana W, Sarker F, Chau T, Mamun KA (2021) Technological evolvement in AAC modalities to Foster communications of verbally challenged ASD children: a systematic review. IEEE Access, 1–18
9. Huang CH, Weng SJ Korla PK, Fang JF, Tsai JJP (2016) Identification and analysis of palindromes for RNA sequences. IEEE 16th international conference on bioinformatics and bioengineering (BIBE), pp 95–100
10. Shah A, Pareek J, Patel H, Panchal N (2013) NLKBIDB—Natural language and keyword based interface to database. International conference on advances in computing, communications and informatics (ICACCI), pp 1569–1576
11. Agrawal R, Chakkarwar A, Choudhary P, Jogalekar UA, Kulkarni DH (2014) DBIQS—an intelligent system for querying and mining databases using NLP. International conference on information systems and computer networks (ISCON), pp 39–44
12. Wu D, Guan Y (2019) Artificial intelligence retrieval algorithm for text data from multiple data sources. Inte J Comput Appl Taylor & Francis, pp 1–5
13. Oladosu JB, Emuoyibofarhe JO (2012) A Yoruba—english language translator for doctor—patient mobile chat application. Int J Comput Appl 34(3):149–156
14. Mukherjee P, Chattopadhyay A, Chakraborty B, Nandi D (2021) Natural language query handling using extended knowledge provider system. Int J Knowl Based Intell Eng Syst 25(1):1–19

# Emotion-Based Music Recommendation

**Akshata Kumble, Sriveda Medatati, and Ashwini Bhatt**

**Abstract** Music-related activities, whether both active and passive, activate various brain regions, boosting well-being and enhancing the quality of life. Through the utilization of rhythms and audio frequency, music therapy is a powerful tool for treating physical problems. According to research, listening to music stimulates the brain's linguistic, motor, and cognitive regions. An emotion-based music recommendation system is proposed to lift up the spirits of the individual with utmost ease. Our goal is to develop a deep learning model that uses neural network models to recommend certain songs based on the user's mood. In order to accomplish this, it is important to classify the speech based on emotion. This paper presents a comparative analysis of emotion-based speech recognition between four models.

**Keywords** Deep learning (DL) · Convolutional neural network (CNN) · Recurrent neural network (RNN) · Long short-term memory (LSTM) · K-nearest neighbor (KNN) · Mel frequency cepstral coeffecients (MFCC)

## 1 Introduction

The practice of examining audio recordings is known as audio classification or sound classification. In contrast to image processing and other classification methods, audio processing is one of the most challenging data science projects. One such application, emotion-based categorization, seeks to group audio recordings into different groups according to their emotional content. Humans are capable of expressing multiple emotions, these are mainly characterized into four types: Anger, Fear, Happy, and

A. Kumble (✉) · S. Medatati · A. Bhatt
Department of Electronics and Communication, PES University, Bangalore, India
e-mail: akshatapraveenkumble@pesu.pes.edu

S. Medatati
e-mail: srivedamedatati@pesu.pes.edu

A. Bhatt
e-mail: ashwinib@pes.edu



S. Borah et al. (eds.), *Advanced Computational and Communication Paradigms*, Lecture Notes in Networks and Systems 535,
https://doi.org/10.1007/978-981-99-4284-8_12

Sad. Music is a form of art that enhances one's mood, calms, and relaxes them. The project's goal is to develop a DL model which suggests music depending on one's feelings by recognizing emotions depending on pitch and other MFCC variables. The implementation deals with a multiclass classification problem with a dataset consisting of audio files (in .wav format) with various emotions such as anger, fear, sadness, surprise, joy, and neutral. Our data set comprises 200 audio files of each emotion. Each file is around 3 s long and pre-processed and the labels are stored in a .json file. Four models were used for audio categorization using (i) basic artificial neural network, (ii) convolutional neural network, (iii) LSTM (long short-term memory)-based recurrent neural network, and (iv) K-nearest neighbors network. Based on the emotions predicted by the models, the user is redirected to a youtube page recommending specific music.

## 2 Related Works

A variety of input formats, machine learning methods, song libraries, and Neural Network (NN) types have all been deployed with varying rates of success. A few of the ML models implemented by the authors in these papers classify music into its many musical genres and also incorporate the analysis of music categorization. The study uses a series of spectrograms generated from temporal slices of music that serve as inputs into the NN [1–3].

The system model implemented in this paper has been trained and classified using a DL technique. Here, training and classification employ CNNs. The most important step in audio analysis is the feature extraction phase. With audio samples, the MFCCs are employed as a feature vector. The suggested technique uses feature vector extraction to categorize music into different genres. Results indicate that this method has an accuracy level of around 76%, which will substantially enhance and ease the automatic classification of musical genres [2, 12].

The feature sets can be categorized as dynamic, spectral, rhythmic, and harmonious. The central moments of each feature up to the fourth order, together with five other statistical characteristics, are taken into account as indicators of the features. In the end, the MRMR algorithm regulates a significant portion of representative attributes.

Expressing from a state of fear, rage, or excitement has a higher, broader range of pitch than speaking from a place of low pitch. Audio files may be used to identify emotions since they contain a variety of characteristics. The features include Tonnetz, MFCC, Mel, and Chroma. MFCC and PITCH can also be used to classify emotions based on their associated vocal speech signals. Here, they used the SVM non-linear classifier and SVM linear classifier for classification [10]. The speech underwent undergo framing and was then passed through Hamming window, after which Fast Fourier Transform (FFT) was perform and the MFCC features were extracted. The mean value of MFCC and standard deviation was found, then thresholding was applied to distinguish between the emotions.

Various emotion recognition approaches that employ linguistic elements are based on a speech emotion recognition study that combines both acoustic and linguistic information. The acoustic characteristics of emotional speech change substantially depending on the kind and strength of the emotion. This research introduces a new emotional speech recognition approach that combines acoustic model and language model adaption to achieve good recognition performance on an emotional speech problem. The system's word recognition accuracy was 82.2%, and recognition mistakes were found. Regardless, the authors show that the combination of lexical and auditory elements is effective for emotion recognition [8, 10, 13, 15].

## 3 Methodology

The first step involves pre-processing the raw audio files into suitable waveforms. Then we transform the waveforms into Mel spectrograms and extract the features. In order to extract features, language input must be recognized and noise must be avoided. Three categories—high-, mid-, and low-level characteristics are used to categorize audio information. (1) High-level elements like chords, rhythms, and melodies are connected to musical lyrics. (2) Pitch-like fluctuation patterns, MFCCs, and beat-level properties are examples of mid-level features. (3) Statistical measurements that are collected from sounds while extracting features include energy and zero-crossing rates (Fig. 1).

So, to create these features, we employ a series of computations that are combined under a single name MFCC, which aids in the extraction of mid-level and low-level audio characteristics. MFCC features are obtained by dividing the audio files into frames and then identifying and extracting different frequencies from each frame. These extracted features are stored in a .json file which is later used for mapping during the training process (Fig. 2).

We have created four models to perform a comparative analysis. These extracted features are fed into each of the models where the performance is measured. The basic steps involved in creating all the models are (a) load data, (b) training and

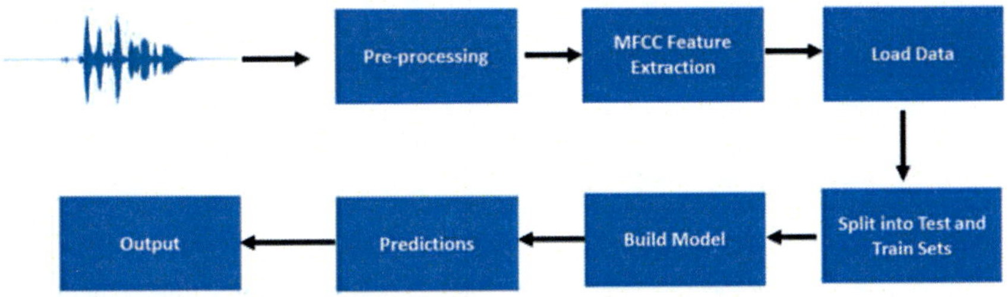


**Fig. 1** Steps involved in the emotion-based speech recognition process

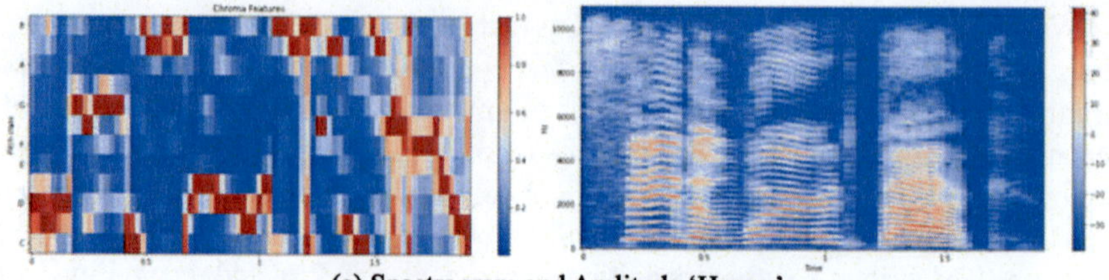

(a) Spectrogram and Amlitude 'Happy'

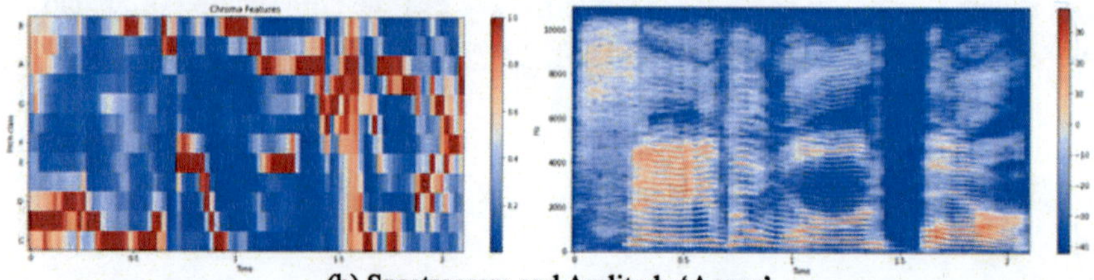

(b) Spectrogram and Amlitude 'Angry'

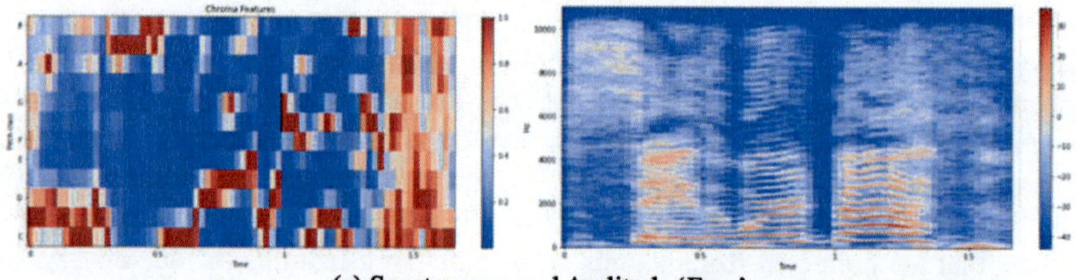

(c) Spectrogram and Amlitude 'Fear'

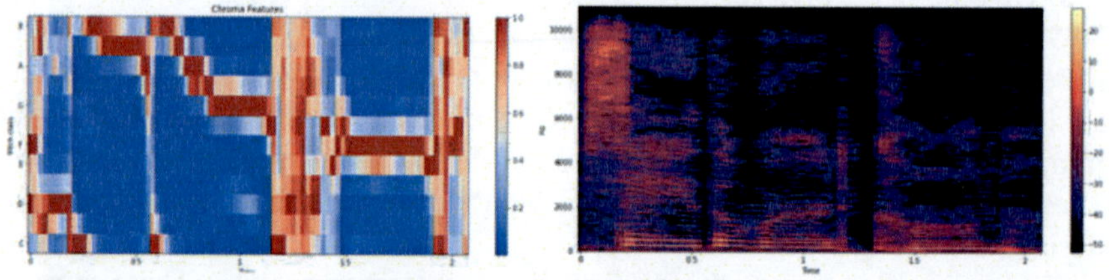

(d) Spectrogram and Amlitude 'Sad'

**Fig. 2** Spectrograms and amplitude plots for sample emotion

testing sets split, (c) build network architecture, (d) train network, and (e) test cases. We look into the methodology and architecture of each of these methods one by one.

## *3.1 Convolutional Neural Networks (CCN)*

CNNs are ideal for audio classification as they can learn translation-invariant patterns with spatial hierarchies. A CCN is often comprised of several convolution and sub-sampling layers, followed by one or more fully connected layers at the end to aid

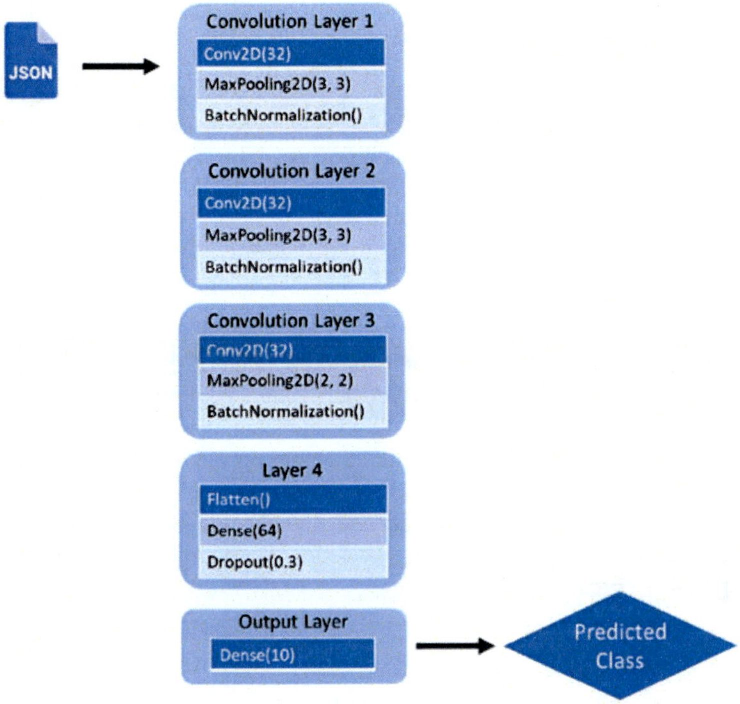


**Fig. 3** Layers present in the CNN model

with output prediction. Initially, a database of audio recordings depicting distinct emotions based on pitch was created, and feature extraction was conducted using the MFCC method. This sequential model is composed of five layers: an input layer, three hidden levels using the ReLu activation function, and an output layer utilizing the Softmax layer. The ReLu activation function is applied to train the network faster, while the Softmax layer is employed to normalize the output and assist prediction (Fig. 3).

Because CNNs outperform multilayer perceptrons (MLPs) despite having fewer parameters, we consider CNNs by perceiving audio files in an image format, i.e., a matrix of pixels, from which we extract features as horizontal and vertical bars, sense all the different components, and extract features using methods such as convolution and pooling. In the convolution procedure, a kernel (filter) associated with learning parameters (weights, bias) is applied to the input image, and an output image is created based on the shape and values of the applied filter in order to preserve certain information from a picture and discard the rest.

Considering that **x** is the original input with **n** elements, the convolution is performed as

$$\mathbf{y} = \mathbf{x} * \mathbf{w} \to \mathbf{y[i]} = \sum_{k=-\infty}^{\infty} \mathbf{x[i-k]w[k]}, \tag{1}$$

While considering a padded input vector $x^p$ with a vector size of $\mathbf{n} + \mathbf{2p}$ and a filter of m elements, the above equation is modified as

$$\mathbf{y} = \mathbf{x} * \mathbf{w} \to \mathbf{y[i]} = \sum_{k=0}^{m-1} \mathbf{x^p[i+m-k]w[k]}, \tag{2}$$

Clusters of pixels are examined in the pooling procedure in order to execute an aggregation over them. Choosing the maximum value of the pixels in the cluster is one of the potential aggregations (known as Max Pooling). A 2D max pooling layer downsamples the input by splitting it into rectangular pooling zones and then calculates the maximum of each region. Computing the average is another standard aggregate (known as average pooling). This approach decreases the amount of information in the audio while retaining and amplifying the valuable characteristics revealed by the filters while convolving the audio. We acquired a precision of 93.53% by using this technique.

## 3.2 LSTM-Based Neural Networks

Recurrent neural networks identify the sequential properties of input and utilize patterns to forecast the next likely situation. The input is a sequence vector, and the output is a single vector. As a result, an LSTM model is implemented since it contains feedback connections and is capable of analyzing not only single data points but also whole sequences of data. LSTMs assign data weights to help RNNs decide whether to accept new information, forget it, or give it enough weight to affect the output. The test accuracy achieved is nearly 78% (Fig. 4).

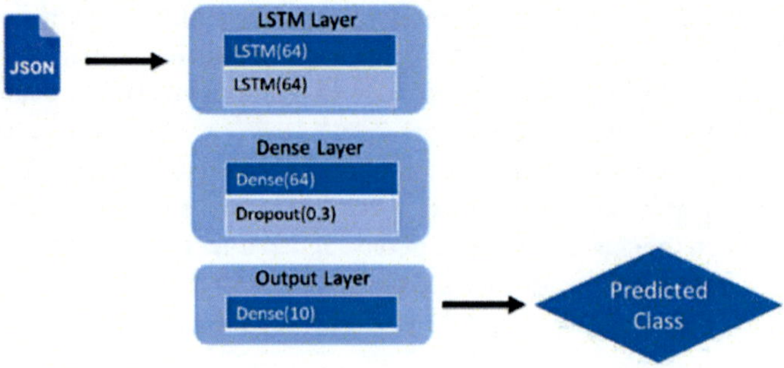


**Fig. 4** Layers present in the RNN-LSTM model

### 3.3 K-Nearest Neighbors Network (K-NN)

Popularly known as the "lazy learner algorithm," KNN is a machine learning technique that is used for classification and regression. It simply applies a distance-based algorithm to determine the K number of identical neighbors to new data and the class to which the majority of neighbors belong, resulting in that class being returned as an output. First, we created a function that would receive training data, current instances, and the number of neighbors required. We calculated the distance between each point in the training data and every other point, then located the K nearest neighbors and returned all neighbors. We developed a dictionary that holds the class and the number of neighbors it has. After producing the frequency map, we sort it in decreasing order depending on the number of neighbors and return the first class. We then extracted the MFCC features and trained the model, using the NumPy library to determine the distance between two locations. The data is then fed into the KNN algorithm, which makes predictions on the testing dataset, achieving a remarkable accuracy of 99.25%.

### 3.4 Multilayer Perceptron Network (MLP)

Speech recognition with a multilayer perceptron for emotion-based classification is used in this technique. A multilayer perceptron is created by connecting many single-layer perceptrons and analyzing their interactions parallelly. A perceptron consists of four basic components: the input value or input layer, the weight, the net summation, and the activation function. This sequential model allows us to develop models in a multilayer perceptron layer by layer; however, it only works for stacks of layers with a single input and output (Fig. 5).

The "Flatten" function flattens the input without modifying the batch size. The activation stage employs the ReLu and Softmax activation functions. The first two thick layers are concealed and serve as the foundation for a fully linked model. The final thick layer, the output layer, has 10 neurons that determine which class the audio sample belongs to.

## 4 Results

Three deep learning methods and one machine learning model have been implemented for emotion-based speech classification. A comparative analysis was performed where the results as depicted in the table were achieved. While keeping the batch size constant and varying the other training parameters, the observations show that the multilayer perceptron network and convolutional neural network obtained comparable results and gave better accuracy over the recurrent neural network. The

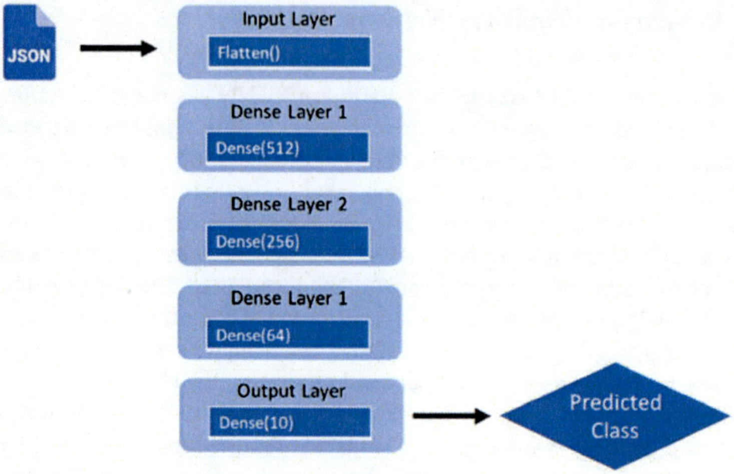


**Fig. 5** Layers present in the multilayer perceptron model

**Table 1** Comparative analysis of the deep learning models using the parameters used for the simulation study

| | CNN | RNN-LSTM | MLP |
|---|---|---|---|
| Accuracy | 0.948 | 0.7800 | 0.9353 |
| Loss | 0.1496 | 0.5068 | 0.1395 |
| Learning Rate | 0.0001 | 0.0001 | 0.0001 |
| Batch Size | 32 | 32 | 32 |
| Epochs | 30 | 30 | 50 |
| Layers | 5 | 3 | 5 |

models were also tested on real-time audio files and it was observed that CNN works well with real-time speech. Based on the emotion detected specific songs are recommended to lift the user's spirits by redirecting the user to a youtube page containing suitable music (Table 1).

## 5 Conclusion

Speech contains rich emotional factors and music therapy has been proven effective over the years. Emotion recognition based on speech is a fascinating subject to explore in the future and can be used to improve one's well-being. We have successfully performed a comparative study using four models and propose that CNN produces efficient results with an accuracy of almost 95%. This framework along with the others can further be extended to provide multilingual emotion and its efficiency can be tested in noisy environments.

## References


1. Pelchat N, Gelowitz CM (2020) Neural network music genre classification. Can J Electr Comput Eng 43(3):170–173
2. Vishnupriya S, Meenakshi K (2018) Automatic music genre classification using convolution neural network. In: International conference on computer communication and informatics (ICCCI), pp 1–4
3. Deshmukh G, Gaonkar A, Golwalkar G, Kulkarni S (2019) Speech based emotion recognition using machine learning. In: 2019 3rd international conference on computing methodologies and communication (ICCMC), pp 812–817
4. Puppala LK, Muvva SSR, Chinige SR, Rajendran P (2021) A novel music genre classification using convolutional neural network. In: 2021 6th international conference on communication and electronics systems (ICCES), pp 1246–1249
5. Elbir A, Çam HB, Emre Iyican M, Öztürk B, Aydin N (2018) Music genre classification and recommendation by using machine learning techniques. In: 2018 innovations in intelligent systems and applications conference (ASYU), pp 1–5
6. Siadat SR, Voronkov IM, Kharlamov AA (2022) Emotion recognition from Persian speech with 1D convolution neural network. In: Fourth international conference neurotechnologies and neurointerfaces (CNN). Kaliningrad, Russian Federation, pp 152–157
7. Yazdani A, Simchi H, Shekofteh Y (2021) Emotion recognition in Persian speech using deep neural networks. In: 2021 11th international conference on computer engineering and knowledge (ICCKE), Mashhad, Iran, Islamic Republic of, pp 374–378
8. Yang Q, Xu F, Ling Z, Li X, Li Y, Fang D (2021) Selecting and analyzing speech features for the screening of mild cognitive impairment. In: 43rd annual international conference of the IEEE engineering in medicine and biology society (EMBC), Mexico, pp 1906–1910
9. Krishna KV, Sainath N, Posonia AM (2022) Speech emotion recognition using machine learning. In: 2022 6th international conference on computing methodologies and communication (ICCMC), Erode, India, pp 1014–1018. https://doi.org/10.1109/ICCMC53470.2022.9753976
10. Sakurai M, Kosaka T (2021) Emotion recognition combining acoustic and linguistic features based on speech recognition results. In: 2021 IEEE 10th global conference on consumer electronics (GCCE), Kyoto, Japan, pp 824–827
11. Jie C (2021) Speech emotion recognition based on convolutional neural network. In: 2021 international conference on networking, communications and information technology (NetCIT), Manchester, United Kingdom, pp 106–109
12. Cai L, Dong J, Wei M (2020) Multi-modal emotion recognition from speech and facial expression based on deep learning. In: Chinese automation congress (CAC), Shanghai, China, pp 5726–5729
13. Ainurrochman, Febriansyah II, Yuhana UL (2021) SER: speech emotion recognition application based on extreme learning machine. In: 2021 13th international conference on information and communication technology and system (ICTS), Surabaya, Indonesia, pp 179–183
14. Wang J, Han Z (2019) Research on speech emotion recognition technology based on deep and shallow neural network. In: Chinese control conference (CCC), Guangzhou, China, pp 3555–3558
15. Assunção G, Menezes P, Perdigão F (2019) Importance of speaker specific speech features for emotion recognition. In: 5th experiment international conference (exp.at'19), Funchal, Portugal, pp 266–267. https://doi.org/10.1109/EXPAT.2019.8876534
16. Wang J, Han Z (2019) Research on speech emotion recognition technology based on deep and shallow neural network. In: Chinese control conference (CCC), Guangzhou, China, pp 3555–3558

# A Comprehensive Study on Natural Language Processing, It's Techniques and Advancements in Nepali Language

**Sital Sharma, Kalpana Sharma, and Biswaraj Sen**

**Abstract** Natural Language Processing is the ability of a system to understand, interpret, and analyze spoken words, text files, etc. There have been plethora of models that have been developed for language processing which consist of rule-based approach, Neural Network approach, and Traditional Machine Learning. NLP has a wide range of applications, including speech recognition, machine translation, sentiment analysis, chatbots, and intelligent personal assistants. In recent years, NLP has made significant progress, thanks to the development of deep learning models, which have greatly improved the performance of NLP systems. To make the machine understand the speech, the machine should be able to understand the sentences, which can only be achieved when the sentences are well-structured, which is ensured by Grammar. In this paper, a survey is done on various natural language processing works that have been done for the Nepali language, different resources, and work available.

**Keywords** Natural language processing · Machine learning · Artificial intelligence · Deep learning · Grammar

## 1 Introduction

The rise of Machine Learning, Artificial Intelligence, and Deep Learning has largely transformed natural language processing and the ability of the machine to understand and better respond to user input commands. Different approaches are used to solve

S. Sharma (✉) · K. Sharma · B. Sen
Sikkim Manipal Institute of Technology, Sikkim Manipal University, Sikkim, India
e-mail: sitalneo@gmail.com

K. Sharma
e-mail: kalpana.s@smit.smu.edu.in

B. Sen
e-mail: biswaraj.sen@gmail.com



S. Borah et al. (eds.), *Advanced Computational and Communication Paradigms*, Lecture Notes in Networks and Systems 535,
https://doi.org/10.1007/978-981-99-4284-8_13

any Natural Language Processing task; the commonly used approaches to solve the task include [1] Rule-based Approach, Neural Network-based Approach, and Traditional Machine Learning.

To make a computer understand a language, grammar plays a pivotal role which can affect both the response as well as information retrieval accuracy of the system. Grammar helps to determine and define rules that are used to form a well-structured sentence. NLP uses mainly three types of grammar which are as follows [2]:

a. Context-Free Grammar, b. Dependency Grammar, and c. Constituency Grammar.

The approaches and grammar play a vital role in NLP and its understanding. In the context of the Nepali language or any language, different phases are used to get an insight which includes text planning and sentence planning. The tree/hierarchy of NLP associated with any language includes the following [3]:

a. Lexical Analysis
b. Syntactic Analysis
c. Semantic Analysis
d. Disclosure Integration
e. Pragmatic Analysis.

Different approaches have been proposed with respect to each of the phases of NLP in the Nepali language, which addresses various domains like POS tagging, Stemming, Dependency Parsing, etc. The components that make natural language processing (in a broad sense) includes

1. Natural Language Understanding: The NLU is used to make the system understand the human language and analyze the language which produces non-linguistic output(s) from a set of natural language inputs.
2. Natural Language Generation: This process includes the generation of language which produces natural language output(s) from a non-linguistic input.

## 2 Approaches in Natural Language Processing

### *2.1 Rule-Based*

It is one of the most used and oldest approaches to NLP. The rule-based approach focuses mainly on pattern matching or parsing; it has low precision and high recall due to which the performance degrades when generalized [4]. Rule-based approach for the Nepali language has been discussed by Tej Bahadur Shahi et al. [5] which is used to predict the parts of speech of words in the Nepali language. The model proposes to go through four stages which are

i. Input user sentence
ii. Tag set

iii. POS tagger module.

The dataset that has been used was The Nepali National Corpus from Nepali Language Resources and Localization for Education and Communication which comprises fourteen million Nepali words [6, 7] which consist of 82% of unambiguous words, 13% unknown words, and 5% unambiguous word. The use of various algorithms is discussed which are SVM and TnT. The system follows the following steps:

- Input of untagged sentence to the SVM/TnT Tagger, the Tagger consists of the following sub-steps which are
  - Tagged corpus
  - Feature extraction and selection
  - SVM/TnT Learning
  - SVM Tagger.

The training data was passed into column format, and the sentence separator used was blank space. The classification was done using One versus rest classification where for every class one classifier was built to distinguish a class. The accuracy obtained for POS tagger for the Nepali language is classified which is classified as known word accuracy, unknown words, and overall accuracy. The overall accuracy for the TnT tagger was 74% with known and unknown words combined and for SVM the overall accuracy was 93.27% [8]. Though the SVM is yielding good results, the training was slow due to the fact that SVM uses different kinds of features so as to construct the feature vectors, which affects the performance of the tagger [9].

## *2.2 Neural Network-Based Approach*

In the Neural Network-based approach, the feature engineering is skipped as the network can learn the important feature; in the neural network, vector representations of the words are fed without engineered features, and the training corpus is generally large. As discussed by Kriti Nemkul and Subarna Shakya [9], English to Nepali sentence translation was done which uses RNN that comprises LSTM with attention. The dataset used here was obtained from the ELRA catalog [10] and the Opus dataset [11] which consist of an English–Nepali parallel corpus. The RNN architecture used was as follows:

- Long Short-Term Memory (LSTM)

The RNN Encoder–Decoder was used using LSTM which consists of bidirectional RNN with forward RNN and backward RNN; the source was scanned using forward RNN and the forward hidden state sequence is calculated. The backward RNN takes inputs as the series in reverse order which give the series of backward hidden states. When the forward and backward states were concatenated, annotations are obtained. For decoding the sentence that has been encoded, the hidden state of the decoder which when provided with encoder annotation the hidden state, proposal hidden gate, output gate, memory, and forget gate is calculated [12]. The score for an individual segment that has been translated is calculated using BLEU [13] score; the score varies from 0 to 1 where 0 indicated that the sentence translated is not accurate while the score 1 indicates that the score is very high and the translation is accurate; generally, the score fluctuates between 0 and 1. The score is converted into the range of 0 to 100 where each sentence is matched with its highest quality reference texts, where a perfect match gives the score of 100, and 0 indicates that there is no match. The model proposed an accuracy (BLEU) of 8.9 which has 256 hidden layers and two layers of neural networks. The main challenge is to handle the unknown words that are not present in the corpora and also the rear words that are often left behind in the corpora which greatly affects the accuracy of the model.

## 2.3 *Traditional Machine Learning*

In the traditional machine learning approach, various classifiers and modeling techniques are used which may consist of probabilistic modeling, linear classifier, and likelihood maximization. The main characterization of such a type of system is that it includes feature engineering which may consist of surrounding words, word type, etc. The models are trained on some parameters which is followed by fitting on test data. In the context of the Nepali language, the model was trained on a large corpus named NepBERTa (Milan Gautam et al. 2022) which is a BERT-based NLU that has been trained with the dataset of 800 million words [14] collected from different news websites. NepBERTa can be used to perform various tasks like POS Tagging and Categorical Pair Similarity. The accuracy greatly improves when NepBERTa is used with the accuracy obtained shown in Table 1.

Models have been proposed using BERT, and on comparing the performance of the existing/pre-proposed model, the said model NepBERTa performs well. The model comparison is as follows in Table 2.

The evaluation metric which was used was Macro-F1. The proposed model has the highest Nep-GLUE score which outperforms many multilingual models like Mbert, etc. The said model can be used to perform various tasks which can yield accurate results.

**Table 1** NepBERTa performance

| No. of parameters | Named entity recognition | Part of speech | Categorical pair similarity | Content classification | Nep-gLUE score |
|---|---|---|---|---|---|
| 110 M | 91.09 | 95.56 | 94.42 | 93.13 | 93.95 |

**Table 2** Performance comparison with different NepBERTa-based multilingual model

| Model | No. of parameters | Named entity recognition | Parts of speech | Categorical pair similarity | Content classification | Nep-gLUE score |
|---|---|---|---|---|---|---|
| NepBERTa | 110 M | 91.09 | 95.56 | 94.42 | 93.13 | 93.95 |
| XLM-Rbase | 270 | 87.59 | 94.88 | 93.65 | 92.33 | 92.11 |
| NepaliBERT | 110 M | 82.45 | 91.67 | 89.46 | 90.10 | 88.42 |

# 3 Grammars in Natural Language Processing

The rule which is used to form a well-structured sentence is known as grammar. For processing any natural language, the syntax of the language plays a vital role which gives an insight into how the sentence is formed and what rules need to be followed in order to have a meaningful sentence. A grammar G can be mathematically represented using 4 tuples which are N, T, S, and P [2], where each letter of the grammar denotes the following:

$V_n$ or $N =$ Indicates all the set of variables or non-terminal symbols.

$\Sigma$ or $T =$ Indicates the set of terminal symbols.

$S =$ Indicates the start symbol where $S \in N$.

$P =$ Indicates the production rules for terminals as well as non-terminal.

Here, it will have a form where $\alpha \to \beta$, which indicates strings $(\alpha, \beta)$ on $V_n \cup \Sigma$ where at least one symbol of the $\alpha$ belongs to $V_{n.}$

In the context of NLP, grammar can be broadly classified into three categories which are as [15, 16]follows.

## 3.1 Context-Free Grammar

Context-free grammar consists of a finite set of grammar rules which have the following components:

- Set of Non-Terminals: The set of non-Terminals is represented by V which are syntactic variables which denote the set of strings which in turn helps in defining the language which is being generated with the help of grammar [17, 18].
- Set of Terminals: It is represented with the help of the symbol $\Sigma$ which is also known as tokens; here the strings are formed with the help of terminal basic symbols.
- Set of Production: It is represented by symbol P; the set of production gives an idea regarding how the terminal and non-terminal can be combined. The set of productions consists of the following components:
  - Non-Terminal
  - Arrows

  - Terminals.

The non-terminals are on the left side of production and the right side of the production is called the terminals.

- Start Symbol: The start symbol is a point from where the production begins, represented by symbol S; the non-terminal symbols are treated as start-symbol [19].

### *3.2 Constituency Grammar*

Also known as Phrase structure grammar, it is based on the constituency relation and it is the opposite of the dependency grammar [18]. In this type of grammar, constituents can be any words, phrases, or any group of words; the main task of the grammar is to organize any sentence into its constituents; this is achieved using POS tagging, verb phrase identification, etc. The grammar organizes any sentence into three constituents which are a subject, context, and object.

Sentence: < subject > < context > < object >

### *3.3 Dependency Grammar (DG)*

Based on the dependency relation, it is the opposite of the constituency grammar; the dependency grammar lacks phrasal nodes [19]; here, the words are connected to each other via a directed link and the verb is considered as the center of the clause structure. It is used to analyze the relationship between words in a sentence; here, the words in the sentences are treated as nodes in a graph, and the dependencies between the words are represented as edges between the nodes. Dependency grammar makes use of graph theory to represent the relationship between words in a sentence, where the nodes represent the words, and the edges between the nodes represent the dependencies between the words [20]. By analyzing the dependencies between the words in a sentence, it is possible to determine the meaning and structure of the sentence [22].

## 4 Grammar in Nepali Language

Nepali follows many of the typical patterns of Indo-Aryan languages [21], which have a rich system of noun declensions, with different endings for nouns depending on their role in the sentence. There are 36 consonants, where 36 are unique consonants and 3 are a combination of the other unique consonants and vowels [22]. The first Nepali lexicon was introduced by Bista et al. [23] which has various file formats

**Table 3** Nepali language processing techniques [8]

| Morphological analysis | Syntax analysis | Semantic analysis |
|---|---|---|
| Stemming | Parsing | Named entity recognition |
| Part-of-speech tagging | Grammar checking | Word sense disambiguation |
| Word segmentation | – | Anaphora resolution |

which consist of root words, parts of speech, synonyms, etc. for each of the words. The first spell checker, machine translation system was proposed/introduced by Madan Puraskar [24]. The Nepali National Corpus [25] was also developed for further developments in Natural Language Processing for the Nepali language. Nepali language processing can be represented by three major parts which consist of [8]

1. Morphological Analysis
2. Syntax Analysis
3. Semantic Analysis.

Here, the morphological analysis consists of three sub-parts, syntax analysis consists of two sub-parts, and semantic analysis consists of three sub-parts which are listed in Table 3.

## 5 Natural Language Pipeline

A natural language pipeline is a series of processes that a natural language processing system uses to analyze and understand human language. These processes typically include tasks such as tokenization, part-of-speech tagging, and syntactic parsing. The goal of a natural language pipeline is to convert raw text into a structured, machine-readable format that can be used by other NLP algorithms or applications. There are various steps shown in Fig. 1 that comes while building the Natural Language Processing Pipeline, which is as follows.

### *5.1 Sentence Segmentation*

Sentence Segmentation is also known as sentence boundary detection; it is the process of dividing a string of written language into its individual sentence; this step allows the individual sentences to be analyzed and processed separately. There are several different approaches to sentence segmentation, but one common method is to use punctuation and other syntactic cues to identify the boundaries between sentences. For example, a period (.), exclamation point (!), or question mark (?) typically signals

**Fig. 1** NLP pipeline

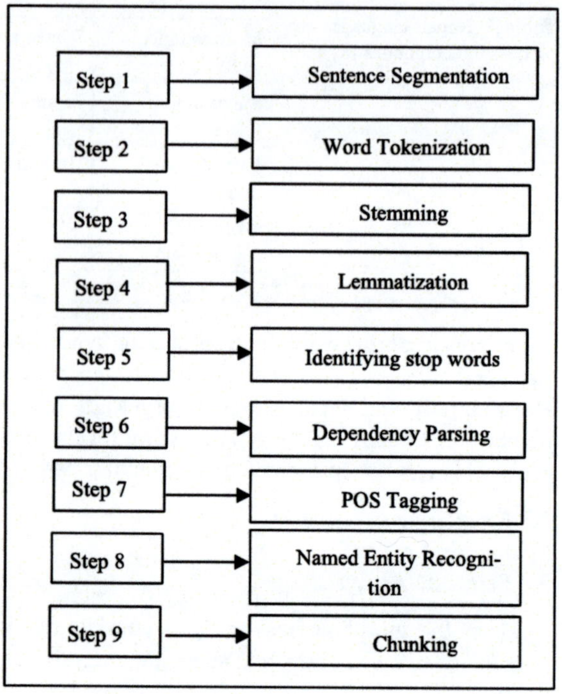


the end of a sentence, while a comma (,) or semicolon (;) may indicate a pause or break within a sentence. In Nepali language, sentence segmentation can be identified differently as compared to the English language; here, the symbol "।" is used to denote the sentence end which is called "PURNA BIRAM" but another symbol "॥" is also used which is denoted by symbol which is used in some special cases mostly in poems' second line which is written on "Chanda" [26], so it becomes a tedious task to segment the Nepali sentence due to the presence of high level of punctuations. Sentence segmentation is a challenging task because written language is often ambiguous and can contain multiple punctuation marks, abbreviations, and other features that make it difficult to automatically identify sentence boundaries. As a result, sentence segmentation algorithms often rely on complex rules and heuristics to accurately identify sentence boundaries in a given piece of text.

## 5.2 *Word Tokenization*

Word tokenization is the process of dividing a string into individual words. The common method that is used for word tokenization is to use whitespace and punctuation to identify word boundaries [27].

## 5.3 Stemming

Stemming is the process of reducing a word to its base form or stem. This is often used in natural language processing to help identify words that have the same basic meaning, even if they have different inflections or spellings. For example, the words "run," "running," and "ran" all have the same basic meaning of moving quickly on foot. However, they are conjugated differently to indicate past, present, and future tenses. A stemming algorithm would reduce these words to their base form of "run," which allows them to be treated as the same word for the purposes of analysis and processing. There are several different algorithms for stemming words, and the specific approach used can vary depending on the language and the desired level of precision. Some common stemming algorithms include the Porter stemmer and the Snowball stemmer [28, 29]. These algorithms typically use a combination of rules and heuristics to identify and remove common word endings such as -ing, -ed, and -s, in order to reduce words to their base form [30]. A new Nepali stemmer has been proposed with a larger set of suffixes which consist of 128 suffixes and is based on the rule-based algorithm for stemming with nearly about 500 words [31]. The stemmer so proposed considered only the suffix of the sentence for stemming which in turn leads to the over-stemming problem. To overcome the shortcomings, i.e., the over-stemming problem, a computational method was proposed for stemming by Sitaula [32], which uses context-free grammar and similarity matching techniques. A finite state machine-based approach was proposed by Borah et al. [33] which was used to perform a morphological analysis of Nepali language; the method was stated to work with minimal resources. Here, the Nepali text was firstly converted into Roman and used a pattern-matching approach which is used to find the location of the noun in a sentence. If there are nouns present in the sentence, then a grammatical rule was used to tag the word which can be before or after the sentence with an adjective. To deal with an inflectional form of a noun, Chhettri et al. [34] proposed a model which was based on the finite state machine method; in the said method, the noun of the sentence is split into its morphemes with the help of a suffix dictionary.

## 5.4 Lemmatization

Lemmatization is a process in natural language processing (NLP) that aims to reduce words to their base forms, called lemmas. This is similar to stemming; while stemming often results in words that are not actual words, lemmatization produces valid words. There are several algorithms for lemmatization, and which one to use can depend on the specific language and context. Some commonly used lemmatization algorithms include the WordNet lemmatizer and the Morpha lemmatizer [35, 36]. Lemmatization can be an important step in the process of natural language processing, as it allows words to be standardized and more easily compared and processed by computer algorithms. Much research has been done in the field of lemmatization as discussed by Ansari Fatima et al. [37] where works have been done for the English language but as contrary to English, there is a lack of Lemmatization techniques for the Nepali language due to which the model that is built for the Nepali language fails to achieve the level of accuracy that is expected and have significantly lower accuracy as compared to its counterparts popular well research language [38, 39].

## 5.5 Identifying Stop Words

Stop words are common words in a language that are often used to connect phrases and sentences, but don't have much meaning on their own. These words are typically considered "noise" when processing natural language and are often removed from the text before further analysis. Examples of stop words include "the," "a," "and," "of," and "to." In many cases, these words can be ignored without affecting the meaning of a sentence. However, this can vary depending on the context, and some stop words may be important in certain situations [40]. There is a pre-defined set of stop words for the research-rich and high resource language like English, as discussed by Tej Bahadur Shahi et al. [8]; most of the works that have been done in the field of Nepali natural language processing have used their own set of stop words which have caused a non-availability of the standardized set of stop words. Due to the non-availability of standard stop words for the Nepali language, it would be a difficult task to process and analyze text. The common and uninformative words are filtered out which can be termed as stop words and to focus only on the important words. The non-presence of standardized stop words makes the algorithm work hard in determining which words in a sentence are important and which words are not, which can result in lesser accurate results. Another major issue of not having standardized stop words is that it makes a difficult task to compare the result across various natural language processing systems, as each system is having its own way of determining which words to be considered as stop words and which words to discard; it may result in the ambiguous result as well as the accuracy of the model is also compromised. Some (not all) of the commonly used stop words are given in Table 4.

## 5.6 Dependency Parsing

Dependency parsing can be classified as syntactic analysis which is used to identify the dependencies between two or more individual words in a sentence; it is basically used to represent the grammatical structure of a sentence which makes it clear which words are dependent on which other words. The relationship or the dependency between the two words is calculated by creating a directed graph, where the nodes represent the word in the sentence and the dependencies between them are represented by the edges [41]. It is basically used in NLP for the tasks such as identifying the verb, subject of the sentence, or can be used to generate a more simplified version

**Table 4** List of Nepali stop words

| | | | | | | | | | |
|---|---|---|---|---|---|---|---|---|---|
| अझै | अरूलाई | उनले | के | गर्नुपर्छ | चाँडे | छैन | जस्तोसुकै | दिए | दुई |
| होइनन् | हामीलाई | सात | स्पष्ट | सबैलाई | हरेक | वरिपरि | शायद | रहेको | राख्छ |
| सट्टा | लगभग | सबैलाई | लगभग | | सक्छ | पर्याप्त | तापनि | देखिन्छ | क्रमशः |
| बन्न | यसबाहेक | दोस्रो | यदि | विरुद्ध | जे | सायद | जबकि | हुनेछ | |
| तत्काल | दिनुभएको | रही | हरेक | यद्यपि | जुन | फेरि | र | प्लस | तिनीहरू |
| निम्नानुसार | होस् | हुन | जताततै | पछिल्लो | गरेर | थियो | दुई | देख्न | ठीक |
| जसबाट | पहिलो | केही | कुनै | चाँडै | | तल | चाहिए | नजिकै | तेस्रो |
| पाँच | म | उहाँलाई | गरेका | अक्सर | भित्री | त्यसकारण | यहाँसम्म | त्यहाँ | छौं |

of the sentence. A Nepali parser has been introduced by Archit Yajnik et al. [42] which uses a simplex method from which the parsing evolves where there are two steps to perform a parser for a given input. First, the given input is tested using the Penalty approach which is also known as the Big M method, and the output from the first step is then fed into the next step which consists of the two-phase simplex method. Here, the number of annotated sentences used was 10 (ten), and the two-phase simplex method yields better results; it is also discussed that whenever a sentence is parsed while only using the Big M method, only six out of the ten sentences were parsed correctly, whereas while using the combined method all the sentences were parsed successfully. Another dependency parser is also developed which was based on annotation projection, where a dependency treebank (word level) and a full parser for the Nepali language were proposed by Pooja Rai et al. [43]. It uses the method that was used to build a tree bank for the Bengali language, i.e., annotation projection to build the tree bank for the Nepali language. The Nepali parser baseline was built using the developed tree bank which uses Maltparser along with the Stack Projective algorithm. The parser so built was based on the Neural Network model which yielded good results in parsing the Nepali language. There are many challenges related to dependency parsing; some of them are ambiguity, idiomatic expression and collocations, noise and error in the input, etc.

## *5.7 POS Tagging*

Part-of-Speech Tagging is the process of tagging individual words in a sentence to its corresponding part of speech where the tags may include "Nouns," "Verbs," "Pronouns," "adjectives," "adverbs," etc. POS tagging is used to identify the structure of a sentence which can be used to improve the NLP task such as sentiment analysis and named entity recognition [44]. There are basically three approaches to performing POS tagging: Machine learning/deep learning-based approach, rule-based approach, and statistical based approach [44]. In the rule-based approach, the rules are pre-defined which is used to classify each word; the rules are used to tag and classify the words with their respective tags like Nouns, Verb, Pronoun, etc. Due to the pre-defined rule where the rules are prewritten and while writing the rule, there are chances of human error and to frame the rule, a language/linguistic expert(s) is needed which make it more time-consuming and also labor-intensive [45]. The rules can be based on the word structure, spelling, pronunciation, or the context of the word with respect to the sentence. The rule-based tagging method may be effective for a small number of datasets but for a larger dataset, it becomes unwieldy and difficult to maintain as when the dataset increases the complexity also increases. In the statistical based approach, a training dataset is used which manually annotates with the correct POS tag. Here, machine learning algorithms are used to train the model in such a way that it learns the pattern and rules that govern the assignment of POS [46]. The advantage of using a statistical based approach is that it can handle a large number of datasets. One of the problems that may occur while using the method is that it is strictly language-based, so it may work for one language which is written on the same script but can generate inaccurate results/tags for another language that is written on the same script, for example, a POS tagger based on a statistical approach which can tag the Hindi language will not work for the Nepali language, though the two languages are written on the same script "Devanagari." The machine learning-based approach can be a supervised and unsupervised learning method, which is used to classify the words; here, various algorithms are used to make the model learn the various rules for assigning POS to various words in any given sentence. The machine learning-based approach is similar to statistical based approach where the former uses various machine learning algorithms to learn the rules for Part-Of-Speech tagging, whereas the latter relies on a pre-defined set of rules [47]. In the context of the Nepali language, the first tagset was introduced in 2008 [8, 48], which was having 112 tags and was used to tag the Nepali National Corpus (NCC). Since the tagset was based on the rule-based approach and was semi-strictly typed which later leads to the increase in the error, to limit the error while handling the large corpus the tags were later reduced to 43. The first rule-based tagger which is known as Unitag was first introduced in 2004 [49]. Since the tagger was rule-based and Nepali being one of the rich languages with a huge number of words and annotations, whenever a new word was introduced, the tagger was not able to tag the new word properly as the word was not present in the corpus which led to defining new rules every time when a word is encountered which is beyond the pre-defined rule. A statistical based tagger was introduced in 2008 [8], which was based on the theory of probability and was based and trained on a large set of corpus which consist of 82 thousand words and was having 42 POS tags. The tagger performed well for known words with an accuracy of 97% and for the unknown words, the accuracy of the model was around 56%. The improvised POS tagger was introduced in 2013 [5] which was based on SVM and was trained on the NCC dataset. The tagger performed well compared to the previously introduced tagger. A Hidden Markov Model and Viterbi algorithm-based tagger was introduced in 2017. The tag set which was used to train the model was self-developed [50]. The accuracy of the model was 95.43%. A similar kind of tagger was also introduced for

**Table 5** Different POS taggers introduced for Nepali language

| Year | Methodology | Accuracy |
|---|---|---|
| 2004 [49] | Rule-Based | N/A |
| 2008 [48] | Manual and Automated | N/A (112 tags, later reduced to 43) |
| 2008 [8] | TnT-statistical tagger | 97% (Known words) and 56% (Unknown), 82,000 words, 42 POS tags |
| 2013[5] | SVM-Based tagger | N/A |
| 2017 [50] | Hidden Markov Model | 95.43% |
| 2018 [51] | Hidden Markov Model, General Regression ANN, Feed Forward ANN, Radial Bias ANN | N/A |
| 2018 [52] | RNN: Vanilla RNN, GRU, LSTM | 99% |

the Nepali language in 2015 [51], which was having 42 tags for POS tagging and 1,50,830 words which are specifically in the Nepali language. A neural network-based POS tagger was introduced in 2018 [52] which has high accuracy of 99%, but the model and the dataset that was introduced are a closed source which makes it difficult to use for further references. Different POS taggers with respect to year and methodologies are listed in Table 5 below.

## 5.8 *Named Entity Recognition*

It is a sub-task of NLP which involves identifying and classifying the named entities in a text which includes recognizing and classifying the proper nouns which represent or refer to a specific individual, location, organization, etc. Named Entity Recognition is very useful in NLP tasks as it helps to understand and interpret text which is done by extracting important information and later organizing it in structural form [53]. Named Entity Recognition can be very useful in information retrieval systems as it helps to better understand what the user is trying to say and extract the resources. In the Nepali language, one of the first implementations was done in 2014 which was created using a custom self-created corpus [54]. In the proposed model, there were certain exceptions which include name ambiguity, words that have been taken from another language, and no capitalization. A similar kind of work was also done in 2014 by Dey et al. [55] which uses HMM model to tag each word and was using a lookup table to find the named entity for a given word. In 2019, a deep learning framework based on Named Entity Recognition was introduced by Singh et al. [56]; the accuracy of the model was 86.89% which uses various features like Grapheme feature and POS features to perform the classification.

### 5.9 Chunking

In chunking, a text/sentence is divided into sub-parts or syntactically correlated parts of a word which are known as chunks. A unit of chunk can be defined as a continuous stretch of words and can be understood as a single entity [57]. Chunking is used to group related words together which makes it easier for interpretation of the text. It is very useful for various NLP tasks such as parsing, language modeling, and part-of-speech tagging. Some of the major challenges in chunking include lack of standardization in chunking conventions, computational complexity, inherent ambiguity of natural language, etc. A model for the Nepali language was proposed by Rupakheti et al. [58] which uses a rule-based algorithm to identify different chunks of words in a sentence; after the chunks are identified the pattern is analyzed. Pointers were used to point to the starting and ending of the tokens and the pattern is searched between the start and end pointers if the relationship is established or found then the entire pattern is kept under one chunk and if there is no relationship then the end pointer is moved to one token from the left side until and unless the rule is established. There were 13 chunk rules which cover the rules nearly 50–60%.

## 6 Phases of Natural Language Processing

There are various phases in Natural Language Processing which are used to develop algorithms and models that can process, analyze, and generate human languages which are shown in Fig. 2.

### 6.1 Lexical Analysis

It is the first step in Natural Language Processing; here, the text is broken down into lexemes or it can be said that here the texts are broken down into individual words which are known as tokens. Here in this step, various tasks are performed which are identifying and extracting of words, punctuation marks, numbers, etc. [59]. Here, Lexicon Normalization is performed; one of the commonly used techniques

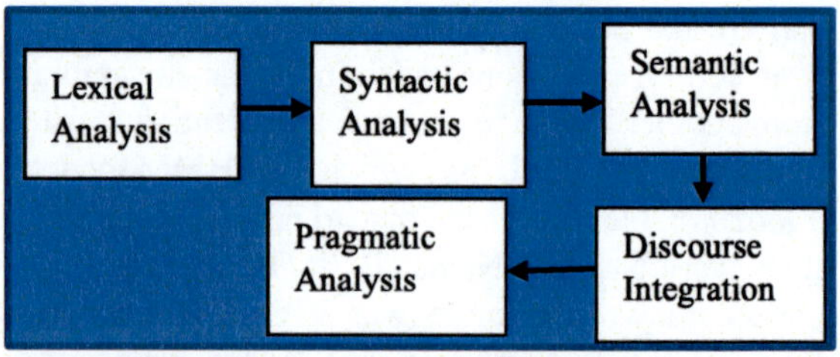


**Fig. 2** Phases of natural language processing

for lexicon analysis is stemming. There are many challenges that are associated with lexical analysis which are handling large volumes of text, supporting multiple languages, identifying and handling complex language constructs, etc.

## 6.2 Syntactic Analysis

It is used to check the grammar of the sentence, the inter-relationship between the two words, and it is also used to check how the words are arranged in a sentence and whether it is making any sense or not; if the sentence is not making any sense, then the sentence is rejected by the syntactic analysis. Here, parts of speech are identified for each word to determine its syntactic roles in the sentence. Some of the major challenges of syntactic analysis are ambiguity, prepositional phrases, idiomatic expressions, and grammatical errors [60]. Some in-depth analyses have been done in the field of syntactic analysis for the Nepali language by P. Chandra et al. (2015) [61], which provides a study which shows how the meaning of the same sentence varies from place and how the same sentence can be spoken differently in different places due to dialectal variations.

## 6.3 Semantic Analysis

Semantic analysis deals with the process of analyzing the meaning of the syntactically correct sentence; it involves understanding the meaning of the words and sentences and determining how they are related to each other which helps to properly interpret the overall meaning of the sentence [62]. One of the key challenges of Semantic Analysis is to deal with ambiguity as sentences can have multiple valid interpretations, and it can be a complex task to determine the exact meaning of the sentence though it may be syntactically correct [63]. It has to deal with a vast number of references of the sentences and to preserve the previously referred word which may be referring to any kind of tuple like name, entity, event, etc.

## 6.4 Discourse Integration

Discourse integration is used to extract the information from a sentence and how a sentence is related to the previous sentence; this is done by reserving the context of the previous sentence [64]. This is one of the challenging tasks in NLP as the context of the sentence frequently changes. Various NLP techniques are used to perform discourse integration which are coherence modeling, discourse parsing, and discourse segmentation [64].

### 6.5 Pragmatic Analysis

It is the last and final phase of NLP, which is used to understand how people communicate with each other and when they are communicating it tries to find the context on which two or more people are talking. It is the study of the ways in which the language is used in real-world situations [65]. There are various factors that need to be considered while doing the analysis such as the speaker's intention, and the relationship that is in between the speaker and the listener. For achieving such things, various NLP techniques are used which are NLU, pragmatic inference, and discourse analysis.

## 7 Conclusions

A detailed study has been provided for natural language processing and its techniques and its advancements in the Nepali language. It has been seen that many works have been done in the field of Nepali NLP such as Sentiment Analysis and Classification problem, but there are only a few works that have been done which aim at building the fundamental structure for NLP task one of which is the non-presence of standardized stop words for the Nepali language; there has been significantly low research and resources available for NLP in Nepali. Most of the NLP tasks are rule-based tasks where the rules are pre-defined which needs to be changed to computational based models and techniques. It has also been observed that the model which is proposed in the Nepali language is not able to perform well when some new words have been added which is not present in the corpora. Nepali language being rich in phonetics makes it difficult to perform the NLP task. There are various challenges that need to be addressed in the field of NLP such as lack of large, high-quality training data, difficulty in handling complex syntactic and semantic structures, limited ability to understand context and intent, ambiguity and variability, and the potential for bias algorithm. With such challenges present in the NLP task and due to the low resource availability for the Nepali language, it becomes quite a challenging task to perform NLP in the Nepali language with a high level of accuracy.

## References

1. Goldberg Y (2017) Neural network methods for natural language processing. Synth Lect Human Language Technol 10(1):1–309
2. Chowdhary K (2020) Natural language processing. Fundamentals of artificial intelligence, 603–649
3. Verma MTSR (2018) Natural language processing (Nlp): a comprehensive study
4. Plisson J, Lavrac N, Mladenic D (2004) A rule based approach to word lemmatization. In: Proceedings of IS, Vol 3, pp 83–86

5. Shahi TB, Dhamala TN, Balami B (2013) Support vector machines based part of speech tagging for Nepali text. Int J Comput Appl 70(24)
6. Lamsal R (2020) A large scale Nepali text corpus. IEEEdataport. https://doi.org/10.21227/jxrd-d245
7. Duwal S, Bal BK (2019) Efforts in the Development of an Aug- mented English–Nepali Parallel Corpus. Technical report, Kathmandu University
8. Shahi TB, Sitaula C (2021) Natural language processing for Nepali text: a review. Artif Intell Rev, 1–29
9. Nemkul K, Shakya S (2021) English to Nepali sentence translation using recurrent neural network with attention. In: 2021 international conference on computing, communication, and intelligent systems (ICCCIS), pp 607–611. IEEE
10. Nemkul K, Shakya S (2021) Low resource English to Nepali sentence translation using RNN—long short-term memory with attention. In: Proceedings of international conference on sustainable expert systems, pp 649–657. Springer, Singapore
11. Tiedemann J (2012) Parallel data, tools and interfaces in OPUS. In: Lrec, Vol 2012, pp 2214–2218
12. Staudemeyer RC, Morris ER (2019) Understanding LSTM--a tutorial into long short-term memory recurrent neural networks. arXiv preprint arXiv:1909.09586
13. Reiter E (2018) A structured review of the validity of BLEU. Comput Linguist 44(3):393–401
14. Timilsina S, Gautam M, Bhattarai B (2022) NepBERTa: Nepali language model trained in a large corpus. In: Proceedings of the 2nd conference of the Asia-pacific chapter of the association for computational linguistics and the 12th international joint conference on natural language processing, pp 273–284
15. Nivre J (2015) Towards a universal grammar for natural language processing. In: International conference on intelligent text processing and computational linguistics, pp 3–16. Springer, Cham
16. Dhanalakshmi V, Rajendran S (2010) Natural language processing tools for tamil grammar learning and teaching. Int J Comput Appl, 0975–8887
17. Triana JG, De Castro R (2019) Grammars and multifactorial numbers. Global J Pure Appl Math 15(3):251–259
18. Klein D, Manning CD (2005) Natural language grammar induction with a generative constituent-context model. Pattern Recogn 38(9):1407–1419
19. Nivre J (2005) Dependency grammar and dependency parsing. MSI report 5133(1959):1–32
20. Debusmann R (2000) An introduction to dependency grammar. Hausarbeit fur das Hauptseminar Dependenzgrammatik SoSe 99(1):16
21. Khatiwada R (2009) Nepali. J Int Phon Assoc 39(3):373–380
22. Matthews D (2013) Course in Nepali. Routledge
23. Bista S, Khatiwada L, Keshari B (2004) Nepali lexicon development. PAN Localization, Working Papers, 2007, 311–15
24. Bal BK, Shrestha P, Pustakalaya MP (2004) Nepali spellchecker. PAN Localization Working Papers, 2007, 316–318
25. Yadava YP, Hardie A, Lohani RR, Regmi BN, Gurung S, Gurung A, ... Hall P (2008) Construction and annotation of a corpus of contemporary Nepali. Corpora 3(2):213–225
26. Bal BK (2004) Structure of Nepali grammar. PAN Localization, Madan Puraskar Pustakalaya, Kathmandu, Nepal, 332–396
27. Jurish B, Würzner KM (2013) Word and Sentence Tokenization with Hidden Markov Models. J Lang Technol Comput Linguist 28(2):61–83
28. Katam S (2014) The porter stemmer. Indiana State University
29. Jivani AG (2011) A comparative study of stemming algorithms. Int J Comp Tech Appl 2(6):1930–1938
30. Khyani D, Siddhartha BS, Niveditha NM, Divya BM (2021) An Interpretation of Lemmatization and Stemming in Natural Language Processing. J Univ Shanghai Sci Technol
31. Shrestha I, Dhakal SS (2021) Fine-grained part-of-speech tagging in Nepali text. Procedia Computer Science 189:300–311

32. Sitaula C (2013) A hybrid algorithm for stemming of Nepali text
33. Borah S, Choden U, Lepcha N (2017) Design of a morph analyzer for non-declinable adjectives of nepali language. In: Proceedings of the 2017 international conference on machine learning and soft computing, pp 126–130
34. Chhetri I, Dey G, Das SK, Borah S (2015) Development of a morph analyser for Nepali noun token. In: 2015 international conference on advances in computer engineering and applications, pp 984–987. IEEE
35. Jayakodi K, Bandara M, Meedeniya D (2016) An automatic classifier for exam questions with WordNet and Cosine similarity. In: 2016 Moratuwa engineering research conference (MERCon), pp 12–17. IEEE
36. Lu X (2014) Lexical annotation. In: Computational methods for corpus annotation and analysis, pp 39–65. Springer, Dordrecht
37. Anees AF, Shaikh A, Shaikh A, Shaikh S (2020) Survey paper on sentiment analysis: techniques and challenges. EasyChair2516–2314
38. Subba S, Paudel N, Shahi TB (2019) Nepali text document classification using deep neural network. Tribhuvan Univ J 33(1):11–22
39. Tripathi M (2021) Sentiment analysis of nepali covid19 tweets using nb svm and lstm. J Artif Intell 3(03):151–168
40. Nothman J, Qin H, Yurchak R (2018) Stop word lists in free open-source software packages. In: Proceedings of workshop for NLP open source software (NLP-OSS), pp 7–12
41. Fernández-González D, Gómez-Rodríguez C (2023) Dependency parsing with bottom-up hierarchical pointer networks. Inf Fusion 91:494–503
42. ArchitYajnik D (2015) Parsing techniques using Paninian framework on Nepali language. DJ J Eng Appl Math 1(1)
43. Rai P, Chatterji S (2022) Annotation projection-based dependency parser development for Nepali. Transactions on asian and low-resource language information processing
44. Chiche A, Yitagesu B (2022) Part of speech tagging: a systematic review of deep learning and machine learning approaches. J Big Data 9(1):1–25
45. Li H, Mao H, Wang J (2021) Part-of-speech tagging with rule-based data preprocessing and transformer. Electronics 11(1):56
46. Zheng X, Chen H, Xu T (2013) Deep learning for Chinese word segmentation and POS tagging. In: Proceedings of the 2013 conference on empirical methods in natural language processing, pp 647–657
47. Marquez L, Padro L, Rodriguez H (2000) A machine learning approach to POS tagging. Mach Learn 39(1):59–91
48. Prasain B, Khatiwada LP, Bal BK, Shrestha P (2008) Part-of-speech Tagset for Nepali. Madan Puraskar Pustakalaya
49. Bal BK, Shrestha P (2004) A morphological analyzer and a stemmer for Nepali. PAN Localization, working papers, 2007, 324–31
50. Yajnik A (2017) Part of speech tagging using statistical approach for Nepali text. Int J Cognit Language Sci 11(1):76–79
51. Paul A, Purkayastha BS, Sarkar S (2015) Hidden Markov model based part of speech tagging for Nepali language. In: 2015 international symposium on advanced computing and communication (ISACC), pp 149–156. IEEE
52. Prabha G, Jyothsna PV, Shahina KK, Premjith B, Soman KP (2018) A deep learning approach for part-of-speech tagging in nepali language. In: 2018 international conference on advances in computing, communications and informatics (ICACCI), pp 1132–1136. IEEE
53. Mohit B (2014) Named entity recognition. In: Natural language processing of semitic languages, pp 221–245. Springer, Berlin, Heidelberg
54. Bam SB, Shahi TB (2014) Named entity recognition for nepali text using support vector machines. Intell Inf Manag
55. Dey A, Paul A, Purkayastha BS (2014) Named entity recognition for nepali language: a semi hybrid approach. Int J Eng Innov Technol (IJEIT) 3:21–25

56. Singh OM, Padia A, Joshi A (2019) Named entity recognition for nepali language. In: 2019 IEEE 5th international conference on collaboration and internet computing (CIC), pp 184–190. IEEE
57. Lee YS, Wu YC (2007) A robust multilingual portable phrase chunking system. Expert Syst Appl 33(3):590–599
58. Rupakheti P, Report on Nepali Computational Grammar Prajwal Rupakheti, Laxmi Prasad Khatiwada Bal Krishna Bal Madan Puraskar Pustakalaya Lalitpur, PatanDhoka, Nepal.
59. Hippisley AR (2010) Lexical analysis
60. Vo AD, Nguyen QP, Ock CY (2020) Semantic and syntactic analysis in learning representation based on a sentiment analysis model. Appl Intell 50(3):663–680
61. Chandra P, Udaar U (2015) Ergative case and verbal agreement: explaining dialectal variations in Nepali. Acta Linguistica 9(1)
62. Goddard C (2011) Semantic analysis: a practical introduction. Oxford University Press
63. Maulud DH, Zeebaree SR, Jacksi K, Sadeeq MAM, Sharif KH (2021) State of art for semantic analysis of natural language processing. Qubahan Acad J 1(2):21–28
64. Meera S, Geerthik S (2022) Natural language processing. Artificial intelligent techniques for wireless communication and networking, 139–153
65. Zhao L, Alhoshan W, Ferrari A, Letsholo KJ, Ajagbe MA, Chioasca EV, Batista-Navarro RT (2021) Natural language processing for requirements engineering: a systematic mapping study. ACM Comput Surv (CSUR) 54(3):1–41

# Quality Evaluation of Enterprise Architecture Frameworks

**Priyanka Chakraborty and Anirban Sarkar**

**Abstract** An Efficient and effective Enterprise Architecture Framework (EAF) design depends on the quality of the EAF and its quality evaluation process. Most of the prior research lacks selecting proper quality factors, which maximize the EAF quality. Moreover, the set of metrics, devised in those works, for the measurement of the quality factors is specific to those frameworks. To resolve these issues, a novel quality evaluation framework is proposed in this paper. In this framework, the quality feature is characterized by several quality factors like reusability, context sensitivity, re-configurability, and structural complexity. Moreover, a set of ten basic quality assessment metrics has also been proposed and validated theoretically.

**Keywords** Enterprise architecture framework · SCORE Framework · Quality metrics · Re-usability · Re-configurability · Context sensitivity

## 1 Introduction

The quality aspect of an EAF has received huge attention from stakeholders in recent years because of its considerable influence on the sound IT decision-making process. Therefore, maintaining a high service standard is nowadays a top priority to develop an EAF that is both effective and efficient. The quality of services plays a key role in reducing both the development and maintenance efforts and costs of a framework. In addition to this, it also gives provision to the stakeholders to include any kind of modifications to the pre-existing framework.

Consequently, a quality assessment framework is required to be established for the measurement of different quality factors like re-usability, context sensitivity, reconfigurability, and structural complexity of an EAF. To measure those quality factors, the proper establishment of several basic quality metrics is also a necessary

P. Chakraborty (✉) · A. Sarkar
Department of Computer Science and Engineering, National Institute of Technology, Durgapur, West Bengal, India
e-mail: priyanka1nitdgp@gmail.com



S. Borah et al. (eds.), *Advanced Computational and Communication Paradigms*, Lecture Notes in Networks and Systems 535,
https://doi.org/10.1007/978-981-99-4284-8_14

step. Thus, several research questions arise like (i) How can an EAF be assessed from the viewpoint of the quality aspect? (ii) What kind of quality metrics are required to be developed for the proper realization of the essence property that focuses on the quality of services in the essence layer of an EAF?

In this context, existing EAFs like Zachman [11], Treasury Enterprise Architecture Framework (TEAF),[1] Generalized Enterprise Reference Architecture and Methodology (GERAM) [2], NATO Architecture Framework (NAF) [9], Federal Enterprise Architecture (FEA) [10], etc., are lacking emphasizing various major quality factors such as reusability, context sensitivity, reconfigurability, and structural complexity.

These are well established and validated in an existing EAF, named SCORE framework [4, 5] which gives a true level of abstractions and the interdependence among the layers is minimized. Consequently, different functionalities like addition, deletion, and modification can be easily performed in this framework. Moreover, the consideration of the context property in the context layer makes this framework context-sensitive while in the object layer, the presence of the re-configurable services facilitates this framework to support the re-configurability property. Finally, the quality of services is focused on the essence layer.

A novel theoretical approach to the quality assessment framework is proposed in this paper. A set of basic quality assessment metrics is also established for the proper measurement and calculation of the quality factors. The proper validation of the proposed set of metrics is necessary to identify any kind of flaws in those metrics and assess whether those metrics are useful in a real-life scenario. In this paper, Briand's Properties of Measures approach [3] is adopted for the theoretical validation of the proposed metrics. Although these quality metrics are devised for the quality assessment in the SCORE framework's essence layer, they can also be used in other quality evaluation frameworks.

## 2 Related Work

Most of the pre-existing EAFs are lacking in emphasizing the quality aspect of services provided by those EAFs. The Zachman framework [11], a pioneer in the domain of EAF, hardly supports any of those four quality factors like (i) Reusability, (ii) Context Sensitivity, (iii) Re-configurability, and (iv) Structural Complexity which is considered in this paper. TEAF, GERAM [2], The Open Group Architecture Framework (TOGAF) [7], British Ministry of Defense Architecture Framework (MODAF) [1], (NAF) [9], Department of Defense Architecture Framework (DODAF)[2], and 4+1 View Model [8] support the reusability property. FEA [10] supports only structural complexity property. All the frameworks except MoDAF hardly support the context sensitivity property, while only the 4+1 View Model supports the re-configurability property. TEAF, NAF, MoDAF, FEA, 4+1 view model, and TOGAF fully satisfy

[1] https://segoldmine.ppi-int.com/node/67633.

[2] https://dodcio.defense.gov/library/dod-architecture-framework/.

the property named structural complexity, while Zachman, GERAM, and DODAF partially support this property. In contrast, the SCORE framework [5] supports those quality attributes entirely.

# 3 Quality Assessment Framework

## 3.1 Proposed Set of Metrics

A set of ten basic quality assessment metrics has been proposed and theoretically validated to measure the four quality factors Reusability, Context Sensitivity, Reconfigurability, and Structural Complexity (*SCM(F)*). This set of metrics is further categorized into three types of metrics such as

(a) ***Construct-Level Quality Metrics***

(i) **Metric 1 (*M1*)**: Number of services (activity objects or user tasks) contained in any process $BP_i$ (*NS(F)*).
Description: If a BP is contained with services $A = \{A_1, A_2, \ldots, A_i, \ldots, A_n\}$. Hence, the total number of services within a BP is $NS = |A|$.
Significance: The larger the value of *NS* means that there are more services in the BP. It increases the structural complexity level of an EAF.

(ii) **Metric 2 ( *M2*)**: Number of Event Object within a BP (*NEO(F)*).
Description: Let a BP is contained with a set of event objects $EO = \{EO_1, EO_2, \ldots, EO_n\}$. The total number of event objects within a BP is $NEO = |EO|$.
Significance: The larger number of *NEO* implies the increasing value of context sensitivity and structural complexity quality factors of a framework.

(iii) **Metric 3 (*M3*)**: Context Sensitivity Metric (*NC(F)*).
Description: *NC* is a set of contexts within a BP.
Significance: The larger value of *NC* implies the higher context sensitivity and structural complexity power exist within the enterprise entities of a framework.

(b) ***Semantic-Level Quality Metrics***

(i) **Metric 4 (*M4*)**: Number of realization relationship among the BP elements ($NRER_{jk}(F)$).
Description: If $RER_{jk}$ is the relationship between two specific business elements *j* and *k*, then the total number of such relationships is denoted as $NRER_{jk}$. Hence, $NRER_{jk}$ is determined as $NRER_{jk} = \sum_{j=1,k=1}^{n} RER_{jk}$.
Significance: The increment of $NRER_{jk}$ value increases the value of dynamicity, context sensitivity, completeness, and analyzability property of a framework and thus, it implies the better quality of the framework.

(ii) **Metric 5** (*M5*): Number of association relationship ($NAR_{jk}$(F)).
Description: If $AR_{jk}$ is an association relationship between two specific business facets *j* and *k*, then the total number of such relationships is denoted as $NAR_{jk}$. Hence, $NAR_{jk}$ is described as $NAR_{jk} = \sum_{j=1,k=1}^{n} AR_{jk}$.
Significance: The higher value of $NAR_{jk}$ signifies the larger value of dynamicity, context sensitivity, structural complexity, completeness, and analyzability.

(iii) **Metric 6** (*M6*): Static Complexity of an EAF (*STC(F)*).
Description: Static complexity is calculated by the summation of *NGT*, *NC*, and *NTO*. $STC = (NGT + NC + NTO)$.
Significance: The higher *STC* value results in greater values of dynamicity, context sensitivity, structural complexity, completeness, and analyzability.

(iv) **Metric 7** (*M7*): Dynamic Complexity of an EAF (*DYC(F)*).
Description: Suppose an EAF is contained with the total number of association relationships (*NAR*), the total number of collaboration relationships (*NCR*), and the total number of realization relationships (*NRER*) among all the business elements. Thus, *DYC* is represented with the summation of all the existing relationships within that framework. $DYC = NAR + NCR + NRER$. Here, $NCR_{jk} = \sum_{j=1,k=1}^{n} CR_{jk}$.
Significance: The larger value of *DYC(F)* implies the greater values of dynamicity, context sensitivity, structural complexity, completeness, and analyzability.

(c) ***Framework-Level Quality Metrics*** Top-level executives deal with the following six framework-level quality metrics.

(i) **Metric 8** (*M8*): Degree of Dynamicity Metric ($D_{DY}(F)$).
Description: $D_{DY}$ is inversely proportional to the summation of *STC* and *DYC*. The higher value of $D_{DY}$ signifies that the architecture is more user-centric. $D_{DY} = 1/(STC + DYC)$.
Significance: The higher value of $D_{DY}(F)$ results in greater values of dynamicity, context sensitivity, structural complexity, completeness, and analyzability.

(ii) **Metric 9** (*M9*): Cost of Re-configurability Metric ($CO_{RCPC}$(F)).
Description: It is related to the degree of re-configurability metric $D_{RCPC}$ [6] value. A higher value of $D_{RCPC}$ signifies a large amount of modification is required in the previous goal and its corresponding BPs to satisfy the new goal and its corresponding BPs. It results in the increasing value of $CO_{RCPC}$, while a lower value of $D_{RCPC}$ means a lower value of $CO_{RCPC}$. Hence, $CO_{RCPC}$ is proportional to $D_{RCPC}$. $CO_{RCPC} \propto D_{RCPC}$.
Significance: The higher value of $CO_{RCPC}(F)$ implies the greater values of the reconfigurability property of an EAF. Thus, the stakeholders decide their business strategies based on the $CO_{RCPC}$ values and their objective is to minimize the $CO_{RCPC}$ values as much as possible.

(iii) **Metric 10** (*M10*): Cost of Re-usability Metric ($CO_{RUPC}(F)$).
Description: It is associated with the degree of re-usability metric $D_{RUPC}$ [6] value. If $D_{RUPC}$ is higher, it means that a greater number of BP components from the previous goal can be utilized to achieve the newly modified goal. Consequently, the value of $CO_{RUPC}$ decreases to a certain extent, while a lower value of $D_{RUPC}$ implies a higher value of $CO_{RUPC}$. Therefore, $CO_{RUPC}$ is inversely proportional to $D_{RUPC}$. $CO_{RUPC} \propto 1/D_{RUPC}$.
Significance: The higher value of $CO_{RUPC}(F)$ implies the lower value of the re-usability property of an EAF.

## *3.2 Framework-Level Quality Measurements*

To make the SCORE framework efficient, its essence layer should be focused on. The primary focus of the essence layer is ensuring the quality of services. Hence, a quality assessment framework needs to be devised for the inclusion of the quality factors like reusability, context sensitivity, re-configurability, and structural complexity.

(i) **Reusability**: This property explains how many portions of an EAF or how many constructs of an EAF can be used in its entirety to similar types of EAFs. This feature is well maintained by a good quality EAF to reduce the development and maintenance cost to a significant extent. Hence, the cost of re-usability metric ($CO_{RUPC}$) (*M10*) is introduced.
(ii) **Context Sensitivity**: The term "context" pertains to any information or knowledge that characterizes the immediate surroundings of the EAF entities. If an EAF can adapt its behavior based on dynamic contextual information, it is considered to be context-sensitive. The more context-sensitive EAF is more expressive and user-centric. Context sensitivity metric (*NC*) (*M3*) is introduced.
(iii) **Re-configurability**: This property implies how many constructs of an EAF need to be modified to satisfy the rapidly changing business requirements and strategies. Highly re-configurable EAF (more modifications are needed) results in a higher value of Cost of Re-configurability Metric ($CO_{RCPC}$) (*M9*).
(iv) **Structural Complexity**: Structural complexity (*SCM(F)*) of an EAF signifies that the level of difficulties exists at the design level in terms of the basic constructs and various kinds of relationships exist among those constructs. Here, $SCM(F) = (NS * W_5) + (NC * W_4) + (NEO * W_3) + (NRER * W_2) + (NAR * W_1)$, where, and $W_1$, $W_2$, $W_3$, $W_4$, and $W_5$ are assigned distinct weightage values and $W_5 > W_4 > W_3 > W_2 > W_1$.

## 4 Theoretical Validation of the Proposed Set of Quality Assessment Metrics

The validation of the proposed set of quality assessment metrics is done with the help of a property-based measurement framework discussed in [3]. Like in [3], the SCORE framework (*F*) is considered as a system and described by a pair $< E, R >$ where E represents the set of all facets that includes the goal, and its corresponding all sub-goals, contexts, objects, events, roles, and essence. Within *F*, the set of all binary relationships on *E* like, realization, association, and collaboration relationships are denoted as *R*. Thus, *R* can be expressed as $R \subseteq E \times E$. Each of the five layers (object layer, role layer, subject layer, context layer, and essence layer) termed as *L* can be expressed as $L =< EL, RL >$, where *EL* represents all the constructs belongs to that layer, while *RL* denotes the set of entire relationships exists among the constructs in *L*. Mathematically, the above convention can be represented as $EL \subseteq E, RL \subseteq R$, and $RL \subseteq EL \times EL$ and should be held for each layer.

Several significant size properties of the framework are discussed below. The function *Size(F)* is used for the calculation of the size of the framework.

(i) **Non-negativity**: Since each service of *F* belongs to a BP that is associated with the goal of *F*, the number of services in the object layer of *F* cannot be negative. Hence, in *F*, the value of *NS* is always greater than or equal to zero. It means $Size(F) >= 0$ or $NS >= 0$.

(ii) **Null Value**: If no service exists in the framework, then $NS = 0$.

(iii) **Module Additivity**: Suppose a layer $L =< EL, RL >$ is comprised of two modules $x_1$ and $x_2$, where $x_1 =< E_{x_1}, R_{x_1} >$ and $x_2 =< E_{x_2}, R_{x_2} >$. Hence, each element of *L* belongs to either $x_1$ or $x_2$ that means $x_1, x_2 \subseteq L, EL = E_{x_1} \cup E_{x_2}$ and $E_{x_1} \cap E_{x_2} = \phi$. Consequently, *Size(L)* is the summation of the size of two modules, $Size(L) = Size(x1) + Size(x2)$ or $NS(L) = NS(x_1) + NS(x_2)$, where and $x_1 \cap x_2 = \phi$.

(iv) **Adding Elements:** The size of *L* cannot be decreased by the addition of an element to it. Hence, $L' =< E', R' >$ and $L'' =< E'', R'' >$ and $E' \subseteq E'' \Rightarrow Size(L') \leq Size(L'')$ or $NS(L') \leq NS(L'')$.

(v) **Sum of Modules**: The size of a layer *L* is always less than any of its pair of modules $y_1$ and $y_2$ where $y_1 =< E_{y_1}, R_{y_1} >$ and $y_2 =< E_{y_2}, R_{y_2} >$ such that any element of L belongs to either $y_1$ or $y_2$ or both ($y_1 \subseteq L$, $y_2 \subseteq L$ and $E = E_{y_1} \cup E_{y_2}$) $\Rightarrow Size(L) \leq Size(y_1) + Size(y_1)$ or $NS(L) \leq NS(y_1) + NS(y_2)$. Therefore, the theoretical validation of the metric *NS* is accomplished. Similarly, *NE*, *NC* and *STC* can also be validated using the size metrics properties. Other semantic level metrics such as $NRER_{jk}$ and $NAR_{jk}$ can be validated using the length properties while all the framework level metrics like *DYC*, $D_{DY}$, $CO_{RCPC}$ and $CO_{RUPC}$ can be validated using complexity properties.

## 5 Illustration of the Proposed Metrics and Quality Factors with a Case Study

A case study on re-configuring Hospital Management System has been adopted from a recent work [6] where the existing goal (*G*) is to provide free medical facilities to all the faculty members while the extended goal (*EG*) is to provide the same to both faculties and students. As described in [6], the goal is realized by six contexts such as Medical Card Submission Time (MCST)-*C1*, Hospital Location (HL)-*C2*, Checking Time (CT)-*C3*, Pressure of Patient (PR)-(*C4*), Weight of Patient (WT)-*C5*, and Number of Medicines (NM)-*C6*.

In the object layer, five activity objects or services exist, Submit Medical Card at Reception (*A1*), Return Card (*A2*), Check Credentials (*A3*), Visit Doctor (*A4*) and Provide Medicines (*A5*).

Furthermore, four structural element objects like a patient (faculty/student) (*S1*), database (*S2*), counter person/attendee (*S3*), and doctor (*S4*) are also introduced to execute the above-mentioned five services.

*A1* is associated with contexts *C1* and *C2*, while *A2*, *A3* are realized by *C2*. *A4* and *A5* are associated with (*C2*, *C3*, *C4*, *C5*) and (*C2*, *C6*), respectively. Moreover, *A1*, *A2*, *A3*, *A4* and *A5* are realized by their corresponding events *A1* [swipe ID-card (*E1*)], *A2* [taking card back (*E2*)], *A3* [checking expiry date on database (*E3*), select suitable appointment date and time (*E4*), pressing submit button (*E5*)], *A4* [taking pressure (*E6*), taking weight (*E7*), giving prescription (*E8*)] and *A5* [taking prescription (*E9*), giving medicines (*E10*)].

In the role layer, *S1* and *S3* play separate roles through the execution of various sets of activities. For example, *S1* and *S3* play roles like service seeker (*R1*) and service provider (*R3*) respectively.

Figure 1 exhibits the diverse types of constructs of the SCORE framework and three types of relationships among them for the Hospital Management System.

The lowermost layer, the essence layer, is concerned with the quality aspect of the services and the overall framework. Hence, this layer answers several questions associated with the framework quality like, "*Whether the hospital management system provides good service in the form of proper treatment?* (*Es1*)"

Table 1 shows values of all the metrics for the above case study. Similarly, the measurement values for four quality factors are depicted in Table 2.

## 6 Conclusion and Future Work

In this paper, a theoretical concept of the quality assessment framework has been presented. This framework possesses several quality factors like reusability, context sensitivity, re-configurability, and structural complexity. A set of ten quality metrics is also developed along with a detailed formulation for the measurement of those four quality factors. The theoretical validation of those quality assessment metrics is done

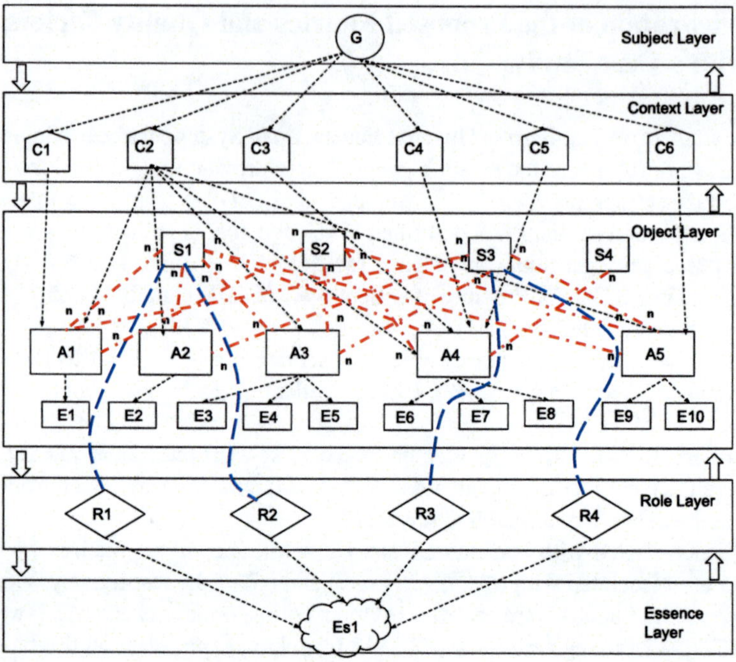


**Fig. 1** The relationships among the different constructs of five layers of SCORE framework for Hospital Management System. Here, the red lines between *S* and *A* signify association relationship, the blue lines between *S* and *R*, are for collaboration relationship, and the remaining lines are for realization relationship

**Table 1** Result of ten quality metrics based on hospital management system

| Metrics | | Value | Unit | Remark |
|---|---|---|---|---|
| NS(F) | (M1) | 5 | Number | Five services within the BP of HMS as described in Fig. 1 |
| NEO(F) | (M2) | 10 | EO | The event objects within the BP of the HMS (Fig. 1) |
| NC(F) | (M3) | 6 | CX | Total number of contexts within the BP, representing the HMS (Fig. 1) |
| NRERjk(F) | (M4) | 30 | RER | According to the realization relationships among the five layers of the SCORE framework described in Fig. 1 |
| NARjk(F) | (M5) | 14 | AR | According to the association relationships in the SCORE framework's object layer described in Fig. 1 |
| STC(F) | (M6) | 27 | Number | The structural complexity of the BP, representing the HMS by considering the total number of services, contexts, and event objects (Fig. 1) |
| DYC(F) | (M7) | 48 | Number | Considering the total number of realization and association relationships within the HMS as described in Fig. 1 |
| DDY(F) | (M8) | 0.013 | Number | Inversely proportional to the total dynamicity (summation of STC and DYC) |
| $CO_{RCPC}$(F) | (M9) | 0.585 | Number | Proportional to $D_{RCPC}$ that is 0.585 for this case study |
| $CO_{RUPC}$(F) | (M10) | 1.056 | Number | Inversely proportional to $D_{RUPC}$ that is 0.947 for the HMS |

**Table 2** The measurement values for four quality factors for the corresponding case study on hospital management system

| Quality factors | Value | Unit | Remark |
|---|---|---|---|
| Reusability | 0.947 | Number | Represents that more than half of the pre-existing enterprise entities can be re-used with its original form to satisfy the ever-changing business requirement related to the HMS |
| Context sensitivity | 6 | CX | This high value of context-sensitivity means the EAF is more expressive and user centric |
| Re-configurability | 0.585 | Number | Less than half of the enterprise entities related to the HMS are required to be modified |
| Structural complexity | 153 | Number | Considering the major basic constructs like services, contexts, and event objects, different kinds of relationships like realization relationship among the constructs of five layers and the association relationships within the enterprise elements in the object layer of the SCORE (Fig. 1) |

using the size, length, and complexity properties of a property-based measurement framework to prove their usefulness in real-life scenarios. The quality framework along with its quality metrics are applicable in the SCORE framework's essence layer as well as useful for other similar types of EAFs. The extension of this work will include more quality attributes and metrics for the proposed quality assessment framework. Another addition would be comparing the proposed framework with some relevant existing frameworks. Moreover, the focus in the future will be on the empirical validation using more real-life case studies to establish the significance of the quality metrics and their associations with the quality factors.

## References


1. Bailey I (2008) Brief introduction to MODAF with v1.2 updates. In: 2008 IET seminar on enterprise architecture frameworks. IET, pp 1–18
2. Bernus P, Noran O, Molina A (2015) Enterprise architecture: twenty years of the GERAM framework. Annu Rev Control 39:83–93
3. Briand LC, Morasca S, Basili VR (1996) Property-based software engineering measurement. IEEE Trans Software Eng 22(1):68–86
4. Chakraborty P, Sarkar A (2017) Context driven approach for enterprise architecture framework. In: IFIP international conference on computer information systems and industrial management. Springer, pp 277–289
5. Chakraborty P, Sarkar A (2019) Score framework: a layered approach for enterprise architecture. Int J Bus Syst Res 13(4):438–467
6. Chakraborty P, Sarkar A (2022) Dynamic context driven re-configurable business process. Ann Emerg Technol Comput (AETiC) 6(3)
7. Edward IYM, Shalannanda W, Agusdian A, Lestariningati SI (2014) E-government master plan design with TOGAF framework. In: 2014 8th international conference on telecommunication systems services and applications (TSSA). IEEE, pp 1–6
8. Kruchten PB (1995) The 4+1 view model of architecture. IEEE Softw 12(6):42–50
9. Ota D, Gerz M (2011) Benefits and challenges of architecture frameworks. Technical report, FGAN-FKIE WACHTBERG (GERMANY)
10. Wang W, Gao C, Lin S, Yin G (2009) Modeling technologies of FEA grid services and grid information management structure for FEA grid portal. In: 2009 WRI world congress on software engineering. vol 1. IEEE, pp 108–113
11. Zachman JA (1987) A framework for information systems architecture. IBM Syst J 26(3):276–292

# Natural Language Query Processing System to Extract Autism Spectrum Disorder Information from Database

**Prasenjit Mukherjee, Nandan Banerji, Kailash Pati Mandal, Manish Godse, and Baisakhi Chakraborty**

**Abstract** This paper proposes a natural language (NL) query-based response system in the Healthcare domain. Today, autism spectrum disorder (ASD) is a big issue among children and most parents are unaware of this disorder. Awareness about this disorder is very important and very much needed for parents to identify ASD in their children at an early stage to start treatment. Our proposed system will extract information about ASD from a database using NL queries. Sometimes the NL query may ask for some aggregate data from the database which may be difficult to process this proposed system is able to handle conventional queries with aggregation and generates responses from the healthcare database. A combinational algorithm has been approached where SQL has been generated from the NL query.

**Keywords** Natural language query · Autism spectrum disorder · Semantic analysis · Combinational words · Structured query language

## 1 Introduction

Autism Spectrum disorder (ASD) is a big issue among children and it is very difficult to manage. Management of ASD is difficult because parents especially in rural areas do not have very good knowledge about this disorder. The main challenge is controlling the disorder with fruitful information. Many parents are unaware of ASD and

P. Mukherjee (✉)
Manipur International University, Imphal, Manipur, India
e-mail: prasen.mscit09@gmail.com

Vodafone Intelligent Solutions, Pune, India

N. Banerji · K. Pati Mandal · B. Chakraborty
National Institute of Technology, Durgapur, India

M. Godse
bizAmica Software, Pune, India



S. Borah et al. (eds.), *Advanced Computational and Communication Paradigms*, Lecture Notes in Networks and Systems 535,
https://doi.org/10.1007/978-981-99-4284-8_15

its symptoms which is very crucial to manage an ASD child. Today, Artificial Intelligence (AI) plays an important role in the healthcare domain where naïve users can grow their knowledge to use such AI services. Natural language processing (NLP) is a part of AI and many NLP interfaces for naïve users have been developed or are in the development process. The main aim of these interfaces is to reduce the gap between humans and the system. Users can extract information easily using NL queries without prior technical knowledge. Information services are one of the most important services and many organizations are working in this area. Many information service architectures have been developed. Recently, Artificial Intelligence (AI) has been adopted by the maximum organizations in this field. NLP and Machine Learning are the two most important components of AI that are being used by organizations in their information services. NLP is a technique that can help to have the ability to alter spoken or written language according to the requirements as in [1]. NLP is a component of artificial intelligence (AI), which can be used to reduce the gap between human and computer interaction. The requirements of automated NL manipulation for spotting, identifying defects, classification and other activities such as isolation and purification are continued by the researchers for more than two decades as in [2]. A huge range of demand has been observed in NLP for its utilization. Maximum NLP-related works were using static data that is related to a corpus, lexicon database and pattern recognition. At present time, the Deep Learning methods are associated with the NLP tools that make better accuracy and efficiency for such NLP tools as in [3]. The main approach of NLP is text analysis which is based on a set of theories and technologies. Some important steps of NLP are morphological, syntactic and semantic analysis. Users mostly interact with NL interfaces to get the desired results. Parts of speech tagging, named-entity recognition and intent analysis are important fields in NL interfaces to get the desired output from NL text or query. One of the most significant areas of study in NLP is sentiment analysis that uses in decision-making processes as in [4].

The proposed research work is focused on ASD-related information extraction where NL queries have been handled from the user side. The most popular steps of NLP have been used like tokenization, POS tagging and substring formation. A combinational algorithm has been applied in the next step to generate various combinations. The SQL from NL query sentences can be generated after processing combinations of words and semantic table generation. SQL with aggregation methods can be generated as per the presence of specifiers, computational adjectives and computational nouns in the NL query sentence. The generated SQL will be processed to produce an answer from the knowledge (default) database by the proposed system.

## 2 Related Works

The long-term health of children is improved by early identification of autism spectrum disorder (ASD). Currently used detection techniques rely on expensive and arbitrary expert assessments. A strategy has been described in [5] that extracts characteristics of ASD from EEG, facial expression and eye fixation data. This system is based on a weighted Naive Bayes algorithm with an accuracy of 87.50%. Eye fixation, facial expression and EEG have a variety of abilities for recognizing children with autism. EEG may be the most discriminating information. The study recommends computational strategy [5] that can improve classification accuracy to detect autism in infants and young children as in [5]. The ASD population exhibits a wide spectrum of failure in language development. Brain imaging modalities and machine learning are now being used in investigations to create objective technology helped by computers to better understand autism. Computer-aided diagnostic methods attempt to classify autistic persons in comparison to their typically developing peers. The main auditory cortex, angular gyrus, cingulate gyri and superior temporal cortex all show signs of hypoactivation as the severity of autism spectrum disorders increases. The accuracy of the second stage, which classifies patients as having moderate or acute autism, is 0.81. The validation findings demonstrate the stability of the suggested architecture for an earlier computer-aided rating scheme as in [6]. Autism, commonly referred to as Autism Spectrum Disorders (ASD) or Autism Spectrum Conditions (ASC), is a neurodevelopmental disorder characterized by recurrent behaviours as well as abnormalities in social interaction and communication. Many autistic persons may have daily challenges, which can manifest as hopelessness, unemployment, or addiction. The patient may receive greater support if intervention starts sooner, which was noted as a key aspect. Article [7] described machine learning techniques to examine this behaviour and categorize participants from a sample of autistic participants in contrast to a control sample of usually developing persons. These tools may have a substantial impact on how mental illnesses are assessed and may help to maintain the development of criteria and categorization as in [7]. The most prevalent form of diagnosis makes use of autistic children's behavioural signals. The improvement of deep learning algorithms and hardware components in recent years has made it possible to utilize intelligent machines to automatically record self-stimulatory behaviours. The key innovation in this work is the use of temporal coherence as unrestricted monitoring between neighbouring frames and the creation of worldwide discrimination on the basis margin to deduce features of self-stimulatory patterns that change slowly. The efficiency of the extracted characteristics has been established by a thorough examination. First, the collected characteristics are generally categorized using the k-means approach to demonstrate the entirely unsupervised categorization of self-stimulation behaviours. The usefulness of features is then assessed using the conditional entropy approach as in [8].

NL query handling is an important part of NLP. Tokenization, POS tagging, grammatical rule creation, permutation-combination, combination creation, deep parsing, word sense disambiguation, semantic representation, SQL generation, etc. are a few important steps to handle NL queries. The WordNet database plays a major role in POS tagging to parse entire query sentences. Word or sentences may be understandable using the WordNet database at the time of word sense disambiguation. The related work has been elaborated on important steps of NL query handling like POS tagging, substring formation, combination creation, semantic analysis and SQL generation. It has been observed that the relationship between dependency parser and POS tagging is close. The ongoing work on template-based joint modelling techniques of constructed manual features faces challenges from the problem of feature sparsity and incompleteness as referred to in [9]. The dependency parsing and POS Tagging have been approached together by the authors using a neural network that is transition based. The solution has been given to shift or reduce, tag and label conflicts where three classifiers based on neural networks have been devised. The performance of this approach is better than earlier attempts on collaborative POS Tagging and dependency parser across numerous different NLs as in [9]. Sindhi is an old dialect that employs certain grammatical rules with Parts-Of-Speech tags. This method is related to proper speech type or measures expression in language processing pronunciation as in [10]. 28,000 words corpus has been implemented using poetry, primary textbook, newspapers and stories. The proposed implemented corpus has been used to present the programmed explanation of POS Tagging based on Sindhi text. A support vector machine (SVM) has been used for tagging sentences in the Sindhi language. The result states better accuracy of SVM tagger than the existing approaches. An accuracy of 97.86% for ambiguous, unambiguous and unidentified words has been found for this method. The accuracy of the POS tagger for recognized words and phrases is higher than the accuracy of the POS tagger for unidentified terms as in [10]. A straightforward but efficient sequence-to-sequence neural architecture has been implemented using a hybrid task-based long- and short-term memory neural network structure based on a clear transition mechanism. Experimentation has been done on five different datasets that proved this model as highly competitive. The performance has been reported in Chinese literature by this proposed neural joint model using well-trained character level embedding as in [11]. The presented corpus is one million annotated words for Arabic text pre-processing and this pre-processing is based on machine learning. POS Tagging, Named Entity Recognition (NER) and word segmentation constitute a few individual sequence labelling tasks as in [12]. Estimates implies that the single-component arrangement produces a quicker operation when it is able to give accuracy and recall. The annotator has configured the outcome using a fine-grained result tag that makes other tasks such as lemmatization easier and this annotator is free trained OpenNLP component as in [12]. Speech processing has been shown to be an effective method of identifying people's varied moods. This study aims to detect the existence of ASD and examine autistic children's vocal expressions of their emotions. By using source factors linked with children's voices, the provided computerized tone-based system can recognize and categorize seven fundamental reactions (anger, disgust, neutral, happy, tranquility, fear and sorrow) that are conveyed by children. To detect potential emotions displayed by the children and gauge their behavioural condition, a Multi-layer Perceptron (MLP) Classifier is trained using a variety of prime speech data, including Mel-frequency Cepstral Coefficients (MFCC) and Spectrogram. In order to determine the appropriate training and care to improve their quality of life, people must first understand their individual needs and limitations [13]. ASD, a neurodevelopmental condition, hampers a person's cognitive, verbal, object identification, communication and social skills. Although early ASD discovery can help with diagnosis and the implementation of appropriate actions for effective mitigation, ASD can be identified sooner with AI approaches than with conventional ones. The objective of this work was to more precisely detect ASD by proposing a machine learning model that examines ASD data at various age levels. In this study[14] collected ASD datasets from toddlers, kids, teens and adults and employed a variety of feature selection strategies. The performance of several classifiers has been evaluated using measures including prediction performance, kappa statistics, the f1-measure and AUROC. A non-parametric statistical test is used to assess each classifier's performance. The Support Vector Machine (SVM) performs better than other classifiers for the small child, adolescent and adult datasets as in [14]. A neurodevelopmental illness called ASD has an impact on people's daily lives. Despite the fact that it is difficult to entirely evade this illness, the extent of the illness can be reduced by beginning early therapies. This study [15] is a useful framework for comparing several computational-based approaches for the early identification of ASD. Four common ASD datasets (Toddlers, Adolescents, Children and Adults) are used in this investigation [15]. The most effective categorization strategies and the top Feature Scaling (FS) strategies for each ASD dataset are determined by comparing the classification results using various statistical analytic techniques. Using normalizer FS scaling for toddlers and children and the Quantile Transformer (QT) FS technique for scaling adults and adolescents, the maximum accuracy levels are attained. In the calculation of the ASD risk factors, the most crucial characteristics are designated. This experimental evaluation [15] suggests that the Machine learning approach is fine-tuned to help predict ASD in individuals of various ages as in [15].

## 3 Architecture

The main approach of this system is to extract ASD information from the healthcare database. The diagram (Fig. 1) of the proposed system has been given where the user will login into the system and post a query in NL. The query will be validated in

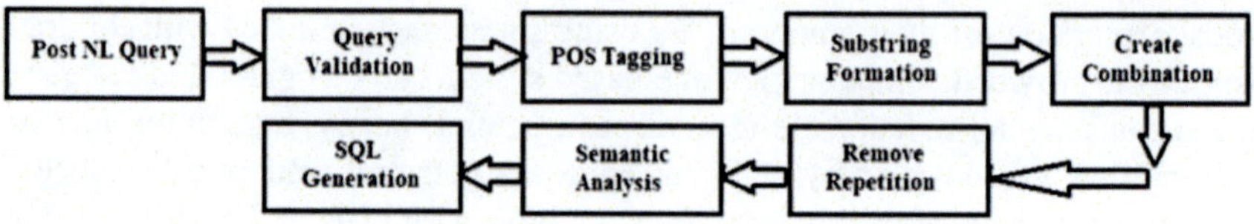


**Fig. 1** Flow Diagram of proposed system

the next step. After completion of validation, the query will be tokenized a token list will be generated. The POS tagger has been developed which is not able to detect noun adjectives, verbs and adverbs. The POS tagger will indicate nouns, verbs, adjectives and adverbs as unknown tokens. Alternatively, a token may be detected as Determiner, pronoun, auxiliary verb, preposition, wh-words or conjunction. If the token is detected by the POS tagger, then it will be removed from the token list and the remaining unknown tokens will be treated as unique tokens. These unique tokens will be used to make substrings. In the next step, the proposed system will create possibilities from each substring, and the occurrence of repetition of possibilities will be removed as in [16]. Now, the semantic table will be prepared using the synonyms database. This semantic table will produce the SQL. The conventional query and conventional query with computation can be handled by this proposed system. The proposed system will handle 'sum', 'count', 'average', 'max', 'min', etc. type aggregation. The step of the proposed system has been given below.

## 3.1 The Proposed System's Steps

### Post NL Query and Verification

The user will log in to the proposed system and this proposed system will read NL query. The POS tagger is used to validate the NL query where the proposed system will check for the unknown token(s). POS tagger follows a list that contains part of speech words except for nouns, verbs, adjectives and adverbs. The POS tagger searches for the unknown word(s) according to the list which means the POS tagger is looking for the word which is not present in the list. This unknown word is termed a unique word or unique token. If the unique token is present in the token list, then the query is validated.

### POS Tagging

The entire sentence is tokenized after completing the validation step. Now, Again POS tagger will match each token with the POS tagger list. If any token is matched, then the matched token will be removed from the list and non-matched tokens will be sent for the next step.

*Pseudo Code:*

```
//Sentences entered by users are splitted and saved in a string array.
(1) Accept Input Sentence S.
(2) Words1[a] is a string array created by splitting input statement S.
   (a) Specify Parts of Speech List: WH1[a] =
{"who","how"...} AUX1[a] ={"was","were","has"...}
PRE1[a] ={"at","into","on","over"...} DET1[a]
={"a","an","the"...} PRO1[a] ={"we","our","they" }
   ....
   Words1[a]=str1.Split();
   FOR x=0 to x<words1.Length FOR y = 0; to y
<Wh1.Length IF(words1[x] == Wh1[y]) THEN
words1[x] = ",";
   FINISH IF FINISH FOR
FINISH FOR
   FOR y=0 to y<Aux1.Length IF(words1[x] ==
Aux1[y]) THEN Search_Aux[z] = words1[x]; words1[x]
= ",";
   z is increased by 1 FINISH IF
   FINISH FOR
   ...
```

Example 1: '*please tell me about the autism problem*'—The NL Query chops the query into a token array (Table 1). After completion of this step, Unique words have been identified (Table 2).

**Substring Formation**

The substring will be formed in this step that has been given in Table 3.

**Table 1** Token array for Example 1

| 0 | 1 | 2 | 3 | 4 | 5 | 6 |
|---|---|---|---|---|---|---|
| Please | Tell | Me | About | The | Autism | Problem |

**Table 2** Unique words of Example 1

| 0 | 1 | 2 | 3 | 4 | 5 | 6 |
|---|---|---|---|---|---|---|
| Please | Tell | | | | Autism | Problem |
| Unique Words | | | | | Unique Words | |

**Table 3** Substring array

| 0 | 1 | 2 | 3 | 4 | 5 |
|---|---|---|---|---|---|
| Please<br>Tell | Autism<br>Problem | Null | Null | Null | Null |

### Create Combination

Now, each substring will create combinations. According to example, combinations are—'please' 'tell', 'please tell', 'autism', 'problem' and 'autism problem'.

```
Pseudo Code:
// Pseudo Code for combination Creation method
  FOR x=0 to x<subs3.Length z=0;
IF(subs3[x] != null) THEN words = subs3[x].Split();
FOR y=0 to y<word.Length
 1=words1[y].CompareTo(" "); IF(x<0) THEN
Do nothing; ELSE
Copywords1[z] = words1[j]; z is incremented by 1
FINISH IF
  FINISH FOR
 CALL GetCombination1 (copywords1, z) Method GetCombination1 (string[] list1, int
m) Method Int x, y, z, vcount1;
 Semantic sm1 = new Semantic(); String calc1 = null;
 Double count1 = Math.Pow(2, m); FOR x=0 to x<=count1
 String str4 = Convert.ToString(x, 2).PadLeft(m, '0'); FOR y = 0 to y<str4.Length
 IF (str4[y] == '1') THEN
 Calc1 = calc1 + list1[y] + " "; FINISH IF
 FINISH FOR
 sm1.InsertCombination(calc1); Method.
 Calc1 = null; END FOR
 END GetCombination1
 .......
```

### Remove Repetition

The repetitive words will be checked by the system in this step. If there are any repetitive words, then they will be removed.

### Semantic Analysis

The proposed system will read each combination and start matching with specifier words and computational adjective words. These specifier words and computational adjective words are used to identify the query type. Examples have been given below.

```
string [ ] specifiers = {"total", "full", "whole" };
string[] comp_adj = {"many", "much", "long", "far", "maximum", "minimum"};
string[] comp_noun = {"number", "amount", "distance", "time", "sum", "average",
"count" };
```

If a specifier word or computational adjective word is present in the user's query, then it will be identified as a conventional query with aggregation and specifier word, and computational noun and computational adjective will be inserted into the semantic table. Without matching of specifier word or computational adjective word, the query will be identified as a conventional query. Next, each substring will go for entity search in the entity table where each entity name, synonym, primary key, foreign key and candidate key is assigned. If a substring matches an entity name or with a synonym, then the matched row will be returned and will be inserted into the semantic table. If a substring finds no match in the entity table, then it will go to the attribute table using the attribute search function where entity name, attribute name, synonym, primary key, foreign key and candidate key are assigned. If there is a match between the substring and the attribute name or synonym, then the selected row will be inserted in the semantic table except for the synonym word. Again, if a substring does not match the attribute table, then the substring will go for the value search in the whole database. If any value is matched with the substring in the database, then the corresponding column name will be selected and this column name will be searched again in the attribute table. The matched row will be selected and will be inserted into the semantic table (Table 4).

***Pseudo Code:***

```
//Semantic Table representation
string [ ] specifiers = {"total", "full", "whole" };
string[] comp_adj = {"many", "much", "long", "far", "maximum", "minimum"};
string[] comp_noun = {"number", "amount", "distance", "time", "sum", "average",
"count" };
FOR i = 0 TO Combination.Length - 1 STEP 1
FOR j = 0 TO specifiers.Length – 1 STEP 1
IF specifiers[j]==Combination[i] THEN
// Insert Combination[i] into the Semantic table ELSE
IF comp_adj[j]==Combination[i] THEN
// Insert Combination[i] into the Semantic table ELSE
IF comp_noun[j]==Combination[i] THEN
// Insert Combination[i] into the Semantic table END IF
```

**Table 4** Semantic table representation

| SID | Specifier | Comp_ Noun | Comp_Adj | Entity | Attributes | PK | FK | CK | Value |
|---|---|---|---|---|---|---|---|---|---|
| 1 | Null | Null | Null | Autism | Autism_ Problem | AID | Null | Null | Null |

*FINISH IF FINISH IF FINISH*
*FOR*
*// If the Combination[i] is not Specifiers, comp_adj and comp_noun then each combination will go for Entity, Attributes and Value check. int Entity=0, Attributes=0, Value=0;*
*Entity=CALL Entity_Search(Combination[i])*
*// If Entity_Search( ) method returns 1, it means Combination[i] is an entity and return value will be stored in Entity variable.*
*IF Entity == 1 THEN*
*// Insert Combination[i] into the Semantic table ELSE*
*Attributes=CALL Attributes_Search(Combination[i]) IF Attributes == 1 THEN*
*// Insert Combination[i] into the Semantic table*
Table 3: Semantic Table for the NL query for example 1
*ELSE*
*Value=CALL Value_Search(Combination[i]) IF Value == 1 THEN*
*CALL Attributes_Search(Combination[i]) END IF*
*FINISH IF FINISH IF*
*FINISH FOR*

## SQL Generation

From the semantic table, the SQL will be constructed where entities will be used as table names and attribute names will be used as attribute names in the SQL. The value will be used in the condition of the SQL. If any specifier word, a computational noun or computational adjective is present, then it will create an aggregation command like count (doctor names). According to the above example, the SQL query will be generated as 'SELECT Autism_Problem from Autism'. The output has been given in Fig. 3. Another example has been given where the NL query is the conventional query with aggregation. The NL query is 'how many doctor names are there in Occupational Therapy'. The semantic table (Table 5) will be.

After completion of semantic table representation, the SQL will be "*SELECT COUNT (DName) FROM Doctor WHERE DSpecialization = 'Occupational Therapy*". The output of these two NL queries can be seen in Figs. 2 and 3.

**Table 5** Semantic table of Conventional query with Aggregation

| SID | Specifier | Comp_ Noun | Comp_ Adj | Entity | Attributes | PK | FK | CK | Value |
|---|---|---|---|---|---|---|---|---|---|
| 1 | How | Many | Null | Doctor | DName | SID | EID | DID | Null |
| 2 | Null | Null | Null | Doctor | DSpecialization | SID | EID | DID | Occupational therapy |

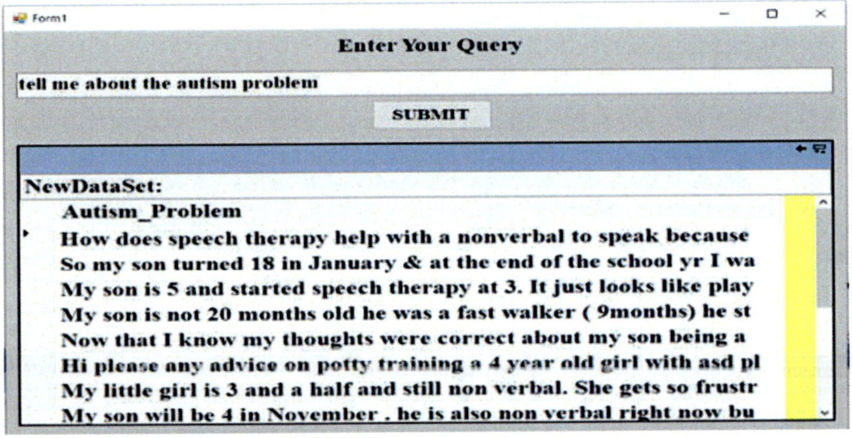


Fig. 2 Output of the first example

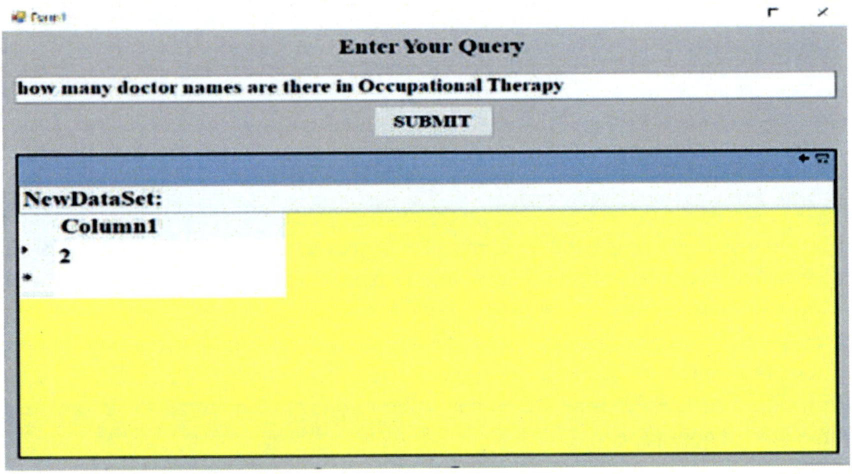


Fig. 3 Output of the second example

# 4 Database

A database has been implemented on autism spectrum disorder where real-life autism problems have been initialized (Table 6). All information regarding autism has been collected from the internet as well as a Doctor database (Table 7) has been implemented where doctors-related information has been given clearly.

**Table 6** Autism Table in the proposed database

| AID | Autism_Problem |
|---|---|
| 1 | I'm confused with my 3-year-old son. All he does is to mumbles only with no proper words but he goes to speech therapy every month. Does it speech therapy help for such patients? |
| 2 | I can't show him how to do potty during the day while his dad is at work. I've realized he needs examples and sounds |
| 3 | Hi! Please advice on potty training for a 4-year-old girl with ASD |
| 4 | I need help for my son who is 2 years old and he has meltdowns but today after every 5 min he is having a meltdown. Can anyone help me with some tips on what to do at this point |

**Table 7** Doctor Table in the proposed system

| SID | DID | DName | DSpecialization |
|---|---|---|---|
| S001 | D001 | Dr. Satyajit Bose | Child Psychiatry |
| S002 | D002 | Dr. Shantanu Das | Clinical Neuroscienes |
| S003 | D003 | Dr. Abhishek Ray | Clinical Psychology |
| S004 | D008 | Dr. Balwant Kumar | Neurology |
| S005 | D006 | Dr. Vinod Rao | Clinical Neuroscienes |

## 5 Limitation

The proposed system is good for conventional queries and conventional queries with aggregation. But, this system is not able to handle time-related queries *like 'what were the doctor names of Occupational Therapy'* or *'how many doctors were in Mission Hospital last year'*. These types of queries are directly related to time and a temporal database needs to be implemented. Time-related queries are difficult to handle because the time frame has to be initialized according to the nature of time-related queries.

## 6 Future Work

The proposed system can manage NL queries with aggregation and the proposed system is flexible to convert these NL queries to SQL. But Time-related query conversion is difficult because responses of time-related queries are different types and responses will be according to the time. The temporal database implementation is needed otherwise it is not possible to extract information from the database according to the time. So, Time-related query handle and temporal database implementation can be considered as future work of this system.

## 7 Conclusion

The proposed system will accept the NL queries for the conventional database as well as can handle NL queries with aggregation. The proposed system will extract computational words, entities and attributes from the NL query using the combinational algorithm and it will prepare the semantic table to generate the SQL for information extraction. The proposed system has the capacity to produce many SQL formations according to the NL queries. This system can be deployed in the healthcare domain to create many services for the naïve user.

## References

1. Mukherjee P, Santra S, Bhowmick S, Paul A, Chatterjee P, Deyasi A (2018) Development of GUI for text-to-speech recognition using natural language processing. 2nd international conference on electronics, materials engineering & nano-technology (IEMENTech), pp 1–4,May
2. Ferrari A (2018) Natural language requirements processing: from research to practice. In: Proceedings of the 40th international conference on software engineering: companion proceedings, 536–537
3. Fahad SA, Yahya AE (2018) Inflectional review of deep learning on natural language processing. In: 2018 international conference on smart computing and electronic enterprise (ICSCEE) IEEE, 1–4
4. Elsaid Moussa M, Hussein Mohamed E, Hassan Haggag M (2021) Opinion mining: a hybrid framework based on lexicon and machine learning approaches. Int J Comput Appl 43(8):786–794
5. Liao M, Duan H, Wang G (2022) Application of machine learning techniques to detect the children with autism spectrum disorder. J Healthcare Eng Hindawi, 1–10
6. Haweel R, Shalaby AM, Mahmoud AH, Ghazal M, Seada N, Ghoniemy S, Casanova M, Barnes GN, El-Baz A (2021).: A novel grading system for autism severity level using task-based functional MRI: a response to speech study. IEEE Access, 100570–100582
7. Robles M, Namdarian N, Otto J, Wassiljew E, Navab N, Falter-Wagner CM, Roth D (2022) A virtual reality based system for the screening and classification of autism. IEEE transactions on visualization and computer graphics, pp 2168–2177
8. Liang S, Sabri AQM, Alnajjar F, Loo CK (2021) A virtual reality based system for the screening and classification of autism. Autism spectrum self-stimulatory behaviors classification using explainable temporal coherency deep features and SVM classifier IEEE Access, pp 34264–34275
9. Yang L, Zhang M, Liu Y, Sun M, Yu N, Fu G (2018) Joint POS tagging and dependence parsing with transition-based neural networks. IEEE/ACM transactions on audio, speech, and language processing, pp 1352–1358
10. Surahio FA, Mahar JA (2018) Prediction system for sindhi parts of speech tags by using support vector machine. International conference on computing, mathematics and engineering technologies (iCoMET), pp 1–6
11. Zhang M, Yu N, Fu G (2018) A simple and effective neural model for joint word segmentation and POS tagging. IEEE/ACM transactions on audio, speech, and language processing, pp 1528–1538
12. Freihat AA, Bella G, Mubarak H, Giunchiglia F (2018) A single-model approach for arabic segmentation POS tagging and named entity recognition. 2nd international conference on natural language and speech processing (ICNLSP), pp 1–8

13. Sukumaran P, Govardhanan K (2021) Towards voice based prediction and analysis of emotions in ASD children. J Intell Fuzzy Syst 41(5):5317–5326
14. Bala M, Ali MH, Satu MS, Hasan KF, Moni MA (2022) Efficient machine learning models for early stage detection of autism spectrum disorder. Algorithms 15(5):1–22
15. Hasan SM, Uddin MP, Al Mamun M, Sharif MI, Ulhaq A, Krishnamoorthy G (2022) A machine learning framework for early-stage detection of autism spectrum disorders. IEEE Access, 15038–15057
16. Mukherjee P, Chattopadhyay A, Chakraborty B, Nandi D (2021) Natural language query handling using extended knowledge provider system. Int J Knowle Based Intell Eng Syst 25(1):1–19

# Spiking Neural Network in Computer Vision: Techniques, Tools and Trends

**Rohit Agarwal, Palash Ghosal, Narayan Murmu, and Debashis Nandi**

**Abstract** Artificial Neural Networks (ANN) have become one of the most powerful machine learning tools that cover a wide range of applications such as surveillance, video and image recognition, medical image analysis, control systems, robotics, bioinformatics, etc. These wide applications have encouraged researchers to search for more realistic neuron models that really mimic human biological neurons. Recent neurological research explored a set of new characteristics of biological neurons which led to generating the third-generation neural networks called Spiking Neural Networks (SNN). SNNs are more faithful to drawing biological properties which makes them energy-efficient and fast with higher processing capability. Many SNN models along with various learning algorithms have been proposed in the last decade to explore fast and efficient machine learning applications. In this paper, we have reviewed recent advancements in SNN architectures, their learning algorithms, applications, advantages, limitations and future scopes. We have also presented a brief comparison among some well-cited state-of-the-art works in terms of performance, advantages and limitations.

**Keywords** Spiking neural networks · SNN learning · SNN coding · Action potential · Integrate-and-fire

---

R. Agarwal (✉) · N. Murmu · D. Nandi
National Institute of Technology, Durgapur 713209, India
e-mail: ra.22cs1102@phd.nitdgp.ac.in

N. Murmu
e-mail: narayan@cse.nitdgp.ac.in

D. Nandi
e-mail: debashis@cse.nitdgp.ac.in

P. Ghosal
Sikkim Manipal Institute of Technology, Majitar 737136, Sikkim, India

 
S. Borah et al. (eds.), *Advanced Computational and Communication Paradigms*,
Lecture Notes in Networks and Systems 535,
https://doi.org/10.1007/978-981-99-4284-8_16

# 1 Introduction

In this fast-paced world, neural network has emerged as one of the most innovative technological developments in various fields like Artificial Intelligence (AI), Internet of Things (IoT) and Computer Vision. It has got thrust for its excellent performance and accuracy. These neural networks are chiefly categorized into three generations based on their evolution. The first generation of neural networks consisted of multilayered perceptrons, Hopfield network [12] and Boltzmann machines which were used to compute Boolean and digital functions. With the increase in data and its feature set, a new advancement in this field was imminent, and as a result, the second generation of neural networks was introduced. This second generation of neural networks applied activation functions and a set of weighted sum inputs. Over the time, this second-generation neural network, ANN, evolved into more powerful and biologically realistic models with an attempt to mimic the human brain. However, in spite of that, we are yet to have complete knowledge of how the human brain solves problems. It has been revealed in neurological research that in numerous parts of the human nervous system information stored in the neurons depends upon the timing of action potentials of each individual neuron giving rise to a new type of neuron called the Spiking neuron. The networks formed by the spiking neurons are termed Spiking Neural Networks (SNN) and are classified to be the third-generation neural networks. The SNNs are biologically more realistic than the traditional ANNs due to the use of discrete spikes for computation and information transmission. Unlike second-generation neural networks, which use backpropagation for learning, SNNs are yet to have a standardized learning method due to the non-differential characteristics of the spikes. So, different kinds of learning techniques are approximated like spike timing-dependent plasticity (STDP) [3], surrogate descent or conversion of analog values of deep neural network to discrete spikes with some additional adjustments.

The history of spiking neural networks began in the mid-twentieth century when Alan Hodgkin and Andrew Huxley first proposed and created the scientific model of SNN. The original inspiration of SNNs is from the neuron's communication process for the transformation of information with the help of discrete spikes. A spike is generated when a threshold is crossed by the sum of the differences in the membrane potential from the pre-synaptic stimulation. These spikes send signals to the intermediate neurons resulting in the increase or decrease in their potential. In the succeeding sections, we will cover the architectures, encoding techniques and learning methods applied to SNNs.

The organization of the rest part of the is as follows: Sect. 2 describes some state-of-the-art architectures of SNN where various models for threshold firing neurons are discussed along with their equations. Section 3 contains few of the coding strategies generally used for encoding. It is followed by Sect. 4 in which the learning methods are discussed, namely, Unsupervised and Supervised Learning in two separate Sects. 4.1 and 4.2, respectively. The applications and comparison with other methods are shown in Sect. 5. The paper is winded up with some future scope in Sect. 6.

## 2 SNN: State-of-the-Art Architectures

Coventionally, when neurons are at rest, the soma, which is the key body of a neuron, is negatively charged, termed resting potential. It is maintained by a certain concentration of sodium and potassium ions inside the cell. With the arrival of currents, the cells initiate depolarization, which triggers the voltage-activated sodium channels allowing the sodium ions to enter the cell, resulting in further depolarization. Depolarization activates the voltage-gated potassium channels and thus potassium ions move out of the cell, leading to the repolarization. The cell is further brought below the resting potential state by the potassium channels, known as hyperpolarization (Fig. 1).

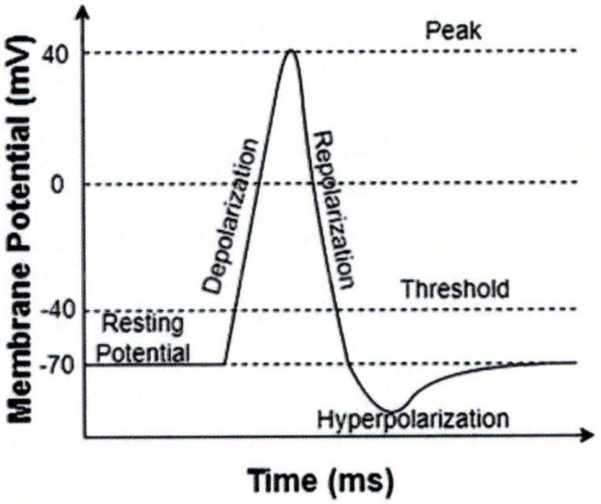


**Fig. 1** Action-potential

Spiking neurons (Fig. 2) and the linking synapses of an SNN are assisted by customizable scalar parameters, known as weights. Initially, encoding of analog data to training spikes is carried out using some state-of-the-art techniques, for instance, rate-based, latency-based, delta modulation-based, temporal and population coding [7]. The adjoining neurons in the network generate spikes $S$ at time interval $t$ when

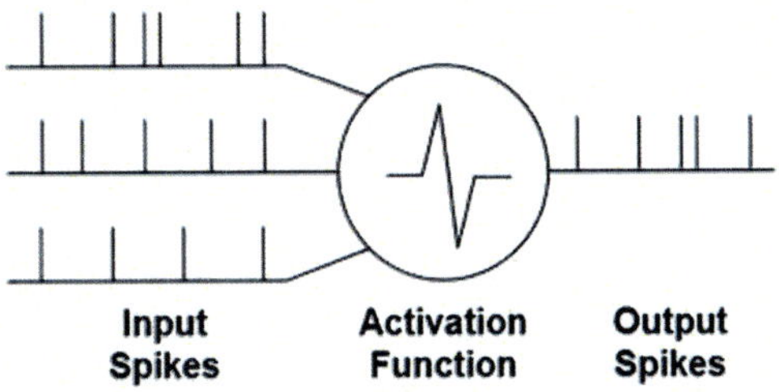


**Fig. 2** Spiking perceptron

the threshold $\theta$ is crossed by the membrane potential $\theta_t$ [7] as shown in Eq. 1:

$$S_t = \begin{cases} 1, & \text{if } \theta \geq \theta_t \\ 0, & \text{otherwise} \end{cases} \tag{1}$$

where 0 and 1 indicate the non-generation and generation of spikes, respectively. With the generation of spikes, the change in the characteristics of the membrane potential is shown in Fig. 1.

In spiking neural network, there are numerous models to implement the threshold firing neurons, for instance, Hodgkin–Huxley [11], Integrate-and-Fire (IF) [1], Leaky Integrate-and-Fire (LIF) [5], FitzHugh–Nagumo (FHN) [9], Izhikevich [13], adaptive Exponential Integrate-and-Fire (aEIF) [4], Spike Response Model (SRM) [10] and more.

Hodgkin–Huxley model [11] laid the founding block of the spiking neural network. This model is described by Eq. 2.

$$I = C_m \frac{dV_m}{dt} + g_K(V_m - V_K) + g_{Na}(V_m - V_{Na}) + g_L(V_m - V_L) \tag{2}$$

where $I$ is the total current passing through the membrane, $C_m$ is the capacitance of the membrane, $g_K$ is the potassium conductance, $g_{Na}$ is the sodium conductance, $g_L$ is the leak conductance, $V_m$ is the membrane voltage, $V_K$ is the potassium reversal potential, $V_{Na}$ is the sodium reversal potential and $V_L$ is the leak reversal potentials.

Owing to the complexity and computational expensiveness of the Hodgkin–Huxley model, the IF model [1] came into the picture of the spiking neural networks. In the model, channel potential was removed from the membrane potential.

$$I = C_m \frac{dV_m}{dt} \tag{3}$$

where $I$ is the input current, $C_m$ is the capacitance of the membrane and $V_m$ is the membrane voltage.

Though the IF model was easy to implement, it lacked the actual behaviour of the biological neurons as various neuronal dynamics aspects were neglected. To make the IF model more realistic, the LIF model [5] came into the scenario in which membrane potential leaks slowly over time, like biological neurons.

$$\tau_m \frac{dV_m}{dt} = V_{rest} - V_m + R_m I \tag{4}$$

where $\tau_m$ is the time constant which is equals to $C_m R_m$, $V_m$ is the membrane voltage, $V_{rest}$ is the resting potential, I is the input current, $R_m$ is the resistance and $C_m$ is the capacity.

FHN model [9] is the simplified version of Hodgkin–Huxley model and is expressed by the following equations:

$$\frac{dv}{dt} = v_t - \frac{v_t^3}{3} - w_t + I_{ext} \tag{5}$$

$$\frac{dw}{dt} = v_t - a - bw_t \tag{6}$$

where $v_t$ is the membrane potential, $w_t$ is the auxiliary function, $I_{ext}$ is the input current and a and b are the controlling parameters for the shape of membrane potential.

Izhikevich model [13] is a combination of biological plausibility of Hodgkin and Huxley and computational efficiency of IF.

$$\frac{dv}{dt} = 0.04v_t^2 + 5v_t + 140 - u_t + I_t \tag{7}$$

$$\frac{du}{dt} = a(bv_t - u_t) \tag{8}$$

where $v_t$ is the membrane potential, $u_t$ is the membrane recovery variable, $a$ controls the decay rate of $u_t$ and $b$ controls the sensitivity of $u_t$.

## 3 Neural Coding Strategies

The spiking neural network models deal with discrete spikes for learning purpose but, usually, the information are in the continuous form. To encode the information into spikes different coding strategies are used such as rate coding, binary coding, temporal coding and population coding.

### 3.1 *Rate Coding*

In rate coding, `spiking frequency` is determined from the input features of the information. This coding technique is inspired by the fact that, in biological neurons, generation of spikes depends upon the intensity of the stimuli, stronger stimuli are more probable to fire a spike rather than a weaker one. Here, firing of spikes is considered as probabilistic events. In general, Poisson or Bernoulli trials are used for the generation of spikes.

### 3.2 *Latency Coding*

In latency coding `spike timing` is determined from the input features of the information. This coding strategy was proposed by Johansson and Birznieks, 2004. Here, the delay between a particular event and the first spike is encoded for a fast response within a few milliseconds.

### 3.3 Temporal Coding

In temporal coding, generation of spikes depends upon the `temporal change` from the input features of the information. Studies in neuroscience suggest that the human brain is more susceptible to changes in nature, that is, more number of spikes are generated if frequent changes are observed in the input and vice versa.

## 4 Learning Methods

Irrespective of the type of network, that is, spiking or non-spiking, learning is achieved by adjusting some scalar parameters between the connections of pre-synaptic and post-synaptic neurons. These scalar parameters are termed synaptic weights. The synaptic weights are fine-tuned with respect to the relative timings of the spikes in a duration of every tens of millions in length roughly. Numerous learning rules have been identified but the most commonly used rules are unsupervised and supervised learning.

### 4.1 Unsupervised Learning

Unsupervised learning simply depends upon the characteristics and behaviour of the input data where no label is provided. In SNN, unsupervised learning can be realized by the Hebbian process [2] and its learning mechanism, which falls under the umbrella of STDP [3]. The adjustments in the synaptic weights depend upon the spike timing difference between the pre-synaptic and post-synaptic neurons. If the post-synaptic neurons fire just after the pre-synaptic neurons, then the synaptic weights are strengthened, which is termed long-term potentiation (LTP), and in the alternate case, the synaptic weights are weakened, which is termed long-term depression (LTD).

### 4.2 Supervised Learning

Supervised learning depends upon the labels of the input data and its objective is to minimize the error by using the backpropagation algorithm. In SNN, the calculation of gradients is insignificant owing to the non-differentiability of the spikes which results in the dead neuron problem, and to overcome this, a modified version of gradient descent, surrogate gradient descent [18], is used. Here, smoothing of the discrete values is performed using smoothing functions like the sigmoid function, which simultaneously smooths out the gradients.

**Table 1** A comparative study of few well-known SNN-derived tasks

| Author | Technique | Dataset | Performance | Advantage | Limitation |
|---|---|---|---|---|---|
| Esser (2015) [8] | Neuromorphic chip (IBM TrueNorth) was used | MNIST | 99.42 | It can be trained using backpropagation | Can only be used for classification and recognition |
| O'Connor (2016) [19] | Fractional SGD | MNIST | 97.93 | Simple addition, indexing and comparison are applied if all the parameters are scaled by the inverse of the learning rate | Low accuracy than the classical ANNs |
| Lee (2016) [16] | Spiking CNN with backpropagation | MNIST | 99.31 | New normalization technique was used for backpropagation | Cannot be used on event-driven data |
| Liu (2017) [17] | SNN with three-layered temporal backpropagating neuromorphic architecture | MNIST | 99.10 | Pliable neural model size reduction maintaining power-efficient information processing | Can only be used for classification neural network |
| Diehl (2015) [6] | Spiking CNN with offline learning and conversion | MNIST | 99.10 | Conversion of ANN to SNN with using ReLU with no bias and normalization to adjust the firing rates | Slow training due to model-based weight normalization |
| Tavanaei (2017) [20] | Spike-timing-dependent plasticity-based backpropagation with three layers | MNIST | 97.20 | Derivatives of activation functions are replaced by spike events | Low accuracy than the classical ANNs |
| Kheradpisheh (2020) [14] | Multilayer SNN with a form of temporal coding | N-Caltech | 99.20 | One spike per stimulus with the latencies carrying information | Undefined latency when threshold is not reached by the neurons |
| Wunderlich (2021) [21] | Spike-timing-dependent based backpropagation without approximations | Ying Yang | 98.10 | Backpropagation without approximations through discrete values are computed | Divergence of the gradient at vital points in the framework |
| Kirkland (2022) [15] | Event-based unsupervised SNN using STDP | N-Caltech | 97.33 | STDP trained Spiking CNN weights are used, for instance, segmentation | Decrease of accuracy with the increase in the number of classes for the same dataset |

## 5 Applications and Comparison Among Different Methods

SNNs are still in the learning phase with no concrete formulation of the learning technique, yet various works have been carried out by the researchers all around the world with some notable results. So far, SNNs have outperformed many of the classical ANN classification models on some commonly used datasets, like MNIST, CIFAR-10, CIFAR-100, Iris and more. Also, SNNs are being implemented on hardwares like Intel's Loihi neuromorphic chip assisted by the Nengo neural network development tool, MT-spike [17] which is a temporal backpropagating neuromorphic chip and many more. A modified version of the ISBI 2D EM Segmentation dataset, N-Caltech, is an instance where image segmentation has been carried out. A comparative study of a few of the well-known SNN-derived tasks are shown in Table 1.

## 6 Conclusion

The authors of this paper have outlined the techniques and tools to implement the spiking neural network along with its trends and applications. SNNs are considered to be the third generation of the neural networks and they are biologically more realistic than the classical ANNs. Also, SNNs are computationally less expensive than other neural networks. There are few challenges in the complete implementation of the SNNs with the segmentation of images being the crucial ones. A huge revamp is possible in all the sectors around the world if accurate implementations of the SNNs can be achieved, especially in the medical field which might save millions of lives due to early and fast detection of disease with high accuracy, which is one of the novel outlines of the said paper.

## References

1. Abbott LF (1999) Lapicque's introduction of the integrate-and-fire model neuron (1907). Brain Res Bull 50:303–304
2. Amit DJ, Brunel N, Tsodyks M (1994) Correlations of cortical Hebbian reverberations: theory versus experiment. J Neurosci 14(11):6435–6445
3. Bi G, Poo M (1998) Synaptic modifications in cultured hippocampal neurons: dependence on spike timing, synaptic strength, and postsynaptic cell type. J Neurosci 18(24):10464–10472
4. Brette R, Gerstner W (2005) Adaptive exponential integrate-and-fire model as an effective description of neuronal activity. J Neurophysiol 94:3637–3642. https://doi.org/10.1152/jn.00686.2005
5. Dayan P, Abbott LF (2005) Theoretical neuroscience: computational and mathematical modeling of neural systems. MIT Press
6. Diehl PU, Neil D, Binas J, Cook M, Liu SC, Pfeiffer M (2015) Fast-classifying, high-accuracy spiking deep networks through weight and threshold balancing. In: 2015 international joint conference on neural networks (IJCNN), pp 1–8. https://doi.org/10.1109/IJCNN.2015.7280696
7. Eshraghian JK, Ward M, Neftci E, Wang X, Lenz G, Dwivedi G, Bennamoun M, Jeong DS, Lu WD (2021) Training spiking neural networks using lessons from deep learning. CoRR. arXiv:2109.12894
8. Esser SK, Merolla PA, Arthur JV, Cassidy AS, Appuswamy R, Andreopoulos A, Berg DJ, McKinstry JL, Melano T, Barch DR, di Nolfo C, Datta P, Amir A, Taba B, Flickner MD, Modha DS (2016) Convolutional networks for fast, energy-efficient neuromorphic computing. Proc Natl Acad Sci 113(41):11441–11446. https://doi.org/10.1073/pnas.1604850113
9. FitzHugh R (1961) Impulses and physiological states in theoretical models of nerve membrane. Biophys J 1(6):445–466. https://doi.org/10.1016/S0006-3495(61)86902-6

10. Gerstner W (2001) A framework for spiking neuron models: the spike response model. In: Handbook of biological physics, vol 4. Elsevier, pp 469–516
11. Hodgkin AL, Huxley AF (1952) A quantitative description of membrane current and its application to conduction and excitation in nerve. J Physiol 117(4):500
12. Hopfield J (1982) Neural networks and physical systems with emergent collective computational abilities. Proc Natl Acad Sci USA 79:2554–2558. https://doi.org/10.1073/pnas.79.8.2554
13. Izhikevich E (2003) Simple model of spiking neurons. IEEE Trans Neural Netw 14(6):1569–1572. https://doi.org/10.1109/TNN.2003.820440
14. Kheradpisheh SR, Masquelier T (2020) Temporal backpropagation for spiking neural networks with one spike per neuron. Int J Neural Syst 30. https://doi.org/10.1142/S0129065720500276
15. Kirkland P, Manna D, Vicente Sola A, Di Caterina G (2022) Unsupervised spiking instance segmentation on event data using STDP features. IEEE Trans Comput 71:1–12. https://doi.org/10.1109/TC.2022.3191968
16. Lee JH, Delbruck T, Pfeiffer M (2016) Training deep spiking neural networks using backpropagation. Front Neurosci 10. https://doi.org/10.3389/fnins.2016.00508, https://www.frontiersin.org/articles/10.3389/fnins.2016.00508
17. Liu T, Liu Z, Lin F, Jin Y, Quan G, Wen W (2017) MT-Spike: a multilayer time-based spiking neuromorphic architecture with temporal error backpropagation. In: Proceedings of the 36th international conference on computer-aided design, ICCAD'17. IEEE Press, pp 450–457
18. Neftci EO, Mostafa H, Zenke F (2019) Surrogate gradient learning in spiking neural networks: Bringing the power of gradient-based optimization to spiking neural networks. IEEE Signal Process Mag 36(6):51–63
19. O'Connor P, Welling M (2016) Deep spiking networks. CoRR. arXiv:1602.08323
20. Tavanaei A, Maida AS (2017) BP-STDP: approximating backpropagation using spike timing dependent plasticity. CoRR. arXiv:1711.04214
21. Wunderlich T, Pehle C (2021) Event-based backpropagation can compute exact gradients for spiking neural networks. Sci Rep 11:12829. https://doi.org/10.1038/s41598-021-91786-z

# A Quantitative Study on the FaceNet System

**Rajesh Gopakumar, Karunakar A. Kotegar, and M. Vishal Anand**

**Abstract** Face recognition is the science of identifying and recognizing human faces in various situations, keeping the constraints like pose variation and occlusion in mind. Due to its impactful applications in safety and security systems, face recognition is becoming extremely popular and is researched extensively even today. The FaceNet method is among the most tested approaches in Deep learning face identification. This method uses a deep convolution neural network for training the data. The face embedding generated can be used to train a face identification system. This study aims to comprehend the FaceNet system, evaluate its performance, and test its accuracy on seven standard datasets. The study also tries to compare how well the FaceNet method works compared to other popular holistic and hybrid methods. From the outcomes of this study, we can conclude that FaceNet showed outstanding results and was better than the other methods. The FaceNet system reached a minimum of 90% accuracy on all standard datasets used on both the pre-trained models, which is a significant number for any face recognition method.

**Keywords** Deep convolutional neural network · FaceNet system · Face recognition · Image processing

R. Gopakumar (✉) · M. Vishal Anand
Computer Science and Engineering, Manipal Institute of Technology, Manipal Academy of Higher Education, Manipal, Karnataka 576104, India
e-mail: rajesh.g@manipal.edu

M. Vishal Anand
e-mail: vishal.anand7@learner.manipal.edu

K. A. Kotegar
Data Science and Computer Applications, Manipal Institute of Technology, Manipal Academy of Higher Education, Manipal, Karnataka 576104, India
e-mail: karunakar.ak@manipal.edu

 
S. Borah et al. (eds.), *Advanced Computational and Communication Paradigms*, Lecture Notes in Networks and Systems 535,
https://doi.org/10.1007/978-981-99-4284-8_17

## 1 Introduction

Due to the large range of commercial law enforcement and security applications that require authentication, face recognition is still a very active topic in computer vision. The non-invasive nature of biometric devices helped them gain popularity. The need to recognize faces is essential in some occupations, and it is proven by research that people vary greatly in their ability to recognize faces. Various constraints, like capture conditions, occlusions, etc., must be addressed to enhance safety and security. For example, various processes like access authority, identity verification, etc. are being made more straightforward due to the advent of face recognition [1].

One of the most popular face recognition methods is the Local Binary Pattern Histograms (LBPH) [2]. The Local Binary Pattern (LBP) labels the pixels in an image by thresholding the area around each pixel and using the result as a binary number. The fundamental method entails segmenting the image containing the face into several local regions, after which LBP texture descriptors are separately retrieved from each part. Concatenating many descriptors results in a general description of the face.

Combining Principal Component Analysis (PCA) and Support Vector Machines (SVM) are other common techniques [3]. PCA is popularly used in exploratory data analysis and for making models for prediction. It works on the principle of dimensional reduction, which is widely used on massive datasets to reduce dimensions. PCA is one of the most widely used techniques due to its simple approach and efficient algorithm [4]. One major disadvantage of this method is the extensive computational effort needed to generate eigenvalues. In addition, the recognition rate decreases significantly in the case of unconstrained images.

One of the prominent face verification and recognition techniques is FaceNet, which is based on deep learning [5]. The Zeiler and Fergus network and the most recent Inception network technique are the two deep learning architectures used by the system [6, 7]. The accuracies achieved for datasets like Labelled Faces in the Wild (LFW) and YouTube Faces are promising 99.63% and 95.12%, respectively [8, 9]. The study aims to test the accuracy of seven standard datasets and compares the results with the other methods. Pre-trained models have been used to simplify the process and reduce training time.

## 2 Related Work

Face recognition systems date back to the early twenty-first century and evolve from extremely low to excellent accuracy, which is applied in real-world scenarios [10]. One of the earliest face recognition methods is using eigenfaces [11]. The eigenface attempts to gather information about the variation in a set of face images, which is used for encoding and comparing the faces. A significant disadvantage of eigenfaces is that it is susceptible to unconstrained constraints like pose variation and low lighting.

A few years later, a concept called Local Binary Pattern Histograms (LBPH) [2] emerged. In this technique, face images are represented in the form of vectors containing data. LBPH is implemented by creating an intermediate image. This image highlights the important characteristics of the face, thereby describing the original input image in a better way. Finally, histograms of each region are extracted, resulting in one ultimate histogram, which is the concatenation of smaller histograms. This ultimate histogram depicts the characteristics of the input image.

Another interesting approach proposed is Fisherfaces [12]. It is better than eigenfaces in classifying the images into different classes. This method removes important features that discriminate one specific person from others through Linear Discriminant Analysis (LDA). The major downside of this method is that it does not capture the constraints like illumination variations quite well.

There are three prominent techniques which employ the Local Binary Patterns (LBP), namely, Ahonen system (AH), Spatial Pyramid Matching (SPM), and the Naive Bayes Nearest Neighbor (NBNN) [13]. The traditional Ahonen approach divides each facial image into non-overlapping, square-shaped mesh areas. A histogram is calculated for each region independently, which, when concatenated, results in a large vector called a "spatially enhanced histogram". Spatial pyramid histogram came into existence due to the need for multi-resolution LBP. Each region is divided into several sub-divisions, and each portion is plotted as a first level histogram until it reaches level L. This technique creates a spatial pyramid histogram with L levels. In the end, all the histograms with multiple resolutions are concatenated.

NBNN offers a more flexible method of spatial matching since local features from one image can be matched to local features in other locations from other photos. This flexibility in spatial matches is beneficial in datasets having pose variations. Since the flexibility is computationally expensive, a variation of NBNN was also proposed, which is called Restricted Naive Bayes Nearest Neighbor (RNBNN). The description matches in the case of RNBNN are limited to coming from the same spot in the image. The computation cost is significantly decreased, comparable to Ahonen's technique.

With the recent developments in deep learning, multiple techniques have come up to solve the complex face recognition problem. Facebook's DeepFace [14] was lauded for achieving excellent accuracy on the well-known Labelled Faces in the Wild (LFW) dataset. With a huge volume of training data and advanced graphic processing units (GPU), face recognition methods using deep learning have significantly improved in performance and were primarily responsible for real-time applications used worldwide. These deep learning techniques use a cascade of numerous layers of processing units for feature extraction and transformation.

The study carried out in [15] on specific models, FaceNet, VCGFace, VCG16, and VCG19, to evaluate the ability of Deep Learning methods in face recognition shows that there is prominent improvement in using deep learning methods. Depending on the face area being compared, the deep learning model accuracy on the whole frontal face and cropped face areas varies. [16] developed a facial recognition system based on transfer learning in fog computing with deep convolutional neural networks for feature extraction. On standard datasets, the system achieved better performance than other algorithms. The method in [17] system uses a pre-trained model VGG16, which works well with medium and smaller face datasets. The image augmentation in the dataset improved the performance and gave a higher accuracy. The ethnicity detection problem in [18] processes the facial images using the intelligent deep learning concept and PCA for dimensionality reduction. The results show a significantly improved performance on benchmark facial datasets. The face-sketch recognition model in [19] is based on transfer learning. A three-channel CNN with triplet loss is used for the recognition. The approach helps limit the dataset to a smaller size and reduces the computing overhead due to the huge dataset usage in common deep learning models.

# 3 Methodology

This study implements and verifies the effectiveness of the FaceNet system based on deep learning for face recognition on various standard face datasets. Deep convolutional neural networks are used by the FaceNet system to optimize its embedding. The one-shot learning technique uses a minimal set of face photos to build the model. In addition, the prior model will be used without training when we need new models to be created.

## *3.1 One-Shot Learning*

One-shot learning is a classification task where an example is given for each class that is used to prepare a model that must predict various unknown examples, which is relatively easy for humans. A person may see a mobile phone one time and will be able to recognize the mobile phone in the future. The deep CNNs are first trained to differentiate between examples of each class. The main idea is to have the models learn the feature vectors, which are effective at extracting features from the input images.

The entire FaceNet system shown in Fig. 1 is comprised of various components. The first component is the images in batches which feed the training and the test data to the FaceNet architecture. The FaceNet architecture is based on the GoogleNet deep CNN, which was trained on a private dataset to build the Siamese network. L2 normalization is used so that the modulus of the embedding generated should be equal to 1. After this stage, we get the final generated embedding in an n-dimensional Euclidean space. FaceNet employs a triplet-loss function to minimize the distance between images of the same identity and maximizes between different characters.

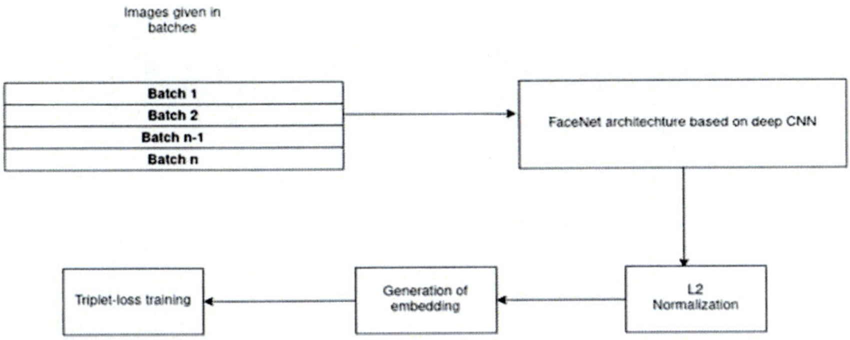


**Fig. 1** Model structure

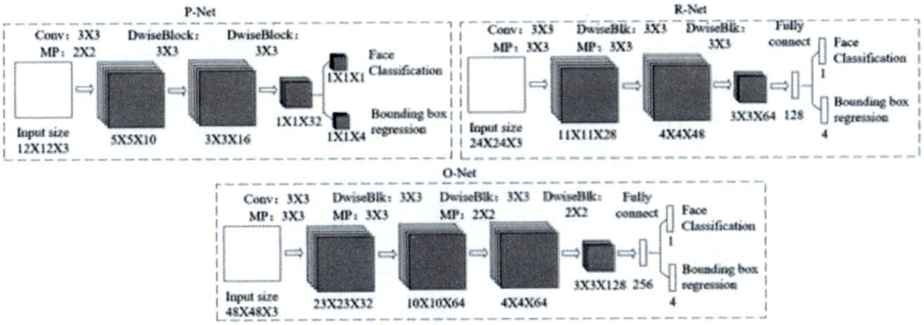


**Fig. 2** Architecture of the various stages of MTCNN

### 3.2 *Multi-task Cascaded Convolution Neural Networks*

Before data training, all the images are aligned and cropped using the Multi-task Cascaded Convolution Neural Networks (MTCNN) [20]. Multi-task Cascaded Convolution Neural Networks (MTCNN) are unified CNNs that work based on the principle of multi-task learning. These CNNs consist of three stages, Fig. 2.

In the first stage, candidate windows are generated using a shallow CNN. Using a somewhat more complicated CNN, a substantial number of non-face windows are rejected in the second stage, which refines the windows. The two eyes, the nose point, and the lips end are the five facial landmarks output in the final stage using a more potent CNN, as shown in Fig. 3.

### 3.3 *Feature Vector Encoding*

Each input for the FaceNet model training phase consists of three pictures of faces. Two of them, the anchor and the positive, show the same person, and the third shows a

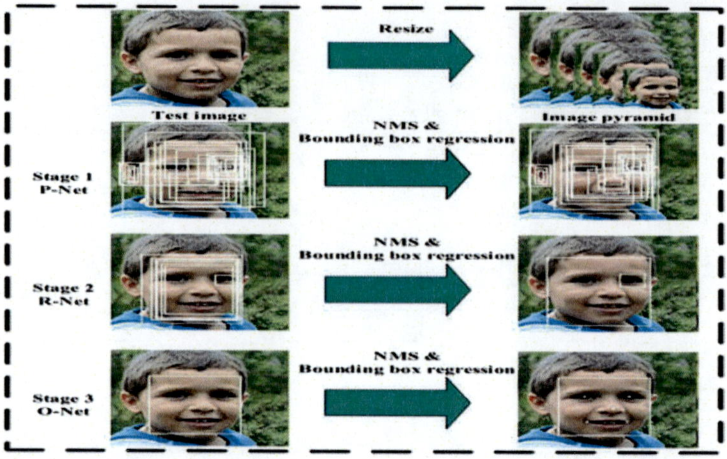


**Fig. 3** Pipeline of the cascaded framework

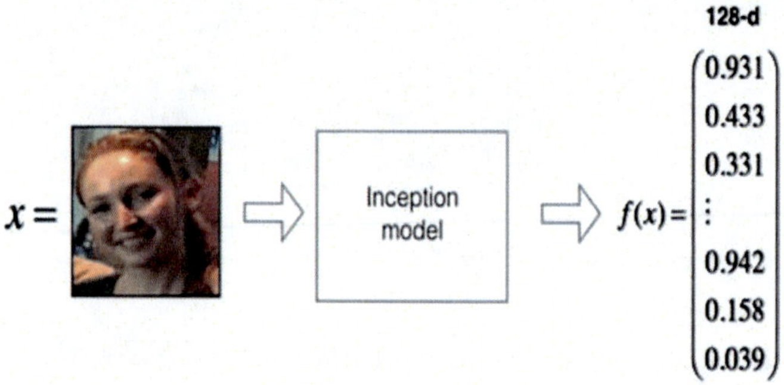


**Fig. 4** Encoding an image to a vector of size 128

different individual (negative). The features of each human face image are processed by the FaceNet model and encoded in a 128-dimensional vector. Figure 4 illustrates the encoding done by the FaceNet system, which generates a 128-dimensional vector.

## *3.4 Triplet Loss Function*

The FaceNet system, which uses the anchor (a), positive (p), and negative (n) inputs, is driven by the triplet-loss function (n). Any random data point can serve as an anchor, and a positive also falls into this category. The third input negative is in a separate class from the positive and anchor.

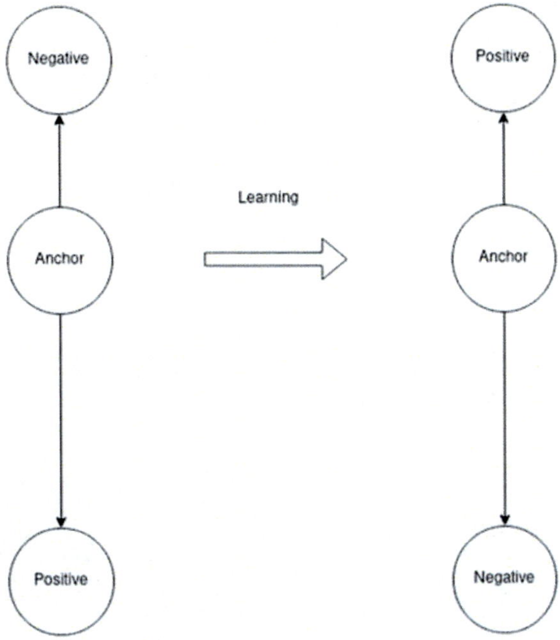


**Fig. 5** The process of minimizing a triplet-loss function

Triplet - loss is mathematically defined as:

$$L = \max(d(a,\ p) - d(a,\ n) + \text{margin}, 0).$$

The distance between the anchor and the positive (d(a, p)) is pushed to 0 and the distance between the anchor and the negative (d(a, n)) is increased above d(a, p) + margin to minimize this loss. After the training, the positive and anchor examples will be closer; otherwise, for the negative and anchor examples, Fig. 5.

The triplet-loss function is the most optimal loss function for training this network. A major challenge of using this loss function is the risk of the model stopping to learn. This happens when the model learns to make the same encoding for different images of the same person, i.e., the distance between positive and anchor will be 0, satisfying the triplet-loss constraint. In these cases, a small margin $\alpha$ is added to create a gap between the anchor and positive versus anchor and negative.

Seven standard datasets have been used for this study, namely, Caltech face dataset, Yale face dataset, YouTube face dataset, GeorgiaTech face dataset, Senthil face dataset, CFP dataset (both frontal and profile considered separately), and Extended Yale B dataset, shown in Table 1. These datasets focus on two major constraints: pose variation and lighting. Since these datasets are widely used to study face recognition, assessing and comparing their performance with the other method will be easier. The samples from the datasets used are shown in Fig. 6.

**Table 1** Datasets used for the study

| Dataset | No. of persons | Total no. of images |
|---|---|---|
| Caltech face [21] | 28 | 450 |
| Yale face [22] | 15 | 166 |
| YouTube face subset [8] | 162 | 464 |
| GeorgiaTech face [23] | 50 | 750 |
| Senthil face [24] | 5 | 80 |
| CFP (only profile images) [25] | 500 | 2000 |
| CFP (only frontal images) [25] | 500 | 5000 |
| Extended Yale B [26] | 28 | 16,588 |

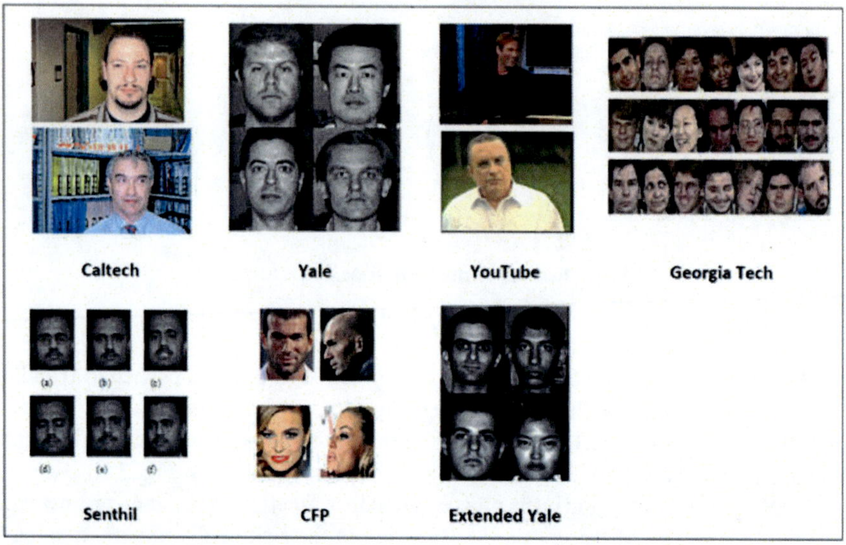


**Fig.6** Samples from datasets used for the study

This study uses the MTCNN to align images fed into the FaceNet system to generate embeddings. All images are aligned to standard 160 × 160 pixels using the three-stage Multi-task Cascaded Convolution Neural Networks (MTCNN).

After training the data, a unified embedding is provided by the FaceNet system. Euclidean space is used to map each face picture in a way that makes the distances in the space correspond to how similar the faces are to one another. In general, the image of a particular person is positioned near to every other image of that person than to any other person in the repository.

Two pre-trained models have been used for this study, Table 2. The pre-trained models are trained on the CASIAWebFace [27] and VGGFace2 [28] datasets. Table

**Table 2** Pre-trained models

| Model name | Training dataset | L F W accuracy | Architecture |
|---|---|---|---|
| 20180408-10 2900 | CASIAWebface | 0.9905 | Inception Res Net v1 |
| 20180402-11 4759 | VGG face2 | 0.9965 | Inception Res Net v1 |

2 summarizes the pre-trained models used and the accuracy of the LFW dataset on them.

The test is now run using the aligned dataset and the selected pre-trained model. In this stage, the model is loaded, and the image pairs will be loaded and parsed. The image pairs are generated randomly from the datasets. The image pairs are used to give the model a pair of photos, and the model answers "yes" or "no". The accuracy is the ratio of the system answer to the real answer of the generated pairs. The embeddings for all images are calculated in the test set before outputting the accuracy and the validation rate. This complete process is repeated for all datasets on the two pre-trained models separately to give out the results.

## 4 Results and Discussion

In this section, the performance and corresponding accuracy of each dataset are determined. The study uses seven public face databases, which have found prominent usage in FR research and are thus used for this study. After preprocessing, the findings are summarized in Table 3.

The findings from evaluating the accuracy of each dataset using pre-trained models are presented in Table 4. From Table 4, it is evident that the majority of the datasets

**Table 3** Number of images aligned for the datasets

| Dataset | Total number of images | Number of images aligned |
|---|---|---|
| Caltech face | 450 | 448 |
| Yale face | 166 | 166 |
| YouTube face subset | 464 | 464 |
| GeorgiaTech face | 750 | 750 |
| Senthil face | 80 | 80 |
| CFP (only profile images) | 2000 | 1911 |
| CFP (only frontal images) | 5000 | 4991 |
| Extended Yale B | 16,588 | 12,552 |

**Table 4** Accuracy (in percentage) of the datasets on facenet system

| Dataset | Accuracy on pre-trained model trained on CasiaWebface dataset | Accuracy on pre-trained model trained on VggFace2 dataset |
|---|---|---|
| Caltech face | 97.20 | 97.93 |
| Yale face | 98.60 | 98.73 |
| YouTube face subset | 89.78 | 90.38 |
| GeorgiaTech face | 99.05 | 98.97 |
| Senthil face | 98.70 | 99.10 |
| CFP (only profile images) | 90.30 | 93.03 |
| CFP (only frontal images) | 97.09 | 97.99 |
| Extended Yale B | 93.73 | 93.96 |

perform better on the pre-trained model trained by VGGFace2. It can also be seen that the accuracy for CFP (only frontal images) is significantly higher than that for CFP (only profile images). The reduced accuracy can be due to more constraints like occlusion while considering the CFP (only profile images) dataset.

It can also be seen that the FaceNet system performs very well (above 95% accuracy) on datasets that have relatively fewer labels with less prominent constraints. In the case of the YouTube Face dataset, accuracy is not so high due to the presence of more prominent pose variation and occlusion. In addition, the presence of dark background and illumination constraint possibly results in the relatively low performance of the FaceNet system on the Extended Yale B database.

The accuracy (in percentage) of the FaceNet system has been tested on the datasets after implementing PCA for dimensionality reduction, shown in Table 5.

It is noted that PCA dimensionality reduction slightly improves the performance of the FaceNet system on all the datasets and reduces the computation time by reducing the number of features for the generated embedding. The advantage is that the number of embeddings can be reduced without losing accuracy, and processes like clustering can be done faster since the files that store the embeddings are smaller.

We compare and contrast the FaceNet system with other methods on three datasets: Yale Face, GeorgiaTech Face, and Extended Yale B, shown in Table 6. After comparing FaceNet with other methods without any illumination normalization performed, FaceNet proved to be the better method overall.

In the case of the Yale Face and the GeorgiaTech Face datasets, which are less limited, FaceNet outperforms all other techniques. However, the FaceNet system performs a little worse than RNBNN in the case of the Extended Yale B dataset. The worse performance is due to the training set enabling each face region to match a comparable face region, inferring a new face by compiling fragments of faces from diverse sources.

**Table 5** Accuracy of the datasets on FaceNet + PCA system

| Dataset | Dataset on which the pre-trained model was trained on | Number of features | | | | |
|---|---|---|---|---|---|---|
| | | 32 | 64 | 128 | 256 | 512 |
| Caltech face | CasiaWebface | 100 | 99.90 | 99.83 | 99.85 | 99.85 |
| | VggFace2 | 99.86 | 99.85 | 99.85 | 99.85 | 99.85 |
| Yale face | CasiaWebface | 99.57 | 99.60 | 99.60 | 99.60 | 99.60 |
| | VggFace2 | 99.83 | 99.80 | 99.80 | 99.80 | 99.80 |
| YouTube face subset | CasiaWebface | 95.18 | 94.82 | 94.75 | 94.75 | 94.75 |
| | VggFace2 | 95.91 | 95.34 | 95.33 | 95.33 | 95.33 |
| GeorgiaTech face | CasiaWebface | 99.92 | 99.85 | 99.78 | 99.78 | 99.78 |
| | VggFace2 | 99.91 | 99.91 | 99.90 | 99.90 | 99.90 |
| Senthil face | CasiaWebface | 97.80 | 97.90 | 97.90 | 97.90 | 97.90 |
| | VggFace2 | 99.60 | 99.60 | 99.60 | 99.60 | 99.60 |
| CFP (only profile images) | CasiaWebface | 93.15 | 93.87 | 93.68 | 93.65 | 93.65 |
| | VggFace2 | 96.33 | 96.57 | 96.47 | 96.47 | 96.47 |
| CFP (only frontal images) | CasiaWebface | 98.03 | 98.18 | 98.01 | 97.99 | 97.98 |
| | VggFace2 | 99.09 | 99.17 | 99.18 | 99.18 | 99.18 |
| Extended Yale B | CasiaWebface | 94.68 | 93.70 | 93.34 | 93.55 | 93.55 |
| | VggFace2 | 94.89 | 93.88 | 93.86 | 93.86 | 93.86 |

**Table 6** Comparison of FaceNet system with other known methods in terms of accuracy

| Dataset | Accuracy (in percentage) of the face recognition method | | | | | |
|---|---|---|---|---|---|---|
| | AH | SPM | NBNN | RNBNN | PCA + SVM | FaceNet |
| Yale face | 84.05 | 82.65 | 86.81 | 88.45 | 93.00 | 98.70 |
| GeorgiaTech face | 75.10 | 77.67 | 92.67 | 81.20 | – | 99.05 |
| Extended Yale B | 73.72 | 72.97 | 93.20 | 94.79 | 81.00 | 93.96 |

## 5 Conclusion

This study uses two pre-trained models to shorten training time and streamline the process, carefully chosen standard datasets to understand the FaceNet approach and assess its accuracy. Seven datasets have been used: Caltech face dataset, Yale face dataset, YouTube Face subset, GeorgiaTech face dataset, Senthil Face dataset, CFP (both frontal and profile considered separately), and the Extended Yale dataset. Two pre-trained models have been trained from CASIAWebFace and VGGFace2 datasets.

Images are aligned using Multi-task Cascaded Convolution Neural Network (MTCNN) before using them for measuring the performance of the FaceNet system.

The embeddings generated after training the data are used for training the multi-class classifier, which is the SVM in this case.

We can conclude from the accuracy testing results that the FaceNet system performs very well and is an efficient face recognition technique. It reaches a minimum of 90% accuracy on all the selected standard datasets. We also noticed that FaceNet is more accurate on the CFP frontal images than the CFP profile images. The decrease in accuracy can be due to more prominent constraints like pose variation and occlusion on the CFP profile images.

FaceNet has evolved to be a better method than the other methods to which it is compared and contrasted. The accuracy of the FaceNet system exceeds 98% on the three datasets used for comparison, and it fares very well against other methods, against all other methods, which achieve an accuracy of not more than 95%.

FaceNet is considered an excellent technique due to a variety of reasons. Traditionally, a classification architecture is presumed to perform well for face recognition applications. The main disadvantage is the presence of a fixed number of classes. FaceNet employs embeddings for training purposes where the embedding of a new class is projected in an embedding space, which could help estimate the Euclidean distance between the new class and the existing class. This results in not needing to retrain the network when a new class is considered. The entire system is an end-to-end training system that further simplifies the process and setup, requiring minimal alignment and optimizing the loss relevant to this method for performance improvement.

**Acknowledgements** The authors profoundly acknowledge the Department of Computer Science and Engineering, Manipal Institute of Technology, Manipal Academy of Higher Education, Manipal, INDIA-576104, for providing a supportive environment to make this work conceivable.

## References

1. Turk M (2006) Eigenfaces and Beyond. Elsevier EBooks, pp 55–86. https://doi.org/10.1016/b978-012088452-0/50003-0
2. Ahonen T, Hadid A, Pietikainen M (2006) Face description with local binary patterns: application to face recognition. IEEE Trans Pattern Anal Mach Intell 28(12):2037–2041
3. Song L, Qiu C (2018) Face recognition by feature extraction and classification. In: 12th IEEE international conference on anti-counterfeiting, security, and identification (ASID). IEEE, pp 43–46
4. Jolliffe IT, Cadima J (2016) Principal component analysis: a review and recent developments. Philos. Trans. R. Soc. A: Math. Phys. Eng. Sci. 374(2065):20150202
5. Schroff F, Kalenichenko D, Philbin J (2015) Facenet: a unified embedding for face recognition and clustering. In: Proceedings of the IEEE conference on computer vision and pattern recognition. pp 815–823
6. Zeiler MD, Fergus R (2014) Visualizing and understanding convolutional networks. In: Computer vision–ECCV 2014: 13th European conference, Zurich, Switzerland, Proceedings, Part I, vol 13. Springer International Publishing, pp 818–833

7. Szegedy C, Liu W, Jia Y, Sermanet P, Reed S, Anguelov D, Rabinovich A (2015) Going deeper with convolutions. In: Proceedings of the IEEE conference on computer vision and pattern recognition. pp 1–9
8. Huang GB, Mattar M, Berg T, Learned-Miller E (2008) Labeled faces in the wild: a database forstudying face recognition in unconstrained environments. In: Workshop on faces in 'Real-Life' Images: detection, alignment, and recognition
9. Wolf L, Hassner T, Maoz I (2011) Face recognition in unconstrained videos with matched background similarity. In: CVPR 2011. IEEE, pp 529–534
10. Bhandiwad V, Tekwani B (2017) Face recognition and detection using neural networks. In: International conference on trends in electronics and informatics (ICEI). IEEE, pp 879–882
11. Turk MA, Pentland AP (1991) Face recognition using eigenfaces. In: Proceedings of IEEE computer society conference on computer vision and pattern recognition. IEEE Computer Society, pp 586–587
12. Belhumeur PN, Hespanha JP, Kriegman DJ (1997) Eigenfaces vs. fisherfaces: recognition using class specific linear projection. IEEE Trans. Pattern Anal. Mach. Intell. 19(7):711–720
13. Maturana D, Mery D, Soto A (2009) Face recognition with local binary patterns, spatial pyramid histograms and Naive Bayes nearest neighbor classification. In: International conference of the chilean computer science society. IEEE, pp 125–132
14. Taigman Y, Yang M, Ranzato MA, Wolf L (2014) Deepface: closing the gap to human-level performance in face verification. In: Proceedings of the IEEE conference on computer vision and pattern recognition. pp 1701–1708
15. Goel R, Mehmood I, Ugail H (2021) A study of deep learning-based face recognition models for sibling identification. Sensors 21(15):5068
16. Salama AbdELminaam D, Almansori AM, Taha M, Badr E (2020) A deep facial recognition system using computational intelligent algorithms. PLoS ONE 15(12):e0242269
17. Saravanan TM, Karthiha K, Kavinkumar R, Gokul S, Mishra JP (2022) A novel machine learning scheme for face mask detection using pretrained convolutional neural network. Mater. Today Proc. 58:150–156
18. Sunitha G, Geetha K, Neelakandan S, Pundir AKS, Hemalatha S, Kumar V (2022) Intelligent deep learning based ethnicity recognition and classification using facial images. Image Vis. Comput. 121:104404
19. Wan W, Gao Y, Lee HJ (2019) Transfer deep feature learning for face sketch recognition. Neural Comput. Appl. 31:9175–9184
20. Chen D, Ren S, Wei Y, Cao X, Sun J (2014) Joint cascade face detection and alignment. In: Computer vision–ECCV 2014: 13th European conference, Zurich, Switzerland, Proceedings, Part VI 13. Springer International Publishing, pp 109–122
21. Angelova A, Abu-Mostafam Y, Perona P (2005) Pruning training sets for learning of object categories. In: IEEE computer society conference on computer vision and pattern recognition (CVPR'05), vol 1. IEEE, pp 494–501
22. Yale face database [Online]. http://cvc.cs.yale.edu/cvc/
23. Georgia Tech Face Database. https://computervisiononline.com/dataset/1105138700
24. Rujirakul K, So-In C, Anonkijpanich B (2014) Weighted histogram equalized PEM-PCA face recognition. In: 2014 International computer science and engineering conference (ICSEC). IEEE, pp 144–150
25. Sengupta S, Chen JC, Castillo C, Patel VM, Chellappa R, Jacobs DW (2016) Frontal to profile face verification in the wild. In: 2016 IEEE winter conference on applications of computer vision (WACV). IEEE, pp 1–9
26. Extended Yale B database, [Online]. https://computervisiononline.com/dataset/1105138686
27. Yi D, Lei Z, Liao S, Li SZ (2014) Learning face representation from scratch. arXiv preprint arXiv:1411.7923
28. Cao Q, Shen L, Xie W, Parkhi OM, Zisserman A (2018) Vggface2: A dataset for recognising faces across pose and age. In: 2018 13th IEEE international conference on automatic face and gesture recognition (FG 2018). IEEE, pp 67–74

# Resampling Strategies for Mitigating Class Imbalance of ASD Dataset on the Performance of Machine Learning Classifiers

**Rahul Kumar Gupta and Kamlesh Dutta**

**Abstract** In the supplied ASD dataset, it is typically seen that there is an extremely large imbalance in the number of samples for two classes, leading to an imbalance. Without addressing this issue, applying binary classification algorithms to such data would produce an extremely biased result. It affects the relationships between features as well. These misclassifications could affect the decision regarding medical treatment and result in a protracted delay for those who urgently require medical intervention. In the current study, we use a variety of resampling strategies to address the issue of class imbalance. Precision, Recall, and F1-score are used as evaluation measures for all models. We have also looked at AUC score, which demonstrates encouraging outcomes for the use of resampling methods for imbalanced dataset.

**Keywords** Autism spectrum disorders · Data imbalance · Random undersampling · Random oversampling · SMOTE · Machine learning

## 1 Introduction

Autism spectrum disorder (ASD) is a developmental impairment caused by differences in the brain. Children with ASD are often troubled with social communications and interaction, and most of the time they are engaged with constrained or consistent actions like stereotypical motor movements (SMMs) [1].

No scientifically proven medical tests for ASD, such as a blood test, are available. Therefore, diagnosis of ASD traits is difficult and significantly depends on subjective

R. K. Gupta (✉)
School Of Technology, Woxsen University, Hydrabad, India
e-mail: rahul.gupta@woxsen.edu.in
URL: https://woxsen.edu.in/

R. K. Gupta · K. Dutta
Department of Computer Science & Engineeing, National Institute of Technology Hamirpur, Hamirpur, HP, India
e-mail: kd@nith.ac.in



S. Borah et al. (eds.), *Advanced Computational and Communication Paradigms*, Lecture Notes in Networks and Systems 535,
https://doi.org/10.1007/978-981-99-4284-8_18

attributes. By the age of two, it is possible to presume a trustworthy diagnosis from a knowledgeable professional [2, 3]. ASD diagnosis is carried out in two steps. In the first stage, using interactive Screening Tool for Autism, designed for those children in case of whom pediatricians as well as parents feel that there could be developmental concerns. The screening task is administered through different activities such as assessing play, communication, and imitation skills and a series of questionnaire asked from parents about their child's specific behavioral aspects. In the second step, a multidisciplinary team conducts a thorough diagnostic evaluation, gathering information using a variety of methods, including observations and interviews, in accordance with DSM-V criteria [4, 5].

Within a given dataset, in case of certain applications, it is possible that one particular class is represented by a larger number of instances as compared to the other classes, the datasets suffer from class imbalance issues. As a result, algorithms frequently become skewed by huge classes and ignore classes with low frequency [6–8]. Standard classifiers typically produce models that are biased toward the majority class because machine learning algorithms' classifiers make the implicit assumption that the training dataset is balanced [9]. In autism-related issues, if the training dataset does not include proper balance of samples, representing all classes, it may lead to a high rate of misclassification, particularly in terms of false negatives [10].

In general, the dominating classes also influence the functioning of machine learning algorithms during the training phase of imbalanced dataset; the classifier may overlook the minority class and produce biased results. In order to have a trust in the performance of the proposed models, the class imbalance issues must be addressed. Techniques such as oversampling or undersampling can be used for data balancing [11]. We specifically use oversampling and undersampling techniques and compare different classification algorithms such as how well they perform when applied to an autism screening dataset [12].

The paper is organized as follows. Section 2 describes various resampling strategies and their applications on various machine learning algorithms for imbalanced dataset [10]. Section 3 presents the methodology focusing on various resampling strategies. The section also analyzes the strategies using various evaluation metrics. Section 4 analyzes the results obtained for various machine learning classifiers before and after resampling and compare the performance. Section 6 summarizes the findings and suggests future directions of research which can make the automated process of autism detection more trustworthy.

## 2 Resampling Strategies

In order to achieve a proper balanced data, so that we can aim at the desired predictive power of machine learning classifiers, various oversampling methods which can create synthetic training data samples for the less represented class or undersampling methods, which can eliminate training data samples for the more represented class, are employed.

By duplicating the samples from those classes which are underrepresented, the dataset can be made relatively balanced. This technique is called Random Oversampling (ROS). However, the application of this technique gives rise to a situation of overfitting of the dataset [11], which in turn result in strong bias during training and high variance during testing, making machine learning models less reliable.

Synthetic Minority Oversampling Technique (SMOTE) operates by clustering the dataset using the K-Nearest Neighbors (KNN) clustering method. The samples are generated at random between neighbors [6]. In the Borderline SMOTE technique, only the minority examples that are close to the borderlines are oversampled. Borderline SMOTE technique outperforms SMOTE and random oversampling methods. The results obtained, True Positive (TP) rate and F-value, are attributed to the sampling of data along the decision boundaries of two classes only, thus focusing on minority classes [13]. SMOTE-SVM combines an ensemble of SVMs with borderline information [14], making ensemble models to capture more complex decision boundaries.

Adaptive Synthetic Sampling Approach (ADASYN) is suitable for those applications where classes with less represented data are of our interest as is in the case of ASD dataset. The technique uses a weighted distribution for various less represented class examples, depending on how challenging it is for them to learn. The ADASYN approach enhances learning and tries to balance diverse data distributions, by minimizing the bias caused by the imbalance in class, and adjusting the classification decision boundary in favor of the difficult examples [15].

Another popular technique is Random Undersampling (RUS). In this technique, the majority class in the sample is truncated and the dataset is resampled using the random undersampling (RUS) approach. This method has a problem that information may be lost if significant undersampling is required [16].

## 2.1 Related Work

Abdelhamid et al. [17] conducted an experiment to determine the effects of the class imbalance problem on the performance of various classifiers, when different approaches to the problem of data imbalance are used in the preprocessing stage. They used RUS, SMOTE, and ROS with various machine learning classifiers. Results indicate that oversampling strategies are preferred to undersampling techniques for the autism dataset.

Rahman et al. [18] tested how well oversampling and undersampling methods perform on cardiovascular data. SMOTE, a well-known oversampling technique, is used, and a few undersampling methods are also investigated. They put forth a modified cluster-based undersampling technique. The proposed method has been reported to perform noticeably better than the current approaches.

In order to balance the data sizes of each class, Li et al. [19] followed bidirectional approach. Authors proposed to minimize the data in the class with better representation and thereafter enhanced the number of synthetic samples in the underrepresented class.

SMOTE was used by El-Sayed et al. [20] to correct for imbalanced data on autism. The data was unbalanced, with 15 non-autistic individuals in the minority class and 100 people with autism in the majority class. The outcomes demonstrated that the model is more trustworthy when imbalanced data are oversampled.

Vakadkar et al. [21] created an automated ASD prediction model. The model was designed using the fewer attributes, taken from each diagnosis dataset [17]. They applied five machine learning models. They used the confusion matrix and F1-score to assess the performance of each model.

Kumar and Das [22] used the dataset [12] to examine machine learning models. Data imbalance mitigation was achieved by employing Recursive Feature Elimination (RFE) approach. Thabtah et al. [23] employed Self-organizing Map (SOM) with classification algorithms. Authors evaluated their models against a real-life autism screening dataset [12, 24].

## 3 Methodology

Five machine learning classifiers were considered for our study. The choice of machine learning classifiers was guided by the research work studied in the related work section. Logistic Regression (LR) was chosen as it had shown promising results earlier. Linear discriminant analysis (LDA) is a classification algorithm. Linear Discriminant Analysis (LDA) allows modeling the class probabilities using the normal assumptions. K-Nearest Neighbors Classifier (KNN) can be chosen for regression as well as classification purpose. Gaussian Naïve Bayes (NB) considers features as independent. Support Vector Machine (SVM), while using non-linear kernel during training, focuses on the possible interactions between features to a certain degree. The classifiers were developed for two scenarios: (i) without resampling the data and generate model for training and testing dataset; thereafter, (ii) resampling using random oversampling, random undersampling, SMOTE, Borderline SMOTE, SVM-SMOTE, and ADASYN, respectively, and apply classifiers.

The dataset we are utilizing for our analysis is the one provided by Thabtah and was obtained from ASDTest [12, 24]. The dataset comprises 1,054 instances with 18 attributes. Authors prepared dataset based on the medical screening questionnaire [5]. Score attribute is the sum of the scores for questions A1 through A10; threshold value of the score for making decision is taken as three. For the score, greater than three, the class indicates that the child has ASD traits. For the score less than three, the class value indicates that the child does not exhibit any ASD traits. The Family ASD attribute contains historical information that reveals whether any of the child's relatives have a history of autism. However, it is to be noted that the author of the dataset has mentioned Autism Spectrum Disorder (ASD) tests not as diagnostic tool but for enabling parents, care-givers, and academic researchers for pinpointing autistic traits through behavioral tests.

The dataset is imbalanced, with 326 instances of the minority class, "No" (children without ASD), and 728 instances of the majority class, "Yes" (children with ASD). During experiment, the initial results "Yes" and "No" are mapped to 1.0 and 0.0, respectively. Features which have types object are converted into int64 by using label encoder. Few attributes which are irrelevant, such as Case No, are dropped from analysis. Thereafter, the dataset is split into training and testing datasets, with the test size of 0.33 proportions.

To evaluate the performance of different classifiers, with various resampling technique, we have used Accuracy, Precision, Recall, F1-score, and AUC-ROC score. When working with a class-imbalanced dataset like this one, where there is a sizable difference between the number of ASD and No ASD labels, accuracy alone is not good enough and other metrics like Precision, Recall, and AUC score are also examined.

## 4 Results

Results produced by various classifiers using the original dataset without applying resampling techniques and after applying ROS, SMOTE, Borderline SMOTE, SVM-SMOTE, ADASYN, and RUS methods are shown in Tables 1, 2, 3, 4, 5, 6, and 7.

When compared to other classifiers, Logistic Regression offers the most accurate results overall. Accuracy, Recall, and F1-score increased while Precision and AUC

**Table 1** Model performance without using resampling technique

| S. no. | Classifier | Accuracy (%) | Precision (%) | Recall (%) | F1-score (%) | AUC score (%) |
|---|---|---|---|---|---|---|
| 1 | Logistic regression | 100.0 | 100.0 | 100.0 | 100.0 | 100.0 |
| 2 | Linear discriminant analysis | 96.68 | 99.12 | 94.58 | 96.80 | 99.60 |
| 3 | K-nearest neighbors classifier | 96.55 | 97.5 | 97.5 | 97.5 | 99.10 |
| 4 | Gaussian Naïve Bayes | 99.97 | 98.29 | 95.83 | 97.04 | 99.78 |
| 5 | Support vector machine | 99.71 | 99.58 | 100.0 | 99.79 | 99.98 |

**Table 2** Model performance using random oversampling technique

| S. no. | Classifier | Accuracy (%) | Precision (%) | Recall (%) | F1-score (%) | AUC score (%) |
|---|---|---|---|---|---|---|
| 1 | Logistic regression | 100.0 | 100.0 | 100.0 | 100.0 | 100.0 |
| 2 | Linear discriminant analysis | 94.25 | 100.0 | 91.66 | 95.65 | 99.68 |
| 3 | K-nearest neighbors classifier | 94.54 | 99.11 | 92.91 | 95.91 | 98.78 |
| 4 | Gaussian Naïve Bayes | 95.68 | 100.0 | 93.75 | 96.77 | 99.83 |
| 5 | Support vector machine | 93.39 | 100.0 | 90.40 | 94.96 | 99.94 |

**Table 3** Model performance using SMOTE sampling technique

| S. no. | Classifier | Accuracy (%) | Precision (%) | Recall (%) | F1-score (%) | AUC score (%) |
|---|---|---|---|---|---|---|
| 1 | Logistic regression | 100.0 | 100.0 | 100.0 | 100.0 | 100.0 |
| 2 | Linear discriminant analysis | 95.40 | 99.55 | 93.75 | 96.56 | 99.52 |
| 3 | K-nearest neighbors classifier | 96.55 | 98.30 | 96.66 | 97.47 | 99.23 |
| 4 | Gaussian Naïve Bayes | 95.40 | 95.16 | 98.33 | 96.72 | 99.41 |
| 5 | Support vector machine | 94.82 | 100.0 | 92.50 | 96.10 | 99.98 |

score, as shown in 1, decreased in SMOTE, as compared to Random Oversampling, for all classifiers except Naïve Bayes and SVM. In Naïve Bayes classifier, accuracy and F1-score decreased while SVM AUC score increased in SMOTE, compared to Random Oversampling.

We can see, the result of Random Undersampling for all classifiers, which is lesser than Random Oversampling except for recall of K-Nearest Neighbors, which is more significant in RUS. In our research, we have also used a few variations of SMOTE such as Borderline SMOTE, SVM-SMOTE, and ADASYN and compared the results of all classifiers. We observe that all these variants of SMOTE produce very little fluctuation in the results, due to smaller dataset size (Fig. 1).

**Table 4** Model performance using borderline SMOTE sampling technique

| S. no. | Classifier | Accuracy (%) | Precision (%) | Recall (%) | F1-score (%) | AUC score (%) |
|---|---|---|---|---|---|---|
| 1 | Logistic regression | 100.0 | 100.0 | 100.0 | 100.0 | 100.0 |
| 2 | Linear discriminant analysis | 95.40 | 99.55 | 93.75 | 96.56 | 99.52 |
| 3 | K-nearest neighbors classifier | 96.55 | 98.30 | 96.66 | 97.47 | 99.23 |
| 4 | Gaussian Naïve Bayes | 95.40 | 95.16 | 98.33 | 96.72 | 99.41 |
| 5 | Support vector machine | 94.82 | 100.0 | 92.50 | 96.10 | 99.98 |

**Table 5** Model performance using SVM-SMOTE sampling technique

| S. no. | Classifier | Accuracy (%) | Precision (%) | Recall (%) | F1-score (%) | AUC score (%) |
|---|---|---|---|---|---|---|
| 1 | Logistic regression | 100.0 | 100.0 | 100.0 | 100.0 | 100.0 |
| 2 | Linear discriminant analysis | 94.54 | 99.55 | 92.50 | 95.89 | 99.54 |
| 3 | K-nearest neighbors classifier | 94.82 | 99.11 | 93.33 | 96.13 | 99.13 |
| 4 | Gaussian Naïve Bayes | 95.68 | 98.23 | 95.41 | 96.82 | 99.68 |
| 5 | Support vector machine | 92.81 | 100.0 | 89.58 | 94.50 | 99.98 |

**Table 6** Model performance using ADASYN sampling technique

| S. no. | Classifier | Accuracy (%) | Precision (%) | Recall (%) | F1-score (%) | AUC score (%) |
|---|---|---|---|---|---|---|
| 1 | Logistic regression | 100.0 | 100.0 | 100.0 | 100.0 | 100.0 |
| 2 | Linear discriminant analysis | 95.40 | 99.55 | 93.75 | 96.56 | 99.56 |
| 3 | K-nearest neighbors classifier | 94.82 | 99.11 | 93.33 | 96.13 | 98.59 |
| 4 | Gaussian Naïve Bayes | 95.97 | 96.31 | 97.91 | 97.10 | 99.44 |
| 5 | Support vector machine | 93.39 | 100.0 | 90.41 | 94.96 | 99.92 |

**Table 7** Model performance using random undersampling technique

| S.no. | Classifier | Accuracy (%) | Precision (%) | Recall (%) | F1-score (%) | AUC score (%) |
|---|---|---|---|---|---|---|
| 1 | Logistic regression | 100.0 | 100.0 | 100.0 | 100.0 | 100.0 |
| 2 | Linear discriminant analysis | 93.96 | 100 | 91.25 | 95.42 | 99.54 |
| 3 | K-nearest neighbors classifier | 92.81 | 98.64 | 93.83 | 94.57 | 98.72 |
| 4 | Gaussian Naïve Bayes | 95.40 | 100.0 | 93.33 | 96.55 | 99.75 |
| 5 | Support vector machine | 92.52 | 100.0 | 89.16 | 94.27 | 99.94 |

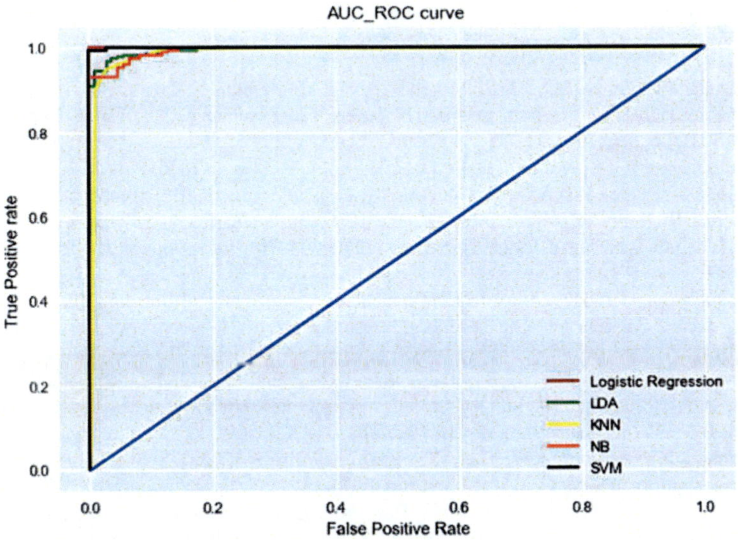


**Fig. 1** AUC-ROC curve for classifiers with SMOTE

## 5 Conclusion

In this paper, we have examined different resampling techniques with various machine learning classifiers and evaluated results in terms of Accuracy, Precision, Recall, and F1-score. Since our dataset is imbalanced, for better results we have also examined AUC score. For Autism Dataset, SMOTE performs better in terms of Accuracy, Recall, and F1-score, while Random Oversampling method gives better results for AUC score and precision. There is a little fall in results when we use Random Undersampling. SMOTE and Borderline SMOTE produce the same results while the results with SVM-SMOTE and ADASYN have shown little improvement.

The findings indicate that we must use resampling techniques to lessen the impact of class imbalance in the dataset related to autism. We could also look into cost-based classification, where we could give selected samples more weight. Individual samples can also be given priority and relevance assignments. Furthermore, a mechanism for segregation is required. Creating machine learning classifiers that are primarily focused on low-frequency data can also be helpful in reducing the impact of dataset imbalance and enhancing prediction abilities.

**Acknowledgements** The work carried out by the first author is supported by the GATE scholarship from Ministry of Education, India.

## References


1. Parellada M, Penzol MJ, Pina L, Moreno C, Gonz´alez-Vioque E, Zalsman G, Arango C (2014) The neurobiology of autism spectrum disorders. Eur Psychiatry 29(1):11–19. https://doi.org/10.1016/j.eurpsy.2013.02.005
2. Lord C, Risi S, DiLavore PS, Shulman C, Thurm A, Pickles A (2006) Autism from 2 to 9 years of age. Arch Gen Psychiatry 63(6):694–701. https://doi.org/10.1001/archpsyc.63.6.694
3. Hyman SL, Levy SE, Myers SM (2020) Identification, evaluation, and management of children with autism spectrum disorder. Pediatrics 145(1):694–701. https://doi.org/10.1542/peds.2019-3447
4. Association AP (2013) Diagnostic and statistical manual of mental disorders, 5th edn. American Psychiatric Association. https://doi.org/ https://doi.org/10.1176/appi.books.9780890425596
5. Allison C, Baron-Cohen S, Wheelwright S, Charman T, Richler J, Pasco G, Brayne C (2008) The q-chat (quantitative checklist for autism in toddlers): a normally distributed quantitative measure of autistic traits at 18–24 months of age: preliminary report. J Autism Dev Disord 38(8):1414–1425. https://doi.org/10.1007/s10803-007-0509-7
6. Chawla NV, Bowyer KW, Hall LO, Kegelmeyer WP (2002) Smote: synthetic minority oversampling technique. J Artif Intell Res 16(8):321–357. https://doi.org/10.1613/jair.953
7. Fern´andez A, del R´ıo S, Chawla NV, Herreral F (2017) An insight into imbalanced big data classification: Outcomes and challenges. Complex Intell Syst 3:105–120. https://doi.org/10.1007/s40747-017-0037-9
8. Abdeljaber F (2019) Detecting autistic traits using computational intelligence and machine learning techniques. Master of research thesis, Psychology Department, School of Health, University of Huddersfield, Huddersfield, UK. http://eprints.hud.ac.uk/id/eprint/34844/
9. Estabrooks A, Jo T, Japkowicz N (2004) A multiple resampling method for learning from imbalanced data sets. Comput Intell 20(1):18–36. https://doi.org/10.1111/j.0824-7935.2004.t01-1-00228.x
10. Thabtah F, Hammoud S, Kamalov F, Gonsalves A (2020) Data imbalance in classification: experimental evaluation. Inf Sci 513:429–441. https://doi.org/10.1016/j.ins.2019.11.004
11. Zheng Z, Cai Y, Li Y (2015) Oversampling method for imbalanced classification. Comput Inform 34(5):1017–1037. https://doi.org/10.1016/j.ins.2019.11.004
12. Thabtah F, Kamalov F, Rajab K (2018) A new computational intelligence approach to detect autistic features for autism screening. Int J Med Inform 117:112–124. https://doi.org/10.1016/j.ijmedinf.2018.06.009
13. Han H, Wang W-Y, Mao B-H (2005) Borderline-smote: a new oversampling method in imbalanced data sets learning. In: Huang D-S, Zhang X-P, Huang G-B (eds) Advances in intelligent computing ICIC. Lecture notes in computer science. Springer, Berlin, Heidelberg, pp 878–887. https://doi.org/10.1007/1153805991
14. Wang Q, Luo Z, Huang J, Feng Y, Liu Z (2017) A novel ensemble method for imbalanced data learning: bagging of extrapolation-smote svm. Comput Intell Neurosci (Article ID 1827016):11 https://doi.org/10.1155/2017/1827016
15. He H, Bai Y, Garcia EA, Li S (2008) Adasyn: adaptive synthetic sampling approach for imbalanced learning. In: 2008 IEEE international joint conference on neural networks, IEEE world congress on computational intelligence. pp 1322–1328. https://doi.org/10.1109/IJCNN.2008.4633969
16. Shelke MS, Deshmukh PR, Shandilya VK (2017) A review on imbalanced data handling using undersampling and oversampling technique. Int. J. Recent Trends Eng Res 3(4):444–449
17. Abdelhamid N, Padmavathy A, Peebles D, Thabtah F, GoulderHorobin D (2020) Data imbalance in autism pre-diagnosis classification systems: an experimental study. J Inf Knowl Manag 19(1). https://doi.org/10.1142/S0219649220400146
18. Rahman MM, Davis DN (2013) Addressing the class imbalance problem in medical datasets. Int J Mach Learn Comput 3(2):224–228. https://doi.org/10.7763/IJMLC.2013.V3.307

19. Li D-C, Liu C-W, Hub CS (2010) A learning method for the class imbalance problem with medical data sets. Comput Biol Med 40(5):509–518. https://doi.org/10.1016/j.compbiomed.2010.03.005
20. El-Sayed AA, Mahmood MAM, Meguid NA, Hefny HA ((2015)) Handling autism imbalanced data using synthetic minority over-sampling technique (smote). In: Third world conference on complex systems (WCCS). IEEE, pp 1–5. https://doi.org/10.1109/ICoCS.2015.7483267
21. Vakadkar K, Purkayastha D, Krishnan D (2021) Detection of autism spectrum disorder in children using machine learning technique. SN Comput Sci 2(5):1–9. https://doi.org/10.1007/s42979-021-00776-5
22. Das PR, Kumar CJ (2021) The diagnosis of asd using multiple machine learning techniques. Int J Dev Disabil. https://doi.org/10.1080/20473869.2021.1933730
23. Thabtah F, Spencer R, Abdelhamid N, Kamalov F, Wentzel C, Ye Y, Dayara T (2022) Autism screening: an unsupervised machine learning approach. Health Inf Sci Syst 10(1):26. https://doi.org/10.1007/s13755-022-00191-x
24. Thabtah F (2019) Machine learning in autistic spectrum disorder behavioral research: a review and ways forward. Inform Health Soc Care 44(3):278–297. https://doi.org/10.1080/17538157.2017.1399132

# Comparative Analysis of Minimax Algorithm with Alpha–Beta Pruning Optimization for Chess Engine

**Harshit Choudhary, Aniket Inamdar, Swapnil Kharade, Anunay Maheshwari, and Preeti Kale**

**Abstract** The chess game engine is introduced which makes use of a minimax algorithm for searching game trees and makes use of alpha–beta pruning algorithm to reduce the search space of game states in the game state tree. The first part of this paper is an expository presentation that demonstrates and explains the minimax algorithm. The work in this paper demonstrates that the alpha–beta pruning method improves the efficiency of the chess engine algorithm based on the results obtained using rigorous experimentation.

**Keywords** Chess engine · Minimax algorithm · Alpha–beta pruning algorithm · Artificial intelligence

## 1 Introduction

Computer programs for playing the chess game usually search a large number of states from the game state tree where each node is a state that represents the state in the game. An algorithm called the "Minimax algorithm" is used to select the best possible state, which may or may not be correct (guarantees a checkmate) but is the best optimal state from the current state to all possible future states. Additionally, the alpha–beta pruning algorithm is used to reduce the number of iterations in the minimax algorithm.

The purpose of this paper is to create and analyze the performance of a chess game engine that makes use of this algorithm. Additionally, we also analyze the performance of the game engine that reduces the search space for the minimax algorithm using the alpha–beta pruning algorithm.

---

H. Choudhary · A. Inamdar (✉) · S. Kharade · A. Maheshwari · P. Kale
Vishwanath Karad MIT World Peace University, Pune 411038, India
e-mail: 1032181634@mitwpu.edu.in

P. Kale
e-mail: preeti.kale@mitwpu.edu.in



S. Borah et al. (eds.), *Advanced Computational and Communication Paradigms*,
Lecture Notes in Networks and Systems 535,
https://doi.org/10.1007/978-981-99-4284-8_19

Section 2 of this document describes the literature review for the minimax algorithm and the Alpha–beta pruning optimization. Section 3 describes the proposed methodology to create the AI chess game engine. Section 4 demonstrates the results along with a comparative analysis of the game engine with and without the alpha–beta pruning function.

## 2 Literature Review

### 2.1 *Minimax Algorithm*

The minimax algorithm [1–4] is a simple recursive algorithm that determines the next state by analyzing the different states in the game tree.

The next state's score is what the maximizing player seeks to increase, while the following state's score is what the minimizing player seeks to decrease. However, since the minimax algorithm cannot search the entire game tree to find the best next state, it uses a fixed depth search to determine the next state.

The minimax algorithm works by producing a game tree up to a certain depth with the current state as the root node. Then we generate a score value for each leaf node using the evaluation function. We choose the value of the parent node based on the category of the parent node (whether it is maximizing or minimizing). If the parent node is a maximizing player, it will consistently select the child node with the highest score value as its state.

Considering the example in Fig. 1, [5, 6] where each node is the state is the game of chess. A white state represents a maximizing player and a black state represents a minimizing player. The leaf nodes are D, E, F, and G which have been assigned score values according to the evaluation function. The current state is the root node A.

Now, considering state B and its child nodes D and E, node B is a minimizing node and hence it will be assigned the minimum value from its child nodes. Hence, node B will be assigned the value 3. Similarly, for the nodes, C, F, and G, node C will be assigned the value 2. Node A is a maximizing node and hence it will select the maximum score value from its child nodes. Hence, node A will be assigned the value 3.

Looking at the above example we can see that the next best state from the current state A is node B. We can assume that the opposing player plays optimally and selects node E.

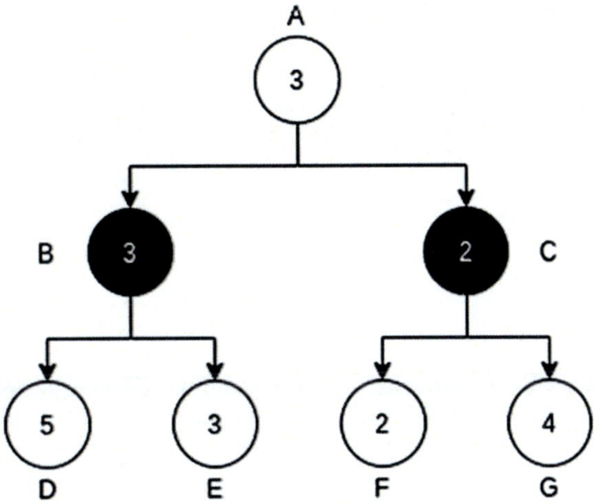


**Fig. 1** Game tree

## 2.2 *Alpha–Beta Pruning Algorithm*

Considering the example in Fig. 2, we can see that node F has a score value of 2. We know that the parent node C is a minimizing node and hence its score value will be at most equal to 2 (from node F). Now considering the left subtree (node B, D, and E), the value of node B will be equal to 3. We can hence determine that node A being a maximizing node will select node B (value 3) over C (at most 2). Thus, we would never have to calculate the score value of node G, since its parent node will always select the minimum value, which can at most be 2.

This method of reducing the search space in the game tree is known as α-β pruning [1, 2, 7, 8], where α and β are the upper and lower bounds. With α-β pruning, we

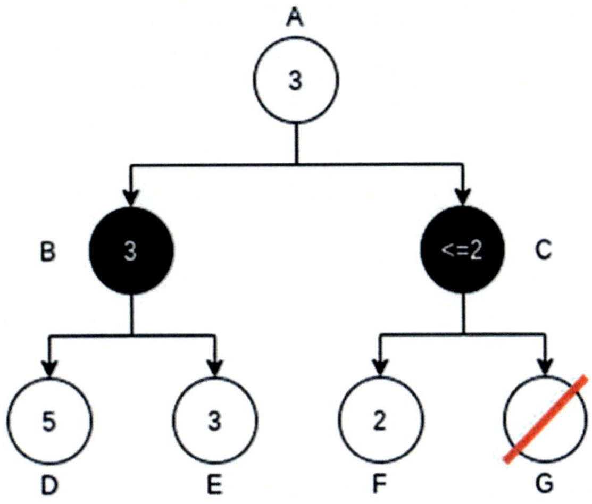


**Fig. 2** Game tree with α-β pruning

**Fig. 3** α-β pruning

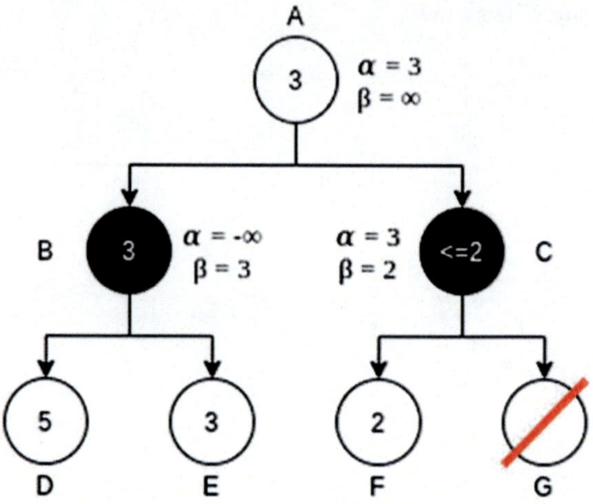


only have to explore the most optimal nodes which can lead to a win (considering the opposing player plays optimally) and we can stop exploring nodes that do not change the score of their parent nodes.

α and β are the best scores that either player can achieve where α corresponds to the maximizing player and β corresponds to the minimizing player. Initially, α is assigned the minimum possible score for the maximizing player which is in $-\infty$ and β is assigned the maximum possible score for the minimizing player which is $\infty$.

Now, consider the example in Fig. 3 using the α and β variables, we can determine which node should be explored further and which can be pruned. The condition to terminate the recursive call is when the value of β is less than or equal to α ($\beta < = \alpha$). For minimizing node B, α would remain the same (-∞) and β would be min (node D, node E) which is 3.

Since the left subtree is now explored, the values of α and β for node A will now be ($\alpha = 3$ and $\beta = \infty$). The algorithm will now start evaluating the right subtree which modifies α and β. Node C is a minimizing node and hence, for node F value of β for node C will be 2 (due to F). As the termination condition is now satisfied the recursive call will be terminated and the current score value will be returned to the parent node.

The sequence in which the child nodes are visited affects alpha–beta pruning. Pruning may not occur if children of a node are visited in the worst possible sequence. For maximizing nodes, we want to visit the child with the highest score value first. Similarly, for the minimizing nodes, we want to visit the nodes, we want to visit the child with the lowest score value first.

### 2.3 Proposed Methodology

The aim of this paper is to evaluate how the performance of the minimax algorithm in chess is improved by using the alpha–beta pruning technique. Furthermore, we will assess the performance results to quantitatively determine the efficacy of the alpha–beta pruning technique.

A. Board representation

The traditional version of the chessboard is used—as an 8 by 8 matrix, each containing a piece or a blank space. For board representation, we use the *chessboard.js* library. This library provides the chessboard layout, its animations, and different API calls to interact with the chessboard.

B. Evaluating the next move

The chess engine uses Flask to interact with the chessboard. We use the *python-chess* library to determine all the legal next moves from the current state. Using the library, we first create a chessboard object with a starting state representing the start of the game. To make a move, the chess object can be passed an alphanumeric string which tells the board to move a piece from position A to position B. If the move is illegal, the method call returns an error. The board object also has a method *legal_moves* which returns an iterable list of all possible next moves. These moves are used by the minimax algorithm to determine the next best state. The board object also implements a method *fen*, which returns a FEN string that represents the state of the board and this string can be used by the *chessboard.js* to update the state to the user.

In order to play against the stockfish game engine, the stockfish python library is used. We pass the FEN string to the stockfish game engine and call the *get_best_move()* which returns the best possible move for the current player. The stockfish game engine has a default configuration having ELO rating of 2000 and depth of 3.

Time required for each move generated by the chess engine is calculated by getting the difference between the timestamp before and after calling *minimax()* and *minimax_ab()* methods which return the move of the chess engine.

C. Heuristic evaluation function

The Minimax algorithm is an educated search algorithm that uses heuristic functions to determine the score of additional nodes needed to get from the current node to the goal node.

In the case of a chess engine, there is no set winning move, the program needs some algorithm to find a desirable move. But, since there are multiple variables, there's no step-by-step algorithm for it. Hence, the algorithm uses such a heuristic function to develop an artificial intelligence of its own and uses the following Table 1 to calculate scores for the next set of possible moves.

Above given are the heuristic evaluation tables that are used to calculate score [9]. The minimax score value for each state will be the sum of values of all the pieces

**Table 1** Heuristic evaluation table

| Black player | Score | White player | Score |
|---|---|---|---|
| Pawn | −10 | Pawn | 10 |
| Bishop | −30 | Bishop | 30 |
| Knight | −30 | Knight | 30 |
| Rook | −50 | Rook | 50 |
| Queen | −90 | Queen | 90 |
| King | −900 | King | 900 |

present on the chessboard. Hence, the following is the equation to calculate the score value,

$$\text{Score} = \left(\sum \text{BLACK}\right) + \left(\sum \text{WHITE}\right)$$

Before calling the minimax function for making a move the $\alpha$ and $\beta$ variables will be initialized to $-\infty$ and $\infty$ respectively. Once the score is calculated for the child node, it will be assigned to the parent node's $\alpha$ or $\beta$ depending on whether it is a maximizing node or a minimizing node. The $\alpha$ and $\beta$ variables are maintained by each state and when the terminating condition is true ($\beta < = \alpha$), the recursive call in the minimax algorithm is terminated.

D. Game over condition

The method *is_game_over* provided by the *python-chess* library is called before every move to determine if the game is over if any of the following conditions happen:

- Checkmate.
- Stalemate (Not in check state but no legal moves are available).
- Insufficient material.

Other conditions for the game to end are not considered (resignation, timeout, repetition, agreement).

## 3 Performance Evaluation

To obtain performance analysis for the minimax algorithm, we considered the time required to make a single move in a game of chess in milliseconds. The minimax algorithm used by the chess engine evaluates the game tree with a depth of 3. The chess engine is played against the Stockfish 14 [10] game engine. The minimax

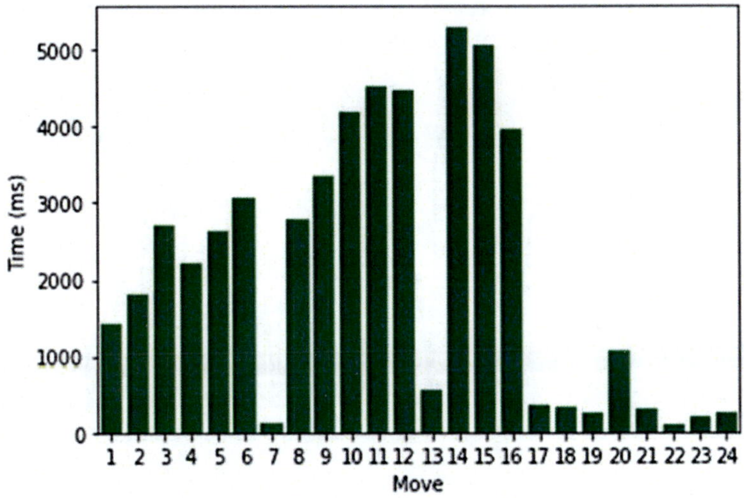


**Fig. 4** Minimax bar graph for 1 game

algorithm plays 10 games with the Stockfish 14 game engine and additionally, 10 games are played with the alpha–beta algorithm.

A. Minimax algorithm

Figure 4 is the bar graph for 1 game played against the Stockfish game engine using the minimax algorithm. It shows that time taken is not proportional to the number of moves and in fact depends on the number of winning moves.

Figure 5 is the line graph for a set of 10 games played against the Stockfish game engine using the minimax algorithm. We could infer from this graph that the starting and concluding moves in a game required time orders of magnitude less than at any other point in a game.

Figure 6 is the bar graph for the number of moves played each game for a set of 10 games against the Stockfish game engine using the minimax algorithm. With varying results, it shows that changing the starting move can in fact change the winning moves and how the game proceeds.

B. Minimax algorithm with alpha–beta pruning

Figure 7 is the bar graph for 1 game played against the Stockfish game engine using the alpha–beta pruning optimization in the minimax algorithm. Compared to Fig. 4, it shows that alpha–beta pruning does help to reduce the number of moves in the game, and also that the highest time taken to find the optimal move was much less with alpha–beta pruning.

Figure 8 is the line graph for a set of 10 games played against the Stockfish game engine using the alpha–beta pruning optimization in the minimax algorithm. The maximum time required for any move throughout all the games did not exceed

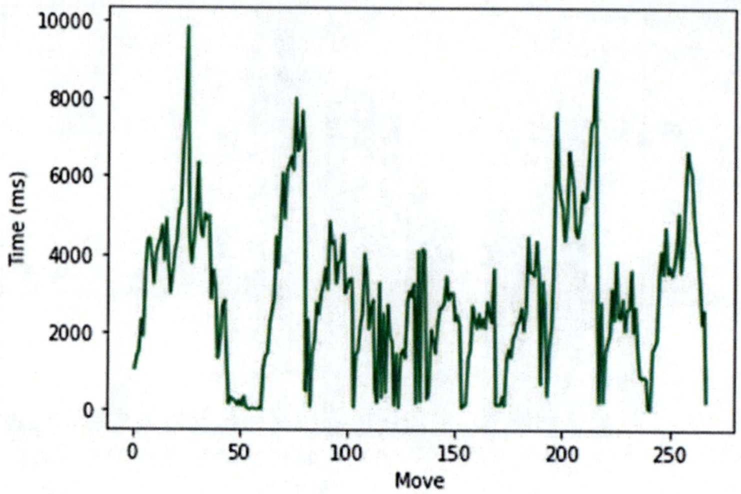


Fig. 5 Minimax line graph plot for 10 games

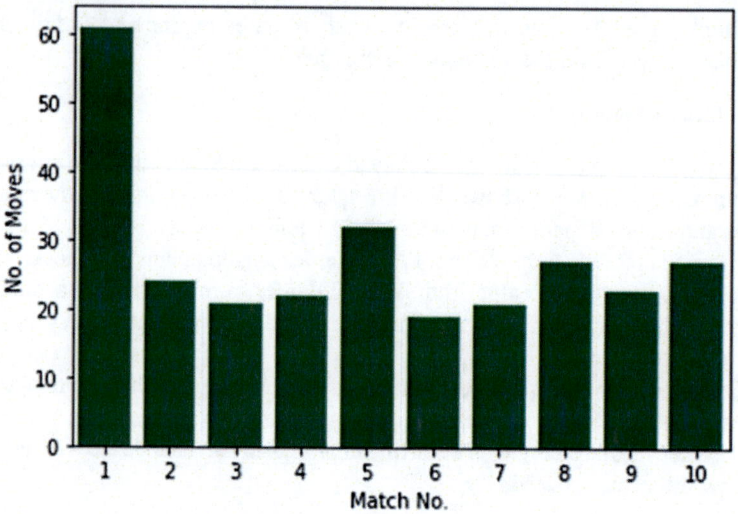


Fig. 6 Minimax bar graph plot for moves in each game

1000 ms. The average time taken for all the moves in a game was also less than the average time required by Minimax without alpha–beta optimization.

Figure 9 is the bar graph for the number of moves played each game for a set of 10 games against the Stockfish game engine using the minimax algorithm with alpha–beta pruning. We observed that the average number of moves required for each game was less as compared to the minimax algorithm without alpha–beta optimization.

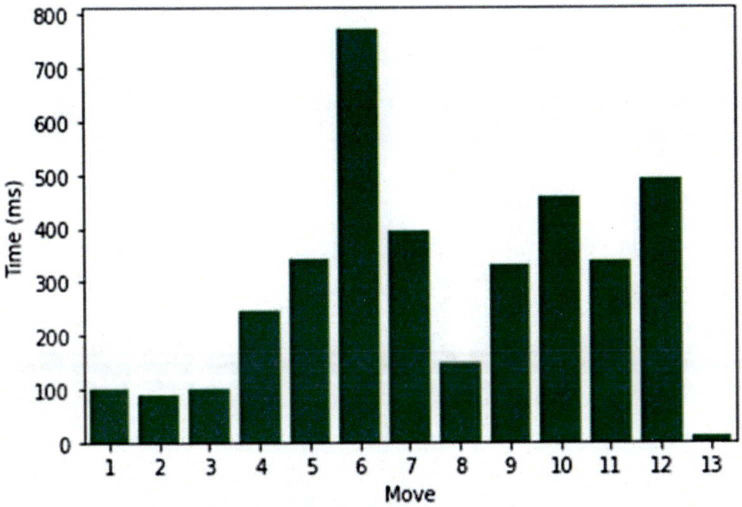


**Fig. 7** Minimax with alpha–beta bar graph plot for 1 game

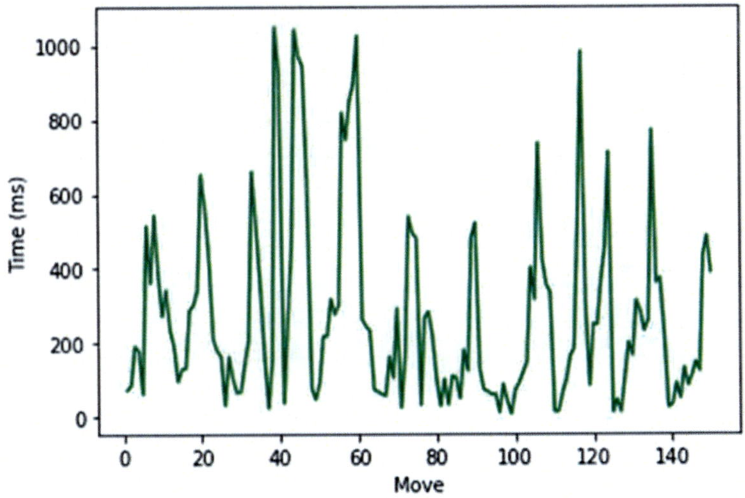


**Fig. 8** Minimax with alpha–beta line graph plot for 10 games

Based on the performance evaluation and analysis of the obtained results as shown in Tables 2 and 3, we infer that the minimax algorithm takes more time to calculate each move since it looks at all the states in the game state tree for a certain depth. Considering the average time taken per move in Tables 2 and 3, the alpha–beta optimization is 90.7455% faster as compared to the minimax algorithm.

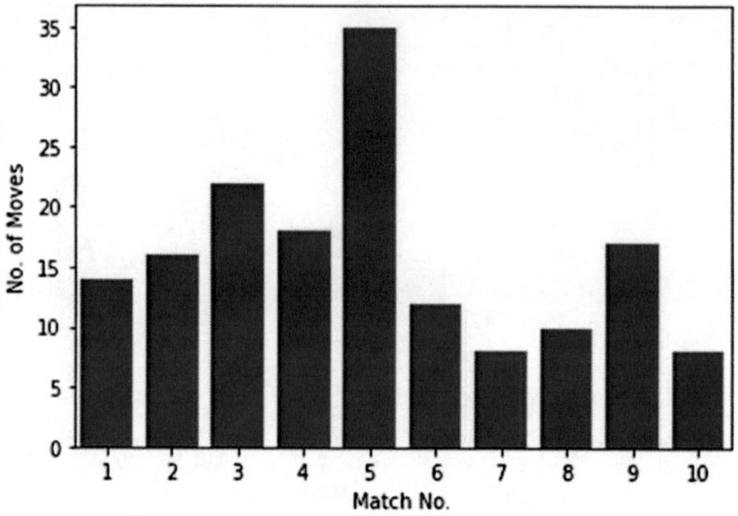


**Fig. 9** Minimax with alpha–beta bar graph plot for moves each game

**Table 2** Minimax algorithm

| No. | Property | Result | Unit |
|---|---|---|---|
| 1 | Average time taken per move for 10 games | 2965.0524 | ms |
| 2 | Average no. of moves for each game | 27.7 | – |

**Table 3** Minimax algorithm with alpha–beta

| No. | Property | Result | Unit |
|---|---|---|---|
| 1 | Average time taken per move for 10 games | 274.4000 | ms |
| 2 | Average no. of moves for each game | 16 | – |

# 4 Conclusion

In this paper, we implemented the minimax algorithm to create a chess engine and also compared its performance with the alpha–beta optimization. The minimax algorithm generated the game tree with a depth of 3. The chess engine was tested against the Stockfish 14 chess engine. The results show that the alpha–beta optimization significantly reduces the time required to determine the next move.

**Acknowledgements** The academic assistance of Dr. Vishwanath Karad MIT World Peace University was crucial for the completion of this work. We are particularly grateful to Dr. Vrushali Kulkarni, who serves as the Head of School for the Computer Engineering and Technology Department, and Dr. Preeti Kale, who taught the Artificial Intelligence course and provided valuable advice and support.

## References


1. Guo T, Qiu H, Tong B, Wang Y (2019) Optimization and comparison of multiple game algorithms in amazon chess. In: 2019 Chinese control and decision conference (CCDC). pp 6299–6304. https://doi.org/10.1109/CCDC.2019.8833089
2. Hu C, Wu X, Qian T, Luo H, Wang J (2020) An improved knowledge base for Chinese chess game. In: IEEE 4th information technology, networking, electronic and automation control conference (ITNEC). pp 1486–1490. https://doi.org/10.1109/ITNEC48623.2020.9085142
3. Vuckovic V (2015) Candidate Moves Method implementation in MiniMax search procedure of the Achilles chess engine. In: 12th International conference on telecommunication in modern satellite, cable and broadcasting services (TELSIKS). pp 314–317. https://doi.org/10.1109/TELSKS.2015.7357795
4. Experiments with Game tree search in real-time strategy games. arXiv:1208.1940v1 [cs.AI] 9 Aug 2012
5. Marsland TA, Bjornsson Y From minimax to manhattan. Games AI Res 5–17
6. Gillogly JJ (1972) The technology chess program. Artif Intell 3:145–163
7. Knuth DE, Moore RW (1975) An Analysis of alpha-beta pruning. Artif Intell 6(4):293–326. ISSN 0004-3702. https://doi.org/10.1016/0004-3702(75)90019-3.
8. Wang XZ, He Y-L, Su P, Li W-L (2009) Two-ply iterative deepening in Chinese-chess computer game. International conference on machine learning and cybernetics. pp 2020–2026. https://doi.org/10.1109/ICMLC.2009.5212141
9. Shannon CE (1950) Programming a computer for playing chess. Phil Mag 41(7):256–275
10. Stockfish—an open source chess engine (2022) http://stockfishchess.org

# Biomedical Informatics

# A Modified Microwave Based System Design for Early-Stage Breast Cancer Detection

**Moutusi Mondal, Palash Ghosal, Amish Kumar, and Debashis Nandi**

**Abstract** This paper applies a modified system architecture for image reconstruction and cancer detection purposes. The proposed method for image reconstruction is based on a Microwave-tomographic system. The data is collected by 12 dipole antennas. The system and the antenna model (2.4 GHz) were designed in HFSS software. The mathematical breast structure is reconstructed with different dielectric properties. After that, the data is applied to Newton–Kantorovich methods using MATLAB environment. Thereafter, the tomographic image is reconstructed. The data set was also implemented for the gradient method in MATLAB. But the reconstructed images are less prominent than the Newton–Kantorovich method. The algorithm helps to detect cancer in different areas and locations. The heterogeneous breast tissues are detected correctly while outperforming some of the state-of-the-art methods.

**Keywords** Cancer detection · Microwave tomography · Microwave imaging · Dielectric breast structure · Newton–Kantorovich method · Gradient method

## 1 Introduction

Breast cancer is the most common cancer among women, and the fatality rate is very high. According to a WHO estimate from 2020, there were 6, 85,000 deaths worldwide and 2.3 million new breast cancer cases in women. Breast cancer will be the most common cancer worldwide by the end of 2020, with 7.8 million women still living with the disease [1]. Improvements came after the 1980s in countries with different modes of treatment. Breast cancer is very much common among Indian women. Studies conducted in 2020 revealed that the survival rate for breast cancer in Indian women is only 66% [2]. Women in nations like the United States and

M. Mondal (✉) · P. Ghosal · D. Nandi
National Institute of Technology, Durgapur 713209, India
e-mail: moutusi.iit@gmail.com

A. Kumar
Zreyas Technology Pvt. Ltd., Kolkata 700118, India

 
S. Borah et al. (eds.), *Advanced Computational and Communication Paradigms*, Lecture Notes in Networks and Systems 535,
https://doi.org/10.1007/978-981-99-4284-8_20

Australia have survival rates for cancer that are almost 90%, according to a study, namely CONCORD-3, a global surveillance project for cancer survival coordinated by the London School of Hygiene and Tropical Medicine [3]. Early detection is an effective process for breast cancer treatment.

The most popular screening method for early breast cancer detection is X-ray mammography. Mammogram images are normally low contrast and noisy. So, the failure rate of breast cancer detection is high [4]. Due to the ionization process in X-ray, it may cause an increase in malignancy. So, to overcome this condition, another non-ionizing microwave imaging system is used for breast cancer detection. The microwave imaging system is of two types: microwave tomography and microwave radar-based systems. For microwave tomography, images are formed by the received scattered waves. Microwave tomography is one of the most popular imaging techniques, and so many research works are there [5–12]. The clinical prototype model has been designed by Jeon et al. [13]. Aldhaeebi et al. [14] discussed the promising works in the microwave tomographic field. Recently, block reconstruction algorithm [15] gives satisfactory results for microwave image formation. Mehedi et al. [16] implemented a back-projection algorithm for intelligent tomographic microwave image formation.

In this paper, we used Newton–Kantorovich and gradient-based microwave image reconstruction techniques and prototype model design proposed for microwave tomographic systems. The results have been compared. The dipole antenna is used for receiving and transmitting data. The mathematical breast structure is proposed through different research works [17–20]. HFSS is used for generating the simulation model. For this purpose, a computational simulation model was designed for image generation purposes. The simulated data are processed by the proposed Newton-based algorithm [11, 21] and gradient-based method [22] Moreover, the advancement of prototype system design, getting a better quality of images, proper shape recognition helps early-stage detection. In these papers [11, 15, 23], advanced tomography-based models have been proposed. The proposed model design is effective, fast, and accurate two-dimensional shape detection is possible using less computational load. In this work, we discussed the dielectric properties of breast structure and described the advanced system design. Then reconstruction algorithms generate images. Here, the two-dimensional image has been reconstructed.

## 2 Dielectric Breast Structure

For breast cancer detection purposes, it is important to know the dielectric properties of breast tissues. The mapping of dielectric values for normal and cancerous tissues of breast organs has been studied for so long [17–20]. The dielectric properties of breast tissues are collected from their measured complex relative permittivity($\varepsilon$).

The Cole–Cole models have commonly been used for frequency-dependent dielectric properties. With the help of single-pole Cole–Cole model, the dielectric data set was established over the wide-band frequency range (from 0.5 to 20 GHz).

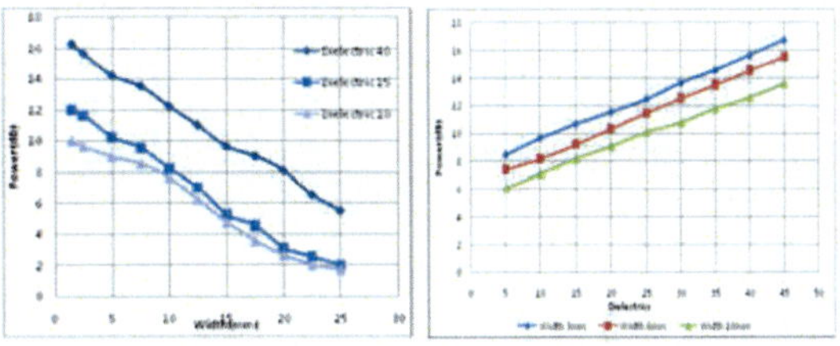

(a) (b)

**Fig. 1** **a** Different dielectric change varies with different widths, frequency fixed at 2.4GHz; **b** dielectric values vary with different widths (3 mm, 6 mm, and 10 mm): frequency fixed at 2.4 GHz

$$\epsilon(\omega) = \epsilon_\infty + \frac{\Delta\epsilon}{(j\omega\tau_c)^{(1-\alpha)}} + \frac{\sigma_c}{j\omega\epsilon_0} \quad (1)$$

In this model, $\epsilon_\infty$, $\alpha$, $\Delta\epsilon$, and $\tau_c$ are variable parameters chosen to fit the experimental data.$\epsilon_\infty$ is the relative permittivity at infinite frequency, $\epsilon_s$ is the static relative permittivity, $\epsilon_0$ is the free space permittivity, $\tau_c$ is the static conductivity, $\omega$ is the angular frequency, and $\tau$ is the relaxation time.

A realistic model design is possible with the help of these parameters. Dielectric properties of tissues, operating frequency, and width help to reconstruct typical breast models. The dielectric contrast between normal and malignant tissue is used for image reconstruction purposes [13]. Around the normal(fatty)tissue region, average static permittivity is $\epsilon_s = 19.8$ ($\epsilon_\infty = 5.57$ and $\sigma_s = 0.52S/m$). In the malignant tissue region, the average static permittivity is$\epsilon_s = 48$ ($\epsilon_\infty = 6.75$ and $\sigma_s = 0.79S/m$). These parametric values help to generate a realistic breast model.

Here in Fig. 1 discussed the dielectric properties with respect to different parameters. It helps to design simulated breast models. Figure 1a shows the changes of different widths along with fixed dielectric value, and Fig. 1b relates the different dielectric values with width. HFSS is used for collecting the data.

## 3 System Design

The antenna is used to scan the breast model at various locations, normally using a dipole antenna (2.4 GHz) for scanning purposes. The 12 dipole antennas are placed in a single line in front of the breast model, and they move at different locations with different angles. Early studies [5–13], researchers followed the circular array system. In this paper, the scanning process covered 180-degree rotation on each side, and the linear antenna array system gradually cover the circular path. In Fig. 2, the model design follows X-ray-type tomographic scanning process. All antennas sequentially transmit one by one, electromagnetic waves which propagate through the breast. The reflected field intensity and phase distributions provide information about breast tissues. The multi-scattered, phase-distributed waves are received by the receiver section of the antenna for imaging. The transmitting and receiving antenna system rotate in an angular fashion with regular intervals. The data obtained from

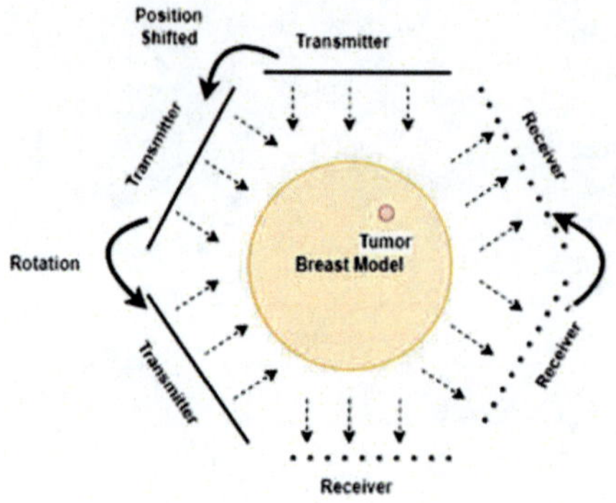


**Fig. 2** Rotational view of scanning object with multiple transmitter and receiver

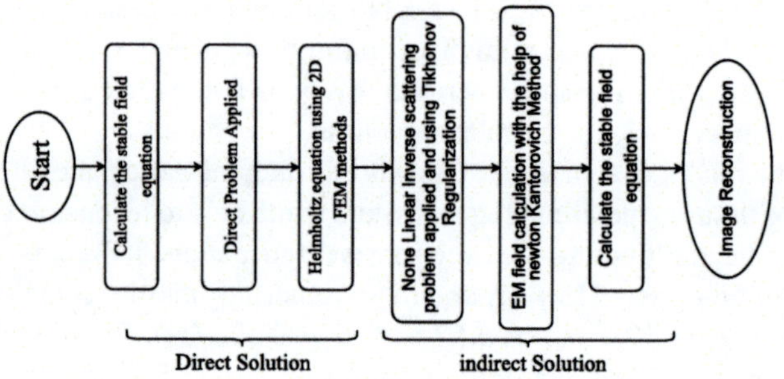


**Fig. 3** Flowchart of the image reconstruction algorithm

the antenna system setup is used for image generation purposes. The system collects the nearest (12 × 35) data points; the scattering parameters are recorded.

## 4 Image Reconstruction Algorithm

Image formation purpose using 2D FEM solver and using Newton–Kantorovich method [11–21]. In Fig. 3, the flowchart gives the idea about reconstruction process. It uses forward solver to calculate Helmholtz equation for electric field calculation. The finite element method used to solve the Helmholtz equation. The incident field $E_0$, which is known, and the scattered field $E_s$ is made by equivalent sources.

$$E_T = E_s + E_o \tag{2}$$

The scalar Helmholtz equation helps to describe the electric field distribution in free space. The scalar approximation is used to develop two-dimensional (2D) tomographic problem. In computation process, the point source act as incident field.

$$\nabla^2 E_s + k^2 E_s - \nabla K E_0 = 0 \tag{3}$$

where $E_s$ is the scattered electrical field, $E_0$ is the incident electric field, k is the wave number, and$\nabla K$ is the difference between the wave number in the homogeneous system. The unknown distribution of permittivity and conductivity is solved by $\nabla K$.

The complex wave number depends on the medium dielectric permittivity and electric conductivity. The unknown function of $\nabla K$ has been generated and the wave equation written in the matrix equation and the Jacobian matrix is used to regularize it. In order to solve the wave equation, the boundary conditions for $E_s$ are applied.

$$E_s = \sum_i (E_s)_i B_i \tag{4}$$

Here, $B_i$ is the basis function and the iteration node is i, which helps to detect the electric field.

The non-linear inverse scattering problem was applied and used linear matrix equation and Tikhonov regularization which help to determine the Newton method.

$$E_0 = \frac{e^{jkr}}{4\pi\epsilon r}\left[Jk^2 r + \frac{1 - Jkr}{r^2}\right] E_i \tag{5}$$

$$E_s = \frac{1}{j\omega\epsilon}\left[\left(\omega^2\mu\epsilon - J\right)\frac{\epsilon^{-\omega^2\mu\epsilon}}{4\pi\epsilon}\right] E_i \tag{6}$$

where $\omega$ is the angular frequency and $\epsilon$ is the reference medium permittivity, $k$ is the wave number, $J$ is the equivalent current source at a particular point, and $r$ denotes the distance position of dipole antennas.

## 5 Results and Discussion

The received signals have been collected at different positions of the antenna. It moves to scan and collect responses from the breast due to the dielectric differences between cancer and its surrounding tissues. An initial set of microwave breast imaging was performed based on the software-based system design tool (HFSS). The background medium for coupling the microwave signals into the breast has nominal relative permittivity and conductivity values as given [13]. The predefined model (HFSS) shown in Fig. 4.

The reconstruction algorithms operated with the different angular positions of antennas organized in a parallel fashion. Here, basically, are followed two most popular methods used in different tomographic microwave-based imaging. We got images using these two methods. So, the data acquisition system design able to reconstruct images and results are compared in Fig. 5. The Newton–Kantorovich

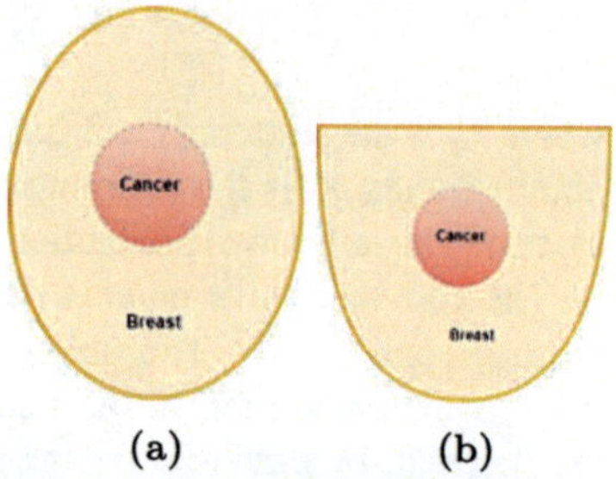


**Fig. 4** The simulated model design for breast cancer **a** top view, **b** front view

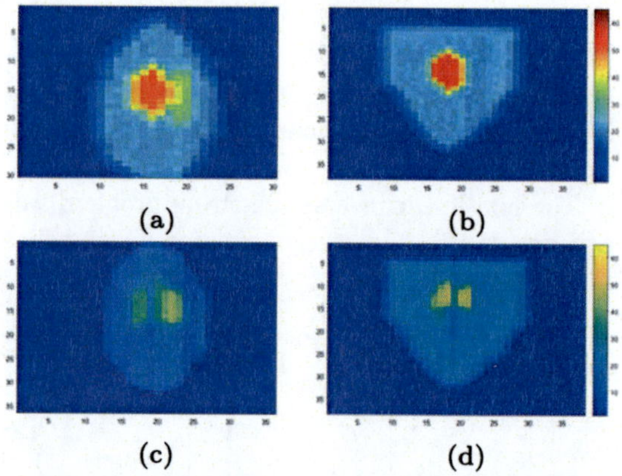


**Fig. 5** The tomographic 2D imaging using Newton–Kantorovich Method—top row (**a**) and (**b**) and Gradient Method—bottom row (**c**) and (**d**). The top view shown in left-hand side and the front view in right-hand side columns

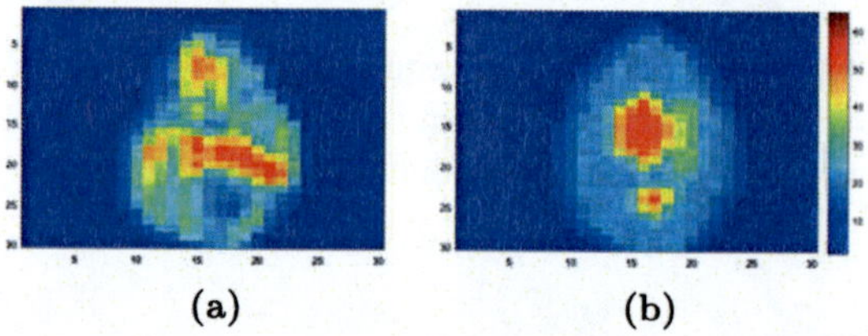


**Fig. 6** **a** and **b** Different cancerous areas detected

method produced better results, it was found that both location and size of the cancer zone are clearly identified. The discrimination between cancerous tissue and fatty tissue is possible in Fig. 5c, d.

In 2.4 GHz frequency, we are getting overall clear images. The cancer zone has been detected with information about the sizes and dielectric properties. The infected cancerous zone and its surrounding fatty tissues are clearly distinguished by the color contrast of dielectric properties shown in Fig. 6.

In Fig. 6a, two different sizes of cancer were also observed prominently. Proper shape detection is also possible. It helps to detect benign and malignant conditions as well as different heterogeneous breast tissues also shown in Fig. 6b. The Newton–Kantorovich and gradient-based methods are the most significant algorithms (qualitative method) employed for microwave imaging. In this work, the

Newton–Kantorovich approach is applied to solve the two-dimensional tomographic problems to get more significant results than the gradient method which suffers from its slower convergence problem.

## 6 Conclusion

According to the findings of this study, microwave imaging may be a valuable method for identifying breast cancer. Cancer can be recognized by the prototype system, which also details its size and dielectric characteristics. The dipole antenna is utilized to serve as the microwave source. The planar configurations detect cancer zone with accurate identification with shapes. The image reconstruction algorithm helps to detect and locate the tumor at an early stage. The remarkable contrast between the different tissue types is existing here and it is well defined. Here, the simulation studies represent some specific case among all possibilities or condition that arises. Future improvement of algorithms along with hardware implementation will enhance early breast cancer detection.

## References

1. Arnold M, Morgan E, Rumgay H, Mafra A, Singh D, Laversanne M, Vignat J, Gralow JR, Cardoso F, Siesling S et al (2022) Current and future burden of breast cancer: global statistics for 2020 and 2040. The Breast 66:15–23
2. Mehrotra R, Yadav K (2022) Breast cancer in India: present scenario and the challenges ahead. World J Clin Oncol 13(3):209
3. Allemani C, Matsuda T, Di Carlo V, Harewood R, Matz M, Nikšić M, Bonaventure A, Valkov M, Johnson CJ, Estève J et al (2018) Global surveillance of trends in cancer survival 2000–14 (concord-3): analysis of individual records for 37 513 025 patients diagnosed with one of 18 cancers from 322 population-based registries in 71 countries. The Lancet 391(10125):1023–1075
4. Huynh PT, Jarolimek AM, Daye S (1998) The false-negative mammogram. Radiographics 18(5):1137–1154
5. Fear E, Stuchly M (2000) Microwave detection of breast cancer. IEEE Trans Microw Theory Tech 48(11):1854–1863
6. Fear EC, Li X, Hagness SC, Stuchly MA (2002) Confocal microwave imaging for breast cancer detection: localization of tumors in three dimensions. IEEE Trans Biomed Eng 49(8):812–822
7. Fear EC, Bourqui J, Curtis C, Mew D, Docktor B, Romano C (2013) Microwave breast imaging with a monostatic radar-based system: a study of application to patients. IEEE Trans Microw Theory Tech 61(5):2119–2128
8. Mehranpour M, Jarchi S, Keshtkar A, Ghorbani A, Araghi A, Yurduseven O, Khalily M (2020) Robust breast cancer imaging based on a hybrid artifact suppression method for early-stage tumor detection. IEEE Access 8:206790–206805
9. Naghibi A, Attari AR (2020) Near-field radar-based microwave imaging for breast cancer detection: a study on resolution and image quality. IEEE Trans Antennas Propag 69(3):1670–1680

10. Bicer MB, Akdagli A (2018) Implementation of the inverse circular radon transform-based imaging approach for breast cancer screening. Int J RF Microw Comput Aided Eng 28(6):e21279
11. Grzegorczyk TM, Meaney PM, Kaufman PA, Paulsen KD et al (2012) Fast 3-d tomographic microwave imaging for breast cancer detection. IEEE Trans Med Imaging 31(8):1584–1592
12. Pagliari DJ, Pulimeno A, Vacca M, Tobon JA, Vipiana F, Casu MR, Solimene R, Carloni LP (2015) A low-cost, fast, and accurate microwave imaging system for breast cancer detection. In: IEEE biomedical circuits and systems conference (BioCAS). IEEE, pp 1–4
13. Jeon S-I, Kim B-R, Son S-H (2016) Clinical trial of microwave tomography imaging. In: URSI Asia-Pacific radio science conference (URSI AP-RASC). IEEE, pp 1–2
14. Aldhaeebi MA, Alzoubi K, Almoneef TS, Bamatraf SM, Attia H, Ramahi OM (2020) Review of microwaves techniques for breast cancer detection. Sensors 20(8):2390
15. Friedrich C, Bourguignon S, Idier J, Goussard Y (2020) Three-dimensional microwave imaging: fast and accurate computations with block resolution algorithms. Sensors 20(21):6282
16. Mehedi IM, Rao KP, Al-Saggaf UM, Alkanfery HM, Bettayeb M, Jannat R (2022) Intelligent tomographic microwave imaging for breast tumor localization. Math Probl Eng
17. Zastrow E, Davis SK, Lazebnik M, Kelcz F, Van Veen BD, Hagness SC (2008) Development of anatomically realistic numerical breast phantoms with accurate dielectric properties for modeling microwave interactions with the human breast. IEEE Trans Biomed Eng 55(12):2792–2800
18. Lazebnik M, Okoniewski M, Booske JH, Hagness SC (2007) Highly accurate debye models for normal and malignant breast tissue dielectric properties at microwave frequencies. IEEE Microwave Wirel Compon Lett 17(12):822–824
19. Lazebnik M, Popovic D, McCartney L, Watkins CB, Lindstrom MJ, Harter J, Sewall S, Ogilvie T, Magliocco A, Breslin TM et al (2007) A large-scale study of the ultrawideband microwave dielectric properties of normal, benign and malignant breast tissues obtained from cancer surgeries. Phys Med Biol 52(20):6093
20. Surowiec AJ, Stuchly SS, Barr JR, Swarup A (1988) Dielectric properties of breast carcinoma and the surrounding tissues. IEEE Trans Biomed Eng 35(4):257–263
21. Ciocan R, Jiang H (2004) Model-based microwave image reconstruction: simulations and experiments: Model-based microwave image reconstruction: simulations and experiments. Med Phys 31(12):3231–3241
22. Semenov SY, Bulyshev AE, Abubakar A, Posukh VG, Sizov YE, Souvorov AE, Van Den Berg PM, Williams TC (2005) Microwave-tomographic imaging of the high dielectric-contrast objects using different image-reconstruction approaches. IEEE Trans Microw Theory Tech 53(7):2284–2294
23. Ozgun O, Kuzuoglu M (2018) Coordinate transformation aided finite element method for contour detection of breast tumors in microwave imaging. Int J Numer Methods Biomed Eng 34(10):e3124

# Research Review and Literature Perception Towards Medicinal Plants Classification Using Deep Learning Techniques

**Himanshu Kumar Diwedi, Anuradha Misra, Amod Kumar Tiwari, and Atif Mahmood**

**Abstract** The pharmaceutical industry is paying more attention to medicinal plants since they are less likely to cause undesirable responses and are less expensive than current medicines. These features have attracted researchers interest in automatic medicinal plant recognition. There are several ways to improve the powerful medicinal plant classifier that can be useful in real-time classification. This work uses several reliable and effective deep-learning techniques for classifying plants based on leaf images. The recent research work is collected systematically and categorized based on their techniques. Further, the dataset details, performance measures used for evaluation, tools utilized for implementation, and chronological review of the collected works are discussed. Finally, this research is concluded with an analysis of eminent ongoing opportunities and research perspectives for improvement in this domain.

**Keywords** Medicinal plants · Classification · Deep learning · Dataset · Performance metrics

H. K. Diwedi · A. Misra (✉)
Computer Science and Engineering, Amity School of Engineering and Technology, Amity University, Lucknow, India
e-mail: amisra@lko.amity.edu

A. K. Tiwari
Computer Science and Engineering, Rajkiya Engineering College, Sonbhadra, India
e-mail: amod.tiwai@recsonbhadra.ac.in

A. Mahmood
Computer Science and Engineering, Pranveer Singh Institute of Technology, Kanpur, India
e-mail: atif.mahmood@psit.ac.in



S. Borah et al. (eds.), *Advanced Computational and Communication Paradigms*, Lecture Notes in Networks and Systems 535,
https://doi.org/10.1007/978-981-99-4284-8_21

## 1 Introduction

By providing oxygen and water for living beings, plants play a major role in protecting life and maintaining biodiversity on earth. A significant class of plants is used as medicines for various diseases. The Previous Generation's knowledge about medicinal plants needs to be maintained and preserved [1]. One of the most difficult tasks is determining a medicinal plant's values. The classification and identification of medicinal plants can be accomplished using image processing technology, pattern recognition, and computer vision. Although herbal medicine has no side effects, if it is incorrectly recognized, the patient's life could be lost [2]. Manual identification is frequently used to categorize medicinal plants. People first use their hands, noses, eyes, or other human organs to detect a plant's texture, taste, color, and shape. Then, based on references or personal experience, they identify the species of medicinal plants.

It is easy and practical to classify medicinal plants using an automatic recognition method with leaf images. The use of supervised approaches to classification is greater than that of unsupervised techniques. Here deep learning technique requires a significant amount of huge data during the training process. The medicinal plant description also uses multi-feature fusion approaches [3]. Several studies used machine learning or pattern matching to train their classifiers. DP, KNN, and SVM are a few examples of typical algorithms.

The most often used machine learning approach for classifying medicinal plant leaves is deep learning. Deep learning techniques are now used to categorize or signify images of medicinal leaves [4]. But there are some limitations to deep learning. High classification accuracy relies on the network having sufficient supervised learning data, which is typically difficult. Only a small number of learning samples are returned. In this, generic deep learning neural networks produce the worst outcomes [5]. Humans can accurately and quickly classify procedures with fewer samples. With the addition of novel samples, computational measurements can make precise judgments. It develops a computation method through the training network so that they may learn from small samples and then use it for autonomous medicinal plant categorization. An automatic discriminative method based on deep learning solves the classification problems by the low inter-class variability of leaf images. Therefore, this survey considers deep learning techniques for determining the best among them that support future research on medicinal plant classification.

This survey has contributions which are explained as follows.

- To review the medicinal plant classification analysis with various information from the existing medicinal plant classification model to classify the medicinal plants accurately.
- To perform a survey based on medicinal plant classification over performance metrics, algorithmic classification, chronological review, and implementation tools.

- To overcome the challenges of the previous medicinal plant classification models, develop an enhanced classification of medicinal plants.

The remaining sections of this survey are described as follows. Section 2 explains the existing works, analysis of datasets, and chronological review of Medicinal Plants Classification. Section 3 describes the implementation of tools, algorithmic classification, performance metrics, and types of medicinal plants used. Section 4 elaborates on the challenges of the existing Medicinal Plants Classification. At last, Sect. 5 summarizes the efficient classification techniques.

# 2 Literature Review on Medicinal Plants Classification

## *2.1 Related Works*

In 2021, Kolhar and Jagtap [6] suggested three distinct approaches that were Convolutional Long Short-Term Memory (ConvLSTM), Convolutional Neural Network (CNN) and 3-dimensional (3-D) CNN, and a vision transformer for achieving the goal of classifying plants based on accession. For evaluating the effectiveness of these techniques, it utilized a dataset for the accession classification of Arabidopsis plants. The neural network model received time series color images of four different Arabidopsis plant accessions as input and recognized the plant accession. These network models could be utilized for determining unidentified plant accessions and forecasting the growth patterns of plants under various climatic circumstances.

In 2022, Puri and Kumar et al. [7] suggested a computer-aided categorization approach for categorizing wide varieties of leaves of medicinal plants. The utilized classification scheme distinguished between the different leaf textures of medicinal plants. Five classes of various plant leaf types were taken for acquiring the desired result, and the texture features of those leaves were extracted using Laws' mask analysis. A support vector machine (SVM) classifier was utilized for the classification operation.

In 2022, Dudi and Rajesh [8] introduced a brand-new model for classifying plant leaves based on deep learning. Grayscale, median filtering, and histogram equalization were utilized for enhancing the image quality required for further processing. The hybrid Shark Smell-Based Whale Optimization Algorithm (SS-WOA) optimized the activation function in maximizing the classification accuracy in CNN. Untrained image classification was a highly difficult task. Hence the improved threshold-based CNN classification was employed. In 2017, Kan and Jin et al. [9] suggested an automatic classification method was introduced based on images of medicinal plant leaves. The leaf images of medicinal plants were first preprocessed. Next, five Texture Characteristics (TF) and ten Shape Features (SF) were computed. At last, the leaves of medicinal plants were classified using the SVM. The outcome suggests that it was possible to automatically classify medicinal plants by combining multi-feature extraction and SVM from leaf images.

In 2019, Dileep and Pournami [10] have suggested categorizing medicinal plants using leaf attributes such as form, size, color, and texture. So, AyurLeaf, a Deep Learning based on the CNN model, was employed. It also suggested a dataset for medicinal plants, which were widespread throughout Kerala. For effective feature extraction from the dataset, a deep neural network based on Alexnet was utilized. AyurLeaf offered a simple way of recognizing and categorizing medicinal plants and aid in preserving traditional medical knowledge. In 2017, Begue and Kowlessur et al. [11] recommended an automated system for medicinal plant identification using machine learning techniques and computer vision. In a scientific environment, a leaf from different species of medicinal plants was collected and photographed. Then, from these properties, several derived features were calculated. With a 10-fold cross-validation method, a random forest classifier produced the best results.

In 2021, Naeem and Ali et al. [12] recommended the classification of medicinal plant leaves using machine learning. The Department of Agriculture at Pakistan's Islamia University in Bahawalpur collected the six different types of medicinal plant leaves that make up the dataset. The names for these plants were Tulsi, Peppermint, Bael, Lemon balm, Catnip, and Stevia. The multispectral and digital image datasets were gathered through a computer vision lab setup. It reduced the leaf's area and converted it to a grayscale format in the preprocessing stage. Second, the Sobel filter performs seed intensity-based edge/line detection and creates five regions of interest. In 2018, Rajani and Veena [13] suggested that plants were essential to human existence since they offered medicine, oxygen, environmental protection, fuel, gums, shelter, and food. The identification of manual medicinal plants takes more time for determining. The automatic identification and classification of medicinal plants were required for addressing this issue. The primary processes in identifying medicinal plants were feature extraction and classification. Many beneficial plant species were being destroyed because of professional secrecy, rising population, global warming, a lack of government backing for research operations, and a lack of knowledge about medicinal plants.

In 2019, Keskar and Maktedar [14] designed a hybrid optimum machine learning based on an Effective Accurate Classifier for Ayurvedic Medical Plant Identification (EAC-AMP) approaches. The type of plants was finally identified using a whale optimization with Deep Neural Network (DNN) classifier. A Spider Optimization Neural Network (SONN) separated leaf regions from an image, and segmentation could be accomplished after preprocessing. The improved edge detection algorithm computed the top and bottom leaf edges after detecting the image corners in EAC-AMP. The Symbolic Accurate Approximation (SAX) was utilized for computing time and frequency domain features. Shape, color, and teeth data were computed using Two-Dimensional Binary Phase Encoding (2DBPE). Understanding and extracting the characteristics of the plant allowed for the accurate identification of ayurvedic plant leaves. In 2019, Dasari and Prasad [15] suggested a cured tobacco leaves-based CNN. The neural networks have four hidden layers, the first three performing convolutions, and pooling, and the fourth layer being completely connected to CNN. The suggested model employed the Max Pooling Technique (MPT) for minimizing the size. The test set consisted of roughly fifteen images from each cluster, and classification was performed using the three major classes of Class-1, Class-2, and Class-3 to achieve overall efficiency.

In 2016, Sladojevic and Arsenovic et al. [16] have concerned with classifying deep convolutional network leaf images to recognize plant diseases. The developed model could be identified various types of plant illnesses. The entire process of this model obtaining images for building a database has been approved by agricultural specialists. In 2019, Kaya and Seydi et al. [17] designed several deep learning algorithms for helping the automatic classification of plant species. DNN was one of the techniques of deep learning that have been utilized on various data sets. However, DNNs have frequently been utilized in isolation and for reusing and transferring the knowledge of other DNN applications.

In 2021, Azadnia and Kheiralipour [18] suggested a new reliable automatic image processing algorithm. The algorithm was applied to the obtained images for extracting texture, color, and form attributes. The most common medicinal herbs were Rumex acetosa, Trigonella foenum-graecum, Pelargonium sidoides, Origanum vulgare, and Falcaria vulgaris. Smartphone vision systems were utilized for capturing plant images. Artificial Neural Networks (ANN) were utilized to categorize different medicinal plant categories. The effective classifiers were chosen based on error, correlation, and accuracy. It has been sufficient for straightforward and precise classification. In 2022, Kyalkond and Aithal et al. [19] designed for overcoming a difficult and time-consuming process and introduced the automated categorization method for medicinal plants. It was making it easier for classifying medicinal plants by utilizing textural characteristics that were essential for recognizing and identifying leaves. It included three main steps that were feature extraction, classification, and image improvement. The traits that could be compared between the leaves images were extracted using digital image processing algorithms after the leaves images using a smartphone.

In 2019, Dudi and Rajesh [20] recommended getting the highest accuracy in the classification and identification processes by utilizing computer vision techniques with deep learning methods. For deep learning of medicinal plant images, the system employed CNN and machine learning techniques. The Flavia leaf dataset from the SourceForge website was utilized. This data was utilized for training the dataset that was based on CNN and machine learning. In 2017, Gao and Lin [21] designed using extraction of vein enhancement in the image was created with an automated segmentation approach for the medicinal plant. By directly solving color RGB images, gradient magnitude, and gradient angle images could be created. The veins in the gradient magnitude images were further enhanced by generating the vein enhancement image by applying the objective criteria. The main veins were recognized using the OTSU method based on this image.

In 2021, Mukherjee and Tudu et al. [22] introduced categorizing medicinal leaves using a computer vision-based technique together with a certain maturity level. The operations were comparatively simpler, quicker, less expensive, and non-invasive. It took images of Medicinal plant leaves, which were then processed and input into an architecture of CNN-based system for categorization. With the use of three well-known species of medicinal plants, namely, Neem, Kalmegh, and Tulsi, the system's potential has been evaluated. The Binary Particle Swarm Optimization (BPSO) was utilized for determining the CNN hyperparameter values. In 2014, Chemburkar and Sartape et al. [23] developed an automated tool that could be identified and categorized plant leaf species after comparison with trained sets. After image processing, an ANN utilized these trained sets. The neural networks have received training for vein and edge identification. Atomizing was challenging, but because of dramatic technological advancements, it was feasible in simple stages. It suggested an automated technique for leaf image recognition in plants. The atomization technique helped for speeding up the time-consuming and manual identification process, which required prior knowledge of the species.

In 2004, Hamilton [24] recommended the most crucial function of medicinal plants in biological conservation for promoting the preservation of natural ecosystems broadly. The results from the unique association's individuals have medicinal plants and their significant impacts on their lives in terms of cultural identity, health support, stability, livelihood, and monetary gain. Medicinal plant conservation and usage with biopiracy or excessive limits on research have started to gain importance in the policy. However, putting this requirement into effect was not always simple. In 2022, Opoku and Twum et al. [25] developed a computer vision system for identifying medicinal plants using CNNs and custom-made filters from Log-Gabor filters. A system was tested using a dataset comprised of 49 plant species created Ghana Dataset. Ten pre-trained networks including Mobilenetv2, GoogLeNet, Alexnet, VGG16, Inceptionv3, VGG19, Resnet50, DenseNet201, Resnet101, and Restnet18 were employed as feature extractors using the transfer learning methods. Other benchmark datasets, including MD2020, Flavia, Folio dataset, and Swedish Leaf, were utilized for evaluating OTAMNet. Finally, a false-positive rate was attained.

## 2.2 *Chronological Review*

Chronologically, various kinds of research are analyzed for the Medicinal Plant Classification in the year-wise implementation that is shown in Fig 1. The review of the Medicinal Plant Classification is taken from the year 2004 to 2022. Nearly 20 papers are reviewed and observed for this survey. The medicinal plant classification is mostly contributed in the year 2019. The contribution level for the years 2004, 2014, 2016, and 2018 is 5%. The contribution level for the year 2017 is 10%. The contribution level for the years 2021 and 2022 is 20%. Finally, the contribution level for the year 2019 is 30% higher than the other levels on this survey.

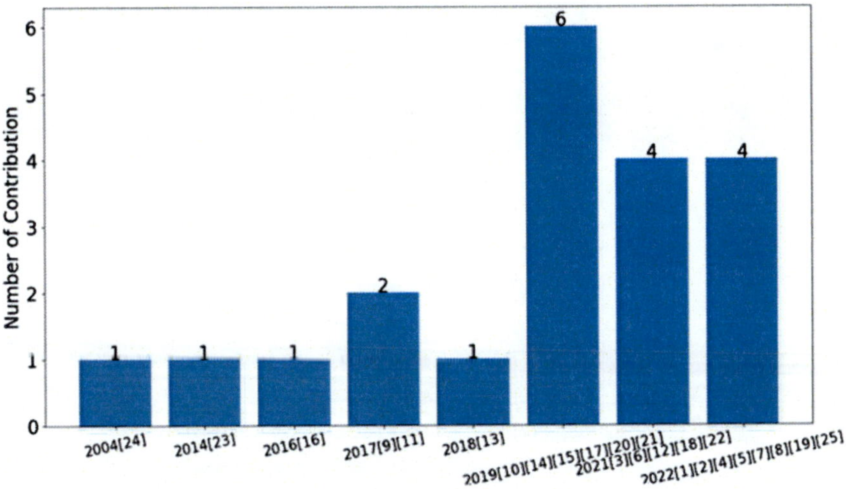


**Fig. 1.** Year-wise implementation for the medicinal plant classification

### *2.3 Medicinal Plant Dataset Analysis*

In medicinal plant classification, a different type of dataset is utilized to classify the medicinal plant, as shown in Table 1. Many researchers utilized various datasets for medicinal plant classification. The online dataset, Flavia and Swedish Leaf datasets, Medicinal plant specimens, and Plant species dataset are the mostly analyzed datasets. Thus, more information about medicinal plant classification is collected from various websites to classify medicinal plants efficiently.

## 3 Short Review on Deep Learning-Based Medicinal Plants Classification Approaches

### *3.1 Algorithmic Classification*

There are various methods and techniques utilized in deep learning models. The techniques related to deep learning and machine learning models are utilized in this survey. To classify the medicinal plants, different classifiers are used by the researchers. Deep learning and machine learning have utilized some techniques in medicinal plant classification, which are explained below.

*Deep learning-based medicinal plant classification*

Deep learning techniques are used in processing, detection, classification, and prediction. The deep learning techniques were analyzed with various algorithmic classifier

**Table 1** Analysis of medicinal plant classification datasets

| Citation | Datasets used |
|---|---|
| [6] | Arabidopsis plant accession classification dataset |
| [7] | Online datasets |
| [8] | Mendeley and Swedish leaf dataset |
| [9] | Medicinal plant specimens |
| [10] | AyurLeaf |
| [11] | The tropical island of mauritius |
| [12] | Department of agriculture |
| [13] | Plants dataset |
| [15] | Test dataset |
| [16] | Online dataset |
| [17] | Flavia and Swedish leaf datasets |
| [18] | Six groups of medicinal plants |
| [19] | Plant species dataset |
| [20] | Flavia dataset |
| [21] | Yaowang mountain dataset |
| [22] | TensorFlow backend library |
| [24] | GWHP dataset |
| [25] | Ghana (MyDataset) |

techniques for medicinal plant classification such as CNN [6, 8, 10, 15, 16, 19–22, 25], CNN-LSTM, and DNN [14, 17, 24].

*Machine learning-based medicinal plant classification:*

Machine learning techniques are also useful in predicting new analyses and classification. The machine learning techniques analyzed with various algorithmic classifier techniques in medicinal plant classification are ANN [13, 18, 23], Multilayer Perceptron (MLP) [11, 12], and SVM [7, 9, 13].

*Difference between the Performance of Deep Learning and Machine Learning Techniques on Medicinal Plant Classification:*

**In deep learning-based medicinal plant classification**, CNN [6, 8, 10] is the most popular choice for an automated classification and feature extraction of images. Especially, the CNN-LSTM [19–21] method has been utilized for extracting the temporal features and for performing classification among the diverse categories. It is an effective neural network method that can provide better classification accuracy with less computational time. However, the LSTM model is more complicated to train the model. It has required sequential input data and also it is utilized to resolve activity recognition problems. DNN is a special type of RNN that can preserve temporal information over a long period [14, 17, 24]. The downsides of the LSTM are it holds few training samples and limited memory resources.

**In machine learning-based medicinal plant classification**, the MLP [11, 12] can produce a better accuracy rate. However, owing to resource constraints, the potential

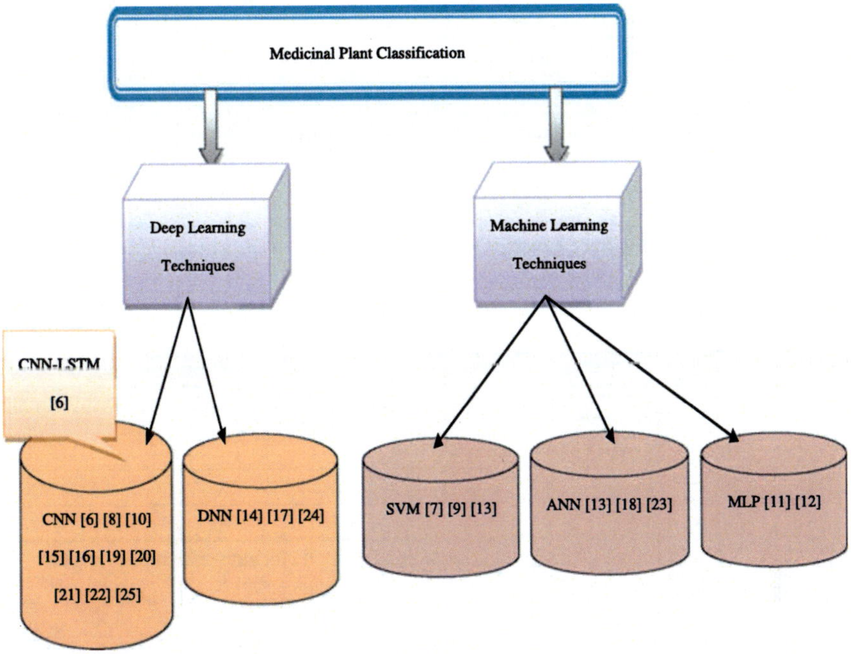


**Fig. 2.** Algorithmic classification on deep learning techniques-based medicinal plant classification models

of neural networks has not been fully utilized and also it more time-consuming. The major scope of ANN [13, 18, 23] is the automatic identification of medicinal plants using image processing approaches. SVM classifier is the better choice for the classification task. Need more investigation about the classification of medicinal plant leaves using SVM [7, 9, 13] models.

The overall review of deep learning algorithms for classifying medicinal plants is represented in Fig 2.

## 3.2 *Types of Medicinal Plants Used in Classification Models*

In medicinal plant classification, there are various medicinal plants utilized in conventional works. Different medicinal plants are reviewed to classify the plants, such as Arabidopsis, Populus tremula, Quercus velutina, Cassia tora, Azadirachta indica, Sushruta, and Vaghbhata, Neem, Origanum vulgare, Brassicaceae, Tulsi, Kalmegh, and Aquilaria malaccensis. The most commonly used medicinal plants are Tulsi, Kalmegh, and Neem. The overall utilized medicinal plants for the classification are shown in Table 2.

**Table 2** Types of medicinal plants used for classification

| Citation | Types of medicinal plants used |
|---|---|
| [6] | Arabidopsis |
| [7] | Asimina triloba, Castanea dentata, Quercus velutina, Salix babylonica and Salix caroliniana |
| [8] | Ulmus glabra, Alnus incana, Sorbus intermedia, Populus tremula, Sorbus aucuparia, Populus, Ulmus carpinifolia, Salix aurita, Betula pubescens Salix alba Sericea, Salix sinerea, Tilia, Acer, Quercus, Fagus silvatica, |
| [9] | Dioscorea bulbifera, Cassia tora, and Coffee Senna |
| [10] | Adalodakam |
| [11] | Antidesma madagascariense, Persea americana, Ayapana triplinervis, Cardiospermum halicacabum, Psiloxylon mauritianum, Azadirachta indica |
| [12] | Lemon balm, Bael, Stevia, Peppermint, Catnip, and Tulsi, |
| [13] | Charka, Sushruta, and Vaghbhata |
| [14] | Neem |
| [15] | Tobacco leaves |
| [18] | Falcaria vulgaris, Pelargonium sidoides, Trigonella foenum-graecum, Origanum vulgare, Rumex acetosa |
| [21] | Apiaceae, Apocinaceae, Asteraceae, Boraginaceae, Brassicaceae, Caryophyllaceae, Cesalpinaceae, Cucurbitaceae |
| [22] | Neem, Tulsi and Kalmegh |
| [23] | Aquilaria malaccensis |
| [24] | Adonis vernalis, Aquilaria malaccensis, Cistanche deserticola, Dioscorea deltoidea, Guaiacum officinal, G. sanctum |
| [25] | Arabidopsis |

### *3.3 Evaluating the Implementation Tools*

In medicinal plant classification, researchers utilized different implementation tools that are plotted in Fig. 3. Some of the implementation tools such as Python [6, 8, 16, 22], MATLAB [9, 14, 18], Google Colab [8, 25], and Java [11, 23]. Based on the overall evaluation, Python and MATLAB are commonly used tools for medicinal plant classification.

### *3.4 Performance Metrics Utilized for Medicinal Plant Classification*

The medicinal plant classification is evaluated with standard performance measures that are shown in Table 3. The performance analysis is analyzed with diverse measures

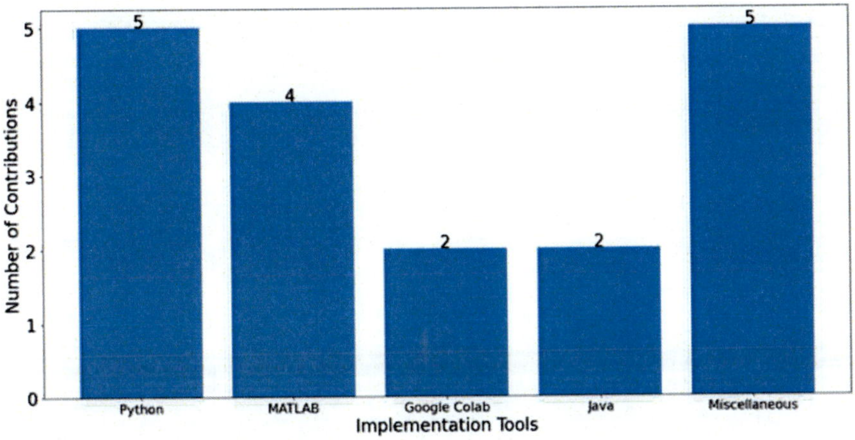


**Fig. 3.** Various implementation tools utilized in medicinal plant classification models

such as precision, specificity, accuracy, recall, False Positive Rate (FPR), F1-score, False Negative Rate (FNP), sensitivity, False Positive (FP), and False Negative [26, 27]. The accuracy, precision, F1-score, FP, and recall are the measures used mostly by the researchers.

# 4 Research Gaps and Challenges on Conventional Medicinal Plants Classification

Certain human diseases can be treated and prevented with the help of medicinal plants. The class of medicinal plants has been used to treat human illnesses. The taxonomy of medicinal plant species is essential for manufacturing and preserving medicines. Local inhabitants lack sufficient awareness about the uses of their urban medicinal plants. The medicinal plants have an impact, including allergic responses, rashes, asthma, headaches, lightheadedness, skin conditions, colds, blood purifier, indigestion, and harmful side effects since it contains toxic chemicals. Therefore it is necessary to identify useful medicinal plants through deep learning techniques. To identify various species, classifying medicinal plants using deep learning high-accuracy image classification algorithms may be helpful. It has many issues when classifying medicinal plants, such as (1) not improving the generalization capacity of the model, (2) it cannot be used with different datasets, (3) it cannot tackle the noisy classification, and the regression problems, (4) advanced monitoring methods are not employed, (5) massive data collections cannot be handled, (6) it is unable to handle leaves that are broken up and have missing leaf pieces, (7) utilizing many distinctive properties of leaves does not enrich, and (8) the edge and leaf vein characteristics are not the main focus. The research gaps in future directions were explained to urge the

**Table 3** Overall performance measures used for medicinal plant classification models

| Citation | Accuracy | Precision | Recall | F1-score | Specificity | Sensitivity | FPR | FNR | Others |
|---|---|---|---|---|---|---|---|---|---|
| [6] | ✓ | ✓ | ✓ | ✓ | – | – | – | – | – |
| [7] | ✓ | – | – | – | – | – | – | – | – |
| [8] | ✓ | ✓ | – | ✓ | ✓ | ✓ | ✓ | ✓ | MCC, NPV, FDR |
| [9] | ✓ | – | – | – | – | – | – | – | – |
| [10] | ✓ | – | – | – | – | – | – | – | – |
| [11] | – | – | ✓ | ✓ | – | – | ✓ | – | ROC, Kappa Statistics, |
| [12] | ✓ | – | – | – | – | – | – | – | – |
| [13] | ✓ | – | – | – | ✓ | ✓ | ✓ | – | True positive |
| [14] | ✓ | – | – | – | – | – | – | – | – |
| [15] | ✓ | – | – | – | – | – | – | – | – |
| [16] | ✓ | – | – | – | – | – | – | – | – |
| [17] | ✓ | – | – | – | – | – | – | – | – |
| [18] | ✓ | – | – | – | – | – | – | – | Mean square |
| [19] | ✓ | – | – | – | – | – | – | – | – |
| [20] | ✓ | ✓ | ✓ | ✓ | – | – | | – | – |
| [21] | – | ✓ | ✓ | ✓ | – | – | – | – | Manhattan, Jaccard |
| [22] | – | ✓ | ✓ | ✓ | ✓ | ✓ | | – | – |
| [23] | ✓ | – | – | – | – | – | – | – | – |
| [24] | ✓ | – | – | – | – | – | – | – | – |
| [25] | – | – | – | – | – | – | ✓ | – | – |

researchers to continue improving the Medicinal Plant Classification models using deep learning networks.

## 5 Conclusion

In addition to exploring several deep learning methodologies, this research survey reviewed the medicinal plant classification technologies which have been utilized previously. This survey provided more information about the recent advancements in medicinal plant classification models. It focused on the types of datasets utilized for the most current advancements in medicinal plant classification approaches as well as the implementation tools being utilized for experiments over the medicinal plant classification approach. Additionally, performance measures have been examined for verifying the model's efficacy. Therefore, this survey motivates the researchers to continue developing the medicinal plant classification models with new efficient deep learning techniques to tackle the current research gaps in the classification of medicinal plants.

## References


1. Sathiya V, Josephine MS, Jeyabalaraja V (2022) An automatic classification and early disease detection technique for herbs plant. Comput Electr Eng 100:108026
2. Shailendra R, Jayapalan A, Velayutham S et al (2022) An IoT and Machine Learning Based Intelligent System for the Classification of Therapeutic Plants. Neural Process Lett 54:4465–4493
3. Roopashree S, Anitha J (2021) DeepHerb: a vision-based system for medicinal plants using xception features. IEEE Access 9:135927–135941
4. Sachar S, Kumar A (2022) Deep ensemble learning for automatic medicinal leaf identification. Intell J Inf Technol 14:3089–3097
5. Wang C, Zhang X, Shi T, Zhang C, Li M (2022) Classification of Medicinal Plants Astragalus Mongholicus Bunge and Sophora Flavescens Aiton Using GaoFen-6 and Multitemporal Sentinel-2 Data. IEEE Geosci Remote Sens Lett 19:1–5
6. Kolhar S, Jagtap J (2021) Spatio-temporal deep neural networks for accession classification of Arabidopsis plants using image sequences. Ecol Inform 64:101334
7. Puri D, Kumar A, Virmani J et al (2022) Classification of leaves of medicinal plants using laws' texture features. Intell J Inf Technol 14:931–942
8. Dudi B, Rajesh V (2022) Optimized threshold-based convolutional neural network for plant leaf classification: a challenge towards untrained data. J Comb Optim 43:312–349
9. Kan HX, Jin L, Zhou FL (2017) Classification of medicinal plant leaf image based on multi-feature extraction. Pattern Recognit Image Anal 27:581–587
10. Dileep MR, Pournami PN (2019) AyurLeaf: a deep learning approach for classification of medicinal plants. IEEE Region TENCON
11. Begue A, Kowlessur V, Singh U, Mahomoodally F, Pudaruth S (2017) Automatic recognition of medicinal plants using machine learning techniques. Int J Adv Comput Sci Appl 8(4)
12. Naeem S, Ali A, Chesneau C, Tahir MH (2021) The classification of medicinal plant leaves based on multispectral and texture feature using machine learning approach. Agronomy 11:263
13. Rajani S, Veena MN (2018) Study on identification and classification of medicinal plants. Int J Adv Sci Eng Technol 6(2):2321–8991
14. Keskar M, Maktedar D (2019) Enhancing classifier accuracy in ayurvedic medicinal plants using WO-DNN. Int J Eng Adv Technol (IJEAT) 9(1):2249–8958
15. Dasari SK, Prasad V (2019) A novel and proposed comprehensive methodology using deep convolutional neural networks for flue-cured tobacco leaves classification. Int J Inf Technol 11:107–117
16. Sladojevic S, Arsenovic M, Anderla A (2016) Deep neural networks based recognition of plant diseases by leaf image classification. Comput Electron Agric 6:1–11
17. Kaya A, Keceli AS, Catal C, Hamdi Yalic, Y., Temucin H and Bedir, T. (2019) ekinerdogan Analysis of transfer learning for deep neural network based plant classification models. Comput Electron Agric 158:20–29
18. Azadnia R, Kheiralipour K (2021) Recognition of leaves of different medicinal plant species using a robust image processing algorithm and artificial neural networks, classifier. J Appl Res Med Aromat Plants 25:100327
19. Kyalkond SA, Aithal SS, Sanjay VM, Kumar PS (2022) A novel approach to classification of Ayurvedic medicinal plants using neural networks. Int J Eng Res Technol (IJERT) 11(01)

20. Dudi B, Rajesh V (2019) Medicinal plant recognition based on CNN and machine learning. Int J Adv Trends Comput Sci Eng 8(4).
21. Gao L, Lin X (2019) Fully automatic segmentation method for medicinal plant leaf images in complex background. Comput Electron Agric 164:104924
22. Mukherjee G, Tudu B, Chatterjee A (2021) A convolutional neural network-driven computer vision system toward the identification of species and maturity stage of medicinal leaves: case studies with neem. Tulsi Kalmegh Leaves Soft Comput 25:14119–14138
23. Chemburkar A, Sartape A, Gawade A, Somawanshi P (2014) Automated tool for plant leaf classification using morphological features. Int J Eng Comput Sci 3:9098–9102
24. Hamilton AC (2004) Medicinal plants, conservation, and livelihoods. Biodivers Conserv 13:1477–1517
25. Oppong SO, Twum F, Hayfron-Acquah J, Missah YM (2022) A novel computer vision model for medicinal plant identification using log-gabor filters and deep learning algorithms. Comput Intell Neurosci 21:1189509
26. Mahmood A, Singh SK, Tiwari AK (2022) Pre-trained deep learning-based classification of jujube fruits according to their maturity level. Neural Comput Appl 34:13925–13935. https://doi.org/10.1007/s00521-022-07213-5
27. Mahmood A, Tiwari AK, Singh SK, Udmale SS (2022) Contemporary machine learning applications in agriculture : Quo Vadis?. https://doi.org/10.1002/(ISSN)1532-0634

# Features Fusion-Based Gait Recognition with Covariate Conditions

**Margaret Kathing, Rishang Kumar Brahma, and Sarat Saharia**

**Abstract** Identification of a person based on their walking pattern is affected by factors such as camera viewing angle, apparel with the subject, holding a bag, walking surface, and complex situations. These are the covariate conditions in human gait analysis. The covariate conditions change an individual's gait pattern, making it difficult to implement gait recognition in a realistic environment. This work addresses the issue of covariate conditions in gait identification through a fusion of features. Using a pre-trained VGG16 model with four fully connected layers, the dynamic features are extracted and merged with HoG (Histogram of Oriented Gradients) features extracted from the raw GEI (Gait Energy Image) gait templates. PCA (Principal Component Analysis) is then used to lower the dimension of the combined features in order to select the discriminant feature vectors. The CASIA-B dataset is used to examine the efficacy of the suggested technique, which employs an MLP (Multi-layer Perceptron) classifier. The findings show that the proposed technique outperforms other existing approaches in terms of accuracy while walking normally, wearing a coat, and carrying a bag under identical viewing conditions.

**Keywords** Gait analysis · Gait recognition · Covariate conditions in gait · Classification · FusionModel · MLP

---

M. Kathing · R. K. Brahma (✉) · S. Saharia
Computer Science and Engineering, Tezpur University, Tezpur, Assam, India
e-mail: rkbrahm@tezu.ernet.in

M. Kathing
e-mail: maggy@tezu.ernet.in

S. Saharia
e-mail: sarat@tezu.ernet.in



S. Borah et al. (eds.), *Advanced Computational and Communication Paradigms*, Lecture Notes in Networks and Systems 535,
https://doi.org/10.1007/978-981-99-4284-8_22

# 1 Introduction

Gait is a pattern of locomotion that encompasses body posture. When a person walks, his or her body posture changes in accordance with the walking motion. Gait recognition is a non-intrusive technique for identifying individuals, which means that data collection and classification can be done without the subjects' interaction or participation. This is why gait has become an attractive modality for person identification. The development of a secure authentication and identification system is challenging with the increasing use of automatic tracking devices in surveillance. In surveillance, one of the most useful biometric cues for distance identification of a person using low-quality images is gait. The collection and processing of gait data depend on the application and functionality. Using temporal and spatial information, gait features are identified. The ideal scenario for gait recognition is when it is covariate-invariant. However, the presence of covariate factors such as wearing coat, carrying bag, walking surface, camera viewpoint, and walking speed with respect to the human subject increases the difficulty of gait recognition. A gait-based human identification using a model-free approach is the most explored and studied with various covariate conditions. In past work on gait recognition, a feature fusion technique employing various CNN network models to handle covariate conditions yielded significant results. In this work, we aim to analyze the factor influencing gait recognition with various covariate conditions.

Shape-based recognition and skeleton-based recognition are frequently used approaches in studies on human gait recognition. In shape-based, silhouette image is considered to represent spatial and temporal parameters. The Gait Energy Image (GEI) a silhouette-based gait recognition is suggested as an efficient method for individual recognition. Subsequently, many other silhouette-based gait identification methods, such as Gaitpart, Chrono-Gait Image (CGI), PoseEnergy, and Gait Entropy Image (GEnI), have since developed with remarkable results.

The following is the paper's structure: Sect. 2 reviews relevant literature, Sect. 3 defines the proposed framework, Sect. 4 analyses the experimental outcomes, and Sect. 5 has the conclusion.

# 2 Related Work

Covariate conditions in human gait have a significant effect on the effectiveness of gait-based human identification in intelligent surveillance monitoring. Model-free gait recognition is one technique that is used mostly in gait recognition. Because it is inexpensive and has low computational complexity. In model-free, gait features are estimated from the whole image or contour image. GEI is a model-free approach to generating enhanced spatial–temporal features. The GEI is calculated by taking the average of a series of binary silhouette images collected over the duration of one gait cycle [1]. The information in the GEI is likely to drastically change the person's appearance under different covariate conditions [2]. In the past, GEI model-free gait templates have been adopted in a number of empirical studies of gait recognition algorithms [3–5]. In our recent work on an effective method to reduce covariate issues in gait-based human identification, an efficient gait representation is suggested [6]. In this work, a good classification accuracy is achieved for multi-view angles for all the covariate conditions by using GEI and GGMI (Gait gradient magnitude image) [7] with a Nearest Centroid (NC) classifier. Many works on gait recognition have indicated that the result analysis was primarily performed on transforming models with known covariate conditions. The uniqueness of gait recognition is that it serves as an effective distance-based human identification system without subject cooperation. So, it is required to evaluate the gait recognition approach with a mixture of gait sequences in various covariate conditions. Bashir et al. [8] work on gait recognition without subject has shown remarkable gait recognition performance for covariate-invariant using GEIn from a side view walking posture of 90 degrees. Using a keyframe extraction method for local feature detection between the frames shows improvement in extracting distinctive features [9] for gait recognition. An effective method for reducing covariate conditions in gait recognition is part-based gait that considers only the less affected part by covariate factors [10].

In recent times, the recognition of gait through the use of Convolutional Neural Network (CNN) has been the main focus of a significant number of cutting-edge human identification systems. Gait recognition based on CNN networks such as Deepgait [11], GEINET [12], GaitPart [13], GaitSet [14], etc., has shown remarkable results. According to the above works, the introduction of deep convolutional network-based feature learning methods has resulted in using pre-trained models for gait recognition. The studies also show that combining the pre-trained model with other static model feature descriptors could be an alternative method for overcoming covariate factor issues in gait recognition.

## 3 Proposed Framework

The proposed framework for reducing the covariate factors in gait-based human identification is described in detail in this section, proposed framework is shown in Fig. 1. The framework is designed to provide a robust and accurate identification system, even in the presence of covariate factors such as clothing types and carrying conditions. The first step in the process is to input video files, which undergo preprocessing to extract silhouette frames. A deep CNN image segmentation model is then used for background subtraction, to further extract the silhouette frames. Once the silhouette frames have been extracted, a GEI gait template is created using the silhouette averaging technique suggested by Han and Bhanu et al. [1]. This GEI gait template is then subjected to feature extraction using HoG (Histogram of Oriented Gradients) of a $16 \times 16$ bin size with nine orientations. The resulting feature vectors are 7056 in number. A VGG16 pre-trained model is then used to extract 1024 feature vectors by unfreezing the last 4 layers with 2048, 2048, 1024, and 1024 neurons, and

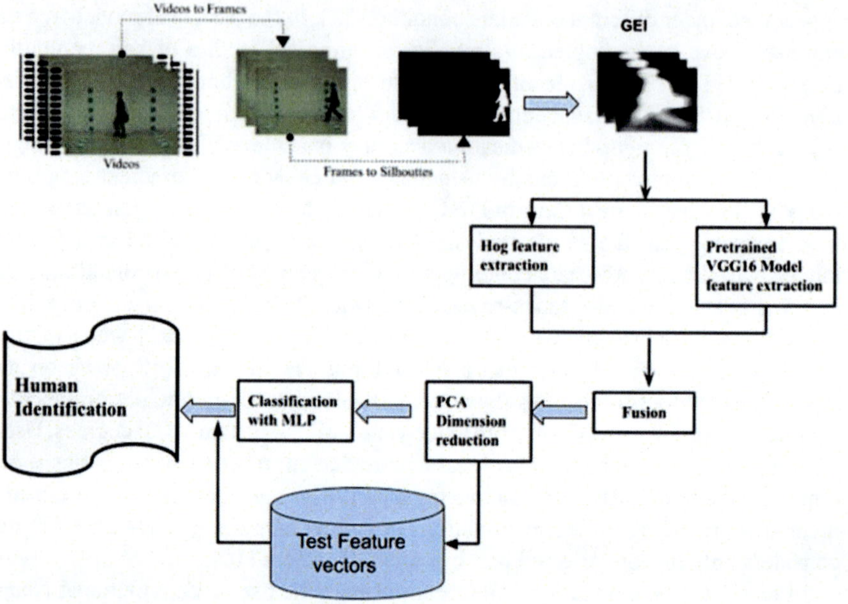


**Fig. 1** The proposed framework

using a Rectified Linear Unit (ReLU) activation function. The extracted features are combined to create fused feature vectors with 8080 features.

To further reduce the feature vectors, PCA (Principal Component Analysis) is applied as a dimension reduction method. The purpose of this is to retain only the discriminant features that characterize the classes. The resulting reduced feature vectors are then used for classification with an MLP (Multilayer Perceptron) classifier. This classifier has three dense layers of 2048, 1024, and 512 neurons, each with a ReLU activation function and an Adam optimizer. The goal of this classifier is to correctly classify the subject in various covariate conditions.

## 4 Experimental and Results Analysis

The CASIA-B standard dataset, which consists of 124 individuals from 11 different view angles with a variation of 18 degrees each, is used for the experiment. The dataset includes 10 different perspectives, with 6 of them being normal walking, 2 being walking while carrying a bag, and 2 being wearing a long or short coat. A sample of each walking normally, carrying a bag, and wearing a coat at a 90-degree angle is shown in Fig. 2.

To evaluate the impact of various covariate conditions on gait-based human identification, the dataset is divided into three parts. For normal walking conditions, six

(a) Normal (b) Bag (c) Coat

**Fig. 2** CASIA-B in 90-degree dataset sample

normal walking sequences were selected and four of them were kept in the gallery set. The gallery set was used to estimate the correct classification for the remaining two sequences of normal walking conditions, as well as for the baggage covariate conditions and clothing covariate conditions.

The Multi-layer Perceptron (MLP) classifier was used for classification and the results obtained from the experiment are shown in Table 1 for different viewing angles.

The analysis reveals that the proposed method demonstrates significant improvement for normal walking and carrying conditions when the gallery and probe angles are identical. However, the clothing covariate condition appears to be a critical issue even with identical viewing angles. Table 2 compares the results with existing works (Fig. 3).

**Table 1** Experimental analysis result indicating normal walk, wearing coat, and carrying bag condition when gallery and probe angle are identical

| Gallery and probe angle | 0° | 18° | 36° | 54° | 72° | 90° | 108° | 126° | 144° | 162° | 180° | Mean |
|---|---|---|---|---|---|---|---|---|---|---|---|---|
| Normal versus normal | 100 | 100 | 97.97 | 96.61 | 99.32 | 97.29 | 98.64 | 96.62 | 98.64 | 100 | 100 | 98.64 |
| Normal versus coat | 63.01 | 66.89 | 64,14 | 69.38 | 73.64 | 74.32 | 68.91 | 63.26 | 67.34 | 59.31 | 67.56 | 67.36 |
| Normal versus bag | 97.26 | 90.54 | 87.16 | 84.35 | 90.54 | 94.59 | 95.27 | 90.47 | 90.47 | 91.09 | 95.91 | 91.6 |

**Table 2** Comparison table with existing methods

| Author | Gait representation | Nm-Nm | Nm-Bg | Nm-Cl | Mean |
|---|---|---|---|---|---|
| Bashir et al. | GEnI | 100 | 78.3 | 44.4 | 74.2 |
| Margaret et al. | GGMI + Hog | 99.18 | 83.06 | 37.77 | 73.33 |
| Rokanujjaman et al. | EnDFT | 97.61 | 83.87 | 51.61 | 77.69 |
| Huang et al. | LFF | 91.9 | 80.3 | 72.3 | 81.5 |
| MuhammadAsi et al. | HoG-GEI | 82.3 | 79.8 | 87.9 | 83.33 |
| Our method | HoG + VGG16 | 97.29 | 94.59 | 74.32 | 88.73 |

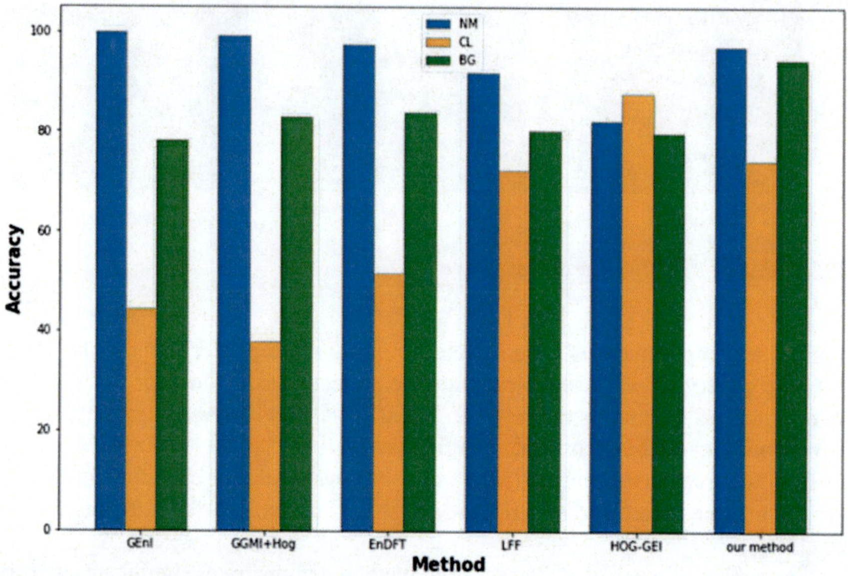


**Fig. 3** Comparison graph with existing methods

# 5 Conclusion

This paper presents a feature fusion approach aimed at minimizing the impact of various covariate conditions on human gait recognition. The evaluation of the proposed method was performed in a realistic scenario where the gallery and probe sets had varying covariate conditions. The CASIA-B dataset, which includes a range of walking conditions, was used to test the method. The results showed that we obtained a better result when both the gallery and probe had identical viewing angles for normal walking and carrying conditions. The proposed method achieved higher classification accuracy compared to current state-of-the-art approaches, with normal walking accuracy at 97.29%, carrying bag accuracy at 94.59%, and wearing coat accuracy at 74.32% at a 90-degree viewing angle. The results indicate that clothing covariate conditions play a significant role in gait recognition, making it a challenge to implement in real-world scenarios. To further improve the robustness of gait recognition and make it suitable for use in automatic human identification systems, future research can focus on developing more robust deep models.

## References


1. Han J, Bhanu B (2005) Individual recognition using gait energy image. IEEE Trans Pattern Anal Mach Intell 28(2):316–322
2. Zhang R, Zhang L, Yu T, Huang J, Yu Y (2021) Feature fusion cnn-lstm network-based gait recognition on covariate of clothing and bag. In: 3rd international academic exchange conference on science and technology innovation (IAECST). IEEE, pp 304–307
3. Tao D, Li X, Wu X, Maybank SJ (2007) General tensor discriminant analysis and gabor features for gait recognition. IEEE Trans Pattern Anal Mach Intell 29(10):1700–1715
4. Xu D, Yan S, Tao D, Lin S, Zhang H-J (2007) Marginal fisher analysis and its variants for human gait recognition and content-based image retrieval. IEEE Trans Image Process 16(11):2811–2821
5. Xu D, Yan S, Tao D, Zhang L, Li X, Zhang HJ (2006) Human gait recognition with matrix representation. IEEE Trans Circuits Syst Video Technol 16(7):896–903
6. Kathing M, Brahma RK, Saharia S (2022) Effective method to reduce covariate issues in gait-based human identification. In: 2nd International conference on intelligent technologies (CONIT). pp 1–5
7. Anusha R, Jaidhar CD (2020) Human gait recognition based on histogram of oriented gradients and haralick texture descriptor. Multimed Tools Appl 79(11):8213–8234
8. Bashir K, Xiang T, Gong S (2010) Gait recognition without subject cooperation. Pattern Recognit Lett 31(13):2052–2060. Meta-heuristic Intelligence Based Image Processing
9. Huang G, Lu Z, Pun CM, Cheng L (2020) Flexible gait recognition based on flow regulation of local features between key frames. IEEE Access 8:75381–75392
10. Rokanujjaman M, Islam M, Hossain M, Makihara Y, Yagi Y, et al (2015) Effective part-based gait identification using frequency-domain gait entropy features. Multimed Tools Appl 74(9):3099–3120
11. Li C, Min X, Sun S, Lin W, Tang Z (2017) Deepgait: a learning deep convolutional representation for view-invariant gait recognition using joint. Bayesian Appl Sci 7(3):210
12. Shiraga K, Makihara Y, Muramatsu D, Echigo T, Yagi Y (2016) Geinet: view-invariant gait recognition using a convolutional neural network. In: International conference on biometrics (ICB). IEEE, pp 1–8
13. Fan C, Peng Y, Cao C, Liu X, Hou S, Chi J, Huang Y, Li Q, He Z (2020) Gaitpart: temporal part-based model for gait recognition. In: Proceedings of the IEEE/CVF conference on computer vision and pattern recognition. pp 14225–14233
14. Chao H, He Y, Zhang J, Feng J (2019) Gaitset: regarding gait as a set for cross-view gait recognition. In: Proceedings of the AAAI conference on artificial intelligence, vol 33. pp 8126–8133

# Computational Modelling of Glucocerebrosidase Signalling Pathways in Parkinson's Disease

**Hemalatha Sasidharakurup, Kasi Viswanadh, Divya M. Sasidharan, Anu Sasidharan, Arushi Tiwari, Devi Krishna, Giovanni Naldi, Egidio D'Angelo, and Shyam Diwakar**

**Abstract** Biological systems are complex dynamical systems that include a diverse range of biomolecular species, as well as their interactions and interconnections. Subcellular systems are even more complicated due to the complex biomolecular interactions such as cross-talks, feedback mechanisms, bifurcations and so on that define the overall system's behaviour. When it comes to complex diseases, understanding these cellular mechanisms is difficult because the causes are not due to a single genetic mutation or altered signalling mechanism. Computational modelling of biological systems aids in the reconstruction of these diseases and to understand clinically or experimentally unanswered questions. In this study, important cellular

H. Sasidharakurup · S. Diwakar (✉)
Amrita Mind Brain Center, Amrita Vishwa Vidyapeetham, Amritapuri Campus, Clappana P.O, Kollam 690525, India
e-mail: shyam@amrita.edu

H. Sasidharakurup
e-mail: hemalathas@am.amrita.edu

H. Sasidharakurup · K. Viswanadh · D. M. Sasidharan · A. Sasidharan · A. Tiwari · D. Krishna
School of Biotechnology, Amrita Vishwa Vidyapeetham, Amritapuri Campus, Clappana P.O, Kollam 690525, India

G. Naldi
Department of Environmental Science and Policy, Università degli Studi di Milano, Via Saldini 50, 20133 Milano, Italy
e-mail: giovanni.naldi@unimi.it

E. D'Angelo
Department of Brain and Behavioral Sciences, University of Pavia, 27100 Pavia, Italy

Brain Connectivity Center, IRCCS Mondino Foundation, 27100 Pavia, Italy

E. D'Angelo
e-mail: dangelo@unipv.it

S. Diwakar
Department of Electronics and Communication Engineering, School of Engineering, Amrita Vishwa Vidyapeetham, Amritapuri Campus, Clappana P.O, Kollam 690525, India

 
S. Borah et al. (eds.), *Advanced Computational and Communication Paradigms*, Lecture Notes in Networks and Systems 535,
https://doi.org/10.1007/978-981-99-4284-8_23

reactions in the GBA (glucocerebrosidase) signalling, one of the critical genetic mutations involved in Parkinson's disease, were reconstructed with kinetic laws and mathematical equations. The model contributes to a better understanding of normal cellular physiology and the alterations that occur in Parkinson's disease. Such models can also be used to make disease predictions, model-based experimental designs and target multiple entities to prevent disease progression.

**Keywords** Systems biology · Parkinson's disease · Mathematical modelling · Computational neuroscience · Biochemical systems theory · GBA Signalling

## 1 Introduction

Biological systems are complicated mechanisms due to the emergence of interactions between molecules and signalling pathways at the molecular and cellular levels. Understanding the complexity of these interactions may be required to comprehend the overall behaviour of the system at the cellular, tissue, organ, individual or population levels. This becomes important when it comes to understanding the mechanisms of complex diseases including cancer, neurodegenerative diseases or infection pandemics like COVID-19. In movement disorders such as Parkinson's disease (PD), the major cause of the disease is the loss of cells that release the neurotransmitter, dopamine which transmits signals in the brain for physical movements [1]. One of the major reasons is that the disease mechanism is not fully understood, especially cellular and molecular mechanisms. Understanding the cellular mechanisms involved in the release of dopamine in PD-affected areas may help to find appropriate therapeutic targets and drug design. Computational modelling of biological systems aids in the reconstruction of complex biological processes at the molecular or organism level that represents our understanding of a system's dynamic behaviour by integrating knowledge and data [2]. One of the strategies for reconstructing biological signalling pathways as compartmental models is to convert biochemical reactions into mathematical equations and represent them as coupled differential equations [3].

This study is an extension of our previous modelling of cellular mechanisms in Parkinson's disease due to mutations in the GBA gene [4]. In this study, a combination of the application of biochemical systems theory (BST) and mathematical modelling has been used to reconstruct some of the cellular mechanisms involved in the GBA pathways related to PD. The computational model helps to identify critical reactions, their interrelation, and cross-talk that led to neuronal death in PD.

## 2 Related Works

Parkinson's disease is an old age-dependent, slow progressive neurodegenerative disorder that affects 2–3% of the world population. Previous studies have reported the role of the GBA gene in PD patients [5, 6]. Although the exact link of how GBA related to PD has not been clear yet, recent studies have shown that mutations in this gene disturb the functions of β-glucocerebrosidase (GCase), a lysosomal enzyme it codes for [7]. GCase is involved in the breakdown of toxic cellular metabolites with the help of lysosomes (an organelle with digestive enzymes). The lysosome is a primary organelle for the degradation of damaged cellular contents such as damaged organelles and any harmful proteins inside the body etc. and degrades them. Mutations in GBA can cause lysosomal damage and accumulation of toxic substances in the brain leading to dopaminergic cell death, resulting in movement problems. In the case of α-synuclein, a critical protein involved in PD, degradation under normal conditions will be proceeded by a lysosomal enzyme called Cathepsin-D, which converts α-synuclein polymers into monomers and eliminated from the body. GBA is an essential gene that encodes for the enzyme ß-glucocerebrosidase, which has an important role in many processes in the body, including the regulation of the amount of Cathepsin-D in lysosomes [8]. The protein α-synuclein, a prominent element of the intraneuronal aggregates known as Lewy bodies, another characteristic of Parkinson's disease, has also been linked to GCase activity. Loss of GCase activity is linked to α-synuclein accumulation and its aggregated forms, which lead to GCase-induced cellular changes and lysosomal dysfunction. It has also been found that the loss of GCase activity in sporadic PD brains. GBA mutations are found in PD patients that decrease enzymatic activity, and mutant proteins result in endoplasmic reticulum-associated degradation and stress. These may play a role in PD development, resulting in lysosomal dysfunction [9]. Understanding how mutations in a GBA gene are interrelated to other cellular reactions and downstream processes involved in dopamine release in the brain may help to understand the crucial mechanisms where researchers could find therapeutic targets. This may also help the clinician to detect, diagnose or even delay the disease progression.

## 3 Methods

The cellular pathways were modelled with the application of biochemical systems theory using the biochemical network modelling tool, CellDesigner™ (www.celldesigner.org) (See Fig. 1), [10]. BST is a mathematical modelling framework to model dynamic systems by using power-law functions and reaction kinetics. This helps to represent all the biochemical reactions with ordinary differential equations. According to BST, in a cellular system, the rate of change of an entity is the difference between the sum of all the reactions that produce it and the sum of all the reactions that degrade it. For example, according to generalized mass action (GMA) kinetics, each reaction can be represented as

$$\frac{dV_i}{dt} = \sum_j \alpha_{ij}.\beta_j \prod_k V_k^{f_{jk}} \tag{1}$$

where $V_i$ represents the concentration of one of the cellular variables (protein, gene and metabolite), $j$ is the type of reaction defining the dynamics of the system, $\alpha_{ij}$ is the stoichiometric coefficient, $\beta_j$, and $f_{jk}$ are the rate constants and order of the reaction, respectively. In this study, the cellular reactions and their initial conditions were taken from previous studies. The reactions were modelled using differential equations using kinetic laws including mass action kinetics, Michaelis–Menten kinetics, convenience kinetics, etc. (see Eq. (1)).

For example, consider the following reaction:

$$\text{GlcCer} \xrightarrow{\text{GCase}} \text{Glucose} + \text{Ceramide}$$

According to simple mass action kinetics,

$$\frac{d[P]}{dt} = k[A][B] \tag{2}$$

where [P] is the concentration of a product, [A] and [B] are the concentrations of reactants and k is the rate constant.

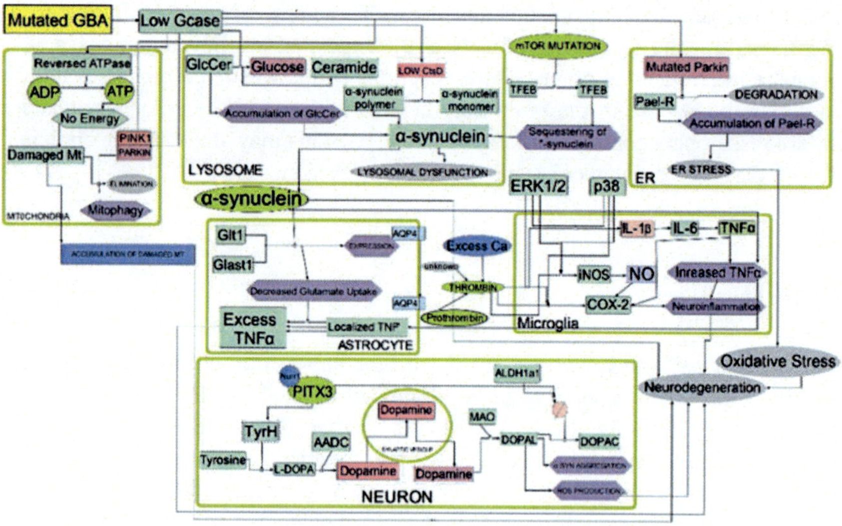


**Fig. 1** Part of the GBA signalling pathway shows reactions leading to oxidative stress, neuroinflammation, and impaired lysosomal mechanisms in PD

By combining Eqs. (1) and (2), the above reaction can be represented as

$$\frac{d[\text{GlcCer}]}{dt} = k_1[\text{GCase}] - k_2([\text{Ceramide}] + [\text{Glucose}]) \tag{3}$$

Equation (3) explains that the rate of change of GlcCer is equal to the concentration of GCase (which is utilized for the production of GlcCer) and the sum of glucose and ceramide produced by the conversion of GlcCer, multiplied by their rate constant $k_1$ and $k_2$. In this way, each biochemical equation has been incorporated with suitable kinetic laws and initial conditions and simulated to acquire a time-series outcome. It helped to determine the rate of each entity along the GBA pathway and analyse its influence on an entire system. The results include a normal state and a diseased state that shows the regulation of species to relative time duration. Some of the initial concentration values were taken from the literature is given in Table 1.

# 4 Results

## *4.1 Cellular Model of GBA Pathway in Parkinson's Condition*

In normal conditions, the GBA gene expresses to form β-glucocerebrosidase enzyme. It ensures the regular activity of PINK1 (PTEN-induced kinase-1) and parkin protein expression for mitophagy and ATPase enzymes in mitochondria to form ATP (Adenosine triphosphate). In contrast, lysosomes hydrolyze GlcCer to glucose and ceramide, and it also helps in the degradation of α-synuclein polymers. In neurons, it helps in the production of dopamine by ensuring the regular expression of the PITX-3 (Pituitary homeobox 3) gene. When the GBA gene gets mutated, it results in lowered production of GCase enzyme, which leads to reversal of ATPase enzyme in mitochondria which will affect the production of ATP (Fig. 1). This leads to lesser energy production and eventually causes mitochondrial damage. The mitophagy process will also get impaired as PINK-1, and parkin proteins are also mutated because of lower GCase levels, whereas, in lysosomes, since GCase production is lowered, it leads GlcCer accumulation and α-synuclein oligomerization resulting in neurodegeneration. In the cytoplasm, lowered GCase levels can lead to mTOR (mammalian target of rapamycin mutation), which inhibits the transport of α-synuclein outside the cell and accumulates inside the cell. In the endoplasmic reticulum (ER), mutated parkin, because of low GCase levels, causes inhibition of the degradation of Pael-R (a putative G protein-coupled transmembrane polypeptide) which leads to the generation of ER stress by accumulation and causes extracellular accumulation of α-synuclein and generation of oxidative stress.

Accumulated extracellular α-synuclein led to a reduction of glutamate uptake and mislocalization of aquaporins. In microglia, lower GCase levels will trigger excessive

**Table 1** The concentration of biomolecules involved in the GBA signalling pathway

| Si No. | Cellular Components | Region | Control | Diseased | Organism | Experiment | Disease stages | Refs. |
|---|---|---|---|---|---|---|---|---|
| 1 | GBA | Post-mortem brain | 55.1 ± 8.5 μmol/L | 50.5 ± 6.6 μmol/L | Human | Lipidomic analyses on post-mortem brain | Early onset | [11] |
| 2 | α-Synuclein | Plasma | 0.42 pg/mL | 1.29 pg/mL | Human | Immunomagnetic reduction-based immunoassay | Progression | [12] |
| 3 | NO | Serum | 7.49 ± 2.573 μmol/L | 29.8 ± 21.631 μmol/L | Human | HPLC | Progression | [13] |
| 4 | GCase | CSF | 0.35 to 2.05 μ/ml | 0.22–1.04 μ/ml | Human | GCase activity assay | Progression | [14] |
| 5 | GlcCer | Brain | 7.66 ± 1.23 μg/mg | 5.25 ± 0.78 μg/mg | Human | Lipidomic analysis and cholesterol measurement | Progression | [15] |

calcium uptake, which leads to the activation of thrombin. Thrombin, in turn, causes the overproduction of proinflammatory cytokines like IL (interleukin)-1β and TNFα (tumor necrosis factor-alpha) which causes neuroinflammation. In contrast, excessive TNFα also triggers astrocytes to produce more amounts of TNFα, which causes accumulation in cells and neurons. GBA gene mutation will trigger the loss of the PITX-3 gene in neurons which decreases dopamine production. Loss of PITX-3 will also affect ALDH1a1 (Aldehyde Dehydrogenase 1 Family Member A1) enzyme activity, which causes lesser degeneration of a highly toxic intermediate DOPAL (3,4-dihydroxyphenylacetaldehyde) which leads to neurodegeneration.

## 4.2 *Mutated GBA May Lead to Oxidative Damage and Excess Release of Proinflammatory Cytokines in PD*

In our previous model [4], simulations of control conditions have already shown high levels of dopamine, glucose or ceramide. And, the GBA and GlcCer levels were decreasing with time. In this model, the predictions of the disease condition have shown a sudden decline in GCase and PITX3 (Fig. 2a). There was an upregulation of aggregated α-synuclein and ROS (reactive oxygen species) which led to a gradual decrease in dopamine levels.

The concentration levels of TNFα, IL-1β and IL-6 levels were increased with high levels of thrombin (Fig. 2b). Simulations showed that changes in the concentration levels of these factors together have led to a decline in dopamine levels.

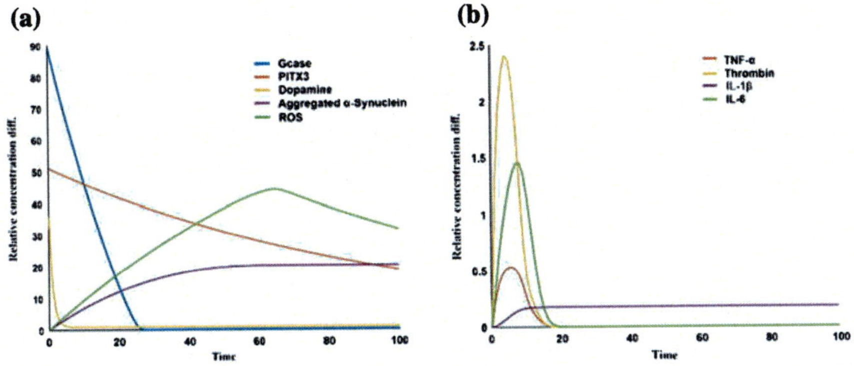


**Fig. 2** Relative concentration differences due to GBA mutation. **a** Low GCase led to decreased PITX3 leading to aggregation of α-synuclein, ROS. **b** Increased thrombin led to the release of inflammatory cytokines

# 5 Discussion

Since the signalling and cross-talk of cellular pathways are accountable for complex diseases, it is very difficult to understand the cellular factors and reactions responsible for them. Modelling these complex systems is one of the ways to scrutinize some of the phenomena to understand the dynamics of the system. In this study, one of the important pathways involved in Parkinson's disease, the GBA signalling pathway has been modelled using BST. Simulations have shown that apart from GBA mutations and changes in α-synuclein and GCase, other processes that lead to oxidative stress and neuroinflammation are also involved in the system. The model predicted that GBA mutations alter the dopamine signalling pathways leading to the accumulation of oxidative agents due to the production of DOPAL. Simulations have shown decreasing levels of PITX3, which involves the degradation of DOPAL. This supports the fact that the loss of functions of factors such as PITX3 and ALDH1A1 leads to the loss of dopamine and the production of ROS inside the neuron. Simulations also have shown high levels of thrombin are one of the reasons for releasing inflammatory cytokines including TNFα, IL-1β and IL-6, which are critical in the degeneration of neurons in PD [16]. The predictions from the model can also be mapped to the clinical symptoms of PD. For example, the abnormal aggregation of α-synuclein is responsible for weight loss and postural tremor in PD patients. Mutation in the GBA gene during the early and late onset of PD leads to memory and cognitive dysfunction. This would help a clinician to track the cellular changes and biomarkers according to the PD symptoms of a patient and also to start appropriate treatment to delay the progression of the disease.

# 6 Conclusion

Computational and mathematical modelling is one of the possible ways to tackle the cellular changes involved in complex diseases and to target them to improve disease conditions. Neuroinflammation and oxidative stress are one of the leading causes of loss of dopamine and neurodegeneration in PD. Our previous GBA pathway model suggested that mutation in the GBA gene leads to dopaminergic cell death through changing the levels of GCase, GlcCer, glucose, ceramide and α-synuclein inside the neuron and by disturbing the autophagy-lysosomal pathway. This model extension has suggested that apart from these processes, GBA also changes the levels of PITX3 and ADLH1a1 along with α-synuclein and GCase which leads to dopamine oxidation and production of DOPAL and other oxidative agents which are difficult to degrade in the absence of PITX3 and ADLH1a1. The model has also shown the role of the GBA gene in regulating inflammatory cytokines and how its mutation leads to excess release. Together, changes in these cellular factors lead to damage in the normal cellular degradation pathway, oxidative stress and neuroinflammation leading to loss of dopamine and neurodegeneration during the progression of PD.

**Acknowledgements** This work derives direction and ideas from the Chancellor of Amrita Vishwa Vidyapeetham, Sri Mata Amritanandamayi Devi. This study was part of the Multiscale Brain Function India–Italy Network of Excellence (MSBFIINE) funded by the Department of Science and Technology, Government of India, and the Ministero degli Affari Esteri e della Cooperazione Internazionale, Italy, and by Embracing the World research for a cause initiative.

## References

1. Emamzadeh FN, Surguchov A (2018) Parkinson's disease: biomarkers, treatment, and risk factors. Front Neurosci 12:612. https://doi.org/10.3389/fnins.2018.00612
2. Fischer HP (2008) Mathematical modeling of complex biological systems: from parts lists to understanding systems behavior. Alcohol Res Heal 31:49–59
3. Voit EO (2013) Biochemical systems theory: a review. ISRN Biomath 2013:1–53. https://doi.org/10.1155/2013/897658
4. Sasidharakurup H, Pisharody D, Bose H, et al (2023) Mathematical modeling of glucocerebrosidase signalling networks linked to neurodegeneration. In: research advances in network technologies. CRC Press, pp 245–256
5. Smith L, Schapira AHV (2022) GBA variants and parkinson disease: mechanisms and treatments. Cells 11
6. Sidransky E, Nalls MA, Aasly JO et al (2009) Multicenter analysis of glucocerebrosidase mutations in Parkinson's disease. N Engl J Med 361:1651–1661. https://doi.org/10.1056/nejmoa0901281
7. Brady RO, Kanfer JN, Bradley RM, Shapiro D (1966) Demonstration of a deficiency of glucocerebroside-cleaving enzyme in Gaucher's disease. J Clin Invest 45:1112–1115. https://doi.org/10.1172/JCI105417
8. Yang S, Gegg M, Chau D, Schapira A (2020) Glucocerebrosidase activity, cathepsin D and monomeric α-synuclein interactions in a stem cell derived neuronal model of a PD associated GBA1 mutation. Neurobiol Dis 134. https://doi.org/10.1016/j.nbd.2019.104620
9. Klein AD, Mazzulli JR (2018) Is Parkinson's disease a lysosomal disorder? Brain 141:2255–2262. https://doi.org/10.1093/brain/awy147
10. Funahashi A, Matsuoka Y, Jouraku A, et al (2006) Celldesigner: a modeling tool for biochemical networks. In: Proceedings—winter simulation conference. pp 1707–1712
11. Muller MVG, Petry A, Vianna LP et al (2010) Quantification of glucosylceramide in plasma of Gaucher disease patients. Brazilian J Pharm Sci 46:644–649. https://doi.org/10.1590/s1984-82502010000400005
12. Lin CH, Yang SY, Horng HE et al (2017) Plasma α-synuclein predicts cognitive decline in Parkinson's disease. J Neurol Neurosurg Psychiatry 88:818–824. https://doi.org/10.1136/jnnp-2016-314857
13. Kouti L, Noroozian M, Akhondzadeh S et al (2013) Nitric oxide and peroxynitrite serum levels in Parkinson's disease: Correlastion of oxidative stress and the severity of the disease. Eur Rev Med Pharmacol Sci 17:964–970
14. Oftedal L, Maple-Grødem J, Førland MGG et al (2020) Validation and assessment of preanalytical factors of a fluorometric in vitro assay for glucocerebrosidase activity in human cerebrospinal fluid. Sci Rep 10:1–8. https://doi.org/10.1038/s41598-020-79104-5
15. Gegg ME, Sweet L, Wang BH et al (2015) No evidence for substrate accumulation in Parkinson brains with *GBA* mutations. Mov Disord 30:1085–1089. https://doi.org/10.1002/mds.26278
16. Sasidharakurup H, Diwakar S (2020) Computational modelling of TNFα related pathways regulated by neuroinflammation, oxidative stress and insulin resistance in neurodegeneration. Appl Netw Sci 5

# COVID-19 Lung Patch Segmentation Using COVSeg-NET

**Vivek Noel Soren, Sakshi Ahuja, B. K. Panigrahi, and Tapan K. Gandhi**

**Abstract** In recent times, the most dangerous and rapidly spreading lung infection disease is COVID-19 infection. Diagnosis of COVID-19 infection from radiographic images is one of the biggest challenges. For COVID-19 infection detection and segmentation, chest CT scans and X-rays are widely used by clinicians. Out of these, Computed Tomography (CT) scans of chest images are one the leading and promising methods to detect COVID-19 patches in the lungs of an infected person. The proposed work aims to automate the process of detection of infection patches with the help of a neural network model, called the COVSeg-Net. The model is trained and validated on a dataset consisting of 2581 CT scan images of volumetric CT scans of 10 patients. The proposed work obtained a Dice coefficient score of 0.783 and a Jaccard index of 0.729 on 381 testing CT scan slices.

**Keywords** COVID-19 · Segmentation · Lung infections · Deep learning · CT scans

## 1 Introduction

Since November 2019, the world has been suffering due to this new respiratory infectious disease called COVID-19. It has affected around 597 M people all around the world while claiming the lives of 6.4 M victims [1]. The US, India, and France

V. N. Soren · S. Ahuja (✉) · B. K. Panigrahi · T. K. Gandhi
Electrical Engineering Department, IITD, New Delhi, India
e-mail: Sakshi.Ahuja@ee.iitd.ac.in

V. N. Soren
e-mail: eet212122@ee.iitd.ac.in

B. K. Panigrahi
e-mail: bkpanigrahi@ee.iitd.ac.in

T. K. Gandhi
e-mail: tgandhi@ee.iitd.ac.in



S. Borah et al. (eds.), *Advanced Computational and Communication Paradigms*,
Lecture Notes in Networks and Systems 535,
https://doi.org/10.1007/978-981-99-4284-8_24

are the most affected nations in this pandemic. One of the challenging tasks faced by the scientific community is to find a way to diagnose the disease in its early stage to prevent the spread of the pandemic. The primary cause of COVID-19 is severe acute respiratory syndrome coronavirus—2 (SARS-COV-2) which is spread via droplets of an infected person's breath, cough, or sneeze. It affects the respiratory organs of an individual, with inflammatory inflation in unilateral or bilateral lungs.

Like other viruses, SARS-COV-2 undergoes mutation to result in new versions or variants of the virus. Omicron is one of such variants that lead to an increase in COVID cases. The Delta variant, first documented in India, was considered a variant of concern by the medical community. Reverse-Transcription Polymerase Chain Reaction (RT-PCR) is the commonly used technique to detect the signs of COVID-19, but it comes with several drawbacks such as a long turnaround time of a few hours to two days. RT-PCR tests also suffer from high false negative results. To overcome these limitations, laboratories are performing radiographic examinations on chest images such as X-rays and CT scans.

Recent studies have found CT scans to be more fruitful than chest X-ray images. They provide better quality and three-dimensional images over the X-ray images and thus help to extract the relevant information from the images. Compared to X-rays, CT scans also capture small bones and soft tissues and provide a detailed view of the lungs. Certain signs of infection from a lung CT scan include patchy, lumpy, segmental ground glass opacity (GGO) at an early stage and pulmonary consolidation at a later stage. Therefore, CT scans are an essential tool for a qualitative analysis of the changes in COVID-19 infection.

The proposed work is based on a model full convolution neural network model named as COVSeg-Net. It comprises of a series of down-sampling operations on the left side of the network with the number of feature maps being doubled on every step of down-sampling. On the right part of the network is a series of up-sampling operations that reduces the number of feature maps by half and doubles the dimension of the feature map with every step of deconvolution. The feature maps on both sides are concatenated, and the image segmentation region is provided as output. The error computed between the predicted segmented region and the ground truth is backpropagated to train the network model.

The rest of the paper is organized as Sect. 2 discusses the literature review; Sect. 3 put forth the proposed methodology; Sect. 4 presents the results and discussions obtained for the proposed methodology. Conclusions and future scope are discussed in Sect. 5.

## 2 Related Work

Several studies have been conducted for efficient and reliable segmentation of the COVID-19 infected lesion from CT scans of lung images. In [2], an edge, auxiliary semantic, and attention fusion supervised module is trained on 4000 CT images and tested on the MosMed dataset. It obtained a Dice score of 0.8743, highlighting low-level boundary features and strengthening high-level semantic information during feature extraction. However, it fails to perform a finer segmentation of the infected region and, thus, cannot estimate the severity of COVID-19. Further, 2D DenseUNet is used to improve the quality and accuracy of segmentation [3]. A Dice score of 0.762 on the uAI dataset and 0.725 on the IS-COVID dataset are achieved. But the architecture fails to avoid the impact of noise annotations.

In [4], lesion areas from chest CT images of COVID-19 infections are segmented using the SRGNet. It obtained a Dice coefficient of 0.83 and a Jaccard Index of 0.71. The proposed work enhanced the local and global information from the lower level layers to the higher level layers. Unfortunately, this model depends on accurate annotations done by clinical experts for better detection. Further, a multi-task learning-based CT imaging analysis is done for the classification and segmentation of COVID-19 that produces an 88.0% of Dice coefficient score [5]. It used a shared encoder that improved the task for segmentation with the less annotated dataset, but the classification methods required many annotated datasets.

A Reliable and Consistent Temporal-Ensembling (RCTE) framework is used for semi-supervised segmentation on the MosMed dataset [6]. It succeeds in efficient segmentation from a limited amount of annotated data and prevents unreliable targets from negatively affecting training performance. However, the RCTE framework is not tested for multi-class segmentations. In [7], an automated tool for segmentation and measurement is proposed using CT scans images of 22 patients. It can achieve a Dice similarity coefficient of 0.71 and detect abnormal regions with low-intensity contrast between lesions and healthy tissues. However, the quality of the segmentation depends on the reliable ground truth of the mask structure.

An image deformation-based segmentation model, called SP-V-Net, is proposed, and provided additional insight into the assessment of infection severity [8]. The proposed model is comprised of a 3D V-Net for CT image segmentation and an STN for output restriction and refinement and achieves a DSC of 0.9796. In [9], pre-trained deep neural network architectures namely VGG16, VGG19, InceptionV3, ResNet50, ResNet50V2, InceptionResNetV2, Xception, and MobileNet are fine-tuned on the Sars-CoV-2 dataset and COVID-CT dataset. It uses an ensemble classifier that learns discriminatory features from CT scans and lung images.

A joint classification and segmentation model (MultiR-Net) is used to identify infected regions [10]. However, it suffers from class imbalance and the high computational cost of 3D input.

Contrast variation of the infected regions is addressed using Inf-Net in [11]. The shortage of annotated data is also handled by a semi-supervised segmentation framework. But it is limited by the absence of a multi-class infection labeling framework. However, to completely alleviate the problem of data annotations, [12] synthesizes artificial lesions into normal lung CT scans and trains Norm-Net to differentiate the infected regions. But it segments all kinds of pulmonary lesions, instead of only COVID-19. A COVID-CT-Mask-Net model comprises Mask RCNN in the first stage to localize and detect lesions and fuse the detection in the second stage to classify the input image [13]. It can train on a small train-test split without any requirement for augmentation tweaks, but it is yet to be generalized in large benchmark datasets.

## 3 Proposed Methodology

The flowchart of the proposed methodology is put forth in Fig. 1. The section provides the brief details of (a) the Chest CT scan dataset used for the proposed methodology, (b) the proposed COVSeg-Net model that proved to be an improvement from the traditional UNet model, (c) post-processing to ensure a further improvement in final scores of the model, and (d) experimental results of the CT images for COVID-19 lesion segmentation that show the scores in terms of Dice Coefficient Score (DSC), Jaccard Index, and several other parameters.

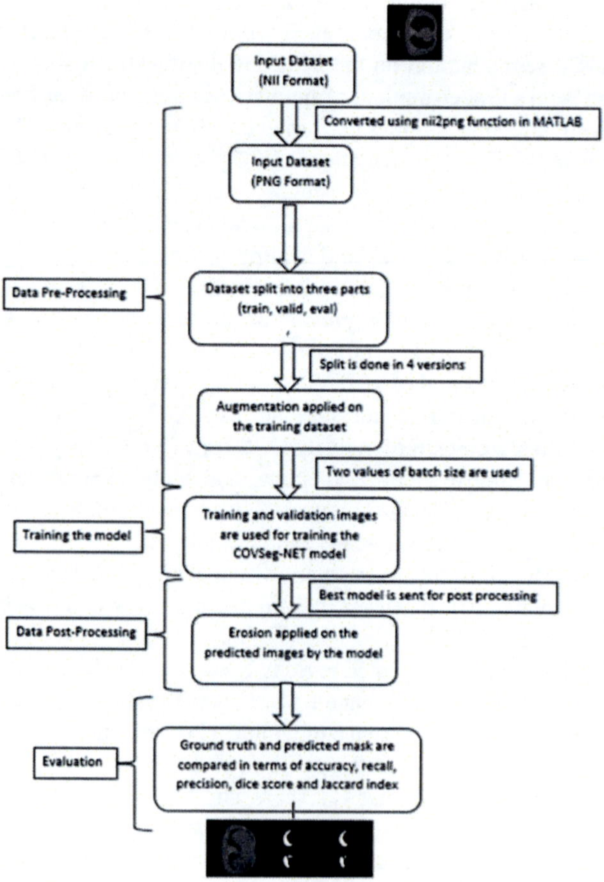


**Fig. 1** Flow diagram to represent the steps of the implementation of the image segmentation

### *3.1 Dataset Pre-processing*

The CT images for the proposed work are collected from [14]. The mentioned dataset contains COVID-19 CT scans of 20 patients. The left lung, right lung, and the infected regions are annotated by experienced radiologists. However, for the paper, we have utilized the COVID-19 CT scans of 10 patients. As for the segmentation task, we have used a mask that highlights the infected regions in the lungs. The CT scans and the mask of each patient are available in NIFTI (NII) format and required to be converted to accessible PNG format. The dataset used in the proposed work has 2581 CT scan images and their corresponding COVID-19-infected region masks. Each slice is $512 \times 512$ pixels.

### *3.2 Data Augmentation*

The proposed methodology is implemented in two phases—firstly, it is conducted with the raw dataset, and then it is performed with an augmented dataset. The augmentations performed on the images of the dataset are horizontal flip, vertical flip, and rotation. The horizontal flip and vertical flip are used with a probability of 1.0 of applying the transform. The rotational transform is also applied with a probability of 1.0 along with a rotational limit of 45°. With augmentation, the size of the training image dataset along with their corresponding masks is increased effectively.

### *3.3 Methodology*

The COVSeg-Net is quite identical to the UNet model as shown in Fig. 2. The COVSeg-NET is an encoder-decoder network architecture, consisting of four encoder blocks on the left hand of the network and four decoder blocks on the right hand of the network. The model comprises a total of 2,441,441 parameters. The encoder and the decoder relate with the help of a bridge which makes it a complete neural network model. The brief details of COVSeg-Net architecture are as follows:

1. Encoder—This network extracts the features through a series of encoder blocks. Every encoder block consists of a $3 \times 3$ convolution layer, followed by a Batch Normalization layer and Rectified Linear Unit (ReLU) activation function. The convolution layer helps in the reduction of the risk of overfitting as well as the computational cost. The output of the ReLU layer acts as a skip connection that is later merged with the corresponding decoder block. In the encoder network, the output is now followed by a $2 \times 2$ max-pooling layer. Here, the dimensions of the feature maps are halved, and the number of feature maps is doubled. This is beneficial for the decrement in the number of trainable parameters.

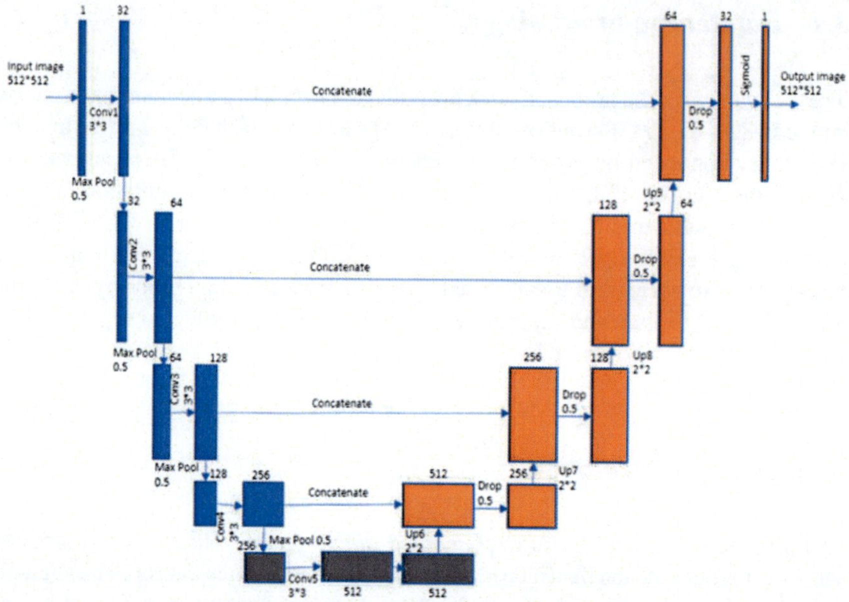


**Fig. 2** COVSeg-NET model structure for the COVID-19 segmentation

2. Dropout layer—A dropout layer is used after the max-pooling layer to force the encoder to learn different representations by ignoring some random neurons in the network. It teaches the model to not rely on specific neurons and thus avoids the overfitting problem.
3. Bridge—The bridge is the connection between the encoder network and the decoder network. It consists of a 3 × 3 convolution layer, followed by a Batch Normalization layer and ReLU activation function. Moreover, a max-pooling layer is also added to the output of the activation function.
4. Decoder—The decoder network uses the abstract representation of the neurons to predict the segmentation mask. Each decoder block comprises a 2 × 2 transpose convolution layer. The output of the transposed convolution layer is concatenated with the skip connections from the corresponding encoder block. Following that, a dropout layer with a parameter of 0.5 is used to generalize the network better. The output of the final decoder block passes through a 1 × 1 convolution block with a Sigmoid activation function. The result from the activation function is the final segmentation mask comprising pixel-wise segmentation.
5. Training parameter—While training the model, the learning rate is set to 0.0001 and Adam optimization is used. It is trained for 50 epochs and since the batch size setting is dependent on the size of memory, the model is trained for both the batch sizes of 8 and 16.

### 3.4 Data Post-processing

The resultant images predicted by the COVSeg-Net model are then subjected to certain morphological operations for an increase in the final scores of the model. For the proposed work, the erosion operation is applied with a 3 × 3 kernel window on best-predicted model results. The output from the post-processing is the result of our segmentation model.

## 4 Result and Discussion

The proposed COVID-19 lung CT image segmentation task is performed on the Kaggle notebook. Kaggle provides free access to 12 GB RAM and NVIDIA Tesla P100 GPUs. The considered COVID-19 dataset is split into three parts—training, validation, and evaluation. The data split is done four times—60:20:20 (SR1), 70:15:15 (SR2), 80:10:10 (SR3), and 90:5:5 (SR4)—and their performance scores are recorded for assessment. The performance evaluation parameters used for the assessment of the model are Accuracy (A), Dice Coefficient (DC), Jaccard Index (JI), Recall value (Re), and Precision (Pr).

The proposed model is trained with augmentation and without augmentation techniques on the considered chest CT dataset. Table 1 shows the evaluation results of the model when it is applied to the raw dataset without any augmentation techniques. Table 1 shows results in terms of the performance parameters with batch size and split ratio varied accordingly. The performance evaluation parameters evaluated with augmentation techniques applied perform better than the segmentation on the raw dataset. Further, the parametric setting of batch size 8 and a split ratio of 70:15:15 gives the best scores in terms of Dice coefficient score and Jaccard index values. With a batch size of 16 and a split ratio of 70:15:15, the model gives the best performance in accuracy, Dice coefficient score, and Jaccard index.

**Table 1** Performance index values obtained by COVSeg-NET without using augmentation techniques

| | Batch size = 16 | | | | Batch size = 8 | | | |
|---|---|---|---|---|---|---|---|---|
| SR | SR1 | SR2 | SR3 | SR4 | SR1 | SR2 | SR3 | SR4 |
| A | 0.269 | 0.991 | 0.991 | 0.996 | 0.995 | 0.996 | **0.996** | 0.995 |
| DI | 0.016 | 0.443 | 0.483 | 0.658 | 0.667 | **0.685** | 0.618 | 0.571 |
| JI | 0.008 | 0.429 | 0.467 | 0.609 | 0.616 | **0.631** | 0.564 | 0.515 |
| Re | 0.9992 | 0.512 | 0.519 | 0.725 | 0.637 | 0.729 | 0.750 | **0.789** |
| Pr | 0.008 | 0.601 | 0.710 | 0.822 | **0.953** | 0.845 | 0.755 | 0.662 |

**Table 2** Performance index values obtained by COVSeg-NET with augmentation techniques

| | Batch size = 16 | | | | Batch size = 8 | | | |
|---|---|---|---|---|---|---|---|---|
| SR | SR1 | SR2 | SR3 | SR4 | SR1 | SR2 | SR3 | SR4 |
| A | 0.996 | **0.997** | 0.997 | 0.997 | 0.996 | 0.996 | 0.997 | 0.996 |
| DI | 0.615 | **0.770** | 0.704 | 0.722 | 0.720 | 0.713 | 0.695 | 0.618 |
| JI | 0.562 | **0.716** | 0.652 | 0.667 | 0.667 | 0.659 | 0.641 | 0.564 |

**Table 3** Performance index values obtained by COVSeg-NET with batch size = 16 and split ratio = 70:15:15

| Performance parameters | Without post-processing | With post-processing |
|---|---|---|
| A | **0.9978** | 0.9977 |
| DI | 0.7708 | **0.7827** |
| Jaccard | 0.7165 | **0.7287** |
| Recall | **0.8500** | 0.8087 |
| Precision | 0.8321 | **0.8920** |

The proposed model performs better than the standard UNet model [15]. This gives an advantage in the form of lower computational time. Despite low computational time when compared to that of UNet, the computational cost is quite high. The proposed methodology fails to perform better in every segmentation scoring parameter unilaterally. It provides an inferior score in terms of accuracy and a recall value. The computational cost being one of the challenging factors, it led to the use of only a fraction of the available dataset. However, with the availability of better processing power, the model could be trained with a larger dataset (Tables 2 and 3).

## 5 Conclusions and Future Scope of Research

This paper proposes a COVSeg-Net model for COVID-19 patch segmentation from chest CT scan slices. The proposed methodology is experimented with different augmentation techniques and different split ratios in its pre-processing phase. For experimentation, COVID-19 CT images from 10 patients are used and a segmentation accuracy of 99.78% is achieved for the validation set. The proposed model produces better segmentation results when appropriate steps are taken to pre-process the dataset and post-process the predicted results properly. The model has a lesser number of trainable parameters compared to the traditional UNet model. In the future, the proposed work needs to be implemented on a larger set of patients, and a cloud computing-based framework is to be explored.

## References

1. https://www.worldometers.info/coronavirus/. Accessed 21 Nov 2022
2. Hu H, Shen L, Guan Q, Li X, Zhou Q, Ruan S (2022) Deep co-supervision and attention fusion strategy for automatic COVID-19 lung infection segmentation on CT images. Pattern Recogn 124:108452
3. Liu X, Yuan Q, Gao Y, He K, Wang S, Tang X, Tang J, Shen D (2022) Weakly supervised segmentation of COVID-19 infection with scribble annotation on CT images. Pattern Recogn 122:108341
4. Wu J, Zhang S, Li X, Chen J, Xu H, Zheng J, Gao Y, Tian Y, Liang Y, Ji R (2021) Joint segmentation and detection of COVID-19 via a sequential region generation network. Pattern Recognit 118:108006
5. Amyar A, Modzelewski R, Li H, Ruan S (2020) Multi-task deep learning based CT imaging analysis for COVID-19 pneumonia: classification and segmentation. Comput Biol Med 126:104037
6. Ding W, Abdel-Basset M, Hawash H (2021) RCTE: a reliable and consistent temporal-ensembling framework for semi-supervised segmentation of COVID-19 lesions. Inf Sci (NY) 578:559–573
7. Oulefki A, Agaian S, Trongtirakul T, Kassah Laouar A (2021) Automatic COVID-19 lung infected region segmentation and measurement using CT-scans images. Pattern Recogn 114:107747
8. Zhao C, Xu Y, He Z, Tang J, Zhang Y, Han J, Shi Y, Zhou W (2021) Lung segmentation and automatic detection of COVID-19 using radiomic features from chest CT images. Pattern Recogn 119:108071
9. Shaik NS, Cherukuri TK (2022) Transfer learning based novel ensemble classifier for COVID-19 detection from chest CT-scans. Comput Biol Med 141
10. Li CF, Xu YD, Ding XH, Zhao JJ, Du RQ, Wu LZ, Sun WP (2022) Multi R-Net: a novel joint learning network for COVID-19 segmentation and classification. Comput Biol Med 144:105340
11. Fan DP et al (2020) Inf-Net: automatic COVID-19 lung infection segmentation from CT images. IEEE Trans Med Imaging 39(8):2626–2637
12. Yao Q, Xiao L, Liu P, Zhou SK (2021) Label-free segmentation of COVID-19 lesions in lung CT. IEEE Trans Med Imaging 40(10):2808–2819
13. Ter-Sarkisov A (2022) COVID-CT-Mask-Net: prediction of COVID-19 from CT scans using regional features. Appl Intell (Dordr) 52(9):9664–9675
14. Ma J, et al COVID-19 CT lung and infection segmentation dataset (version 1.0). Zenodo https://doi.org/10.5281/zenodo.3757476
15. Zhang XQ, Wang GY, Zhao S-G (2021) COVSeg-NET: A deep convolution neural network for COVID-19 lung CT image segmentation. Int J Imaging Syst Technol 31(3):1071–1086. https://doi.org/10.1002/ima.22611

# Research Study on Leukemia Detection Using Deep Learning Techniques

**Rohan Raj Maram, Sasibhushana Rao Pappu, Aravapalli Rama Satish, and Balajee Maram**

**Abstract** Leukemia is a type of cancer found in blood, which commonly starts within the bone marrow and impacts the capacity of the body to combat infection. In a country like India, less than 1000 Lakh instances having leukemia are stated each time. Several of the not uncommon place signs and symptoms are pores and membrane outbreaks, hemorrhage, sensing tiredness, infection, and accelerated hazard of infections. It could be extremely critical to come across leukemia in the initial phases. Conventional techniques, along with limited studies on plasma stains, for detecting Leukemia are no longer cost-effective and rely heavily on skilled scientific personnel. To overcome these limitations, this study proposes the implementation of automated algorithms using image processing and the MATLAB software for the recognition and categorization of Leukemia.

**Keywords** Leukemia · Segmentation · Augmentation · MATLAB · Deep learning

R. R. Maram
Chitkara University, Baddi, HP, India
e-mail: rohan1315.be22@chitkarauniversity.edu.in

S. R. Pappu
CSE Department, GITAM(Deemed to Be University), GITAM School of Technology, Visakhapatnam, India

A. R. Satish
School of Computer Science and Engineering, VIT-AP University, Vijayawada, Andhra Pradesh, India
e-mail: rama.satish@vitap.ac.in

B. Maram (✉)
Apex Institute of Technology—Computer Science and Engineering, Chandigarh University, Mohali, India
e-mail: maram.e15007@cumail.in



S. Borah et al. (eds.), *Advanced Computational and Communication Paradigms*, Lecture Notes in Networks and Systems 535,
https://doi.org/10.1007/978-981-99-4284-8_25

# 1 Introduction

Leukemia is a plasma cancer that affects blood cells. Because leukemia spreads quickly, it must be observed as soon as probable. The issue begins in the bone marrow, while the bone marrow generates an abnormally huge amount of white blood cells. Acute Lymphocytic Leukemia (ALL), Acute Myelogenous Leukemia (AML), Chronic Lymphocytic Leukemia (CLL), and Chronic Myelogenous Leukemia (CML) are the four major types of leukemia [1]. Experts use a microscope to manually detect leukemia. The manual assessment technique is costly, time-consuming, and heavily reliant on the expertise and experience of the worker. The above issues are overcome by detection via digital image processing. Digital image processing involves pictures as the main feedback and does non necessitate the use of an expensive laboratory.

In [1], Agaian et al. suggested using the k-means segmentation algorithm to routinely identify and segment AML in blood stains. A technique for identifying Acute lymphoblastic leukemia was proposed by Bhattacharjee and Saini in [2]. A set of features of a cell, including its Area, Perimeter, Circularity, and Form Factor, is created for classification. In [3], Mashiat Fatma and Jaya Sharma propose a neural network for classification.

In [4–10], many image-processing algorithms have been developed to aid in the detection of leukemia. As input, this system uses microscopic blood smear images. Different image handling methods have been applied to reach the required production due to the type and quality of the image. To detect and identify leukemia types, several methods, for example, image improvement, separation, feature abstraction, and categorization are applied. All this research indicates that picture separation is an extremely important stage in leukemia image handling. The precision of the following steps (feature extraction and classification) is determined by the separation output.

This paper has the following sections. Section 2 has Process Overview. Section 3 has Experimental Results. Simple block diagram of suggested system is shown in Fig. 1.

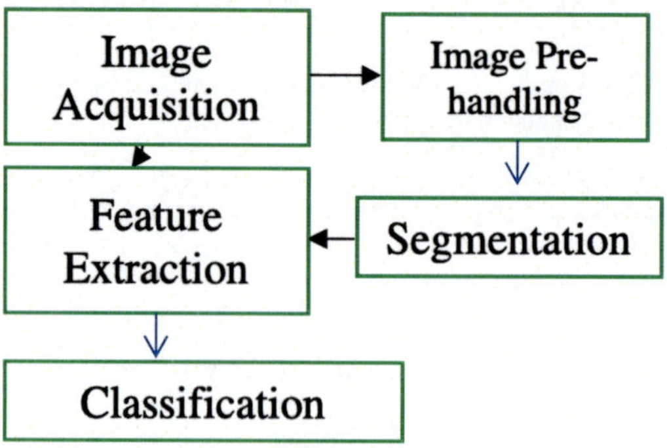


**Fig. 1** Simple block diagram of suggested system

## 2 Process Overview

### 2.1 *Image Acquisition*

The planned method receives microscopic pictures of blood smears as input. The method is skilled in images of various kinds of leukemia. The RGB pictures are shown here. RGB images are saved as three-dimensional arrays in MATLAB. Functionality with this matrix data in MATLAB is thus simplified. The images in the database are from an online database of various types of leukemia.

### 2.2 *Image Pre-handling*

Pre-processing is the image processing move that prepares the picture for the subsequent managing stages. These pictures have been enhanced. Many pre-processing methodologies can be applied to remove noise, enhance picture quality, adjust differences, and so on, depending on the results required. It is the color conversion in this case. The RGB image prototype is illustrated in the HSI design on the way to improve and differentiate the different colors in the picture.

### 2.3 *Segmentation*

The method of recognizing bodies, colors, shapes, or surfaces inside an illustration is known as segmentation. The primary goal of the classification stage in image processing is to extract specific objects or features from the entire image. The segmented output influences the accuracy of feature extraction and classification. Here come about several separation algorithms, although the k-means grouping algorithm is very popular. The k-means system is utilized in this planned scheme. In this case, the number of clusters (k) is set to four.

### 2.4 *Feature Extraction*

Various types of features from the segmented image are obtained in this step. To achieve the desired result, features containing critical information are chosen. Multiple features are required for better classification in this proposed system. This project extracts features such as strength, association, sum information, difference information, information, data rate of association, contrast, association, and so on.

### 2.5 Classification

The classifier is trained on features extracted from various types of leukemia. Adoption of data individuality to a stored picture is established on the skilled identified programs. A Neural Network is used to perform classification in this proposed system.

## 3 Experimental Results

As input images, leukemia images from an electronic archive and Goa Medical College are utilized. These pictures are also subjected to pre-processing and segmentation. GUIs are designed to make it easier for users to execute code step by step. The RGB picture is first translated keenly on an HSI model. Next, k-means algorithm is utilized to portion the picture into different components. Figure 2 depicts a flow graph of the actions taken.

A graphical user interface was created for implementing this paper. It has four controls: choose picture, RGB to HSI, k-means cluster, and diseased collection. Initially, simply "select image" is allowed. It permits you to browse as well as select the preferred picture. When selecting an image, the option to convert it from RGB to HSI is available.

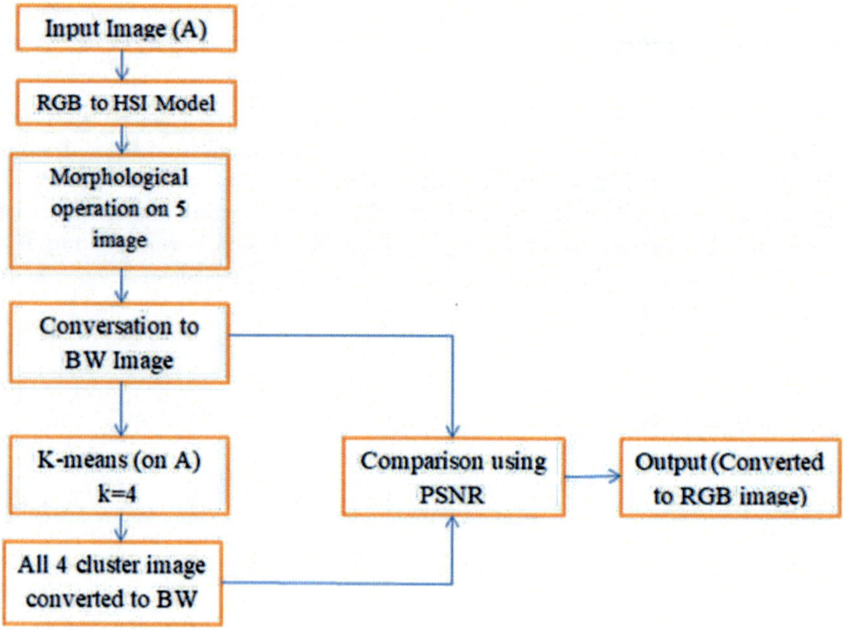


**Fig. 2** Flow graph of the planned structure

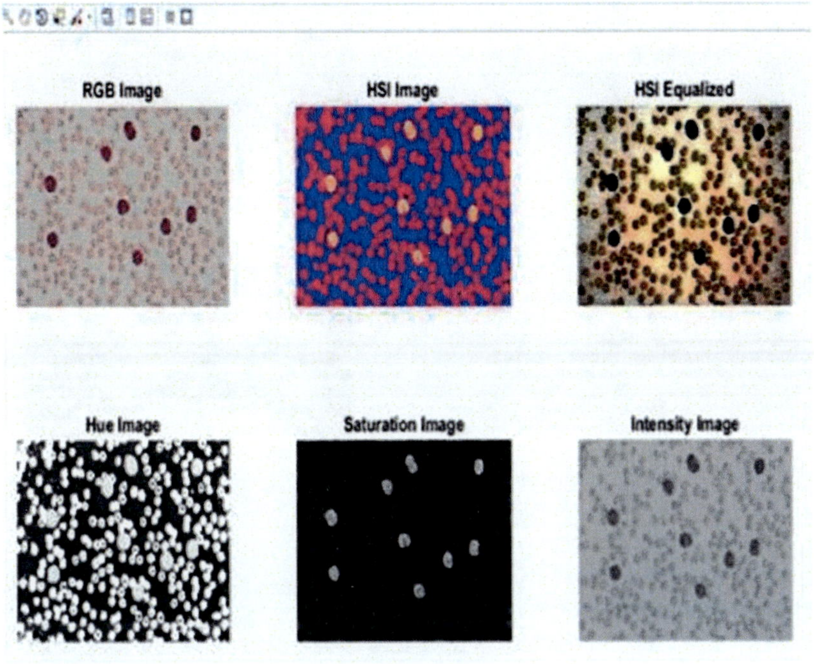


**Fig. 3** The conversion to HSI from RSB of selected image

We get the output shown in Fig. 3 after clicking on "RGB to HSI." It displays images of the chosen leukemia with HSI equalization, hue, saturation, and intensity. Diffusion picture improves and then reflects the requisite cells as the whole picture. The diffusion picture is applied to locate an appropriate collection.

Once choosing "k-mean," the output displayed in Fig. 4 is drawn. Figure 4 depicts a segment outcome of the chosen image. The picture is divided into four (k = 4) groups. By conducting experiments and checking all the images in the dataset, the value 4 is chosen. The "infected cluster" button is enabled after this step. Once choosing "diseased group," it returns the exact group for additional categorization. The results are shown in Figs. 5 and 6.

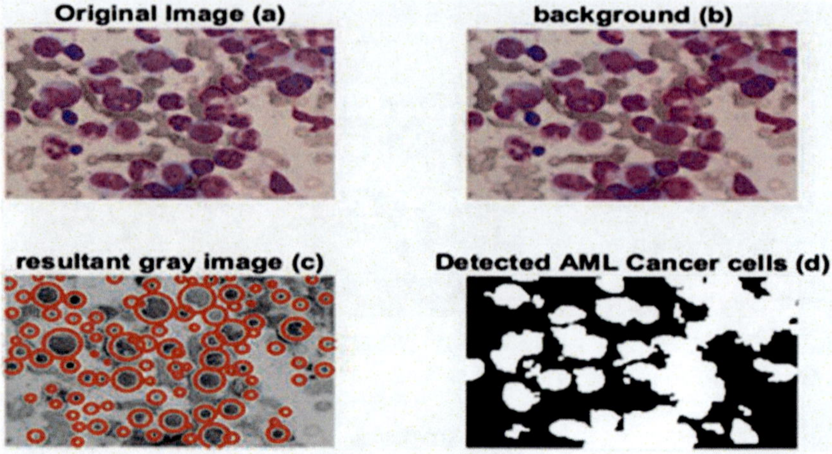


**Fig. 4** For k = 4, k-means clustering segmentation output

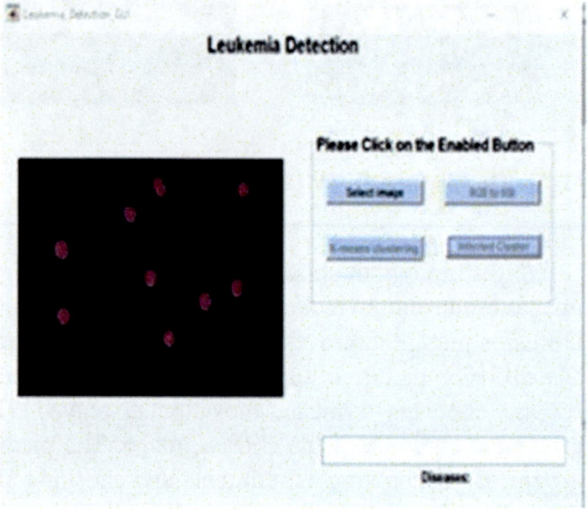


**Fig. 5** Identified group of Acute Myelogenous Leukemia (AML) cells

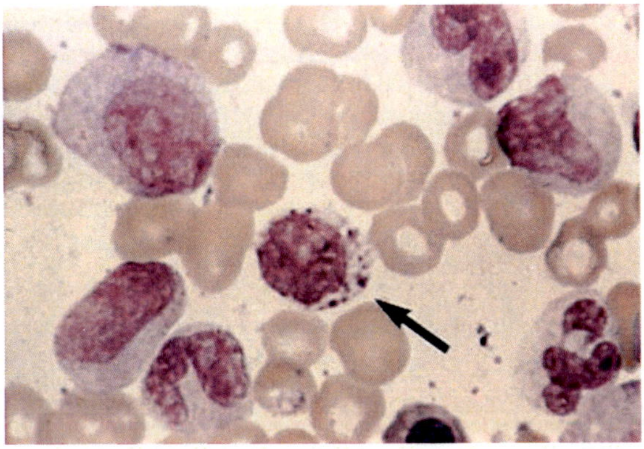

**Fig. 6** Identified group of Chronic Myelogenous Leukemia (CML), showing the identified group of CML image

# 4 Conclusion

This article describes the development of an automation system used for the identification along with tracking of leukemia cells. The planned method consists of the quartet most important stages: pre-treating, dissection, feature abstraction, and categorization. It is conducted on 12 leukemia-described images from Goa Medical College (GMC) as well as several leukemia images obtained from an available dataset. For better segmentation, pictures are transformed into the HSI color prototype. For the databases used in the segmentation process, the K-means algorithm produces accurate results. It correctly segregates all four types of cells. More experiments with algorithms can be performed to achieve better results. More studies can be performed to instantaneously select the amount of groups.

# References


1. Agaian SS, Madhukar M, Chronopoulos AT (2014) Automated screening system for acute myelogenous leukemia detection in blood microscopic image. IEEE Syst J IEEE
2. Bhattacharjee R, Saini LM, (2015) Robust Technique for the detection of acute lymphoblastic Leukemia. In: IEEE power, communication and information technology conference (PCITC). Siksha O Anusandhan University, Bhubaneswar, India
3. Fatma M, Sharma J (2014) Identification and classification of acute leukemia using neural network In: International conference on medical imaging, m-health and emerging communication systems (MedCom)
4. Rejintal A, Aswini N (2016) Image processing based leukemia cancer cell detection. In: IEEE international conference on recent trends in electronics information communication technology. India
5. Sandhu RK, Maini R (2017) Automated detection of Leukaemia. Int J Adv Res Comput Sci 8(5)
6. Singh M, Bansal M (2015) A survey on various K-means algorithms for clustering. IJCSNS Int J Comput Sci Netw Secur 15(6)
7. Pavithra S, Bagyamani J (2015) White blood cell analysis using watershed and circular Hough transform technique. Int J Comput Intell Inform 5(2)

8. Nanda I, Adikari N (2022) Design and analysis of reconfigurable computing of SoC for IoT applications, Chapter 2, 1st edn. Handbook of research for big data. Taylor Francis https://doi.org/10.1201/9781003144526, eBook, ISBN9781003144526
9. Vimala S, Madhusudhana Rao TV, Balajee Maram BA Taylor-Student psychology based optimization integrated deep learning in IoT application for plant disease classification. Wireless Networks. Springer Publications. Electronic ISSN: 1572-8196, Print ISSN: 1022-0038
10. Rajeswari R, Gampala V, Maram B, Cristin R (2022) FwLICM-Deep learning: Fuzzy weighted Local Information C-Means Clustering based lung lobe segmentation with deep learning for COVID-19 detection. J Digit Imaging, Impact Factor: 4.056, h-index:63

# Automated Classification of Diabetic Retinopathy Using Deep Learning Architecture

Noel D'Souza, P. C. Siddalingaswamy, and N. V. Subba Reddy

**Abstract** Diabetic Retinopathy (DR) is an eye disease that can lead to blindness. It is one of the leading causes of vision loss in the world. Early detection of DR is critical to prevent permanent vision loss. DR is diagnosed by a trained ophthalmologist. In many developing countries, access to ophthalmologists for regular screening is often difficult. Using Machine Learning to overcome the need for skilled personnel to diagnose DR is a viable alternative. To automate the diagnosis of DR, we preprocessed our images to obtain 6 channels before training a modified Inception-v3 model on retinal fundus images obtained from Kaggle. We then trained a second model, combining data from both eyes, obtained as outputs from Inception-v3 to diagnose the stage of DR. Using this architecture, we were able to obtain a Kappa score of 0.821 on the dataset provided by Kaggle.

**Keywords** Diabetic retinopathy · Color model · Deep learning · Inception-v3 · Fundus image

## 1 Introduction

Diabetic Retinopathy (DR) leads to vision loss in certain diabetic patients. A long history of high blood sugar can lead to DR as it is one of the leading causes of blindness in developed countries and it is fast increasing in developing countries. Globally, DR cases are expected to increase from 126 million in the year 2010 to 191

N. D'Souza
Language Technologies Institute, Carnegie Mellon University, Pittsburgh, USA
e-mail: ndsouza@andrew.cmu.edu

P. C. Siddalingaswamy (✉) · N. V. S. Reddy
Department of Computer Science and Engineering, Manipal Institute of Technology, Manipal Academy of Higher Education, Manipal, Karnataka, India
e-mail: pcs.swamy@manipal.edu

N. V. S. Reddy
e-mail: nvs.reddy@manipal.edu

 
S. Borah et al. (eds.), *Advanced Computational and Communication Paradigms*, Lecture Notes in Networks and Systems 535,
https://doi.org/10.1007/978-981-99-4284-8_26

million by 2030 [1]. There are currently 463 million cases of diabetes worldwide, with India and China alone accounting for 77 and 116 million cases, respectively. Among the top 10 countries with diabetes worldwide, 8 are developing nations. The overall prevalence of DR among diabetic patients was estimated to be 35%. [2]. DR may go undiagnosed in developing countries where access to trained healthcare professionals may be inadequate. In the US, at the time of diagnosis of type-2 diabetes, 20–40% of people already have DR [3].

DR happens due to prolonged high blood sugar levels, blood vessels in the retina can swell and leak, and they can also become blocked; sometimes, this can lead to new abnormal growth of blood vessels on the retina leading to vision loss. Diabetic Retinopathy can be divided into 5 stages according to the International Clinical Diabetic Retinopathy Disease Severity Scale [4]. The first stage is no Diabetic Retinopathy. The next four stages can be divided into two classes, Non-Proliferate Diabetic Retinopathy (NPDR) and Proliferate Diabetic Retinopathy (PDR). The former can further be divided into 3 sub-stages namely mild, moderate, and severe NPDR. Proliferate Diabetic Retinopathy is the advanced stage of DR; it is caused when the retina starts growing new blood vessels. It is categorized by Neovascularization and preretinal hemorrhage. PDR can lead to vision loss. The chances of developing PDR from NPDR increase as the disease progresses; with mild NPDR, the chances of getting PDR within a year are 5%; this increases to 53% with severe NPDR [3].

Diabetic Retinopathy is diagnosed primarily by retinal fundus imaging. Here, an ophthalmologist usually dilates the pupils with drops before imaging. Another technique called fluorescein angiography can reveal changes in the structure of blood vessels on the retina; a special dye is used to highlight these retinal blood vessels. This technique however is a costly and time-consuming procedure. In many developing countries, not everyone has access to trained ophthalmologists; sometimes even when they do, the wait times can be long. An automatic DR grading system could provide a quick diagnosis and course of treatment to a large number of people. It would also provide economically challenged people access to cheap medical care.

A lot of research has been done in the automatic diagnosis of DR. Traditional methods work by employing several feature extractors to identify parts like blood vessels, micro aneurysms, etc. These feature extractors need to be hand-crafted to identify the underlying feature. Deep Learning using Neural Networks has made a lot of progress this decade and has broken several state-of-the-art benchmarks in various fields. They have achieved or exceeded human-level skills in tasks ranging from image recognition and object detection to video games such as Atari. They have even performed remarkably well in NLP tasks with language models GPT-2 generating extremely convincing human-like text [5, 6]. With the power of Deep Learning, a lot of progress has been made in automatically classifying DR images.

In the proposed work we train a deep convolutional neural network to grade DR images from Kaggle [7] into one of five classes. The proposed algorithm is evaluated on the test set provided by the competition, achieving results in the top 10 of the competition leaderboard; in addition, due to our preprocessing technique, we can make it work with a moderately sized neural network. We achieve much better results compared to recent works in the field.

## 2 Related Work

Various techniques have been developed to automatically diagnose Diabetic Retinopathy. The earlier techniques used manually designed feature extractors to identify the underlying features. The detected features are then inputted to classifiers like K-Means, Random Forest, etc. to group and classify DR. In recent years, deep learning techniques have been deployed to identify and classify DR. These employ Neural Networks that learn the underlying features. There also exist different datasets that have been used to train DR algorithms with a large variation in the number of images in these datasets, ranging from a few hundred to tens of thousands. Here, we summarize some recent work in automatically detecting DR as well as highlight some traditional techniques.

Traditional techniques rely on domain knowledge and handcrafted feature selection. Designing these feature extractors is often time-consuming. Agurto et al. [8] used a technique that defined and characterized features based on their pixel intensity, size, and geometry. A dataset comprising images from 822 patients was used. An area under the curve of 0.89 was obtained for the detection of retinopathy. M. Usman Akram et al. [9] detect microaneurysms based on their shape, color, etc. These features are extracted using various filters and morphological transforms and are then classified using a hybrid classifier. They achieved 99% sensitivity and 99% specificity on the DIARETDB0 and DIARETDB1 datasets, comprising a total of 219 images. Harini et al. [10] extracted blood vessels, exudates, microaneurysms, etc. using an FCM classifier and morphological transforms, and classified them based on their area using an SVM. They too used the DIARETDB0 and DIARETDB1 datasets and achieved a sensitivity of 100%, 95.83% specificity, and accuracy of 96.67%.

Several recent works using deep convolutional neural networks have achieved remarkable results in identifying DR. Gulshan et al. [11] used a private dataset consisting of more than 128 k images, to train an Inception-v3 [12] model to classify DR images. The algorithm was evaluated at two operating levels, one selected for high sensitivity and another for high specificity.

Gao et al. [13] used a novel dataset consisting of 4476 images labeled into 4 classes. They divide each image into four parts and use a different Inception-v3 network for each. The output features are merged into a single vector and fed to a softmax layer for classification. They achieved accuracy, precision, and recall of 88.72%, 95.77%, and 94.84%, respectively. Graham [14] was the winner of the Kaggle Diabetic Retinopathy competition with a Kappa score Of 0.849, making use of fractional max pooling and an ensemble of 3 models. The 5-class softmax output for an eye was combined with that of the other and fed to a Random Forest. Arcadu et al. [15] used a deep learning approach to predict future DR progression over the course of 6, 12, and 24 months based on fundus images acquired during a single visit.

They achieved an AUC of 0.79 for a prognosis of 12 months. Yang et al. [16] used the Kaggle dataset along with a two-stage CNN to classify severity. Each image was divided into patches and graded by a network; these gradings were used to generate weighted lesion maps that were combined with the input image and fed to another network for final class DR prediction. They achieved an AUC of 0.959. Ni et al. [17] used an Inception-v3 model trained on the dataset from Kaggle to classify retinal fundus images into 2 classes and 5 classes. They used selective data sampling to overcome data imbalance among the classes in the dataset. On the 2-classs and 5-class problems, they achieved accuracies of 92% and 87.2% respectively, and Kappa scores of 0.744 and 0.806, respectively. Wan et al. [18] used the Kaggle dataset for training. They trained and compared several model architectures, VggNet-s, AlexNet, VggNet-16, GoogleNet, VggNet-19, and ResNet. Transfer learning and hyperparameter tuning were used to overcome overfitting. They used five-fold cross validation to compare the accuracy of the different architectures. The highest accuracy obtained was 95.68% with VggNet.

# 3 Methodology

## *3.1 Dataset*

The dataset was obtained from Kaggle; it consists of images from EyePacs in collaboration with the California Healthcare Foundation. The images were captured under various lighting conditions and cameras. In addition to this, some images are overexposed, some are underexposed, while others may be out of focus or contain artifacts. Thus, an algorithm that is robust to such variance in the input is needed. Each pair of images has a unique ID along with a label specifying either the left or right eye. The images are graded into 5 categories. Furthermore, the number of samples present in each class is extremely unbalanced, with class 0 containing roughly 73.5% of the 35,126 images. In addition to these images, another dataset consisting of 53,576 was obtained from Kaggle. Our model was evaluated on this dataset; the details are given in Table 1. The retinal dataset was divided into a training set and a validation set. Since the number of images per class was extremely unbalanced, we chose to have a validation set that represented balanced classes; to achieve this, we sampled each class randomly and extracted 150 images from each. This gave us a validation set with a total of 750 images. These images were preprocessed according to techniques we describe later and saved. Figure 1 illustrates the sample images from the dataset.

**Table 1** DR grades and their sample counts for train and test sets

| Class name | Train set | Test set |
|---|---|---|
| 0—No DR | 25,810 | 39,533 |
| 1—Mild | 2,443 | 3,762 |
| 2—Moderate | 5,292 | 7,861 |
| 3—Severe | 873 | 1,214 |
| 4—Proliferative DR | 708 | 1,206 |

### *3.2 Data Preprocessing and Augmentation*

Due to high variability in image size, composition, and contrast brought about by different lighting conditions and cameras used, the image needs to be transformed to a uniform format. This way, camera-specific artifacts and lighting conditions can be removed, allowing the to model to generalize to underlying image information rather than camera-specific data. Figure 2 illustrates the various images after the following preprocessing steps:

(1) *Resize to specific radius*: The image is first resized such that the retina is a fixed radius; we experimented with various values, finally settling on 250 pixels.
(2) *Subtract mean local color*: The mean local color was subtracted from the image to correct for any differences in lighting, creating uniform standard images.
(3) *Clip Radius*: The outer 10% of the radius is clipped to remove any external edge effects.
(4) *CLAHE*: It is a technique used to improve contrast in an image. It calculates different histograms over different sections of the image, improving local contrast. It also clips contrast amplification above a certain threshold.
(5) *HSV and LAB transforms*: This image is then converted to HSV (hue, saturation, value) and LAB (CIELAB) formats; the V channel from HSV is extracted; it contains data in the range (0, 1); it is multiplied by 255 to bring it to the standard (0, 255) range. The AB channels from LAB are also extracted; these channels contain values in the range (−128, 127). They are converted to the standard (0, 255) range by adding 128, giving a new VAB image.
(6) *Resizing*: The newly created RGB and VAB images are both cropped to leave behind only the retina; this new image is then padded with a value of 128 to make it square. And finally resized down to 300 × 300.

### *3.3 Model*

Neural Networks are a set of supervised learning algorithms that attempt to map information from an input space to an output space. They are loosely modeled after biological neurons in the brain. They perform tasks without being explicitly programmed to do so. Every ANN is made up of individual neurons, with input, output, and optional

**Fig. 1** Image samples from each class: **a** Normal, **b** Mild, **c** Moderate, **d** Severe, and **e** affected

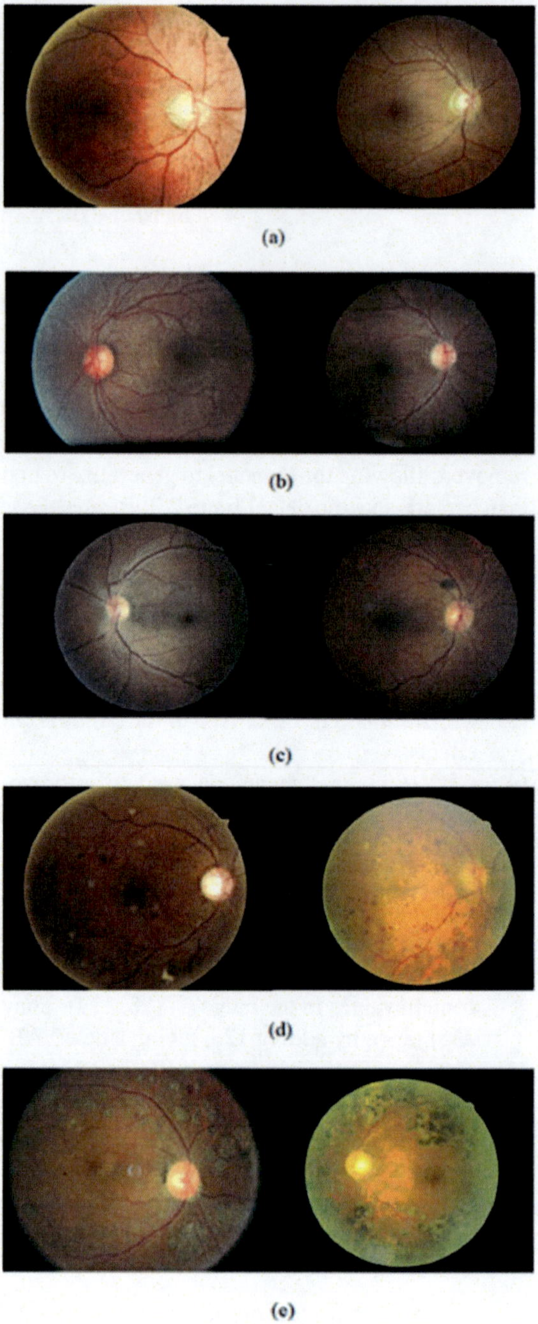

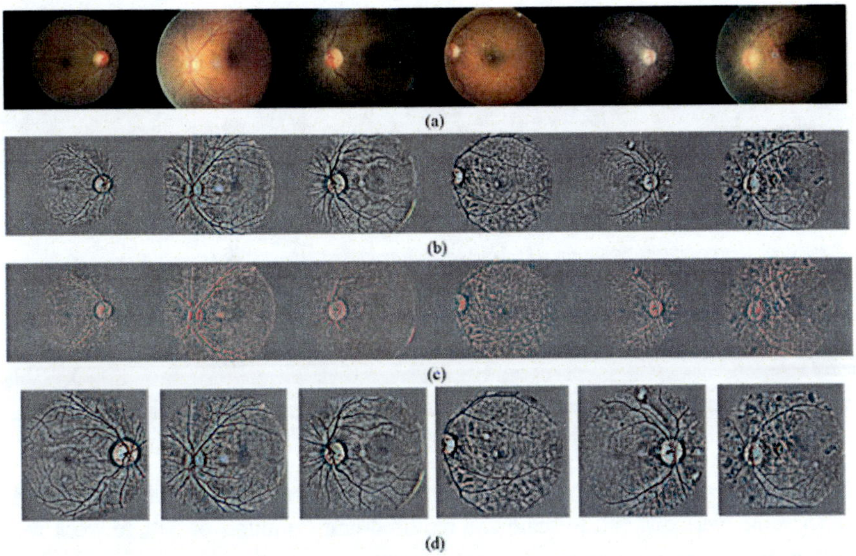

Fig. 2 The various images after preprocessing steps: **a** Original Image, **b** The image after Radius Normalization, color Normalization, and CLAHE, **c** The image after converting to VAB channels, and **d** The image after cropping, padding, and resizing

intermediate hidden layers. Each neuron in one layer is connected to every neuron in the next layer. The input consists of real numbered data that passes through the network, and is multiplied by the neuron's weights; the output is then compared to the true output with the help of a loss function; the weights of the model are then adjusted based on the output of this loss function. The Back-Propagation algorithm is used to calculate how much each weight should be changed based on the output; it does so by calculating the gradient of the loss function with regard to individual network weights. Convolutional Neural Networks are a type of neural network typically employed for visual data. They generally consist of convolution operations, activation functions, and pooling layers.

The convolution layers work by computing the convolution with a fixed-sized filter over the input. The weights of this filter are learnt by the model. The outputs from a filter are passed to an activation function before moving on to further layers. Rectified Linear Unit (RELU) activation function is typically employed in CNNs. Pooling operations down-sample the input reducing the number of computations; they also make the network robust to translational variance in the image. Max pooling is a type of pooling operation that is commonly used in CNNs; it works by sliding a window over the input at selecting the maximum value within that window. Using a combination of these three properties, convolutional networks of arbitrary shapes and sizes can be built. Figure 3 illustrates the CNN architecture of the overall prediction pipeline, from raw image input to output class prediction. Tremendous breakthroughs have been made using deeper and deeper convolutional networks. Starting with LeNet

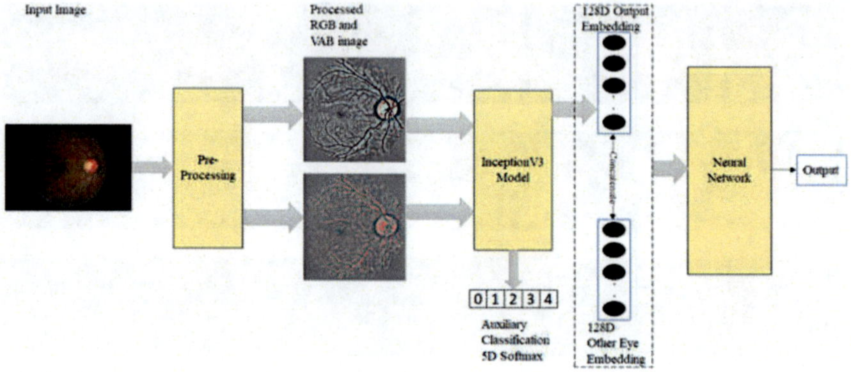


**Fig. 3** The overall prediction pipeline, from raw image input to output class prediction

in 1998, this conv net could recognize handwritten digits [19]. However, the networks of the time were difficult to scale and required large amounts of data to work well. Import breakthroughs came in the last decade, starting with AlexNet [20] in 2012, which broke the then benchmark on ImageNet [21], owing its success to being deeper and using GPUs for quick training. VGGNet [22], ResNet [23], and Inception Net [12] are some of the networks that improved on the ImageNet benchmark, with each of these networks having several different iterations based on parameters like neural structure.

Inception Net is a Deep Learning image classifier built by Google. It was built to classify images on the ImageNet [21] classification task where it achieved a new benchmark in terms of accuracy. Since images can have a large variation on where certain objects of the same type are located, choosing appropriate filter sizes becomes difficult, Inception Net solves this by using filters of different sizes at the same level. The network has several such predefined blocks consisting of different filter sizes that are stacked on top of each other. Inception-v3 [12] was the third iteration in the Inception Net architecture; it had several changes in filter sizes to improve both training times and accuracy over previous versions. We used this version of Inception Net to build our model.

We used the inception-v3 model with weights that were pre-trained on ImageNet [21]. Since our input consisted of 6 channels rather than 3 channels that the Inception Net was designed for originally, a few modifications to its structure were required. The new model had two parallel input layers rather than one earlier. The input to these layers were our two $300 \times 300 \times 3$ images. The output from the input layers was concatenated to give an image with 6 channels. Since the original network had only 3 channels, the pre-trained weights from the first convolution layer were appended together to match the dimensions of the new input shape. Thus, it was not necessary to initialize the weights of this layer from scratch.

The last convolution layer of the network was followed by a Global Average Pooling Layer and a dropout of 30% to prevent overfitting. We experimented with various fully connected layer configurations, finally settling on a single 128D fully

**Table 2** Network structure of second neural network

| Layer name | Output shape | Layer name |
|---|---|---|
| Input layer | 256 | Input layer |
| Layer 2 | 256 | Layer 2 |
| Layer 3 | 256 | Layer 3 |
| Layer 4 | 128 | Layer 4 |
| Layer 5 | 128 | Layer 5 |

connected layer, with RELU activation. The output layer has 5 neurons with a softmax activation function. The weights for the last two layers were initialized randomly using the Xavier initializer. The categorical cross-entropy function was used to calculate the loss.

During the evaluation, the final layer of the model was removed and the 128-dimensional output from the second last layer was saved. This output embedding was calculated for all fundus images. The second neural network was trained to predict the class label. The input to this network was the embedding from a pair of eyes, thus requiring a 256-dimensional vector as input. The neural network structure is described in Table 2.

## 3.4 *Training the Model*

We trained the main model using the Adam optimizer [24]. Since the samples per class were extremely unbalanced, class weights were used to prevent overfitting to any one class. These class weights were set according to the following formula as given in (1).

Let $n_i$ denote the number of samples in class $i$, $\forall i \in [0, 4]$ $[0, 4]$.

Then,

$$t = 1 + n_0\left(\frac{1}{n_1} + \frac{1}{n_2} + \frac{1}{n_3} + \frac{1}{n_4}\right) \quad (1)$$

If $w_i$ denotes the weight of class $i$, $\forall i \in [0, 4]$ then,

$$w_0 = \frac{1}{t}$$

$$w_1 = \frac{n_0/n_1}{t}, w_2 = \frac{n_0/n_2}{t}, w_3 = \frac{n_0/n_3}{t}, w_4 = \frac{n_0/n_4}{t}$$

Thus, for the dataset of 35,126 images, the class weights calculated are given in Table 3.

**Table 3** Class weights used during training

| Class | Weight |
|---|---|
| 0 | 0.01212686 |
| 1 | 0.12811882 |
| 2 | 0.05914480 |
| 3 | 0.35852722 |
| 4 | 0.44208230 |

Initially, only the first convolution layer and the last two fully connected layers were trained, with the rest of the model frozen for training. Following this, the entire model was trained, starting with a learning rate of 6e-4 and ending with a learning rate of 2e-5. The learning rate was manually decreased in steps at certain epochs when the validation loss started to stagnate. The learning rate along with accuracy versus epochs is plotted below.

The weight of class 0 was then increased to 0.025 and the model was further trained; the accuracy on the training set greatly increased, however, the validation accuracy was observed to marginally drop, suggesting that the model was not overfitting on the training data or any particular class. The change in weight is denoted by the red line at epoch 46. Figure 4 illustrates accuracy versus number of epochs on training and validation sets.

The 128-D output embedding from the Inception Net was combined with that from the other eye. This was then passed to our second neural network. We trained the second network as a regressor for the 5-class problem, and final class labels were arrived at after thresholding. For the 2-class problem, we trained the second model as

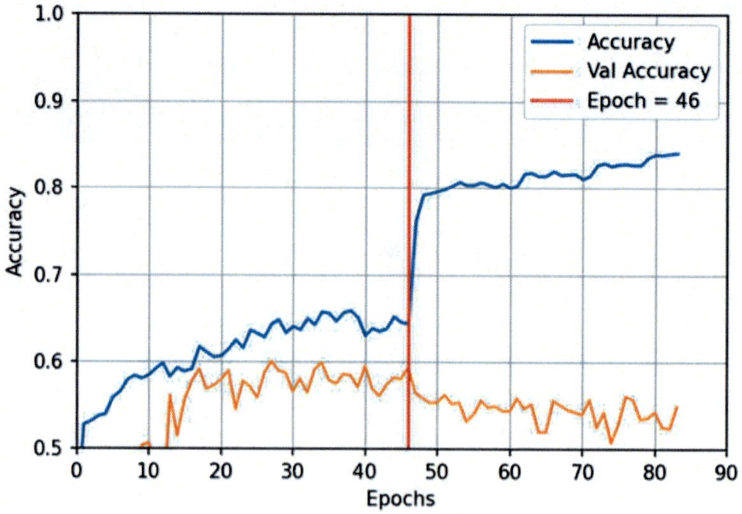


**Fig. 4** Accuracy versus number of epochs on training and validation sets

a binary classifier, even though thresholding the regressor network into two classes gave similar results. This was done in order to get the probability scores for the ROC curve.

## 4 Results and Discussion

We focused on the 5-class and 2-class problems. In the latter, we clubbed classes 0, 1, and classes 2, 3, and 4 as negative and positive, respectively. With our second network as a regressor and binary classifier for the 5-class problem and 2-class problems respectively, predictions for both problems were obtained. We used a technique called Test Time Augmentation (TTA) while evaluating the Inception model on the training and testing sets. TTA works by evaluating the same image several times by randomly augmenting it before each run through the model.

Let $m_i$ be the $i$th image in the dataset, and let $f(x)$ be a function that returns an augmented image $x\prime$; $K$ number of calls to this function return $K$ images together denoted by $m_{i,k}\prime$; each of these images is fed to our classifier to get an output 128-D embedding vector denoted by $p_{i,k}$. The output class probabilities for each set of augmented images are then averaged as shown below where $P_i$ represents the new probability vector for image $m_i$ as given in (2):

$$P_i = \frac{1}{K}\sum_{k=1}^{K} p_{i,k} \tag{2}$$

Random augmentations were performed on the image prior to prediction, Width/Height shift in a (0, 10)% range and Rotation between (0,180) degrees. The number of samples $K$ generated was set to 7. The above values were arrived at by trying various permutations and combinations of parameters. Using this technique gave us a boost in overall accuracy.

The metrics we use to evaluate our model are as follows:

(1) Accuracy
(2) Quadratic Weighted Cohen Kappa for the 5 classes
(3) Sensitivity and Specificity
(4) ROC curve.

The outputs from the 5-class problem were thresholded to give the final labels. The threshold values we used were 0.47, 1.25, 2.04, and 3.29. Based on this, we calculated the accuracy (Table 4).

Sensitivity and Specificity were obtained on the 2-class problem as 0.752 and 0.976, respectively. The ROC curve for the 2-class problem is shown in Fig. 5.

Table 4 Result metric score

| | Accuracy | Kappa |
|---|---|---|
| 5-class | 0.80 | 0.821 |
| 2-class | 0.933 | 0.771 |

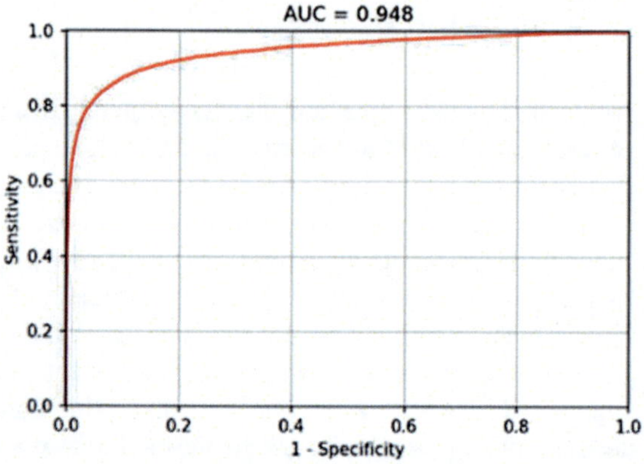


Fig. 5 ROC curve for testing set with an AUC of 0.948

# 5 Conclusion

In this paper, an Inception-v3 model was trained on retinal fundus images obtained from Kaggle. We focused on the 2-class and 5-class problems where we obtained accuracies of 0.933% and 0.80%, and Kappa scores of 0.821 and 0.771, respectively. By using class weights, we avoid the model overfitting on any one class. The results of the method described are comparatively better than the existing methods, thus proving the efficacy of adopting the proposed method in a clinical setting.

# References

1. Zheng Y, He M, Congdon N (2012) The worldwide epidemic of diabetic retinopathy. Indian J Ophthalmol 60(5):428
2. Williams R, Colagiuri S, Chan J, Gregg E, Ke C, Lim LL, Yang X (2019) IDF Atlas 9th
3. American Optometric Association (2014) Evidence-based clinical practice guideline: eye care of the patient with diabetes mellitus. St. Louis (MO), American Optometric Association
4. American Academy of Ophthalmology (2002) International clinical diabetic retinopathy disease severity scale, detailed table
5. Mnih V, Kavukcuoglu K, Silver D, Graves A, Antonoglou I, Wierstra D, Riedmiller M (2013) Playing atari with deep reinforcement learning. arXiv preprint arXiv:1312.5602.
6. Radford A, Wu J, Child R, Luan D, Amodei D, Sutskever I (2019) Language models are unsupervised multitask learners. OpenAI blog 1(8):9
7. 'Diabetic Retinopathy Detection' (2015) Kaggle. https://kaggle.com/competitions/diabetic-retinopathy-detection

8. Agurto C, Barriga ES, Murray V, Nemeth S, Crammer R, Bauman W, Zamora G, Pattichis MS, Soliz P (2011) Automatic detection of diabetic retinopathy and age-related macular degeneration in digital fundus images. Invest Ophthalmol Vis Sci 52(8):5862–5871
9. Akram MU, Khalid S, Khan SA (2013) Identification and classification of microaneurysms for early detection of diabetic retinopathy. Pattern Recogn 46(1):107–116
10. Harini R, Sheela N (2016) Feature extraction and classification of retinal images for automated detection of diabetic retinopathy. In: Second International conference on cognitive computing and information processing (CCIP). IEEE, pp 1–4
11. Gulshan V, Peng L, Coram M, Stumpe MC, Wu D, Narayanaswamy A, Venugopalan S, Widner K, Madams T, Cuadros J, Kim R (2016) Development and validation of a deep learning algorithm for detection of diabetic retinopathy in retinal fundus photographs. JAMA 316(22):2402–2410
12. Szegedy C, Vanhoucke V, Ioffe S, Shlens J, Wojna Z (2016) Rethinking the inception architecture for computer vision. In: Proceedings of the IEEE conference on computer vision and pattern recognition. pp 2818–2826
13. Gao Z, Li J, Guo J, Chen Y, Yi Z, Zhong J (2018) Diagnosis of diabetic retinopathy using deep neural networks. IEEE Access 7:3360–3370
14. Graham B Kaggle. [Online] https://www.kaggle.com/c/diabetic-retinopathy-detection/discussion/15801. Accessed 6 Sept 2020
15. Arcadu F, Benmansour F, Maunz A, Willis J, Haskova Z, Prunotto M (2019) Deep learning algorithm predicts diabetic retinopathy progression in individual patients. NPJ Digit Med 2(1):1–9
16. Yang Y, Li T, Li W, Wu H, Fan W, Zhang W (2017) September. Lesion detection and grading of diabetic retinopathy via two-stages deep convolutional neural networks. In: International conference on medical image computing and computer-assisted intervention. Springer, Cham, pp 533–540
17. Ni J, Chen Q, Liu C, Wang H, Cao Y, Liu B (2019) An effective CNN approach for diabetic retinopathy stage classification with dual inputs and selective data sampling. In: 18th IEEE international conference on machine learning and applications (ICMLA). IEEE, pp 1578–1584
18. Wan S, Liang Y, Zhang Y (2018) Deep convolutional neural networks for diabetic retinopathy detection by image classification. Comput Electr Eng 72:274–282
19. LeCun Y, Boser B, Denker JS, Henderson D, Howard RE, Hubbard W, Jackel LD (1989) Backpropagation applied to handwritten zip code recognition. Neural Comput 1(4):541–551
20. Krizhevsky A, Sutskever I, Hinton GE (2017) ImageNet classification with deep convolutional neural networks. Commun ACM 60(6):84–90
21. Deng J, Dong W, Socher R, Li L-J, Li K, Fei-Fei L (2009) ImageNet: a large-scale hierarchical image database. IEEE Conf Comput Vis Pattern Recognit
22. Simonyan K, Zisserman A (2014) Very deep convolutional networks for large-scale image recognition. arXiv preprint arXiv:1409.1556
23. He K, Zhang X, Ren S, Sun J (2016) Deep residual learning for image recognition. In: Proceedings of the IEEE conference on computer vision and pattern recognition. pp 770–778
24. Kingma DP, Ba J (2015) Adam: a method for stochastic optimization. ICLR. arXiv preprint arXiv:1412.6980, 9
25. Nguyen QH, Muthuraman R, Singh L, Sen G, Tran AC, Nguyen BP, Chua M (2020) Diabetic retinopathy detection using deep learning. In: Proceedings of the 4th international conference on machine learning and soft computing. pp 103–107

# Deep Learning-Based Tumor Segmentation from CT Images

**Nitya Jitani, Anup Basumatary, and Rosy Sarmah**

**Abstract** The segmentation or delineation of gallbladder malignancies and liver tumors is paramount in the prognosis, treatment, and diagnosis of gallbladder and liver diseases. The segmentation of liver tumors and malignant regions indicative of gallbladder cancer from abdominal 2D CT scan images has been accomplished by implementing a deep learning algorithm. An enhanced U-net (EU-Net) model has been introduced, obtained by training on a liver tumor (LITS) dataset and Gallbladder cancer (GBC) dataset for the segmentation of various malignancies. We conducted a comparison of the performances of the EU-Net model to that of the U-Net, U-Net++, and Attention U-Net models and found that Eu-Net outperformed the latest state-of-the-art deep learning models with respect to IOU and F1-score.

**Keywords** CT scan images · Deep learning · Gallbladder cancer · Medical image segmentation · Liver tumor · U-Net

## 1 Introduction

For radiologists to diagnose, treat, plan for, and monitor cancer patients, it is essential to delineate or segment the tumor region. Medical image segmentation refers to classification of image pixels based on their local properties or texture attributes, which are critical for understanding the image content (or what is where in an image), mostly the shape and volumes of the various organs or malignancies, from medical images. The liver tumor or GBC is generally identified from Computed Tomography (CT) images, based on the intensity differences or density of the regions. However, in many cases, the differences in feature values such as intensity and contrast, between the tumorous regions and the healthy region is very minimal, making it difficult to

N. Jitani · A. Basumatary · R. Sarmah (✉)
Department of Computer Science and Engineering, Tezpur University, Assam 784028, India
e-mail: rosy8@tezu.ernet.in

N. Jitani
e-mail: jitani@tezu.ernet.in



S. Borah et al. (eds.), *Advanced Computational and Communication Paradigms*,
Lecture Notes in Networks and Systems 535,
https://doi.org/10.1007/978-981-99-4284-8_27

segment using traditional image processing methods such as manual delineation or semi-automatic delineation methods. Also, the manual segmentation is quite laborious and time-consuming, considering the huge number of slices (around 300–400) per patient. Deep Learning (DL)-based automatic segmentation approaches are being designed for a variety of imaging modalities and diseases to address this issue. The automatic detection of a liver tumor or Gallbladder Cancer (GBC) may be the best alternative for speeding up the search for answers to the tumor's cure.

GBC is a highly lethal malignancy affecting a vast number of people in North and North-East India. Late detection of GBC is a major contributing factor to fatality. Therefore, early detection of GBC is paramount to its prognosis. Here, computational methods will greatly help in early disease diagnosis. To date, not much work has been done for GBC disease diagnosis from CT images. Therefore, this has emerged as a motivating factor for us to tackle the problem of GBC diagnosis from a computational point of view which will aid in early detection and assist clinicians. DL is advancing as the state of the art, with improved performances across diverse applications in medicine. Among the various DL models, Convolutional Neural Networks (CNN) and their extensions are the most frequently used. Although there are numerous competing DL frameworks to choose from, U-Net model, which is also a variant of CNN architecture, shows tremendous performance for segmentation or delineation of medical images. This study emphasizes the U-Net architecture as the fundamental model for liver tumor and GBC segmentation. This work's primary contributions are outlined as follows:

- An effective segmentation method for liver tumor and malignant areas reflecting GBC from CT images.
- Performance evaluation of proposed EU-Net model in comparison to latest DL methods. The outcomes support the EU-Net model's claimed effectiveness.

## 2 Literature Review

In image analysis for medical applications, segmentation plays a vital role [1]. Accurate structural delineation of various organs and malignant regions is essential for further feature extraction and diagnosis. DL models are extensively being used for segmentation, prediction, and classification problems for medical images [2]. CNNs are the best-known DL models with varied applications [3]. Researchers have mainly focused on CNN and its variations, U-Net models and fully connected CNN for various image segmentation tasks for medical purpose. Roth et al. [4] developed a 3D CNN model that is fully connected for multi-class segmentation of 331 liver, spleen, and pancreas clinical CT images. Ronneberger et al. [5] developed the U-Net model for biomedical image segmentation using a contracting and an expanding path. Zhou et al. [6] developed a UNet++ architecture for segmentation of diverse biomedical images, using nested, dense skip paths to connect the encoder and decoder networks. Budak et al. [7] developed a cascaded classification framework using deep CNN to

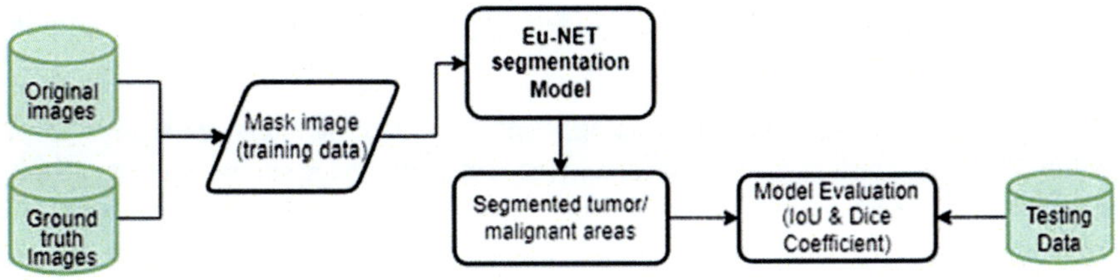


**Fig. 1** Proposed model workflow (Color figure online)

segment liver tumors. Xiaomeng et al. [8] assessed the results of the LITS dataset for segmentation of liver tumors using 24 recent algorithms, applied to 131 CT scan images. Tran et al. [9] built a multi-layer Un-Net framework employing CT scan images for liver and its tumors segmentation. Manjunath et al. [10] evaluated a modified U-Net architecture to segment CT images of liver and its tumors. Kushnure et al. [11] suggested a multi-scale feature-based CNN model, MS-UNet, for automatically segmenting the liver and its tumor incorporating CT scan images. Although several methods have already been developed for CT image segmentation of liver and its associated tumors; however, there is still scope for improvement in terms of various performance measures. A major driving force for this research is the observation from numerous studies that the segmentation of gallbladder cancer has not received much attention. By utilizing the U-Net architecture, we propose an Eu-Net model for segmenting liver tumor and malignant areas reflecting GBC.

## 3 Method

We first process the CT images and then apply the proposed EU-Net DL architecture for segmentation as discussed next. TensorFlow and Keras were used for the DL model implementations. TensorBoard, a collection of visualization tools for TensorFlow algorithms, was used to track validation loss and other validation metrics such as IoU, validation accuracy, F1-score, etc. Furthermore, all the experiments were carried out on Google Colaboratory Pro using Python version 3. The process of the suggested methodology is shown in Fig. 1.

**Datasets**: The GBC dataset comprises of CT scan images of abdominal region of patients, acquired from Dr. Bhubaneshwar Borooah Cancer Institute, Guwahati, India. This dataset contains 71 CT volumes of different patients in Nrrd format, leading to 2000 2D image slices in total, along with the ground truth images, which were segmented manually under the supervision of the domain experts (radiologists), with the corresponding labels, respectively.

The liver tumor dataset is taken from a publicly available dataset, The Liver Tumor Segmentation Benchmark (LITS) [12]. It consists of a total of 131 CT volumes in

Nifti file format, which contains 7190 2D image slices along with the corresponding labels and ground truth images.

**Data preparation and pre-processing**: Since the patient image data were in 3D volumes, the first task involved converting them into 2D image slices for compatibility with the DL model. Although each patient has around 300–400 slices in total, we extracted only those slices which include the region of interest (tumor). For training the model for tumor segmentation, the input images are the mask images, generated by performing a logical AND operation between the 2D CT scan image slices and the corresponding ground truth image slices.

A simple normalization step was also performed by dividing the values of each image pixel by 255 to ensure that all inputs were in a comparable range and had small values, i.e., ranging from 0 to 1 before feeding to the neural network. We performed an 80/20 split on the dataset, i.e., 80% (training), 10% (test), and other 10% for validation by random sampling. So, the training set had 5752 images, 719 images were in the validation dataset, and the remaining 719 were for testing with their corresponding masks. The original image were of dimension $512 \times 512$.

**Proposed EU-Net Model**: The U-Net model served as the foundation for liver tumor and GBC segmentation. The proposed network architecture is shown in Fig. 2. Like the standard U-Net model, the EU-Net also has an encoder path, also known as downsampling path, an upsampling path or the decoder, and a bottleneck path. Four convolutional blocks make up the encoder, also referred to as the contracting path. There are two $3 \times 3$ convolution layers in each block, with filter sizes of 16, 32, 64, and 128, respectively, following batch normalization, activation is performed using ReLU and $2 \times 2$ Max Pooling operations, and each layer has a 0.1 dropout rate. The objective of the contraction path is to obtain input characteristics/metadata for segmentation. As the depth of the encoder increases, the image sizes are gradually decreased. The convolutional and pooling layers are used for high-level or deep feature extraction from the input images. The feature map undergoes upsampling in the expansive path (decoder), a $2 \times 2$ convolution to minimize the feature channels, and concatenated with the matching cropped feature map from the encoder path, followed by two convolutions of kernel size as $3 \times 3$ each, activation function as ReLU, batch normalization and a dropout of 0.1. A convolution of kernel size $1 \times 1$ maps each of the feature vectors to the required classes in the fully connected layer. The network consists of 23 convolutional layers in total. We employ skip connections layers in subsequent decoder stages to combine the feature map obtained from the encoder with the output of the transposed convolutional layer so as to acquire more exact position information. The ADAM optimizer algorithm is employed, incorporating learning rate as 0.001. For the output layer, the activation function used is sigmoid. A binary cross-entropy loss function has been applied for the classification problem.

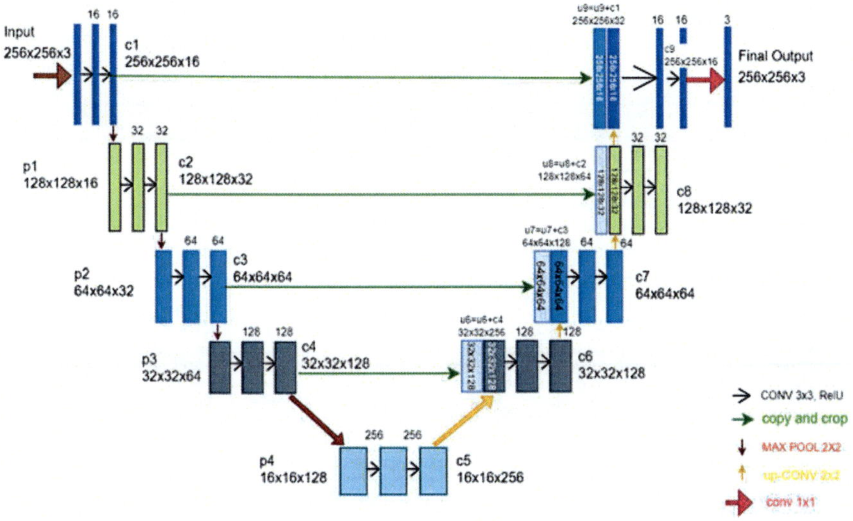


**Fig. 2** Proposed EU-Net model (Color figure online)

## 4 Experimental Results and Discussion

The experiments have been performed on Google Colaboratory Pro using Keras version 2.8.0 in a Python version 3.7.13 programming environment. Two different models with comparable architectures have been employed for training. The training for the first model was performed by working on the CT scan images of liver tumor (LITS dataset) with their respective tumor class labels for liver tumor segmentation, whereas the second model training was done using GBC CT scan images (GBC dataset) with corresponding tumor class labels for GBC segmentation. The processing time was significantly reduced by decreasing the size of the images from 512 × 512 to a dimension of 256 × 256.

**Validation Metrics**: For performance evaluation of the proposed method, we have incorporated the two commonly used metrics, Intersection over Union (IoU score) and Dice coefficient (F1-score). Let the segmented and ground truth images be $A$ and $B$, respectively.

- **Intersection-Over-Union** (IoU, Jaccard Index) [2]: The formula for calculating IoU is as $Jacc(A, B) = \frac{|A \bigcap B|}{|A \bigcup B|}$. IoU score is positively correlated and varies between 0 and 1.
- **Dice coefficient** (F1-Score) [2]: It is computed as $Dice(A, B) = \frac{2 \times |A \bigcap B|}{|A| + |B|}$. The F1-score also has a positive correlation and lies in the range of 0–1, with 1 indicating the greatest similarity between the predicted and ground truth images.

**Table 1** Comparison table for LITS dataset

| Model | IoU | Dice coefficient |
|---|---|---|
| UN | 0.8845 | 0.9286 |
| UN++ | 0.8267 | 0.9049 |
| A-UN | 0.8824 | 0.9374 |
| EU-Net | 0.9013 | 0.9480 |

**Table 2** Comparison table for GBC dataset

| Model | IoU | Dice coefficient |
|---|---|---|
| UN | 0.7102 | 0.8265 |
| UN++ | 0.6385 | 0.7777 |
| A-UN | 0.7352 | 0.8465 |
| EU-Net | 0.7400 | 0.8494 |

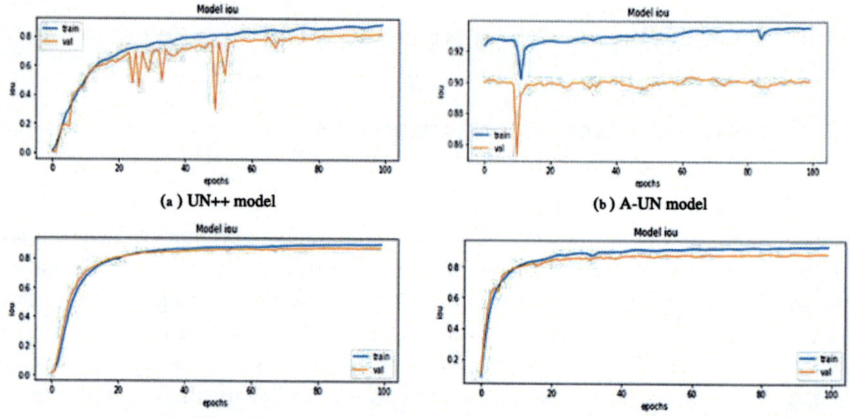


(a) UN++ model (b) A-UN model

(c) UN model (d) Eu-Net model

**Fig. 3** Training and validation curves for IoU of LITS dataset. Blue represents training IoU and orange represents validation IoU (Color figure online)

## *4.1 Segmentation of LITS Dataset*

For liver tumor segmentation, the LITS dataset comprises of 7190 2D CT scan images along with the corresponding masks for each label. The performance evaluation of the EU-Net model has been done with three pre-trained models, namely, U-Net (UN) [5], Attention U-Net (A-UN) [13], and U-Net++ (UN++) [6] models.

Table 1 depicts the IoU and dice coefficient score for validation of the segmentation, using all the four DL models. The model has been tested on 100 images and the results are reported. The proposed EU-Net model showcases the best IoU score

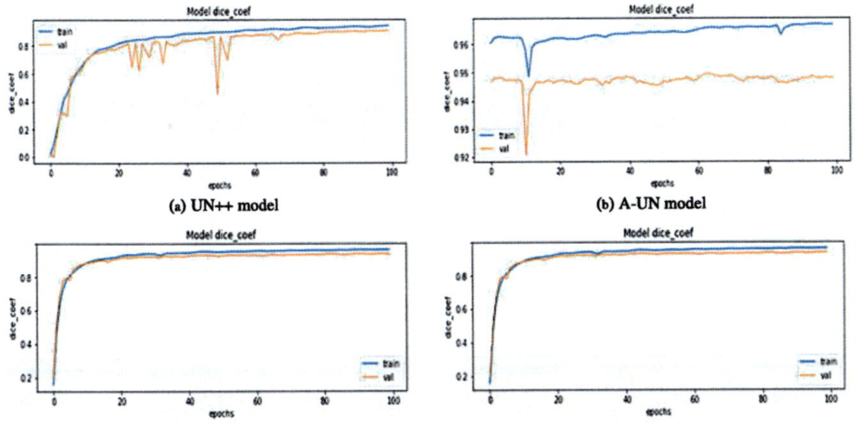


(a) UN++ model (b) A-UN model (c) UN model (d) Eu-Net model

**Fig. 4** Training and validation curves for Dice Coefficient of LITS dataset. Blue represents training scores and orange represents validation scores (Color figure online)

of 90.13% and dice coefficient score of 94.80%, which is far higher than the other models.

The training and validation accuracy curves based on the IoU score for the LITS dataset are shown in Fig. 3. The plot for training and validation accuracy for UN and EU-Net models is comparable, it increases initially for 20 epochs and then becomes almost constant, depicting a good curve. However, for UN++ model, the validation curve has too many fluctuations for up to 70 epochs. Also, the training and validation graphs for A-UN take a downward peak after 10 epochs, after which it is almost constant. Figure 4 infers the training and validation accuracy based on dice coefficient for the LITS dataset. We can see that the plots for accuracy based on dice coefficient are very similar to the IoU curves, for all the respective models. However, we see a rise in the accuracy based on dice coefficient for EU-Net model by 0.4%.

Figure 5 shows the segmentation outcomes for LITS dataset by deploying (A) UN++ model, (B) A-UN model, (C) UN model, and (D) EU-Net model. The first column shows the input CT images, the second column depicts the ground truth tumor mask images, and the third column depicts the resultant predicted tumor masks.

## *4.2 Segmentation of GBC Dataset*

For GBC segmentation, the dataset comprises of 2000 2D CT scan images along with the corresponding ground truth mask images. The EU-Net model's performance has been compared with other three pre-trained models, namely, UN model, UN++ model, and A-UN model. Table 2 shows the IoU and dice coefficient score comparison

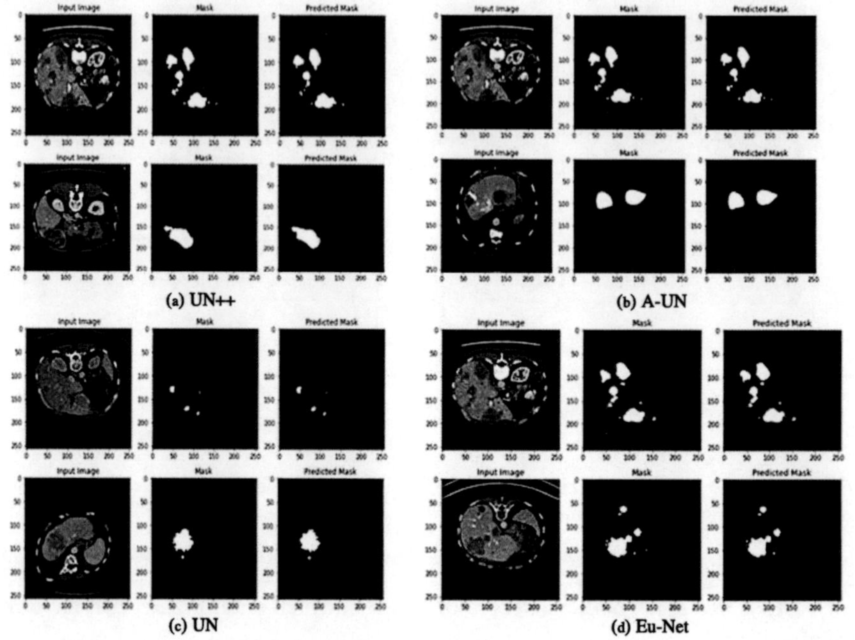


**Fig. 5** Segmentation results for LITS dataset

for the EU-Net model with the other three state-of-the-art models. The IoU score obtained is 74% which is significantly higher than the Iou score for UN++ model, i.e., 63.85%. Also, the IoU score for UN model is 71.02% and for A-UN model is 73.52%. The dice coefficient score obtained using the EU-Net model is 84.94%, which is again better than the dice coefficient scores obtained by UN, UN++, and A-UN models. Hence, it can be concluded that the proposed EU-Net Model gives significantly better performance than the comparative methods for GBC segmentation based on IoU and dice coefficient values.

The training and validation accuracy plots based on IoU score are depicted in Fig. 6 and the training and validation accuracy plots based on the F1-score are depicted in Fig. 7 for the four different models for GBC segmentation respectively. The training and validation plots for the proposed EU-Net model for both IoU and dice coefficient show a gradual increasing pattern up to 40 epochs, after which it's a smooth constant curve. The training and validation curves for UN++ model show an increasing trend up to almost 100 epochs but with lots of fluctuations. The validation curve for the Attention UN model although in an increasing trend shows high downward peaks up to 40 epochs, after which it attains a smooth curve, for both IoU and dice coefficient plots. The validation curve based on both IoU and dice coefficient score for UN model, however, shows an unusual downward peak after 80 epochs although it attained a smooth consistent curve after 20 epochs. From the inferences obtained from these

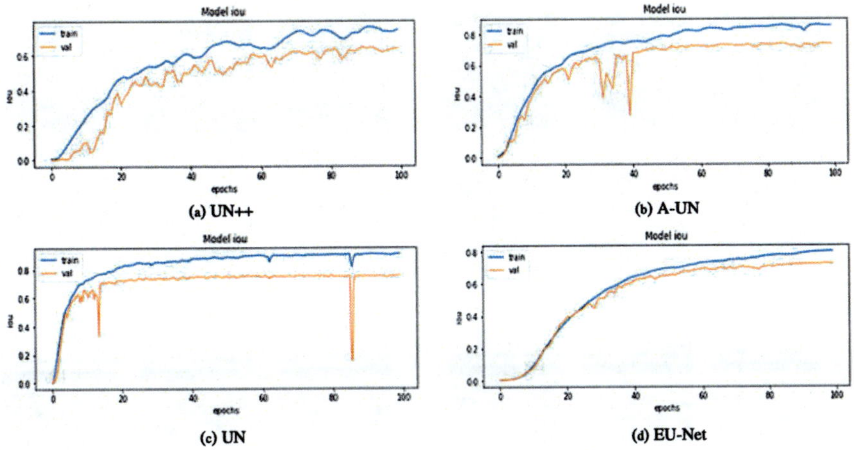


(a) UN++ (b) A-UN (c) UN (d) EU-Net

**Fig. 6** Training and validation curves for IoU of GBC dataset (Color figure online)

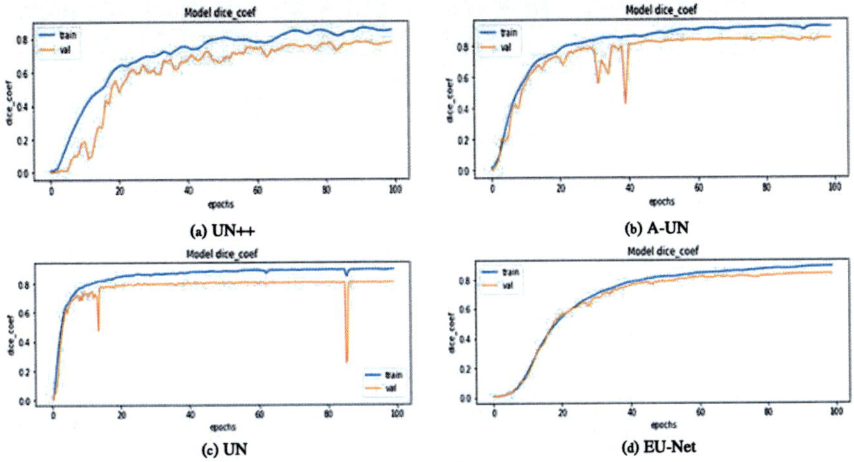


(a) UN++ (b) A-UN (c) UN (d) EU-Net

**Fig. 7** Training and validation curves for dice coefficient of GBC dataset (Color figure online)

plots, it is evident that the EU-Net model performs better than the other models for GBC segmentation.

Figure 8 depicts the segmentation results for GBC dataset using (A) UN++ model, (B) A-UN model, (C) UN model, and (D) EU-Net model. The first column depicts the input CT images, the ground truth tumor mask images are depicted in the second column and the third column depicts the resultant predicted tumor masks.

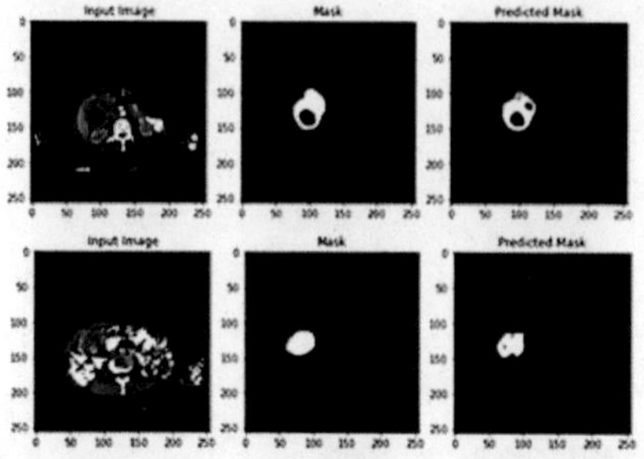


(a) UN++

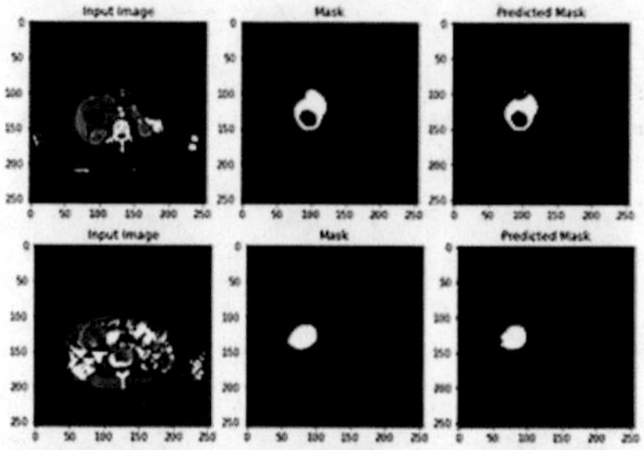


(b) A-UN

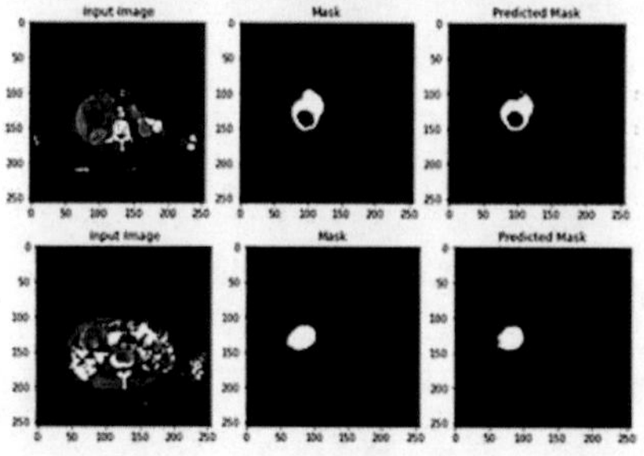


(c) UN

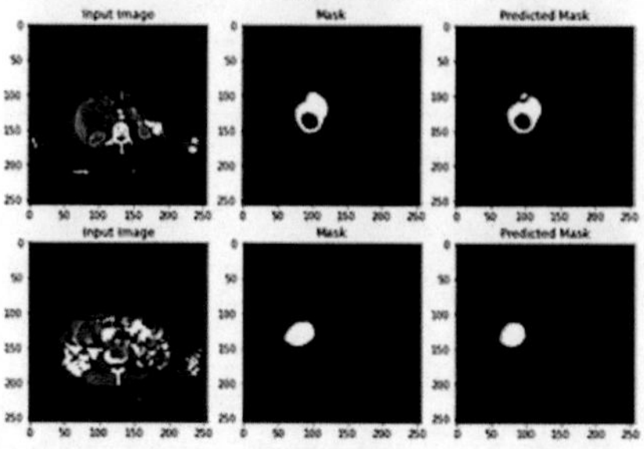


(d) EU-Net

**Fig. 8** Segmentation results for GBC dataset

# 5 Conclusion

This paper presents a DL-based EU-Net model for delineation of liver tumor and malignant areas reflecting GBC from 2D CT images, by training on liver tumor (LITS) dataset for segmentation of liver tumor and GBC dataset for segmentation of various gallbladder malignancies. The results attained using the proposed method were compared with three state-of-the-art methods, viz., UN, UN++, and A-UN, and showed significantly better performance based on the validation metrics, IoU, and dice coefficient scores. In the future, we can assess how well the proposed method segments 3D medical CT scan images.

**Declaration** This work is an outcome of a research project titled "Radiomics with machine learning methods towards prediction of gallbladder cancer", supported by Indian Council of Medical Research, New Delhi.

## References


1. Jitani N, Singha B, Barman G, Talukdar A, Choudhury BK, Sarmah R, Bhattacharyya DK (2022) Gallbladder CT image segmentation by integrating rough entropy thresholding with contours. In: Advanced computational paradigms and hybrid intelligent computing: proceedings of ICACCP 2021. Springer, pp 651–659
2. Haque IRI, Neubert J (2020) Deep learning approaches to biomedical image segmentation. Inf Med Unlocked 18:100297
3. Yu H, Yang LT, Zhang Q, Armstrong D, Deen MJ (2021) Convolutional neural networks for medical image analysis: state-of-the-art, comparisons, improvement and perspectives. Neurocomputing 444:92–110
4. Roth HR, Oda H, Zhou X, Shimizu N, Yang Y, Hayashi Y, Oda M, Fujiwara M, Misawa K, Mori K (2018) An application of cascaded 3d fully convolutional networks for medical image segmentation. Comput Med Imaging Graph 66:90–99
5. Ronneberger O, Brox T, Fischer P (2015) U-net: convolutional networks for biomedical image segmentation. In: International conference on medical image computing and computer-assisted intervention. Springer, pp 234–241
6. Zhou Z, Siddiquee Md MR, Tajbakhsh N, Liang J (2018) Unet++: a nested u-net architecture for medical image segmentation. In: Deep learning in medical image analysis and multimodal learning for clinical decision support. Springer, pp 3–11
7. Budak Ü, Guo Y, Tanyildizi E, Şengür A (2020) Cascaded deep convolutional encoder-decoder neural networks for efficient liver tumor segmentation. Med Hypotheses 134:109431
8. Bilic P, Christ P, Li HB, Vorontsov E, Ben-Cohen A, Kaissis G, Szeskin A, Jacobs C, Mamani GEH, Chartrand G et al (2022) The liver tumor segmentation benchmark (LiTS). Med Image Anal 102680
9. Tran S, Cheng C, Liu D (2020) A multiple layer u-net, un-net, for liver and liver tumor segmentation in CT. IEEE Access 1–1
10. Manjunath RV, Kwadiki K (2022) Modified u-net on CT images for automatic segmentation of liver and its tumor. Biomed Eng Adv 100043
11. Kushnure DT, Talbar SN (2021) Ms-unet: a multi-scale unet with feature recalibration approach for automatic liver and tumor segmentation in CT images. Comput Med Imaging Graph 89:101885
12. Bilic P, Christ PF, Vorontsov E et al (2019) The liver tumor segmentation benchmark (LiTS)
13. Oktay O, Schlemper J, Lee M, Le Folgoc L, Heinrich M, Misawa K, Mori K, McDonagh S, Hammerla NY, Kainz B et al (2018) Attention u-net: learning where to look for the pancreas. arXiv preprint arXiv:1804.03999

# Image and Video Processing

# DIVS: A Real-Time Video Stabilization Framework for Consumer Camera

**Varanasi L. V. S. K. B. Kasyap, Sumathi, Abhijit Adhikari, and V. S. Bhagavan**

**Abstract** Video stabilization is one of the most interesting areas of research that can be implemented in consumer camera technology. Various works have been proposed to overcome this problem. However, there are many practical difficulties in deploying them in real time. Many frames are taken as an input to generate a stabilized output, which leads to high computation of the image sensing hardware. A novel approach to solve these challenges is proposed in this paper using the Hybrid Deep Neural Network Model (HDNNM) that uses gyroscopic sensor data and the optical flow analysis. This network is based on the deep unsupervised learning techniques, which require less computation and low buffer memory. The workflow of the proposed model is based on the integration of the optical flow with the joint motion representation to maintain a correspondence between the frames and the poses of scene objects. The LSTM module in the network wraps the grid and stabilizes the video frames. The relative motion along with the multi-stage training strategies was adopted to make the model unsupervised. The HDNNM is validated using the ablative dataset collected from various scenes, and the results of the model surpass the existing model's performance metrics. To the best of the authors' knowledge, this is the first unsupervised approach to overcome the challenges in video stabilization.

**Keywords** Pose estimation · Video processing · Optical flow analysis · Shutter compensation · Real-time stabilization

V. L. V. S. K. B. Kasyap (✉) · Sumathi · A. Adhikari
School of Computer Science and Engineering, VIT-AP University, Inavolu, India
e-mail: varanasi.20bce7315@vitap.ac.in

Sumathi
e-mail: sumathi.d@vitap.ac.in

V. S. Bhagavan
Department of Engineering Mathematics, Koneru Lakshmaiah Education Foundation, Vaddeswaram, India



S. Borah et al. (eds.), *Advanced Computational and Communication Paradigms*, Lecture Notes in Networks and Systems 535,
https://doi.org/10.1007/978-981-99-4284-8_28

## 1 Introduction

A few videos are often shaky when captured using hand-held devices. The growing usage of smartphones for live streaming, video recordings, and short-film shooting makes the video stabilization with high quality inevitable. However, this remains a hard challenge in real time because of the complex camera moments, illuminations in the frame, and the variations in scene. The existing video stabilization methods can be diversified into two categories, i.e., Image-based and Sensor-based. The Image-based approach outputs video using the smooth camera moments from the sparse frame features that are extracted from the input video or by dense optical flow estimation. This approach gives non-linear flexibility on the motion compensations. Though these methods often fail when there are complex camera moments such as parallax, pedestal, tracking shot or if there are non-reliable features in the frame, they produce the visible non-taut distortions and artifacts in the output due to the rigid reliable constraints. However, these types of methods are used in smartphones because the threshold value of shakiness from the mobile camera is relatively lower, and these methods are able to stabilize that. The Sensor-based approach uses the motion-sensor data, i.e., accelerometer and gyroscope, to enhance accurate motion. This approach is untied from the scene contents and achieves noteworthy stabilization results with efficacious corrections. Sensor-based approaches are widely used in professional camera shooting for effective results, since this is a tedious approach for video stabilization. However, the proposed method bestows homographies to stabilize the hyper-plane and does not adopt the frame depth leading parallax moments for short scenes. In the proposed work, a novel and efficient Deep Integrated Video Stabilization (DIVS) framework is used to integrate two motion sources of Image and Sensor and acquires advantages from both sides. Hence, the network outputs uni-camera pose with dense warping flow, and the video is stabilized by warping sensor-based camera pose and real camera motion toward the virtual pose. In this approach, the motion is naturally saved and the spin shutter distortion is evaluated using artifacts like wobbling. The network learns to mutually minimize both, camera pose smoothness and optical flow losses. Thus, this method automatically adjusts to different scenes like depth variations and light variations to minimize residual motions. This method is predesigned with different loss functions and a multi-stage training approach as shown in Fig. 1.

This shows the overview of convolutional methods for recent learning-based methods and the usage of DIVS. As the datasets available do not record the sensor data, a new dataset that contains videos with the gyroscopic data and the OIS data is curated by capturing various scenes in the surroundings and used for both training and evaluation. This dataset derives scenarios with different illumination conditions like the camera moments. The sensor data and the captured video frames are accurately aligned through estimation. This model is evaluated both subjectively and objectively, and the results obtained outperform the state-of-the-art methods achieving the best results. The contributions of the proposed paper are

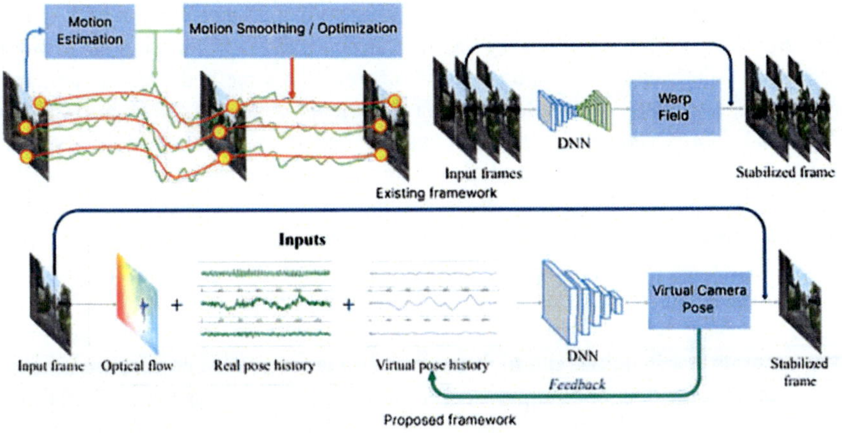


**Fig. 1** Comparisons of existing video stabilization methods and the proposed methods

- A novel approach of video stabilization using the optical flow of the images and the motion sensor data.
- A multi-stage training approach and relative motion representation of camera and optical flow using unsupervised learning process.

A benchmark dataset that contains videos with both gyroscopic data and OIS sensor data is collected from various scenarios. Both the source code and dataset are uploaded in GitHub.

## 2 Related Works

### *2.1 Prosaic Methods*

The classic video stabilization algorithms integrate motion estimation, video rendering or wrapping steps, and camera path smoothing [1–4]. The shutter rolling distortions are also some solutions for video stabilization [5]. These methods can be grouped under 3D, 2D, 4D, and 2.5D approaches depending on their motion estimation. The 3D approaches compute the camera poses and estimate the smooth virtual camera trajectory in the 3D space [6, 7]. Several techniques were proposed to find the 6DoF camera poses including the 3D subjective and projective reconstruction, motion structure, depth camera, and illuminated light field [8]. The 3D reconstruction can handle tracking shot that produces the high-quality results, often necessitating computation cost that requires analogous hardware devices. 2D approaches represent that estimate motions of the camera perform perspective or affine transformations. The outlier rejection and the feature tracking mechanism give robustness to the model and are being used to the accurate estimation of camera poses. Nie et al. [3] proposed a method to replace the feature trajectories using optical flow estimation with spatial-variant motions. The methodology behind this approach is to compute the low-pass filter for smooth motion of individual parameters. This method uses the and Joint optimization packed with the local camera moments. A few of the Hybrid 2D- and 3D-based approaches interpolate the subspace constraints with epiploic geometry. Fang et al. [7] proposed a smooth 3D rotation technique based on the image feature and texture matching. The above methods compute the whole process when video is offline, which cannot be used for the live streaming of videos and mobile usage. Raut et al. [9] proposed a MeshFlow mechanism with only one frame latency rate for the video stabilization online. This method is built using both OIS and EIS data. However, the frame rate of this method cannot be extended therefore giving less accurate stabilization, and high computation is required. The proposed method DIVS utilizes the data of both gyroscope and OIS with the latency rate of 10 frames enabling for video stabilization without any other optimization in less time compared to pre-existing works.

## 2.2 *Training-Based Methods*

The growth of deep learning techniques in all the neighboring areas of computer vision [10] and achieving state-of-the-art results enable to use the visualization techniques to overcome video stabilization challenges. However, deep learning-based methods with noteworthy results are not yet proposed for this problem. Zhe Xu et al. [11] tuned the dataset with 70 pairs of both stable and unstable videos and trained Convolutional Neural Network (CNN) to wrap the mesh-grids and predict them in the video. PWStabNet [12] is proposed by Wang et al. that can learn from the 2D wrapping fields and stabilizes the video. Minh et al. [13] trained a Generative Adversarial Network (GAN) to generate the steady frames in the video using spatial transformer network, extracting the affine transforms for video wrapping in frames. Huang et al. [14] used the optical flow and optimized the weights of the network that generates wrap field of the specific fields. The further stabilization is achieved using the optimization techniques [15–17]. Grundmann et al. [18] proposed a model that learns the frame interpolation with the iterative stabilizing approach without cropping the results. The training-based methods [19–21] learn to stabilize the video from the video content using the optical flow. The performance of this methods is highly dependent on the training data. However, these methods suffer from the problem of the large motion visible distortions [22, 23]. In the DIVS, the gyroscope is used to observe the complex camera motions that utilize the optical flow mechanism to evaluate the residual motions from the scene epiploic geometry.

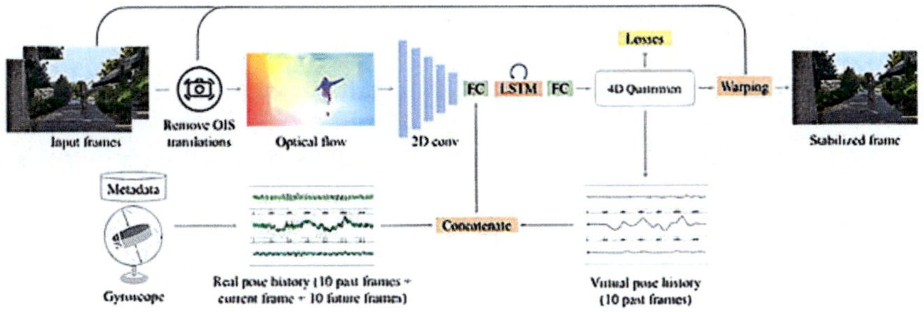


**Fig. 2** The workflow of the proposed framework

## 3 Deep Integrated Video Stabilization (DIVS)

The advantages of both Image-based approach and Sensor-based approach are there in the proposed framework, DIVS. The overview of the network architecture is given in Fig. 2. The gyroscopic data and the OIS reading are processed initially so that both intrinsic and extrinsic moments at arbitrary time stamps are analyzed. After this, OIS translations of the input video are removed to extract the optical flow from the raw video frames.

These optical flows are encoded with a latent space of 2D layers of convolution. This is concatenated with the original camera pose in a temporal window and the virtual camera pose in the motion representation jointly. In the next step, this representation is fed as input into the LSTM cell and other fully connected layers to predict the virtual camera pose at the present time stamp. As the last step, the grid-based wrapping method is used to wrap the video frame for stabilized rolling shutter domain with the input camera rotations, OIS data, and the predicted virtual camera poses. The proposed solution, DIVS, stabilizes the video in the queue of frames and is also suitable for online processing. In the training stage of the model, long subsequences of the training videos were randomly selected and evaluated on the network with the loss function of virtual camera pose for super vision ground-truth videos. The training of the model is stabilized using the multi-stage training strategy that strains a solution space.

### *3.1 OIS Processing and Gyroscopic Analysis*

In the dataset, the gyroscopic coordinates and optical image stabilizer translation sampled 240 Hz, where is the angular velocity and are the OIS translative moments. The camera rotations are integrated by, where I is the sample interval (3 ms). This rotation is represented as the 4D quaternion and stored in the buffer memory. To get the camera rotation at the arbitrary time stamp, the consecutive gyroscopic samples are selected such that they are used as function input by applying spherical linear interpolation (SLERP) as given in (1):

$$Ro(ts) = SLERP(Ro(t1), Ro(t2), (t2 - ts)/(t2 - t1)) \tag{1}$$

Similarly, O $(\boldsymbol{ts})$ is calculated by the linear interpolation between O $(\boldsymbol{ta})$ and O $(\boldsymbol{t_b})$.

## 3.2 Camera Pose Representation

The camera pose is estimated as the P = $(Ro, 0)$ in the 2D offset camera principal point (u, v). From the given 3D coordinates of the world $X_w$, the 2D image at the given time t is projected as shown in (2):

$$x = M(T)Ro(T)X_{\mathrm{w}} \tag{2}$$

Here, M(T) is the intrinsic matrix of focal length f. In the original camera pose, $P_C = (R_o, O)$ and the virtual camera pose is $(R_v, O)$; the transformation from original pose to the virtual pose is given in (3):

$$x_v = M(T)Ro(T)Rr(T)^{-1}Kr(T)^{-1} \tag{3}$$

## 3.3 Unsupervised Learning for Senior Fusion

This section discusses the Deep Integrated Video Stabilization network as shown in Fig. 2. DIVS consists of a sequence of 2D convolutional layers for the optical flow analysis along with LSTM cells to integrate the temporal information of the scene and latent motion representations using a fully connected layer network to decode. The OIS-optical flow is extracted first from the input frames and the OIS data, mapped to a low-dimensional representation $Z_{LD}$. Parallelly, the camera motion history is extracted to define motion representations of the frame jointly as $[Z_{LD}, Ht_r, Ht_v]$ and fed into an LSTM cell for predicting incremental rotation to the past virtual pose in the history. The frame wrapping steps and virtual pose prediction can be interpreted as the decoder module that maps current motion state as $[Z_{LD}, Ht_r, Ht_v]$ in the stabilized video frame.

## 3.4 Free Optical Flow in OIS

The input video camera motions are replaced with the OIS that significantly reduces the motion blur. However, the OIS moments depend on the hand motion of the user.

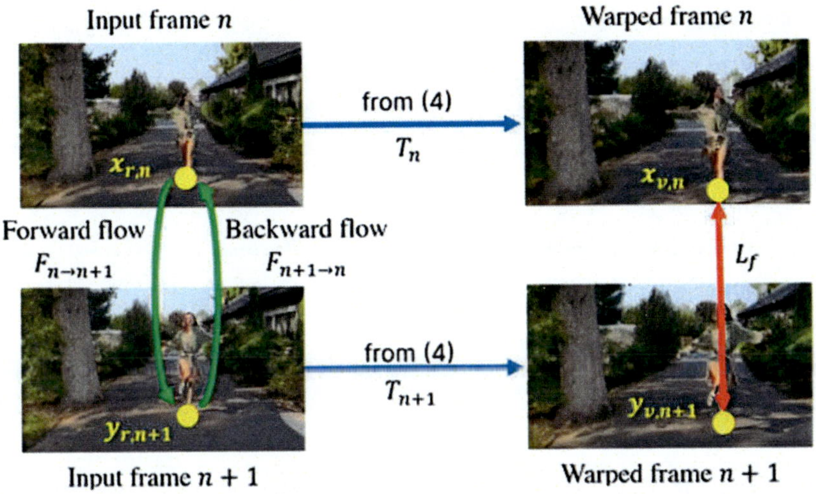


**Fig. 3** Loss function estimations of the optical flow in a video

At each scan line, the offset is different because of the rolling shutter and the random noise. It is not essentially required for the network to associate the local offset with the focal point changes. To overcome this issue, the OIS motions in the input are ignored while estimating the optical flow, giving only camera and object motions. The forward optical flow can be estimated as given in (4):

$$OF_m^{m+1} = y_{n+1,r} - x_{n,r} \tag{4}$$

From the OIS moments of the pixel arbitrary time stamp, at $y_{n+1,r}, x_{n,r}$ that are mapped to $x_{n,r} - O(t_{x,n,r})$ and $y_{n,r} - O(t_{y,n,r})$, respectively, the optical flow in the forward direction is adjusted to (5) as shown in Fig. 3.

$$\begin{aligned} OF_m^{m+1} &= (y_{n+1,r} - O(t_{y,n+1,r})) - (x_{n,r} - O(t_{x,n,r})) \\ &= OF_m^{m+1} - (O(t_{y,n+1,r}) - O(t_{x,n,r})) \end{aligned} \tag{5}$$

The backward optical flow estimation is adjusted correspondingly. The FlowNet [1] architecture is adopted to extract the optical flows in the training data. In this step, the proposed model will directly handle when the OIS is not buttressed.

## 3.5 Motion Based on Relativity

The original and virtual pose $[H_o, H_v]$ at the arbitrary time stamp t. For the original pose. The sample N from the queue in gyroscopic data to get the absolute camera moments as given in (6):

$$H_{abs,original} = (Ro_{org}(T_S - N\nabla T_S), \ldots, Ro_{org}(T_S), \ldots, Ro_{org}(T_S + N\nabla T_S)) \tag{6}$$

In parallel, virtual pose is sampled from the queue in the gyroscopic data to get the absolute camera moments as given in (7):

$$H_{abs,virtual} = (Ro_{virtual}(T_S - N\nabla T_S, \ldots, Ro_{virtual}(T_S + N\nabla T_S)) \tag{7}$$

The output of the network is relative to the previous virtual camera pose. So, the proposed model learns the first order pose and is invariable to the absolute poses. The experiments show that the relative representation gives the stabilized video frames and gives better generalization.

### 3.6 Loss Function

The loss functions used to evaluate the model do not need any ground-truth images. The time stamp is omitted in the loss function to maintain the simplicity. **Smoothness Loss Function**: The $S^0$ and $S^1$ are used to measure the loss function of the smoothness in the camera poses as given in (8) and (9):

$$L_{S^0} = ||Ro_{virtual}(T) - Ro_{org}(T_S - \nabla T_S)||^2 \tag{8}$$

$$\begin{aligned} L_{S^1} = ||&Ro_{virtual}(T)Ro_{virtual}(T)^{-1}(T_S - \nabla T_S) - Ro_{virtual} \\ &(T_S - T_S)Ro_{virtual}(T)^{-1}(T_S - 2\nabla T_S)||^2 \end{aligned} \tag{9}$$

These losses make the camera to take stabilized videos and smoothly.

**Protuberance Loss Function**: The noisy undefined areas and imprudent cropping in the frames are avoided in the stabilized video through this method. This is measured using the protrudes in the wrapped video frame of original boundary as given in (10):

$$L_{pt} = \Sigma_{j=0}^{N} GW_{p,i}||min(proturb(Pt_{virtual}(T_S), Pt_{org}(T_S + j\nabla T_S))/\alpha, 1||^2 \tag{10}$$

Here, N is the total count of look-ahead video frames in the input video. $GW_{p_i}$ is the generalized Gaussian Weights with the mean $\mu$ and standard deviation $\sigma$. The angle of reference protuberance is given as $\alpha$ that can be indulged. To evaluate the proturb () function, with the ratio $\beta$ each side of the original camera pose is projected and max generalized signed distance is measured between the four wrapped video frames with the ratio $\gamma$. The $\beta$ and $\gamma$ are used to control the camera motion sensitivity. The protuberance value is modified to 1 disregard with the large motion and stabilized balance. The parameter values are set to $\sigma = 2.4$, $N = 15$, $\alpha = 0.3$, $\beta = 0.07$, and $\gamma = 0.03$.

**Contortion Loss Function**: The wrapping contortion in the video is estimated using (11):

$$L_c = \delta(Ro_{virtual}, Ro_{org})/(1 + e^{(-\beta_1(\delta(Ro_{virtual}, Ro_{org}) - \beta_0)} \quad (11)$$

In (11), spherical angle in between the original camera and virtual poses is given by $\delta(Ro_{virtual}, Ro_{org})$, and $\beta_0$ and $\beta_1$ are the threshold parameters to tune the shape of logistic regression. The loss function is effective if the angle of deviation is greater than the threshold value; the empirical values of $\beta_0$ and $\beta_1$ are 8° and 120.

**Optical Flow Loss Function**: This loss function is similar to minimizing the pixel motion in the adjoint video frames. In Fig. 3, $y_{n+1,r}$, $x_{n,r}$ resembles the correspondence frames between n and n+1 frames. To get the virtual camera pose as T from the original camera space to obtain $x_{n,virtual} = T_s(x_{n,org})$ and $y_{n+1.org} = T_{s+1}(y_{n+1,org})$, from the forward flow and backward flow incorporation $F_m^{m+1}$, the warped pixel representation is given in (12) and (13):

$$x_{n,virtual} = T_S(x_{n,org}) = T_S(y_{n+1,r} + OF_m^{m+1}) \quad (12)$$

$$y_{n+1,virtual} = T_{S+1}(y_{n+1,virtual}) = T_{S+1}(x_{n,org} + OF_m^{m+1}) \quad (13)$$

The final goal is to decrease the $||x_{n,virtual} - x_{n+1,virtual}||^2$ and can be measured using (14):

$$L_{ofl} = |x_n|^{-1}\Sigma_{X_n}||x_{n,virtual} - T_{S+1}(x_{n,org} + OF_m^{m+1})||^2 + |x_{n+1}|^{-1} \Sigma_{X_{n+1}}||y_{n+1,virtual} - T_S(y_{n+1,org} + OF_m^{m+1}||^2 \quad (14)$$

$x_n$ is the group of different pixel positions in the frame n that are in the wrapped video's undefined regions.

**General Loss Function**: At the arbitrary time stamp $T_s$, the final loss function is the weighted sum of all the loss functions:

$$L = W_{0L_{s0}} + W_{1L_{s1}} + W_{2L_{pt}} + W_{3L_c} + W_{4L_{ofl}} \quad (15)$$

### 3.7 Multi-stage Training

In the virtual camera pose, there always exists a trade-off between the respective original camera pose and stability. Even though the loss function of the model is given to constrain the solution space, the network faces difficulty in learning non-linearity from the training data that cannot be converged to optimize the loss function. The multi-stage training is adopted to overcome this problem. In the early stage of the model, $L_{S^0}$, $L_{S^1}$, $L_{pt}$, $L_c$, $L_{ofl}$ are minimized to generate the meaningful camera position. In the next stages, $L_p$ is used in the network to scale down the undefined

**Table 1** Comparison of the proposed model with existing methods based on loss functions

| Method | Contortion | Associativity | Stabilization | FOV |
|---|---|---|---|---|
| [9] | 0.932 | 0.921 | 0.834 | 0.907 |
| [24] | 0.955 | 0.949 | 0.901 | 0.894 |
| [13] | 0.852 | 0.841 | 0.859 | 0.847 |
| [15] | 0.869 | 0.851 | 0.802 | 0.846 |
| [25] | 0.871 | 0.862 | 0.881 | 0.854 |
| **DIVS** | **0.978** | **0.969** | **0.938** | **0.949** |

image regions in the video frame. In the final stages, $L_f$ is used that can enhance the output video frame quality. The model is trained on approximately 500 iterations in each stage. The generalization of the model is improved by applying the image augmentations with an angular change of $\pm$ 10 degrees on the input video data in the test sequence for virtual original pose deviation. The model is compared with the existing techniques based on the above-discussed loss functions and the results are given in Table 1.

## 4 Experiments and Results

In this sub-section of the paper, the DIVS module is proposed that achieves state-of-the-art results in the quantitative and qualitative analysis and the experimental study. The proposed model is tested on its effectiveness of the proposed framework. The dataset used in building the DIVS module contains 265 videos in various surrounding conditions like illuminations, occlusions, random motions, and people in the frame. The video dataset that is curated is captured using the Google Pixel 5 camera with 45 fps. The resolution of the input video is 2160 $\times$ 840 pixels. The dataset is split into three parts in the ratio of 80%, 10%, 10% for training, validation, and testing, respectively. The visual comparison of the DIVS with other existing works in the complex camera motions is given in Table 2. Many recent algorithms are unable to give high-quality videos after video stabilization. The existing works are post-processing and hardware-based techniques that are computationally intensive. It is very difficult to embed the existing hardware-based devices in a consumer-based camera. Real-time processing-based techniques are essential for a consumer-based camera that can stabilize the video at that instance. DIVS is a exemplar software-based module that can process in real time taking very less computation and can also be deployed easily into the consumer camera. The comparative outputs of different models are shown in Fig. 4.

**Fig. 4** Sample video frame output of the proposed model

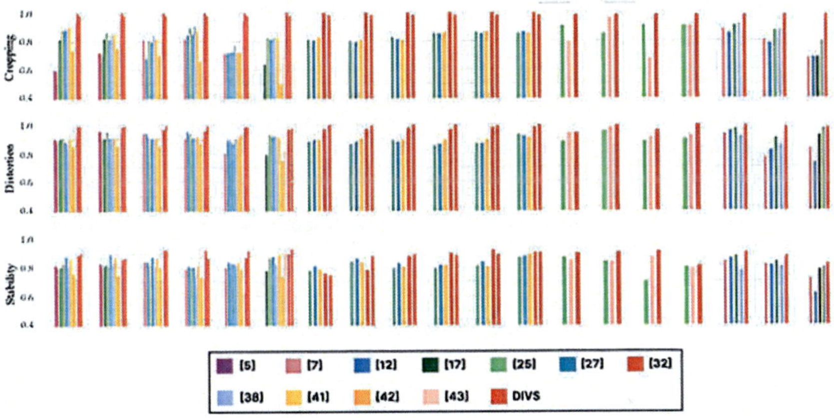


**Fig. 5** Comparative graphical representation of accuracies

As the motion patterns in the video can be converted into the kindred relative poses, this will be easier for the model to be robust from the rotation deviations and the absolute poses. The quantitative comparison of the existing works with the proposed model is shown in Fig. 5. Three different metrics (cropping, stability, distortion) are considered for comparison. The higher the metric value, the better the model.

## 5 Conclusion and Future Works

In this study, the DIVS module is proposed for video stabilization based on the optical flow analysis and the gyroscopic data. The proposed module is a software-based model that can be easily deployable in the consumer camera. The proposed model is capable of stabilizing the high-quality video outputs. The proposed module DIVS achieves high performance because of the effective usage of the joint motion representation. The relative motion along with multi-stage training in the DIVS makes it robust to stabilize videos in diverse surroundings too. In contrast to the image-based methods, this proposed framework uses the external sensor data that is most commonly available in present day devices. The experiments conducted on this model show the robustness and effective video stabilization capability. The FOV representation is not supported by the DIVS module. This can be surpassed by adjusting the loss function accordingly. Extending the video stabilization to the relative motion of parallax posing can be an interesting future scope of work. However, this work can also be extended in such a way in future. This proposed model architecture can also be trained using H&E-stained images. However, this paper focuses on the early detection of gastric cancer, so such an approach is not included here. The computation time required by this model is 2.3 s which is significantly less than the existing works. This model is independent of the quality of WLE image, since the patch-based classification is carried out.

## References


1. Guilluy W, Oudre L, Beghdadi A (2021) Video stabilization: overview, challenges and perspectives. Signal Process Image Commun 90:116015. https://doi.org/10.1016/j.image.2020.116015
2. Greeta S, Raajan NR, Meenu M (2019) Video stabilization. In: 2019 innovations in power and advanced computing technologies (i-PACT), pp 1–4. https://doi.org/10.1109/iPACT44901.2019.8959991
3. Lee YG (2022) Low memory access video stabilization for low-cost camera SoC. Sensors (Basel). 22(6):2341. https://doi.org/10.3390/s22062341. PMID: 35336512; PMCID: PMC8949514
4. Survey on digital video stabilization: concepts, methods, and challenges. ACM Comput Surv (2023). https://dl.acm.org/doi/10.1145/3494525
5. Yu J, Ramamoorthi R (2019) Robust video stabilization by optimization in CNN weight space. In: CVPR
6. Bross B, Wang Y-K, Ye Y, Liu S, Chen J, Sullivan GJ, Ohm J-R (2021) Overview of the versatile video coding (VVC) standard and its applications. IEEE Trans Circuits Syst Video Technol 31:3736–3764
7. Lee YG (2019) Fast global motion estimation on single instruction multiple data processors for real-time devices. Electron Imaging 58:113105
8. Guo H, Liu S, Zhu S, Shen HT, Zeng B (2018) View-consistent MeshFlow for stereo-scopic video stabilization. IEEE Trans Comput Imaging 4(4):573–584
9. Raut S et al (2019) Real-time high-resolution video stabilization using high-frame-rate jitter sensing. ROBOMECH J (SpringerOpen). https://doi.org/10.1186/s40648-019-0144-z
10. Nie Y, Su T, Zhang Z, Sun H, Li G (2018) Dynamic video stitching via shakiness removing. IEEE Trans Image Process 27(1):164–178
11. Sen-Zhe X, Jun H, Wang M, Tai-Jiang M, Shi-Min H (2018) Deep video stabilization using adversarial networks. Comput Graph Forum 37:267–276
12. Wang M, Yang G-Y, Lin J-K, Zhang S-H, Shamir A, Lu S-P, Hu S-M (2019) Deep online video stabilization with multi-grid warping transformation learning. IEEE Trans Image Process

13. Wang Z, Zhang L, Huang H (2018) High-quality real-time video stabilization using trajectory smoothing and mesh-based warping. IEEE Access 6:25157–25166
14. Raptis GE, Fidas C, Avouris N (2018) Effects of mixed-reality on players behaviour and immersion in a cultural tourism game: a cognitive processing perspective. Int J Human Comput Stud 114:69–79
15. Kamranian Z, Sadeghian H, Nilchi ARN, Mehran M (2021) Fast, yet robust end-to-end camera pose estimation for robotic applications. Appl Intell 51(6):3581–3599
16. Walter H, Li R, Munafo J, Curry C, Peterson N, Stoffregen T (2019) APAL coupling study
17. Jia C, Evans BL (2012) Probabilistic 3-D motion estimation for rolling shutter video rectification from visual and inertial measurements. In: Proceedings of the IEEE international workshop on multimedia signal processing (MMSP), pp 203–208
18. Grundmann M, Kwatra V, Essa I (2011) Auto-directed video stabilization with robust L1 optimal camera paths. In: Proceedings of the IEEE conference on computer vision and pattern recognition (CVPR), pp 225–232
19. Aguilar WG, Angulo C (2016) Real-time model-based video stabilization for microaerial vehicles. Neural Process Lett 43(2):459–477
20. Tareen SAK, Saleem Z (2018) A comparative analysis of SIFT SURF KAZE AKAZE ORB and BRISK. In: 2018 international conference on computing mathematics and engineering technologies (iCoMET), pp 1–10
21. Noble FK (2016) Comparison of opencv's feature detectors and feature matchers. In: 2016 23rd international conference on mechatronics and machine vision in practice (M2VIP), pp 1–6
22. Casser V, Pirk S, Mahjourian R, Angelova A (2019) Depth prediction without the sensors: leveraging structure for unsupervised learning from monocular videos. AAAI 33:8001–8008
23. Gordon A, Li H, Jonschkowski R, Angelova A (2019) Depth from videos in the wild: unsupervised monocular depth learning from unknown cameras. In: ICCV, pp 8977–8986
24. Wang M, Yang G-Y, Lin J-K, Zhang S-H, Shamir A, Lu S-P, Hu S-M (2018) Deep online video stabilization with multi-grid warping transformation learning. In: IEEE TIP
25. Guervós E, Ruiz JJ, Pérez P, Muñoz JA, Díaz C, García N (2019) Using 360 VR video to improve the learning experience in veterinary medicine university degree. In: Electronic imaging. Society for imaging science and technology, pp 217-1–217-7
26. Fang M, Li H, Si S (2018) A video stabilization algorithm based on affine SIFT. In: 2018 international conference on computing, mathematics and engineering technologies (iCoMET), Sukkur, pp 1–4
27. Minh T, Hong M (2018) Video stabilization using feature-based classification. In: 2018 IEEE international conference on consumer electronics-Asia (ICCE-Asia), Jeju, 2018, pp 206–212
28. Huang TS, Netravali AN (2002) Motion and structure from feature correspondences: a review. In: Advances in image processing and understanding: a Festschrift for Thomas S Huang, pp 331–347
29. Luchetti A, Tomasin P, Fornaser A, Tallarico P, Bosetti P, De Cecco M (2019) The human being at the center of smart factories thanks to augmented reality. In: 2019 IEEE 5th international forum on research and technology for society and industry (RTSI), pp 51–56
30. Milgram P, Kishino F (1994) A taxonomy of mixed reality visual displays. IEICE Trans Inf Syst 77:1321–1329
31. Butaslac IIII, Luchetti A, Parolin E, Fujimoto Y, Kanbara M, De Cecco M, Kato H (2020) The feasibility of augmented reality as a support tool for motor rehabilitation. In: International conference on augmented reality, virtual reality and computer graphics. Springer, pp 165–173
32. Litleskare S, Calogiuri G (2019) Camera stabilization in 360° videos and its impact on cyber sickness, environmental perceptions, and psychophysiological responses to a simulated nature walk: a singleblinded randomized trial. Front Psychol 10:2436
33. Zhang G, Qin X, Hua W, Wong T-T, Heng P-A, Bao H (2007) Robust metric reconstruction from challenging video sequences. In: Proceedings of the IEEE conference on computer vision and pattern recognition (CVPR), pp 1–8

34. Yan C, Xie H, Yang D, Yin J, Zhang Y, Dai Q (2018) Supervised hash coding with deep neural network for environment perception of intelligent vehicles. IEEE Trans Intell Transp Syst 19(1):284–295
35. Dong J, Liu H (2017) Video stabilization for strict real-time applications. IEEE Trans Circuits Syst Video Technol 27(4):716–724
36. Kejriwal L, Singh I (2016) A hybrid filtering approach of digital video stabilization for UAV using Kalman and low pass filter. Procedia Comput Sci 93:359–366
37. Bian J, Li Z, Wang N, Zhan H, Shen C, Cheng M-M et al (2019) Unsupervised scale-consistent depth and egomotion learning from monocular video. NeurIPS, pp 35–45
38. Choi J, Kweon IS (2019) Deep iterative frame interpolation for full-frame video stabilization. In: SIGGRAPH Asia
39. He K, Zhang X, Ren S, Sun J (2016) Deep residual learning for image recognition. In: CVPR, pp 770–778
40. Liu S, Li M, Zhu S, Zeng B (2017) Codingflow: enable video coding for video stabilization. IEEE TIP 26(7):3291–3302
41. Liu S, Tan P, Yuan L, Sun J, Zeng B (2016) Meshflow: minimum latency online video stabilization. In: ECCV, pp 800–815
42. Luo X, Huang J-B, Szeliski R, Matzen K, Kopf J (2020) Consistent video depth estimation. ACM TOG 39(4)
43. Ranjan A, Jampani V, Balles L, Kim K, Sun D, Wulff J et al (2019) Competitive collaboration: joint unsupervised learning of depth camera motion optical flow and motion segmentation. In: CVPR 2019-June, no. 1, pp 12232–12241

# Saliency-Driven 'Evidence CNN' for Fine-Grained Recognition of Twins

**Razia de Loyola Furtado e Sardinha**

**Abstract** The extremely subtle differences between a pair of twins make their individual recognition an interesting fine-grained classification challenge. When the images are clicked in the wild, the confidence score of predictions is generally low. Most of the previous work in this domain focuses on improving the accuracy of the model. There is no check on whether the correct evidence was considered while making a prediction. Without such a check, the predictions cannot be trusted, especially when the confidence score is low. The objective of our research is to use visual attribution techniques like image saliency to understand, guide and improve a model's recognition. We propose a saliency-driven fine-grained classifier that learns the most discriminative evidence for each twin during training, and uses it to train an 'Evidence CNN'. At test-time, the Evidence CNN validates the predictions, by matching the coherence of the evidence used to make the prediction, with the evidence detected for correct predictions of the same class during training. We used the Guided Grad-CAM method for generating saliency evidence, and provide the theoretical justification for selecting this technique. We tested our model on a dataset of 700 public images of the celebrity Olsen twins, 'Ashley' and 'Mary-Kate'. With saliency-driven prediction refinement, our classifier showed a significant improvement in the test accuracy. Since our predictions are grounded on explainable evidence that is cross-checked at test-time, our model can be considered trustworthy.

**Keyword** Fine-grained recognition · Twins · Saliency · Explainable predictions

R. de Loyola Furtado e Sardinha (✉)
Department of Information Technology, Padre Conceicao College of Engineering, Verna, Goa, India
e-mail: razia@pccegoa.edu.in



S. Borah et al. (eds.), *Advanced Computational and Communication Paradigms*,
Lecture Notes in Networks and Systems 535,
https://doi.org/10.1007/978-981-99-4284-8_29

# 1 Introduction

Identifying twins in natural images is an interesting computer vision challenge. It is a fine-grained recognition task where the inter-class differences between the objects (i.e. twins) are extremely small. Deep learning models have been popularly used for fine-grained object recognition [1, 2]. However, when natural images are involved, the accuracy drastically drops [2]. Convolutional Neural Networks (CNN) learn features in a hierarchical, feed-forward fashion and hence critical base-level features become indistinct as the learning progresses [3, 4]. With fine-grained objects, this can harshly affect the accuracy of the model. It is thus important to capture the features at different layers and at multiple scales. Consequently, this can lead to overfitting, especially if the training dataset is small or not representative of the important features of all its classes [5].

Besides accuracy, the trust factor of the CNN is also important, particularly when dealing with applications like auto-tagging photos, crime prevention, activity monitoring, etc. [6] Trust is best ascertained when there is a good human understanding of why a prediction was made [6, 7]. It thus makes sense to investigate a test prediction for conclusive evidence, instead of blindly accepting it [8]. Attribution techniques like saliency maps help to visually understand how each input pixel contributes to a particular prediction [9, 10].

In this paper, we aim to provide 'correct' and 'trustworthy' fine-grained recognition of twins. The general working of our model is illustrated in Fig. 1. We first train a conventional *'Base CNN'* with the training set of the twins' images. For each correct prediction, we detect the corresponding image's salient region as evidence. This information is used to train an *'Evidence CNN'*. The training evidence is augmented by progressively mining the most salient feature regions by improvising on adversarial erasing [11]. At test-time, the Evidence CNN verifies the predictions by matching the coherence between the saliency for the test image, and the corresponding training evidence for correct predictions of that class.

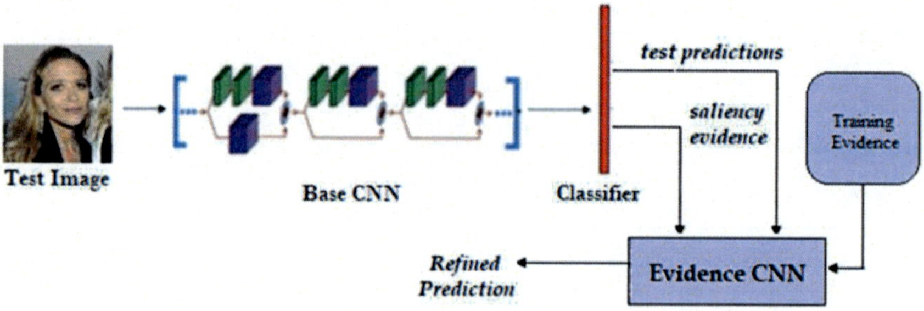


**Fig. 1** General outline. The *Base CNN* is trained to recognize the *Olsen* twins. The *Evidence CNN* validates and refines the test predictions based on saliency evidence

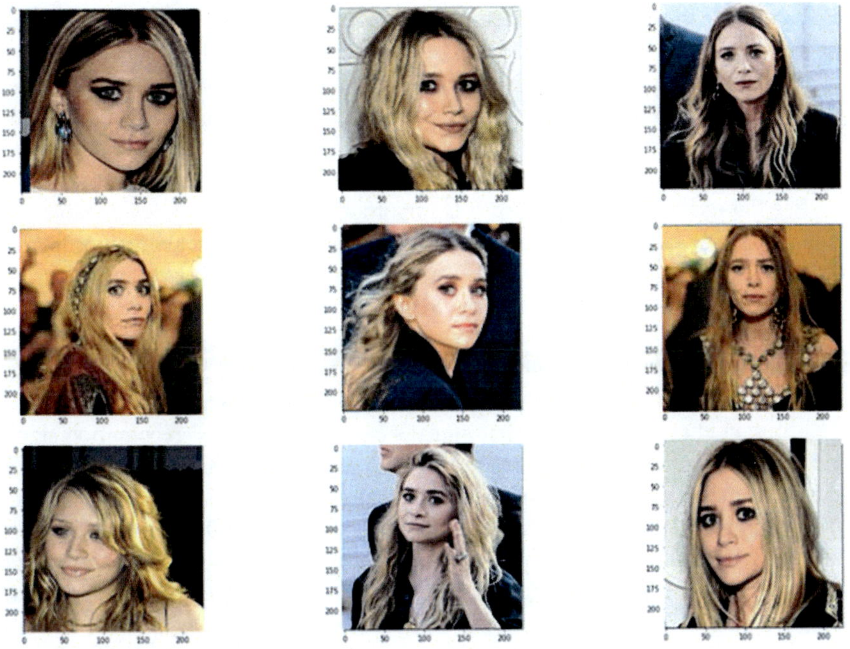

**Fig. 2** Dataset images of '*Olsen* twins'

Majority of the research involving image saliency focuses on generating precise and minimal saliency maps that best explain a prediction [9]. Our research explores the end-to-end functionality of saliency maps, from training the model, to improving fine-grained recognition, to providing justifiable test predictions. Our model is simple to implement and does not require either part-annotated training images or hard-coded attention modules. To train our model, we annotated 700 images from the public digital photos of the celebrity twins Ashley and Mary-Kate Olsen. The images are natural and the twins are clicked at different ages, in varying poses and with diverse backgrounds. We only considered images where the twins are aged approximately sixteen years and above. A few images from our dataset are shown in Fig. 2. The dataset is uploaded on Kaggle and is available by request.

## 2 Related Work

Most classifiers for twins' recognition use facial similarity as the deciding factor. Some notable methods have been proposed in [12–14] where non-CNN approaches use a fusion of facial feature extractors like Principal Component Analysis (PCA), Histogram of Gradients (HOG), Local Binary Patterns (LBP), etc. These methods perform poorly when the faces are clicked at odd angles. LBP considers small

$3 \times 3$ neighbourhoods to capture similarity and cannot be used to capture the dominant features at larger scales. CNN-based feature extraction is a more generic and effective approach. The survey by Hu et al. [15] discusses the various aspects of face recognition using CNNs and highlights the mechanisms used to control the explosion of parameters and overfitting. Many of these solutions come at the cost of accuracy, especially when natural images are used.

In a fine-grained task like ours, it is important that critical low-level features are not lost during training. Rather, they must be learned. Multi-scale feature learning techniques are explored in [16, 17]. Part-based models [2, 18] work well for tasks like classification of bird species, dog breeds, etc. where the individual features like beak length, head shape, etc. can be visualized as separatable parts. In our task, the individual physiological features are generally similar for both classes. It is more important to learn the correlation between the features of each twin [13].

Mousavi et al. [19] observe that humans can perform twins' recognition significantly well if given enough time. Their study suggests that humans look for facial marks like moles or scars, nose length, eye colour, etc. and fail in their predictions if the major discriminating features are not visible. In [3, 20] the authors propose human-in-the-loop, guided training models where the CNN is fine-tuned after manual segregation of the images where the distinct discriminating features are prominently visible. Besides being tedious to implement and prone to human errors, there is no provision to verify whether a prediction occurs based on the discriminative features identified by the human.

For improved learning, Lin et al. [21] propose a bilinear CNN model. Sun et al. [22] implemented an attention module that learns to localize and correlate different object parts during training. In Fu et al. [23], the model learns where to focus by recurrently zooming into a location by using a recurrent-attention CNN. Although effective in terms of accuracy, these approaches do not correlate the test predictions with the training evidence for correct predictions.

Our objective is to improve fine-grained recognition with explainable predictions. Saliency maps highlight the most-informative pixels in an image and can be visualized and interpreted [9, 10]. Flores et al. [24] show how better saliency maps yield an improved prediction accuracy. Class Activation Mapping (CAM) and its generalizations like Guided Grad-CAM [25] combine the fine-grained details of Guided Backpropagation with the course localization advantages of Grad-CAM. We use Guided Grad-CAM for the saliency detection. The main advantage of our implementation is that along with saliency-driven model training, we will also provide saliency-validated and refined test-time predictions.

## 3 Methodology

Our implementation is logically divided into two main modules, one for training the *Base CNN* and detecting image saliency, and the other for training the *Evidence CNN* for test prediction validation and refinement.

### 3.1 Detecting Saliency Evidence

The *Base CNN* is trained with the training set images. The output is a Boolean $\in \{0,1\}$ and specifies the class conditional probability of the prediction. We find the top-k saliency regions for the correct predictions using Guided Grad-CAM [25] technique.

Guided Grad-CAM combines Guided Backpropagation [26] with up-sampled Grad-CAM attributions. Guided Backpropagation computes the gradient of the output $f_c(x)$ w.r.t. the input, except that, when propagating through the ReLU functions, only non-negative gradients are backpropagated. Grad-CAM computes the gradients of the output $f_c(x)$ with respect to the feature map activations at a layer. The gradients are averaged for each channel, along with the width and height dimensions, to compute the importance weights. The average gradient for each channel is multiplied by the layer activations and the results are summed over all channels. Thus, Grad-CAM highlights coarse regions of the image that have a positive contribution to $f_c(x)$. As seen in Eq.1, Guided Grad-CAM is the element-wise product of up-sampled Grad-CAM attributions and Guided Backpropagation;

$$\Phi_{GuidedGrad-CAM}(f_c, x) = \Phi_{Grad-CAM}(f_c, x) \odot \Phi_{GuidedBackprop}(f_c, x) \quad (1)$$

Hence, we combine the fine-grained details of Guided Backpropagation with the course localization advantages of Grad-CAM for the saliency detection.

### 3.2 Prediction Refinement

Once the peak saliency is detected, we extract a small pixel patch from the image, around the peak saliency. This patch contains the most class-discriminative evidence. Since the next most-discriminative patches may also provide important evidence, we use iterative adversarial erasing [11] to augment the reference pool of evidence. Additional attention maps are obtained for each prediction by adding black pixel squares to the previous most salient region. This method results in 'implicit parts localization' and creates richer evidence for each twin. All the evidence patches are given the same ground-truth label as the original image. We then train the *Evidence CNN* with the weighted evidence patches for both the classes.

At test-time, it can happen that the test prediction score is not very high. This is particularly true when the critical features are not clearly visible. For validating test predictions, we extract a pixel patch around the peak saliency of the test image and use iterative adversarial erasing to acquire more evidence. We then run these weighted evidence patches through the Evidence CNN.

The Evidence CNN refines the top-1 prediction based on the training and test-time evidence. Figure 3 illustrates the prediction refinement step. When a test image is passed through the Base CNN, we get a pair of class conditional probabilities for the test prediction. We detect the saliency evidence for both classes and pass their attention maps through the Evidence CNN. The refined prediction will be inclined toward the best-match class by an amount proportional to how coherent its evidence is, with the evidence for correct predictions during training. For prediction refinement, we aggregate the class conditional probability vectors with weights proportional to the saliency of each attention map. The class having the maximum aggregate prediction for the weighted combination is selected as the 'refined' class prediction.

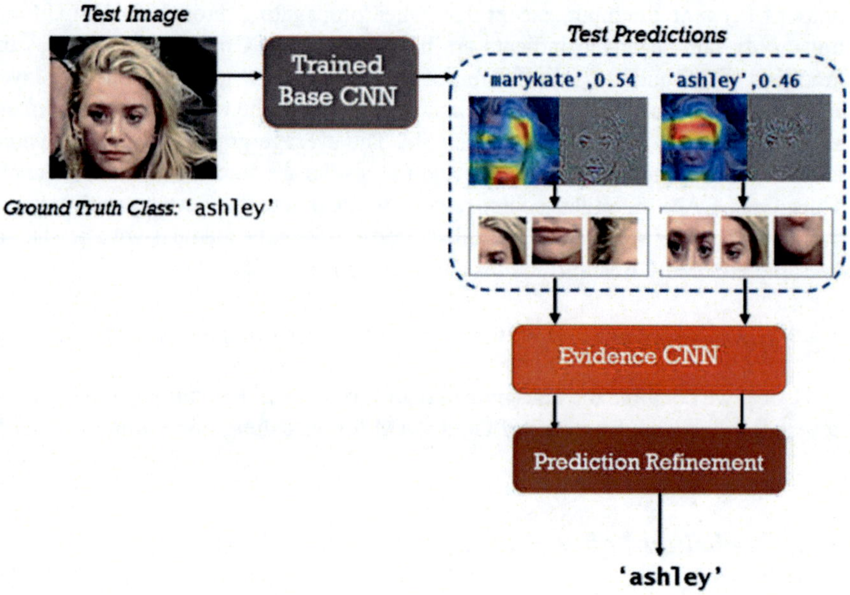


**Fig. 3** *Evidence CNN* refines test predictions based on the coherence between the test prediction's saliency and training evidence

## 4 Experiments

We created an annotated dataset of 700 publicly available, natural images of the celebrity twins Ashley and Mary-Kate Olsen. We segregated the images into 480 for training, 120 for validation and 100 for testing.

The Olsen twins are not exactly identical but sororal twins, i.e. the female version of fraternal twins [3]. By studying the twins' images, we observed that there were a few striking differences between them. However, our dataset had very few images in which the differences were clearly visible. The number of easy examples was small compared to the hard examples. We decided to use the saliency evidence of the strong predictions to fine-tune the model. This stops the hard examples from overwhelming the easy examples, and consequently enhances the learning of critical differences.

We used MobileNet_v2 [27] for the *Base* and *Evidence CNNs,* as it learns faster than other CNNs, with comparable performance. We trained the Base CNN with images resized to 224x224 pixels, with Adam adaptive optimizer, and batch size of 32 for 50 epochs. We found that Adam works best for our dataset as the learning rate automatically adapts to the parameters by performing small updates for frequently occurring features and larger updates for the rarer ones. The network can thus capture information from features that are not frequent, and use it as evidence by assigning an appropriate weight. We identified the top-gain images based on binary cross-entropy loss and segregated them. We detected the salient regions in these images using Guided Grad-CAM and extracted 150x150 pixel patches of the peak-saliency regions. We augmented the salient image patches, and used them for fine-tuning the *Base CNN*. With saliency-guided fine-tuning, we were able to achieve an improvement in the train and validation accuracies as seen in Table 1.

We applied the same training parameters for the Evidence CNN. We used 150x150 pixel, peak-saliency patches of the correct predictions. We augmented the network evidence by performing three rounds of weighted adversarial erasing on the salient patches, giving highest weight to the most salient region, and reducing it as the iterations progressed, i.e. w1=0.3, w2=0.2 and w3=0.1. This step implicitly enforces part-annotations as shown in Fig. 4.

We perform test prediction validation by matching the coherence of the test prediction's saliencies for both classes, with their corresponding training evidence. The weighted combinations of three saliency maps for each class are passed through the Evidence CNN. The class with the maximum aggregate prediction is selected as the final prediction.

**Table 1** Accuracy of base CNN

| Method | Train accuracy (%) | Validation accuracy (%) |
|---|---|---|
| Simple training | 70.2 | 61.23 |
| Training after saliency-guided fine-tuning | 75.41 | 69.91 |

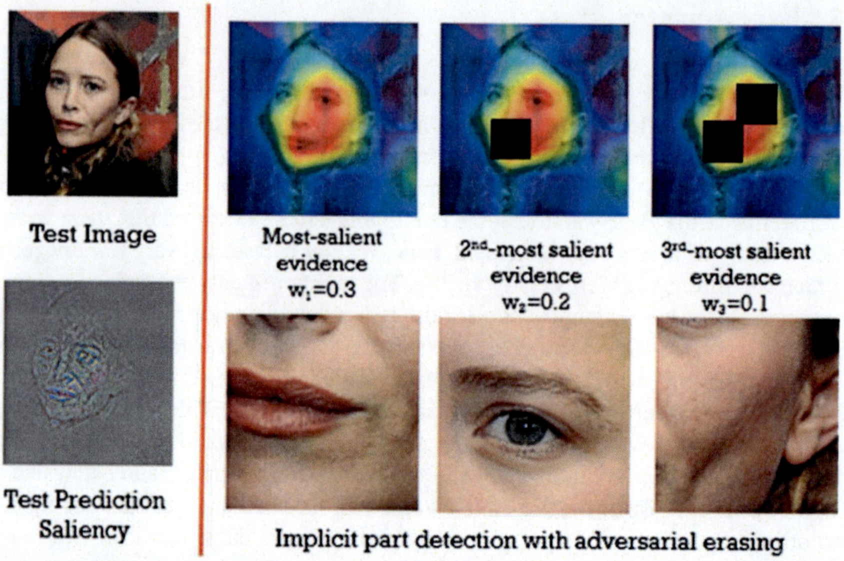


**Fig. 4** Extraction of evidence patches during adversarial erasing [11] and consequent implicit part-annotation

# 5 Results

The test predictions for two test images with are shown in Fig. 5. We can observe the change in confidence score after decision refinement.

The performance of our model before and after decision refinement is tabulated in Table 2. Decision refinement improves the test accuracy by over 4%. The mAP for *Ashley* is higher than that of *Mary-kate* as the dataset has relatively easier examples for Ashley. After decision refinement, we see a significant increase in the mAP for Mary-kate. This can be attributed to the implicit parts recognition after adversarial

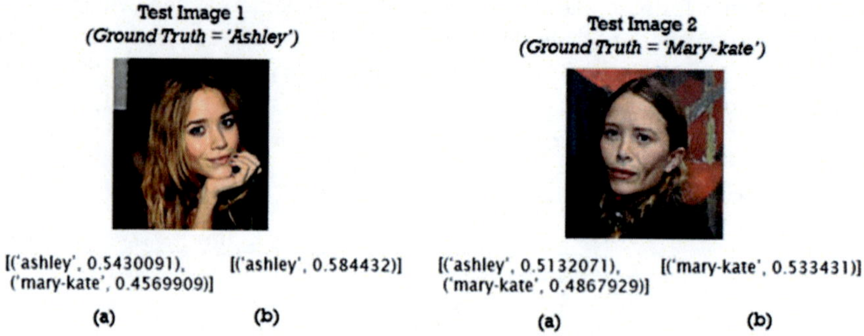


**Fig. 5** Test prediction examples (**a**) Before decision refinement (**b**) After decision refinement

**Table 2** Test accuracy

| Method | Test accuracy (%) | Mean average precision (mAP) | |
|---|---|---|---|
| | | 'Ashley' | 'Mary-kate' |
| Without decision refinement | 62% | 0.6414 | 0.5955 |
| With decision refinement | 66% | 0.6851 | 0.6616 |

erasing. Mary-kate's features are more distinct, and hence get better highlighted. A few test results before and after decision refinement are presented in Fig. 6. We can observe a noteworthy improvement in the conditional probability of the correct class.

## 6 Conclusion and Future Scope

Our paper proposes an explainable saliency-driven model for recognition of identical twins. We have a *Base CNN* that is fine-tuned using the salient regions of the top-gain training images. Fine-tuning improves the train and validation accuracy of the *Base CNN*. The salient image regions are also used to train our *Evidence CNN*. At test-time, the test predictions from the Base CNN are refined, by matching the coherence of their evidence, with the training evidence for correct predictions. We theoretically justify the use of Guided Grad-CAM [25] for saliency detection and adversarial erasing [11] for augmenting the network evidence. We created our own dataset and conducted an empirical study using a simple MobileNetv2 for both, the *Base* and *Evidence CNN*s. Our model showed a significant improvement in accuracy after evidence-driven prediction refinement. By using complex CNN architectures and experimenting with larger datasets, we can evaluate the generic performance of our model for other real-world applications that require reliable predictions. This is the future scope of our research.

(a)

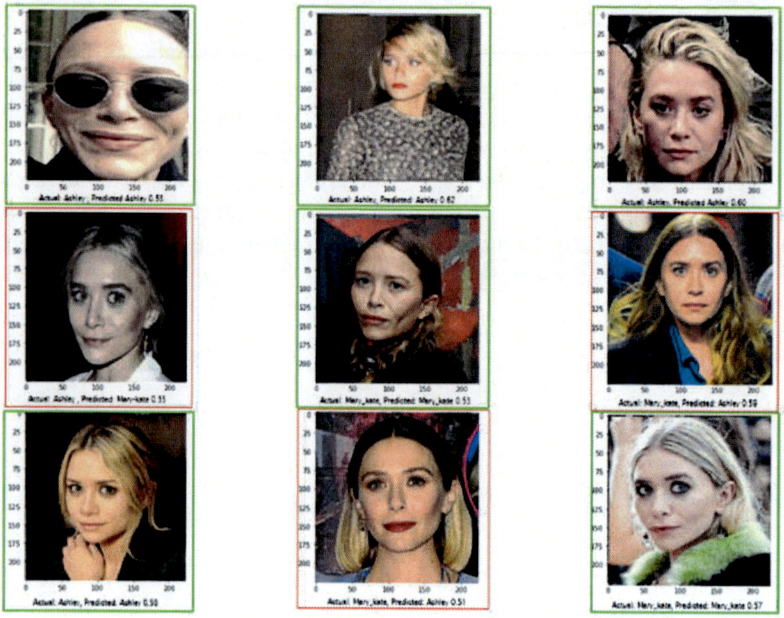


(b)

**Fig. 6** Test predictions (**a**) Before decision refinement (**b**) After decision refinement. The correct predictions are marked with green boxes and wrong predictions in red boxes

## References


1. Wang D, Shen Z, Shao J, Zhang W, Xue X, Zhang Z (2015) Multiple granularity descriptors for fine-grained categorization. In: Proceedings of the IEEE international conference on computer vision. pp 2399–2406
2. Wang M, Deng W (2018) Deep face recognition: a survey. arXiv preprint arXiv:1804.06655
3. Krause J, Gebru T, Deng J, Li LJ, Fei-Fei L (2014) Learning features and parts for fine-grained recognition. In: 2014 22nd International conference on pattern recognition. IEEE, pp 26–33
4. e Sardinha RDLF (2019) Twins' identification using single-shot detector. In: 2019 Fifth international conference on image information processing (ICIIP). IEEE, pp 521–525
5. Cai S, Zuo W, Zhang L (2017) Higher-order integration of hierarchical convolutional activations for fine-grained visual categorization. In: Proceedings of the IEEE international conference on computer vision. pp 511–520
6. Hu T, Qi H, Huang Q, Lu Y (2019) See better before looking closer: Weakly supervised data augmentation network for fine-grained visual classification. arXiv preprint arXiv:1901.09891
7. RichardWebster B, Hu B, Fieldhouse K, Hoogs A (2022) Doppelganger saliency: towards more ethical person re-identification. In: Proceedings of the IEEE/CVF conference on computer vision and pattern recognition. pp 2847–2857
8. Xu F, Uszkoreit H, Du Y, Fan W, Zhao D, Zhu J (2019) Explainable AI: a brief survey on history, research areas, approaches and challenges. In: CCF international conference on natural language processing and Chinese computing. Springer, Cham, pp 563–574
9. Bargal SA, Zunino A, Petsiuk V, Zhang J, Saenko K, Murino V, Sclaroff S (2018) Guided zoom: questioning network evidence for fine-grained classification. arXiv preprint arXiv:1812.02626
10. Bylinskii Z, Recasens A, Borji A, Oliva A, Torralba A, Durand F (2016) Where should saliency models look next?. In: European conference on computer vision. Springer, Cham, pp 809–824
11. Recasens A, Kellnhofer P, Stent S, Matusik W, Torralba A (2018) Learning to zoom: a saliency-based sampling layer for neural networks. In: Proceedings of the European conference on computer vision (ECCV). pp 51–66
12. Wei Y, Feng J, Liang X, Cheng MM, Zhao Y, Yan S (2017) Object region mining with adversarial erasing: a simple classification to semantic segmentation approach. In: Proceedings of the IEEE conference on computer vision and pattern recognition. pp 1568–1576
13. Phillips PJ, Flynn PJ, Bowyer KW, Bruegge RWV, Grother PJ, Quinn GW, Pruitt M (2011) Distinguishing identical twins by face recognition. In: 2011 IEEE International conference on automatic face & gesture recognition (FG). IEEE, pp 185–192
14. Paone JR, Flynn PJ, Philips PJ, Bowyer KW, Bruegge RWV, Grother PJ, Quinn GW, Pruitt MT, Grant JM (2014) Double trouble: Differentiating identical twins by face recognition. IEEE Trans Inf Forensics Secur 9(2):285–295
15. Afaneh A, Noroozi F, Toygar Ö (2017) Recognition of identical twins using fusion of various facial feature extractors. EURASIP J Image Video Proc 2017(1):1–14
16. Hu G, Yang Y, Yi D, Kittler J, Christmas W, Li SZ, Hospedales T (2015) When face recognition meets with deep learning: an evaluation of convolutional neural networks for face recognition. In: Proceedings of the IEEE international conference on computer vision workshops. pp 142–150
17. Duan Y, Lu J, Feng J, Zhou J (2017) Context-aware local binary feature learning for face recognition. IEEE Trans Pattern Anal Mach Intell 40(5):1139–1153
18. Guo C, Fan B, Zhang Q, Xiang S, Pan C (2020) Augfpn: Improving multi-scale feature learning for object detection. In: Proceedings of the IEEE/CVF conference on computer vision and pattern recognition. pp 12595–12604
19. Wang M, Deng W (2021) Deep face recognition: a survey. Neurocomputing 429:215–244
20. Mousavi S, Charmi M, Hassanpoor H (2021) Recognition of identical twins based on the most distinctive region of the face: human criteria and machine processing approaches. Multimedia Tools Appl 80(10):15765–15802

21. Biswas S, Bowyer KW, Flynn PJ (2011) A study of face recognition of identical twins by humans. In: 2011 IEEE international workshop on information forensics and security. IEEE, pp 1–6
22. Lin TY, RoyChowdhury A, Maji S (2015) Bilinear CNN models for fine-grained visual recognition. In: Proceedings of the IEEE international conference on computer vision. pp 1449–1457
23. Sun M, Yuan Y, Zhou F, Ding E (2018) Multi-attention multi-class constraint for fine-grained image recognition. In: Proceedings of the European conference on computer vision (ECCV). pp 805–821
24. Fu J, Zheng H, Mei T (2017) Look closer to see better: Recurrent attention convolutional neural network for fine-grained image recognition. In: Proceedings of the IEEE conference on computer vision and pattern recognition. pp 4438–4446
25. Flores CF, Gonzalez-Garcia A, van de Weijer J, Raducanu B (2019) Saliency for fine-grained object recognition in domains with scarce training data. Pattern Recogn 94:62–73
26. Xu S, Chang D, Xie J, Ma Z (2021) Grad-CAM guided channel-spatial attention module for fine-grained visual classification. In: 2021 IEEE 31st International Workshop on Machine Learning for Signal Processing (MLSP). IEEE, pp 1–6
27. Sandler M, Howard A, Zhu M, Zhmoginov A, Chen LC (2018) Mobilenetv2: Inverted residuals and linear bottlenecks. In: Proceedings of the IEEE conference on computer vision and pattern recognition. pp 4510–4520

# IoT-Based Landslide Monitoring System Using LiDAR Sensors: A Review

**Bishal Gupta, Kalpana Sharma, Tawal Koirala, and Sital Sharma**

**Abstract** Landslides are gravitational movements of rock and soil down slopes that have an effect on the ecology. They seriously harm both property and human life and are a typical natural occurrence on the globe. The most widely used sensor technology is Wireless Sensor Network (WSN), which enables broad monitoring and landslide prediction. In this study, we reviewed a number of publications on ground monitoring methods using a variety of sensors, including LiDAR sensors, as well as some significant advancements that help to improve the architecture and satisfy user demands. A cost-effective ground monitoring technique has been studied for several landslide warning systems.

**Keywords** WSN · LiDAR · Landslide · Sensors · Monitoring

## 1 Introduction

In the Himalayas and other mountainous regions, landslides are one of the natural risks. They are described as the occurrences of land or rock sliding down a slope, which are caused by the disruption of soil stability and are one of the regular natural disasters during the wet season. The activity of structural, lithological, geomorpho-logical, climatic, environmental, hydrological, seismological, and other factors in the

B. Gupta · K. Sharma (✉) · T. Koirala (✉) · S. Sharma (✉)
Department of Computer Science and Engineering, Sikkim Manipal Institute of Technology, Sikkim Manipal University, Majhitar, Sikkim 737136, India
e-mail: kalpana.s@smit.smu.edu.in

T. Koirala
e-mail: koirala01@gmail.com

S. Sharma
e-mail: sitalneo@gmail.com

B. Gupta
e-mail: bishal_202127004@smit.smu.edu.in



S. Borah et al. (eds.), *Advanced Computational and Communication Paradigms*, Lecture Notes in Networks and Systems 535,
https://doi.org/10.1007/978-981-99-4284-8_30

**Table 1** Landslide monitoring techniques [2]

| Techniques for monitoring landslides | | |
|---|---|---|
| Earth-based sensing | | Remote sensing |
| S. No. | Methods | Methods |
| 1 | Rain gauge | UAV-DP |
| 2 | Accelerometer/Inclinometer | SAR |
| 3 | Wire extensometer | GB-INSAR |
| 4 | GPS | IRT |
| 5 | Piezometers | TLS |
| 6 | Acoustic | Optical VHR |
| 7 | Laser | |
| 8 | Pressure/Strain gauge | |

affected region may also produce landslides [1]. Due to the unpredictability and short lifespan of landslides, technology that allows those in charge of disaster mitigation to continually monitor the elements that cause landslides is essential [2]. For various case studies throughout the globe, researchers have done landslide prediction, detection, and monitoring. Landslides may be found using a variety of approaches, such as visual examination using image/video processing, satellite remote sensing, statistical techniques, or machine learning algorithms. Data-driven methods for detecting landslides may be implemented via wireless sensor networks [3]. In landslide-prone locations, it is necessary to set up Early Warning Systems that can recognise and warn people of the potential for a landslide. Numerous methods have been developed and proposed to do this. Thanks to its network of wireless sensors, a system based on a wireless sensor network may detect any anomaly in a specific area [4]. LiDAR models the surface and looks for recent landslide activity using topographical imagery [5]. The condition of mine slopes has been closely monitored, and warnings regarding rock falls have been sent frequently using digital video cameras. As shown in Table 1, there are two major categories for landslide monitoring: ground-based sensing and remote-based sensing. It is extremely difficult for developing countries to use most of these landslide monitoring systems since they demand expensive instrumentation and upkeep. This demonstrates the necessity for an inexpensive technology that can identify landslides in real time and alert those in danger so that appropriate action can be taken [6].

## *1.1 Detection and Ranging Based on Laser*

LiDAR uses a laser to target an item or a surface, then measures how long the reflected light takes to reach the receiver. By altering the light's wavelength, it may also be used to create digital 3D images of regions on the surface of the earth and the ocean's bottom that are intertidal and close to the shore. It may be used for terrestrial, aerial, and mobile purposes [7]. It is a unique process that combines 3D scanning with laser scanning and is known as 3D laser scanning. To create high-resolution maps, LiDAR is often employed. To image things, LiDAR employs ultraviolet, visible, or near-infrared light. It might be directed towards non-metallic items, rocks, rain, aerosols, chemical substances, clouds, or even single molecules. In order to determine the distance to a surface or object, a LiDAR uses the formula $D = c.t/2$ where c is the speed of light, d is the distance between the detector and the surface or object being detected, and t is the amount of time it takes for the laser light to travel to the target and return to the detector. Narrow laser beams enable the creation of high-resolution maps of physical characteristics [8].

Remote sensing has several advantages over traditional survey methods, which is why it is increasingly being used in the discipline of geological engineering. They reduce danger, save time and effort, enhance the quality and dependability of data, and even save the resultant scans or photographs for future use [9]. Remote sensing technology has advanced tremendously during the last several decades. Recent advancements in scanning data processing and scanning sensor electronics have increased the applicability of these methods [10]. One of the two main remote sensing methods that is quickly growing in the study of landslides and rock falls is interferometric synthetic aperture radar (InSAR), along with light detection and ranging (LiDAR) [11].

InSAR is a geodesy radar technique for remote sensing. This geodesic approach produces maps of surface deformation or digital elevation based on variations in the phase of the waves returning to the satellite using two or more synthetic aperture radar (SAR) photos. A target is irradiated with pulsed laser light in the LiDAR surveying technique, and a sensor records the amount of time it takes for the pulses that are reflected to return to the source. Both of these technologies have the potential to significantly advance geological hazard research and mitigation [12].

LiDAR uses ALS and TLS [13]. The laser scanner sensor location classifies LiDAR. Laser scanner sensor position categorises LiDAR. LiDAR technology uses phase-shift or time-of-flight sensors to build a 3D point cloud.

1. Time of flight LiDAR calculates the distance between the laser source and scanning targets by measuring the light pulse's transit time. LiDAR technology also uses phase shift, which involves emitting modulated light and scanning the light's reflected wave to determine the phase change. The laser scanner then calculates the scanner-target distance [14].

LiDAR is a more accurate long-distance measurement method. A scanning range of several hundred metres allows a single measurement accuracy of 6 mm and a sample resolution of 1 mm. The laser can measure up to 1,000,000 points per second, creating a scan point cloud with millions of data points to represent the surface of scanned objects (Fig. 1).

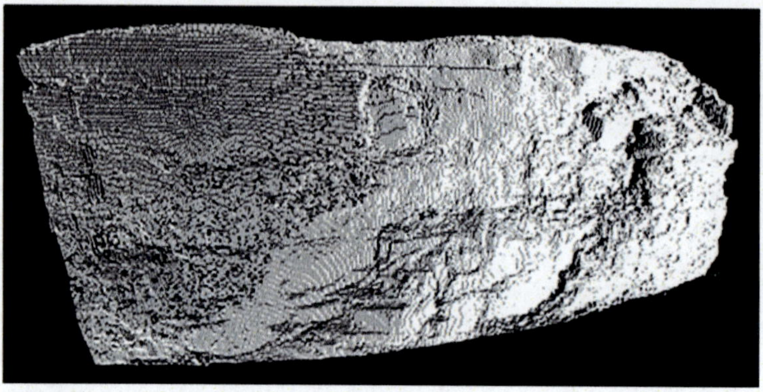

**Fig. 1** A LiDAR scan point cloud example (Kassebaum [15])

## 1.2 Categorisation of the Principal Applications

TLS and ALS can be applied to a variety of investigational tasks, which can be further divided into four categories (Table 1): (1) landslide detection and characterisation; (2) hazard assessment and susceptibility mapping; (3) modelling; and (4) monitoring. This categorisation allows us to classify the study's main objectives, even though there may be overlaps.

## 1.3 Background and Previous Work

LiDAR's landslide studies and modelling are described here. Geological engineering is adopting LiDAR. TLS and ALS are growing [13]. Hazard assessments, hydropower, mining, forensics, history, and geological mapping utilise TLS. LiDAR sensor has been used to study landslides in four areas: scanning movements or volume changes; hazard assessment, susceptibility mapping, and modelling; forecasting and monitoring; and integrating LiDAR with computer simulation methodologies [12].

## 1.4 Scanning Rockfall and Landslide Displacement

The characterisation of landslides using LiDAR has been the subject of many research teams' study initiatives in recent years [16]. These researchers show how the use of LiDAR technology might be highly beneficial for understanding landslide causes, defining borders, and estimating volume change [17]. The majority of study focuses on the deformation of landslide surfaces since LiDAR can only detect surface data.

A circular landslide on silty riverbanks was reported by Michel Jaboyedoff [12]. (Switzerland). This result shows the immense potential of LiDAR technology in the study of landslides.

Miller [18] employed the HDS2500 scanner, which had a 6 mm precision, to detect small changes in slope surfaces (differences of less than 8 cm could be detected over a six-week period). TLS is a useful approach for evaluating embankment stability at small scales, according to the study's conclusions. TLS is a good complement to LiDAR and photogrammetry for integrated risk assessment because of its high spatial and temporal resolutions [18]. The study results presented in this publication have been useful in determining the likely features of early slope deformation. A variety of scanning objectives were set up by the author to monitor changes in volume and displacement and guarantee the validity of the findings. Akca [19] built a reality model with scanning targets to detect slope movement using IDS uEye UI-6240 C gigabit Internet video cameras. The author found that small tennis balls pushed into the ground may quantify surface deformation. Photogrammetry improves landslide dynamics understanding. Artificial targets can monitor landslide surface changes [19]. Barbarella and Fiani [20] suggested GPS receivers and laser scanners (TLS). The study employed Optech Ilris 3D, Leica Scan station C10, and Riegl VZ400. All surveys employed the same absolute reference system using Cyclone, Leica, and RiscanPro to edit scanning data. A 4-month scanning technique estimated the landslide volume change in Pisciotta, Campania, Italy [20].

Maerz and his research team have used Leica Scan Station II and FARO Focus3D in addition to LiDAR. The research showed how LiDAR scanning technology might be used to monitor volume changes and deformation in outcrops [15].

Franz [21] gave a presentation on his 2016 study on the use of LiDAR to track landslides. The inquiry is centred on the covert Peney landslide that occurred close to Geneva, Switzerland. Within and outside of the landslide zone, polystyrene targets were scanned using LiDAR. Within were floating points, while outside were control points. InnovMetric PolyWorks v.12.1's IMAlign modules were utilised by the author to analyse the data, and MATLAB was used to remove vegetation since volume change is difficult to estimate when vegetation is present.

There are several types of studies for which TLS and ALS may be used, and these investigations may be further divided into four categories as indicated in Table 2.

## 1.5 Risk Assessment

Airborne LiDAR has been used to analyse landslide mass movement for a long time due to its accuracy, precision, and large scanning range. ALS is less precise and efficient than terrestrial LiDAR for landslide analysis. The International Earth Rotation and Reference Systems Services (IERS) uses laser range to accurately assess geodetic satellites over 3500 miles distance [22]. TLS reaches an accuracy of 0.3mm [23]. The site-specific modelling is enhanced by TLS. An earth-destabilised flow

**Table 2** A classification of several LiDAR uses in landslide investigations

| | Landslide | Rockfall | Debrisflow |
|---|---|---|---|
| Landslide characterisation and detection | Geomorphic feature mapping | Characterisation and imaging of the rockface. Determine the direction of a discontinuity | Mobilizable volume detection. A hydromorphic description |
| Risk evaluation and vulnerability mapping | Mostly as assistance in mapping | Some hazard and susceptibility mapping efforts (not yet achieved) | Input for hazard mapping using a geomorphological technique |
| Modelling | Large 3D information density is currently beyond the capabilities of traditional modelling tools. HRDEM enhances the geometrical characterisation of landslides, allowing for more precise modelling | DEM has high resolution for modelling trajectories | A modelling input for spreading |
| Monitoring | Tracking surface displacements. Budget for the volume | Tracking changes in surface displacements. detection of displacements prior to failure (under development). the measurement of rockfall activity (volumes) | Budget for emulsion. tracking the channel's morphological changes |

zone was identified by a researcher using a Bishop analysis and a TLS-DEM slope profile [12].

TLS is a useful technique for finding out how slowly landslides are moving, according to Travellettia [24], who came to this result after conducting a well-thought-out experiment. The 120 $m^2$ study landslide for the controlled rain experiment was conducted in the Super-Sauze mudslide in the South French Alps. From July 10 through July 14, 2007, five days of observation produced five LiDAR scanning point clouds. Measurements produced utilising benchmark and shaded relief image correlation methods, among other measuring approaches, were compared to these scanning data. The comparison's results showed that the laser scanner can identify landslides moving in a direction of 27°/30° at an average speed of 3.2 cm/day.

LiDAR's potential for monitoring slowly moving landslides was assessed in 2009 by Prokop and Panholzer [25]. TLS was used to track the landslide's volume change. TLS's DEM can rapidly estimate the point cloud's mass displacement by comparing scanning results with orthophotos. However, the results showed that the LiDAR scanner utilised in the study was unable to establish movement rates less than 50mm in each period due to the precise limitations of the laser scanner and monitoring method employed. Changes smaller than 100mm may still be recognised if the scanning point density's quality and the region's very low point density are both known.

Kasperski [26] employed TLS technology to track the course of the Séchilienne landslide in Isère, France, from 2004 to 2007. The research team can track the 3D movement of the study site, thanks to the time series of data from terrestrial laser scanning. Although there was some homogeneous data in the scanning point cloud, the results were trustworthy. The findings of additional monitoring contributed to the argument that shifts in scanning locations and DEM may be used to infer changes in landslide volume. The study demonstrated how well LiDAR technology classifies landslides.

Carrea's study titled "Oren-Nayar reflectance model correction of terrestrial LiDAR intensity channel: An application to lithological differentiation" [27]. The strength and accuracy of the laser scanner were assessed across a range of scan distances and laser incidence angles using black-and-white matte paper. The author then conducted a second test utilising actual rock blocks with various lithologies and surface conditions after the first test. The findings of the two studies in question demonstrated that the suggested correction strategy may effectively address the unequal diffuse reflectance of rock surfaces. The approach presented in this paper may enhance LiDAR technology's capacity to determine lithological surface properties for geological mapping.

Wang [28] employed TLS to identify the characteristics of shallow landslides and explain the underground deformation by rotating the shapes of the tree stems. The author released a novel monitoring technique that uses LiDAR to gather point cloud data and determine the volume and stem inclination angle. Tree characterisation showed 1.6 cm DBH and 8.2° inclination. Consequently, this information may be used to forecast the activity of shallow landslides.

## 1.6 *Forecasting and Monitoring Techniques*

A range of technologies, such as GIS, mathematical models, computer simulations, borehole measurements, and others, are used to anticipate and monitor landslides. One of the most effective research tools is LiDAR. Many kinds of landslides have been predicted using semi-empirical prediction functions [12].

LiDAR technology has been used for continuous monitoring [25] and landslide prediction [29].

Oppikofer obtained a reliable TLS monitoring result [24]. The study's slope was the eastern Eiger (Switzerland). The slope shifted more than 70 cm daily during paroxysmal activity, according to monitor observations. Prokop and Panholzer [25] used TLS to collect slope point cloud data for a 3D model. For geohazard mitigation, LiDAR's 3D model helps collect and monitor landslide data.

Abellán [30] proposed the TLS methodology as a method to recognise and foresee rockfalls. The main scarp of the 1881 Puigcercós landslide is being studied near

Pallars Jussà, Catalonia. After studying rockfall volume change, the author utilised pre-failure deformation to forecast rockfall geographically.

Benni Thiebes [31] summarised the Landslide Early Warning Systems (LEWS) conceptual approaches and concluded that TLS technology aids LEWS monitoring. Engineering journals barely address LiDAR. Laser scanners are costly, hence LEWS rarely shows TLS.

Michel Jaboyedoff [12] found that TLS's monitor function outperforms ALS's despite using the same monitoring methods. This is because TLS has more accuracy and efficiency. The accuracy of TLS was just 1.5 cm, but the precision of ALS was 15 cm. The author also argued that rockfall displacements may be seen as alterations to rigid bodies, which would make rockfall monitoring far simpler than landslide monitoring.

There are two main precursory signs that LiDAR can be used to monitor landslides and rockfall: (1) The volume increase indicates that there were more frequent rockfalls before the ultimate collapse [32]. (2) The Eiger rockslide's 170,000 $m^3$ partial collapse was preceded by a 125cm/day displacement [24].

## 1.7 Discussion

Research on landslides first used TLS technology in 2000, based on the findings of the aforementioned background research. The ALS was utilised for the geohazard research since there wasn't an original application specifically designed for processing 3D data; TLS was subsequently developed. To aid in the scanning process, previous researchers have created a broad variety of extension techniques, with uses ranging from spotting slope movement and volume changes to monitoring and defining landslide features. How to employ LiDAR to analyse the subterranean aspects of landslides, however, is the most urgent issue [23]. To increase the use of LiDAR, several researchers worked to create fresh strategies. Franz [21] employed artificial scanning targets to avoid vegetation and enhance scanning findings, whereas Wang [28] used tree stem inclination angle and stem volume to characterise landslide subsurface characteristics despite the results being imprecise. Additionally, some scientists are aiming to enhance LiDAR's functionality so they may use it to examine subsurface deformation. Table 3 includes a list of the sensors and monitoring parameters for landslides.

## 1.8 Related Works

In this paper, in order to give geolocation, Fekih's wireless sensor network for landslide prevention [40] has shown how LoRa (Long Range) radio transmission technology may be used. The main goal is to use as little money and energy as feasible while still having the sensor detect field movements. A sensor network with fixed

**Table 3** Various network analyses for monitoring landslides [10]

| Ref. No. | Region | Methods | Network | Analysis | Proposed work | Future directions |
|---|---|---|---|---|---|---|
| [6] | Testbed experiment | Camera sensor | IoT | Raspberry Pi, camera sensor, and Android app | Employing vision-based algorithms to capture video transmission and find landslides | AI may increase productivity but requires a lot of data storage |
| [33] | Testbed experiment | GPRS and WSN were used to build multiple network help | WSN | Humidity, displacement, and slope angle | Low power usage and wide coverage | The use of energy-saving routing methods is lacking |
| [34] | Test bed experiment | Zigbee and a low-power camera sensor are used to cover a small region | Heterogeneous network | Image compression, object detection, slope movement, slope schedule | In order to verify anomaly changes and inter-layer validation, a two-layer heterogeneous network is used | The prototype allows for lab tests. Lacks the capacity to scale under challenging conditions |
| [35] | Kerala (India) | Energy optimisation algorithms | Heteroge neous network | Moisture, pore pressure, acceleration, and rainfall | A thorough examination of energy conservation and data compression for landslide detection | Adverse environmental changes cut down on network performance |
| [36] | Korea | Low coverage communication (Bluetooth) | WSN | Acceleration, Ground vibration, slope angle | Landslide detection prototype utilising the test bed | Lack of alarm control and short battery life |
| [37] | Torgiovan netto (Italy) | Event-driven mechanism | WSN | Stain, wind speed, tensile, rainfall, temperature, moisture, humidity | A fault-tolerant, self-organising network with poor link quality and low battery usage is proposed | The difficulties of real-time deployment are absent |
| [38] | Nusa Tenggara (Indonesia) | Vibration sensors | Ad-hoc | Ground vibration measurement | Prototype for measuring ground motion induced vibration | Public warning system absent |
| [39] | Super Sauze (France) | Sensor-based network monitoring | Ad-hoc | Accelerometer, barometric pressure, and tilting | Model for research, analysis, and data gathering | Unorganized system for handling warnings |

(non-mobile) nodes and anchors is a frequent way to geolocate. One approach sends data from other anchors to a "master" anchor.

In this paper, using a wireless ZICM2410-based sensor, Wang's [41] study of a wireless sensor network for monitoring landslides employs this technique. The devastating landslide that struck the Chinese province of Guizhou served as the impetus for the study. The risk is greater after a downpour. The remainder of this system consists of a PC, a base station, a monitoring host, and acquisition nodes. WSN nodes capture data on landslide factors such as displacement, soil temperature and moisture, tilt angle, and nearby precipitation. The Zigbee ZICM2410 chip-based wireless sensor network is cost-effective and efficient. It supports mesh, point-to-point, and point-to-multipoint networks. Data from far areas is sent to the viewer via GSM technology at the base station. The monitor is inside and updates the PC through an RS232 interface with information obtained from the worldwide system for mobile communications (GSM), enabling landslide monitoring using PC-based software. PC administration software must include both a database management system and a real-time graphical user interface. The system's user interface (UI) was developed using VC++ tools, while the database was constructed using the SQL Anywhere 11 platform. A landslide's forces must be accommodated in the design of the wireless sensor network (WSN) used for landslide monitoring.

In this research, Kebaili describes the creation of a prototype early warning system for landslide risk utilising Web of Things technology and MongoDB [42]. The created prototype receives and shows real-time sensor telemetry data, such as data on rainfall, soil movement, and soil moisture. As soon as specific criteria are surpassed, built-in alarms ring.

In this work, they proposed a mechanism known as SMARTCONE to continue operating even if the power goes out. Lee, the creator of the Open-Source wireless slope monitoring system, first proposed it [43]. Using a wireless sensor to monitor environmental factors outside requires constant power. Since it's in the mountains, replacing the sensor's battery is difficult. While consuming minimal power, the sensor must swiftly detect and relay data. To meet these needs, they developed the SMARTCONE active slope monitoring system. When not searching for events, it uses minimal electricity.

Giorgetti's [44] uses wireless sensor network for landslide risk analysis. Design, development, and testing employs WSN to collect data on battery life, radio connections, and paths. The WSN utilised in a landslide monitoring system must be constructed to withstand the negative effects of a landslide, run for a considerable amount of time on a single charge, and maintain optimal performance during that time. Sensor nodes and communication protocols provide self-organisation, fault tolerance, and situation adaption. Because of the system's network protocols, self-organisation, node failures, low connection quality, and unexpected battery life are all handled efficiently. For a few months, this technique was utilised experimentally on a landslide near Torgiovanetto, Italy, with little assistance from humans. It proved to have a very high degree of resilience to the destructive landslide, which made it a superb option for harsh landslide circumstances. A central point receives information to process and analyse.

Badru [45] researched the Bududa region in eastern Uganda to design a web-based landslide detection and alarm system that could monitor landslide-prone regions and inform residents. They investigated elements including pore pressure, vibration, movement, rainfall intensity, and humidity using quantitative experimental designs. They are being monitored by sensors, and the system analyses their values to improve itself. These measurements are sent online to the control centre, where they are evaluated before being used to send a "landslide detection alert" by GSM text message. It is believed that the findings would improve disaster preparedness and reduce the number of fatalities and property loss. "Landslide Monitoring System Implementing IoT Using Video Camera" author Aggarwal claims that eye inspection, image/video processing, satellite remote sensing, statistical approaches, and machine learning algorithms may predict landslides [6]. Machine learning, deep learning, and other technologies are used to identify landslides. In this study, they deal with landslides in India's Konkan area, where soil movement destroys railway lines, using image processing. A landslide detection system gathers sample video frames from low-resolution cameras. In MATLAB, many methods for detection include the Hamming distance, entropy, Euclidean distance, correlation, block processing, etc. With a threshold margin of 80.24%, the average system efficiency for the examined images was 86.67%.

In this paper, the author Susanto [46] has proposed a design for an IoT-integrated landslide disaster-monitoring system in Slope, Humidity, and Vibration Sensor Performance for Landslide Monitoring System. Many online, real-time monitoring experiments of landslide detection have focused on the soil displacement caused by earthquakes and man-made precipitation. In order to better understand how to measure soil movement, slope, humidity, and vibration sensor performance were investigated. Sensing data shows that any quantifiable variable may be helpful in landslide monitoring. A system was built to process data in the cloud with the aid of sensors for detection. For this, a number of sensors are used, including a rain gauge, moisture sensor, piezometer, tiltmeter, geophone, and strain gauge. Data from rain gauge sensors, which gather samples of rainfall over time, may help better understand a landslide tragedy where precipitation had a crucial impact. The efficiency of sensors employed in systems designed to identify and prevent landslides is the main topic of this research by the authors. Three sensors—a slope sensor, a moisture sensor, and a vibration sensor based on a piezo sensor—support soil movement. For landslide monitoring systems up to this point, Wireless Sensor Network (WSN) has been rated as one of the best methodologies and technologies.

## 2 System Architecture

As illustrated in Fig. 2, the system comprises the base station for the acquisition nodes, the monitor host, and the computer with administration software.

A wireless sensor network (WSN) is often made up of a number of acquisition nodes that collectively gather data on landslide monitoring parameters such as

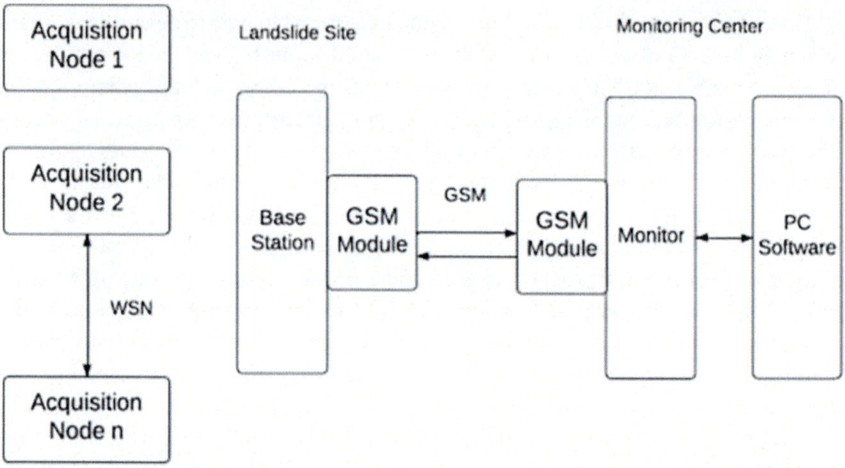


**Fig. 2** System architecture diagram [35]

displacement, temperature, moisture, tilt angle, and local rainfall. One GSM module built within the base station may wirelessly transmit data to the monitor. In addition, the monitor host is constantly inside and connected to the GSM network to receive monitoring data, which it then uploads to the PC. The landslide monitoring management software that is installed on the PC allows users to search for monitoring data.

## 3 System Operation

Landslides may be detected and people may be informed using wireless sensor technology. There is no physical labour involved in landslide monitoring, nor do any other surveying instruments. In landslide-prone areas, it is feasible to independently install sensors like LiDAR. In order for the microcontroller to process the data further, these sensors collect signals and transform them to digital form. The data analysis centre receives the processed figures over the Internet. Each station's findings are processed and sent by a station microprocessor. The public is informed of the potential landslide hazard very quickly and in advance. There is still time to protect time, space, and property as a consequence [2].

## 4 Advantages of IoT-Based Landslide Monitoring System

IoT-based technology has the capacity to be widely deployed and to detect landslide losses instantly. IoT-based networks can detect even the tiniest changes in unstable ground or slopes due to a number of causes, including pore pressure, dielectric moisture, and other elements that may occur during a landslide. It has the ability of automatic tools to acquire and elaborate data independently. The key objectives of an Internet of Things-based device for landslide monitoring systems are low cost, robustness, and delay effectiveness.

## 5 Conclusion and Future Work

This research illustrates the development of several landslide monitoring systems using sensors and a stable wireless network. The monitoring of landslides in limited regions may be done at a low cost to society by using wireless networks and smart sensors. One of the problems with the deployment of sensor networks is using the appropriate energy-saving design or algorithms to minimise the energy consumption of sensor nodes. Over the last several decades, a number of methods and tools for analysing landslides have been created, including LiDAR technology, modelling software, and methods for landslide detection and susceptibility mapping. But there are obvious problems with the program and the monitoring hardware. The laser scanner can only detect surface changes in the item it is scanning, while having excellent scanning quality and precision. The internal state of a landslide may be examined using computer simulation methods, although the results of the simulation are limited by erroneous input settings. Combining high-precision LiDAR equipment with computer modelling techniques is the preferred approach for studying the landslide. Table 4 shows how different WSN models, approaches, assessments, suggested action plans, and possible future directions for landslide monitoring vary from one another. WSN-based systems for landslide detection and monitoring are contrasted in Table 4. Future work on the WSN-based system may include employing deep learning to improve the performance and capability of the landslide monitoring strategy.

**Table 4** Landslide monitoring and analysis using WSN

| Sl. No. | Title of paper | Landslide detection based on | Communication technology or standard | Radio module/MCU | Sensors | Merits |
|---|---|---|---|---|---|---|
| 1 | Rihab Fekih Romdhane [40] | WSN | LoRa | iM880A/ STM 32L511Cx | 3 axis capacitive accelerometer, 3 axis magnetometer, pressure sensor, 3 axis angular rate sensor | Low Power, low cost |
| 2 | Wang Honghui [41] | WSN | Zigbee (IEEE 802.15.4), GSM | ZICM2410/ 8051 | Soil moisture, soil temperature, displacement, tilt angle, rainfall | Low cost, low power, small size, efficient |
| 3 | Mohamed Omar Kebaili [42] | WSN | Wi-Fi/Zigbee/ GPRS/ Bluetooth/BLE/ RFID-NFC | Waspmote (ATmega 1281) | Meshlium sink, precipitation sensor, accelerometer sensor, humidity sensor | Low energy consumption due to external database |
| 4 | Huang-Chen Lee [43] | WSN-based (SMART CONE) | GPS trajectories | Nordic NRF24L01/ Nuvoton NANO102SC 2AN | Battery fuel gauge, adxl 345, temperature and humidity sensor | Low power |
| 5 | Giorgetti Andrea [44] | WSN | IEEE 802.15.4 compliant, GPRS | Texas CC2530 Tx/ Rx | Air thermometer, rain gauge, air hygrometer, wind gauge, soil hygrometer, extensometer, tiltmeter | Low power |

## References

1. Parkash S (2013) Landslides caused by earthquakes in the Indian Himalaya: past experience and future implications. Landslide Sci Pract Complex Environ **5**, 327–334. https://doi.org/10.1007/978-3-642-31427-8-42

2. Bagwari S, Gehlot A, Singh R, Priyadarshi N, Khan B (2021) Review of low-cost sensor-based and LoRaWAN opportunities for IoT platform landslide monitoring systems. IEEEAccess 1 https://doi.org/10.1109/ACCESS.2021.3137841
3. Pawar P, Patil A, Rathod H, Hadale R, Kharche S (2019) IoT based landslide monitoring and detection
4. Giri P, Ng K, Phillips W (2019) Wireless sensor network system for landslide monitoring and warning. IEEE Trans Instrum Meas 68(4):1210–1220
5. Chengxun L (2019) Studying the interaction between the soil and LiDAR scanning targets and using LiDAR to determine subsurface movement of slowmoving landslides. Doctoral Dissertations, 2809. https://scholarsmine.mst.edu/doctoraldissertations/2809
6. Aggarwal S, Mishra PK, Sumakar KVS, Chaturvedi P (2018) Landslide monitoring system using IOT and video camera. In: 2018 3rd International conference for convergence in technology (I2CT). pp 1–4. https://doi.org/10.1109/I2CT.2018.8529424
7. Abellán A, Jaboyedoff M, Oppikofer T et al (2012) Review on the use of LIDAR in landslide investigations. Safety Risks 61:5–28. https://doi.org/10.1007/s11069-010-9634-2
8. Shan J, Toth CK (2008) Principles and processing for topographic laser ranging and scanning. CRC Press
9. Fruneau B, Achache J, Delacourt C (1996) Observation and modelling of the Saint-Étienne-de-Tinée landslide using SAR interferometry. Tectonophysics 265(3–4):181–190. ISSN 0040–1951. http://www.sciencedirect.com/science/article/pii/S0040195196000479
10. Tarnopolsky MA, Rennie C, Carter SL (2001) Utilization of substances during endurance exercise in both men and women after endurance training. Am J Physiol Endocrinol Metab 280(6):E898–907. PMID: 11350771. https://doi.org/10.1152/ajpendo.2001.280.6.E898
11. Haugerud RA, Harding D (2001) Some methods for LiDAR topographic survey data virtual deforestation (VDF). J Int Assoc Photogram Remote Sens
12. Jaboyedoff M, Oppikofer T, Abellan A, Derron M-H, Loye A, Metzger R, Pedrazzini A (2012) A review of the use of LIDAR in landslide investigations. Na. Hazards 5:28. https://doi.org/10.1007/s11069-010-9634_2
13. Rutzinger M (2009) Topographic laser ranging and scanning—principles and processing by J. Shan and C.K. Toth. Int J Appl Earth Obs Geoinf 11:368–369. https://doi.org/10.1016/j.jag.2009.05.001
14. Woodbury N, Brubacher M, Woodbury JR (2003) Noninvasive tank gauging using frequency modulated laser ranging sensors. 27–31
15. Duan Y, Maerz NH, Youssef AM, Otoo JN, Kassebaum TJ (2013) A simple method for measuring discontinuity orientations from terrestrial LIDAR data. Environ Eng Geosci 19(2):185–194. https://doi.org/10.2113/gseegeosci.19.2.185
16. Rowlands KA, Jones LD, Whitworth M (2003) Landslide laser scanning: a fresh perspective on an old issue. Quart J Eng Geol Hydrogeol 36:155–157. https://doi.org/10.1144/1470-9236/2003-08
17. Borgatti L, Cervi F, Dahne A, Ronchetti F, Sterzai P, Corsini A (2009) Estimating mass-wasting processes in active earth slides and flows using time-series of high-resolution DEMs from photogrammetry and aerial LiDAR. Nat Hazards Earth Syst Sci 9:433–439. https://doi.org/10.5194/nhess-9-433-2009
18. Miller P, Barr S, Parkin G, Clarke B, Stephanie G (2008) Assessing the risk of slope instability along transport corridors using terrestrial laser scanning
19. Akca D, Gruen A, Askarinejad A, Springman S (2011) Monitoring of a simulated landslide using photogrammetry. https://doi.org/10.13140/RG.2.1.3653.8967
20. Fiani M, Barbarella M (2013) Monitoring of big landslides by terrestrial laser scanning techniques: field data collection and processing. Europ J Remote Sens 46(1):126–151. https://doi.org/10.5721/EuJRS20134608
21. Franz M, Carrea D, Abellan A, Derron M-H, Jaboyedoff M (2016) Tracking 3D displacements with targets in heavily forested regions susceptible by landslides. 13:821–831. https://doi.org/10.1007/s10346-016-0685-7

22. Renslow MS (2012) Topographic lidar airborne manual. Am Soc Photogramm Remote Sens 246
23. Maerz NH, et al (2015) Evaluation of rock fall hazards using LiDAR technology. J Civil Eng Architect 9(1):80–89. David Publishing
24. Travelletti J, Oppikofer T, Delacourt C, Malet JP, Jaboyedoff M (2008) Monitoring landslide displacements using a long-range Terrestrial Laser Scanning during a controlled rain experiment (TLS). Int Arch Photogramm Remote Sens Spat Inform Sci 37
25. Alexander Prokop and H. Panholzer (2009). Evaluating the effectiveness of terrestrial laser scanning for keeping an eye on landslides that are moving slowly. Nat Haz Earth Syst Sci https://doi.org/10.5194/nhess-9-1921-2009
26. Kasperski J, Delacourt C, Allemand P, Potherat P, Jaud M, Varrel E (2010) Study of the Séchilienne landslide (Isère, France) using a terrestrial laser scanner (TLS). Remote Sens https://doi.org/10.3390/rs122785
27. Carrea D, et al (2015) Automatic rockfalls volume estimation based on terrestrial laser scanning data. The second volume of Engineering Geology for Society and Territory. Geophys Res Lett 425–428
28. Di Wang M, Hollaus EP, Pfeifer N (2016) Stem reconstruction in forests affected by landslides: automatic and self-adaptive. Satel Imaging 8:974. https://doi.org/10.3390/rs8120974
29. Biasion A, Bornaz L, Rinaudo F (2005) Applications for disaster management using laser scanning. In: Fendel EM, Zlatanova S, van Oosterom P (eds) Geospatial data for emergency preparedness. Berlin, Heidelberg: Springer. https://doi.org/10.1007/3-540-27468-52
30. Lim M, Jaboyedoff M, Rosser NJ, Abellán A, Oppikofer T, Lato MJ (2014) Rock slope instability detection using terrestrial laser scanning. Earth Surf. Process Landforms 39: 80–97. https://doi.org/10.1002/esp.3493
31. Darvishi, Schlögel R, Kofler C, Cuozzo G, Rutzinger M, Zieher T, Toschi I, Remondino F, Mejia Aguilar A, Thiebes B, Bruzzone L (2018) Sentinel-1 and ground-based sensors for ongoing corvara landslide monitoring (South Tyrol, Italy). Remote Sensing 10:1781. https://doi.org/10.3390/rs10111781
32. Rosser N, Lim M, Petley D, Stuart D, Allison R (2007) Patterns of rockfall that occurs before a slope fails. J Geophys Reshttps://doi.org/10.1029/2006JF000642
33. Jian L, Li C, Li K, Liu Y (2014) Internet of things-based landslide monitoring and early warning system design. Appl Mater Mech. https://doi.org/10.4028/www.scientific.net/AMM.511-512.197.511-512.197-201
34. Rangan VP, Kumar S, Duttagupta S (2021) Voice channels for resilient green cellular IoT for landslide monitoring. J Sens Actuat Netw 10(3):59. https://doi.org/10.3390/jsan10030059
35. Ramesh MV (2014) Design, development, and implementation of a wire-less sensor network for detection of landslides. Ad Hoc Netw 13(Part A):2–18. ISSN1570–8705. https://doi.org/10.1016/j.adhoc.2012.09.002
36. Song YS, Chae BG, Kim KS, Park JY, Oh HJ, Jeong SW (2021) Development and use in hazard assessments of a landslide monitoring system for natural terrain in Korea. Sensors (Basel) (Basel) 21(9):3040. PMC8123579 and PMID 33926052
37. Intrieri E, Gigli G, Mugnai F, Fanti R, Casagli N (2012) Design and implementation of a landslide early warning system. Eng Geol 147–148:124–136, ISSN 0013–7952. https://doi.org/10.1016/j.enggeo.2012.07.017
38. Duc-Tuyen T, Duc-Nghia CN, Tran D-T (2015) Development of a wireless accelerometer network for a rainfall-triggered landslide system. Int J Comp Technol Adv 7:14
39. Benoit L, Briole P, Martin O, Thom C, Malet J-P, Ulrich P (2015) Monitoring landslide displacements using the Geocube wireless network of low-cost GPS. Eng Geol 195:111–121. ISSN 0013–7952. https://doi.org/10.1016/j.enggeo.2015.05.020
40. Romdhane RF, Lami Y, Genon-Catalot D, Fourty N, Lagreze A, Jongmans D, Baillet L (2017) Computational intelligence and virtual environments for measurement systems and applications (CIVEMSA) IEEE 2017 IEEE International Conference on Annecy, France (2017.6.26–2017.6.28)] Wireless sensors network for landslide prevention. In: 2017 IEEE international conference on computational intelligence and virtual environments for measurement systems and applications (CIVEMSA). pp 222–227

41. Zhang Z, Tuo X, Xi D, Fan L, Zhang G, Hao S, Wang H (2011) [2011.09.23–2011.09.25] [IEEE 2011 7th International conference on wireless communications, networking and mobile computing (WiCOM) Wuhan, China] Research on One ZICM2410-based wireless sensor network for landslide monitoring. In: 2011 7th international conference on wireless communications, networking, and mobile computing. 14. https://doi.org/10.1109/wicom.2011.6040373
42. Kebaili M, Karim F, Karim F, Ali F, Tahar E, Claramunt C (2016) Early warning system for landslides using MongoDB and web of things technologies. Proc Comp Sci 98:578–583. https://doi.org/10.1016/j.procs.2016.09.090
43. Lee H-C, Ke K-H, Fang Y-M, Lee B-J, Chan T-C (2017) Wireless open-source sensor system for long-term slope movement monitoring. IEEE Trans Instrument Meas 66(4):767–776. https://doi.org/10.1109/TIM.2017.2657838
44. Giorgetti A, Lucchi M, Tavelli E, Barla M, Gigli G, Casagli N, Chiani M, Dardari D (2016) System design, deployment, and field testing of a robust wireless sensor network for landslide. Risk Anal. https://doi.org/10.1109/jsen.2016.2579263IEEESensorsJournal,(),1-1
45. Badru L, Semwogerere T, Victoria O, Gilibrays G (2022) A case study of the Bududa District in Eastern Uganda's Hilly areas using an IOT based landslide detection and early warning system. East Afric J Inform Technol 5:31–38. https://doi.org/10.37284/eajit.5.1.588
46. Susanto E, Budiman F, Mukhtar H, Latief M (2019) Performance of slope, humidity, and vibration sensors for monitoring system of landslides. 139–142. https://doi.org/10.1109/APWiMob48441.2019.8964155
47. Colesanti C, Wasowski J (2006) Using synthetic aperture radar (SAR) interferometry to study landslides. Eng Geol 88:173–199. https://doi.org/10.1016/j.enggeo.2006.09.013
48. Allemand P, Delacourt C, Squarzoni C (2003) SAR interferometry has tracked the La Valette landslide's nine-year geographical and temporal progression. Eng Geol 68:53–66. https://doi.org/10.1016/S0013-7952(02)00198-9
49. Priori SG, Blomström-Lundqvist C, Mazzanti A, Blom N, Borggrefe M, Camm J, Elliott PM, Fitzsimons D, Hatala R, Hindricks G, Kirchhof P, Kjeldsen K, Kuck KH, Hernandez-Madrid A, Nikolaou N, Norekvl TM, Spaulding C, Van Veldhuisen DJ (2015) ESC scientific document 2015 ESC guidelines for the treatment of ventricular arrhythmias and sudden cardiac death european society of cardiology task force on ventricular arrhythmias and sudden cardiac death (ESC). Assoc Europ Pediatr Cong Cardiol (AEPC) 36(41):2793–2867. PMID: 26320108. https://doi.org/10.1093/eurheartj/ehv316
50. Slob S, Hack R (2013) 3D terrestrial laser scanning as a new field measurement and monitoring technique. In: Hack R, Azzam R, Charlier R (eds) Engineering geology for infrastructure planning in Europe. 104th volume of Earth Sciences Lecture Notes. Berlin-based Springer, Heidelberg. https://doi.org/10.1007/978-3-540-39918-6_22
51. Wang YY, Wu YL, Duan ZP, Jia ZH, Zhang J, Liu S, Chen Y, Liang LC, Zhang CQ, Zhang Z, Sun Y, Zhang SQ, Zhang J (2011) Herbal natural remedies A randomised, double-blind, positive controlled clinical study using the Lianhuaqingwen pill as an anti-influenza A (H1N1) treatment. Chem Med J (Engl) 124(18):2925–2933. PubMed: 22040504
52. Oppikofer T, Penna I, Carrea D, Derron M-H, Jaboyedoff M, Benjamin R, Jaboyedoff M (2020) A overview of techniques for calculating the depths and volumes of the first landslide failure surface. Eng Geol 267:105478. https://doi.org/10.1016/j.enggeo.2020.105478
53. Jaboyedoff M, Robert M, Oppikofer T, Couture R, Derron M-H, Locat J, Turmel D (2007) COLTOP-3D programme provides new insights into the analysis of rock-slope relief utilising DEM and 3D-imaging cloud points. https://doi.org/10.1201/NOE0415444019-c8
54. Aryal A, Brooks BA, Reid ME, Bawden GW, Pawlak GR (2012) Displacement fields from point cloud data: application of particle imaging velocimetry to landslide geodesy. J Geophys Res 117:F01029. https://doi.org/10.1029/2011JF002161
55. Oppikofer T, Jaboyedoff M, Blikra L, Derron M-H, Metzger R (2009) employing terrestrial laser scanning to characterise and monitor the Knes rockslide. Earth Syst Sci Environ Haz 9:1003–1019. https://doi.org/10.5194/nhess-9-1003-2009

# Change Detection in LULC for Durgapur: An Efficient GIS-Based Framework

**Kazi Amrin Kabir and Parag Kumar Guha Thakurta**

**Abstract** A geographical information system (GIS)-based framework on the changes in land use land cover (LULC) is proposed in this paper for efficient urban planning and management. The proposed work compares two classified LULC maps of Durgapur city for the years 2014 and 2022. The proposed methodology downloads the satellite images of the study area from USGS EarthExplorer for the year 2014 and 2022, and it is explored into a GIS framework. The study area is clipped and preprocessed into the GIS framework. A classification algorithm is applied to the preprocessed image. The classified output image is obtained in terms of different classes, such as waterbody, vegetation, agricultural land, built-up area and mining. A higher accuracy obtained from the classification for the proposed work highlights that it can efficiently detect the changes in LULC from 2014 to 2022 based on the classes.

**Keywords** Land use land cover (LULC) · Classification · Geographical information system (GIS) · Accuracy · Change detection · Satellite image

## 1 Introduction

In a large urban centre, land cover is continually changed with respect to time and space. The local government can update their database to include those changes in the land use [1]. Thus, the monitoring of the changes in urban land can aid efficient urban planning and development [2]. The gap between supply and demand for basic services and infrastructure [3] defines the positive aspects of urbanization. An uncontrolled urbanization becomes responsible for creating problems related to land use land cover (LULC) [4]. A process of the changes in LULC that ensure a result of human

K. A. Kabir (✉) · P. K. G. Thakurta
Department of Computer Science and Engineering, NIT Durgapur, Durgapur, West Bengal 713209, India
e-mail: kaziamrin17@gmail.com



S. Borah et al. (eds.), *Advanced Computational and Communication Paradigms*, Lecture Notes in Networks and Systems 535,
https://doi.org/10.1007/978-981-99-4284-8_31

interaction with the physical environment [5] also explain either shifting to a new type of land use or intensifying the use of the existing type [6].

The resulting negative impacts of the changes in LULC on ecosystems and human well-being are containing erosion, excess flooding, loss of water resources, degrading water quality, etc. [7]. In order to minimize the physical hazards in terms of social and environmental issues due to uncontrolled urbanization, modern technology uses remote sensing data as a source to obtain urban LULC mapping [8]. Here, the geographic information system (GIS) as a computer-based information system capable of integrating data from various sources is used for effective decision-making in urban planning based on remote sensing data [4]. The importance of GIS capabilities is handling digital cartographic data and the associated databases with attribute information for map features [2]. These features are used to obtain the LULC maps which can feed into the QGIS [9] software for the next levels of analysis.

In this context, a GIS-based framework to detect the changes in LULC is proposed in this paper for efficient urban planning and management. The proposed work compares two classified LULC maps of a city obtained from satellite images in two different years. The proposed methodology considers the satellite images of the "Durgapur" city for the year 2014 and 2022. These images are downloaded from USGS EarthExplorer and subsequently, explored into a GIS framework. Here, the study area is clipped and preprocessed into the GIS framework. It is important to note that the latitude and longitude values are the same for both the years 2014 and 2022. Furthermore, a classification algorithm is applied to the preprocessed image. As a result, it produces the classified output image where different classes, such as waterbody, vegetation, agricultural land, built-up area and mining, are visualized. In order to obtain the accuracy of the proposed classification, the Google satellite image is used to consider the ground truth data. Furthermore, the values of various accuracy parameters are obtained depending on the classes. A higher accuracy of the proposed work highlights that it can detect the changes in LULC efficiently from 2014 to 2022 based on the classes.

The rest sections are organized as follows: materials and methods are discussed in Sect. 2. Simulation results are shown in Sect. 3. Conclusion is discussed in Sect. 4.

## 2 Materials and Methods

### *2.1 Study Area*

The city "Durgapur" is located in Burdwan, West Bengal, India, which is shown in Fig. 1. This city is considered as the study area for the proposed work due to its large population growth and massive mining activities which may affect the environment. The latitude and longitude of Durgapur are 23.533440 and 87.321930, respectively.

It has an area around 154 $km^2$ and the elevation is 65 meters. The average temperature in Durgapur is 29 °C and has 81% humidity.

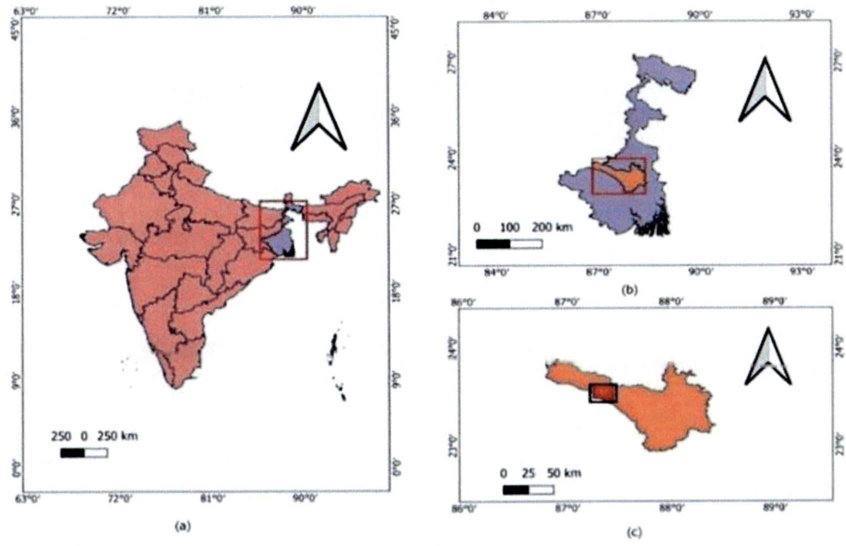


**Fig. 1** A view of the study area: **a** West Bengal in India, **b** Burdwan in West Bengal and **c** Durgapur in Burdwan

**Table 1** Information of the satellite image of the study area

| Acquisition date | Satellite | Sensor | Path/row |
|---|---|---|---|
| 2022/04/11 | Landsat 8 | OLI/TIRS | 139/044 |
| 2014/04/13 | Landsat 8 | OLI/TIRS | 139/044 |

## 2.2 Data Collection

Landsat 8 OLI/TIRS C1 Level 1 satellite images, for the years 2014 and 2022, of the study area, are freely downloaded from the USGS website (http://earthexplorer.usgs.gov/). These satellite images are utilized to explore the changes in LULC of the study area for 8 years period from 2014 to 2022. For better visibility of the satellite image, less than 10 % cloud-free imagery is used. The characteristics of the input data for the proposed work are shown in Table 1.

## 2.3 Proposed Methodology

In order to detect the changes in LULC in the study area from 2014 to 2022, an efficient GIS-based framework is proposed. Hence, a flowchart of the proposed methodology is shown in Fig. 2.

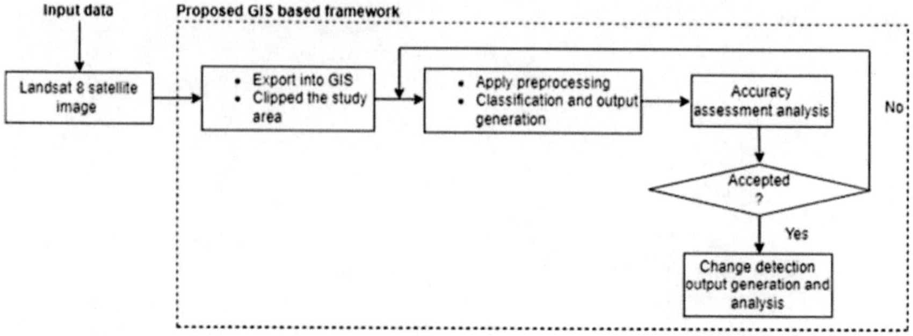


**Fig. 2** Block diagram of proposed methodology

In the proposed methodology, the satellite image of Durgapur is searched from the USGS website through its longitude and latitude values. Next, the image processing tasks such as radiometric calibration, atmospheric correction and the cloud cover deduction are applied to the input satellite image. For further processing, the satellite image of Durgapur is downloaded, and it is exported into QGIS. Afterwards, the study area is clipped from the downloaded satellite image. By using the clipped layers of GIS, the map layers are preprocessed to obtain the corresponding reflectance (RT) layers with suitable RGB (Red, Green, Blue) values. Subsequently, the band set is defined for those RT layers and the training input file of the image is obtained. The regions of interest (ROIs) on the training input file are selected in accordance with the RGB values. Now, the minimum distance method is used for training the input file having the selected ROIs earlier as input. As a result, it produces the output in terms of classified images. For post-processing, the accuracy of the classified image is measured by comparing it with the ground truth values on the base map of Google satellite. In the accuracy analysis, if the overall accuracy and the kappa value cannot be efficient enough for the classification, then preprocessing is applied again on the image layers. Otherwise, the proposed work can detect the changes in LULC from the year 2014 to 2022.

## 3 Simulation Results

### *3.1 Classification Results*

In order to identify the changes in LULC from 2014 to 2022, the classified output image for both years is shown in Fig. 3a, b. The different classes obtained after classification by the minimum distance method are denoted by different colours in the output images. It shows that the blue-coloured, green-coloured, yellow-coloured, red-coloured and pink-coloured pixels can indicate the waterbody, vegetation area, agricultural land, built-up area and mining, respectively. For better interpretation, the classified output images from 3 (a) and Fig. 3b are presented by their corresponding pie charts in Fig. 4a, b, respectively. For example, in Fig. 4a, 2, 14, 47, 28 and 9% pixels are used in the case of waterbody, vegetation, agricultural land, built-up area and mining, respectively. Similarly, 4, 11, 7, 71 and 7% pixels are used in Fig. 4b to show the waterbody, vegetation, agricultural land, built-up area and mining, respectively.

The classification report generated after applying the minimum distance method for both years is shown in Table 2. Here, both the sum of the pixels (PixelSum) and the percentage of coverage (%_coverage) for each class as well as the area in square kilo meter ($km^2$) covered by those pixels in the corresponding classes are high-lighted. For example, in Table 2 for the year 2014, 9014.4, 50295.6, 169418.7, 103623.3 and 31483.8 sq km area are covered in case of waterbody, vegetation, agricultural land, built-up area and mining, respectively. However, for the year 2022, 10939.5, 32266.8, 18777.6, 199599.3 and 19953.9 sq km area are covered in case of waterbody, vegetation, agricultural land, built-up area and mining, respectively.

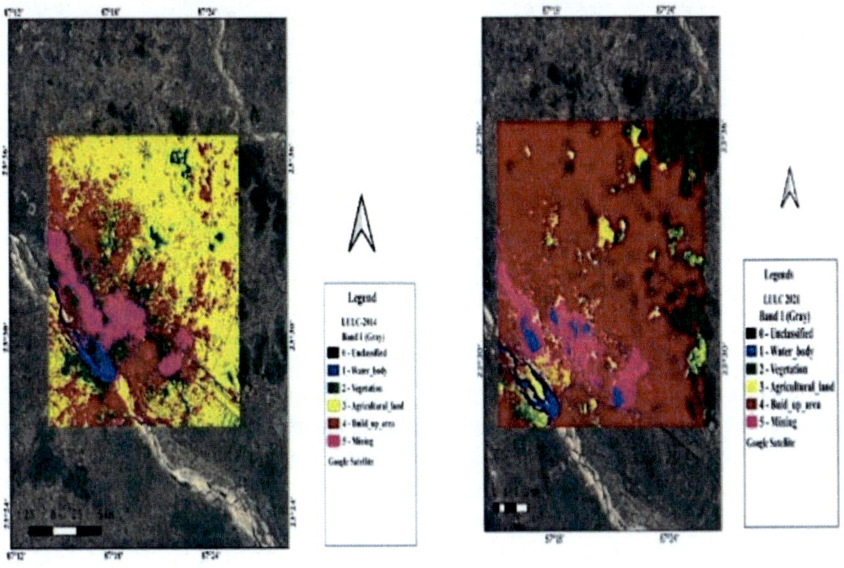

(a) (b)

**Fig. 3** Classified image for **a** year 2014 and **b** year 2022

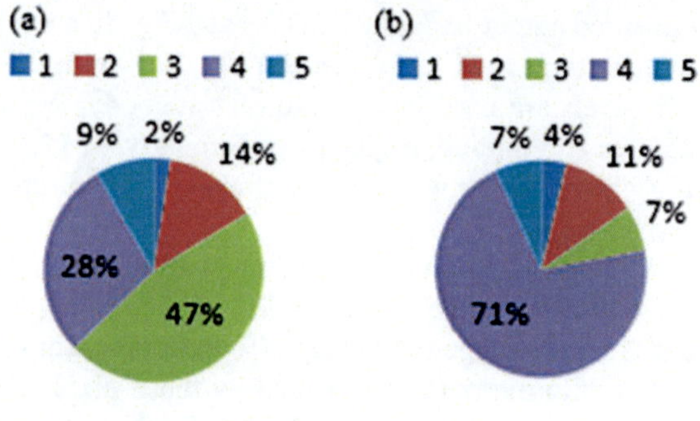


**Fig. 4** Pie chart corresponding to the classification for **a** 2014, **b** 2022

**Table 2** Classification report for the years 2014 and 2022

| | 2014 | | | 2022 | | |
|---|---|---|---|---|---|---|
| Class | PixelSum | %_coverage | Area (km$^2$) | PixelSum | %_coverage | Area (km$^2$) |
| Water_body | 10,016 | 2.477 | 9014.4 | 12,155 | 3.885 | 10,939.5 |
| Vegetation | 55,884 | 13.823 | 50,295.6 | 35,852 | 11.460 | 32,266.8 |
| Agricultural_ land | 188,243 | 46.564 | 169,418.7 | 20,864 | 6.669 | 18,777.6 |
| Built_up_area | 115,137 | 28.480 | 103,623.3 | 221,777 | 70.896 | 199,599.3 |
| Mining | 34,982 | 8.653 | 31,483.8 | 22,171 | 7.087 | 19,953.9 |

## *3.2 Accuracy Assessment*

In order to assess the accuracy of the classified image, a base map image such as a Google Earth image is used for post-processing. The Google Earth satellite image as the reference is downloaded by the HCMGIS plugin. The confusion matrix based on the area covered by each class, for the years 2014 and 2022, is shown in Fig. 5a, b. Here, the diagonal elements and off-diagonal elements of the error matrix denote the correctly classified pixels and the misclassified pixels, respectively. The values of overall accuracy and overall kappa hat are shown in Table 3. For the year 2014, the values of overall accuracy and overall kappa hat are 91.9055 and 0.8804, respectively. On the other hand, the values of those measures obtained for 2022 are 98.1448 and 0.9614, respectively. Hence, the satellite image of Durgapur in 2022 provides a better classified image than in 2014.

## *3.3 Change Detection Analysis*

The detection of LULC changes for each class from 2014 to 2022 is shown in Fig. 6. Here, the blue-coloured legend indicates the waterbodies which are not changed, vegetation to waterbody, agricultural land to waterbody. The green-coloured legend

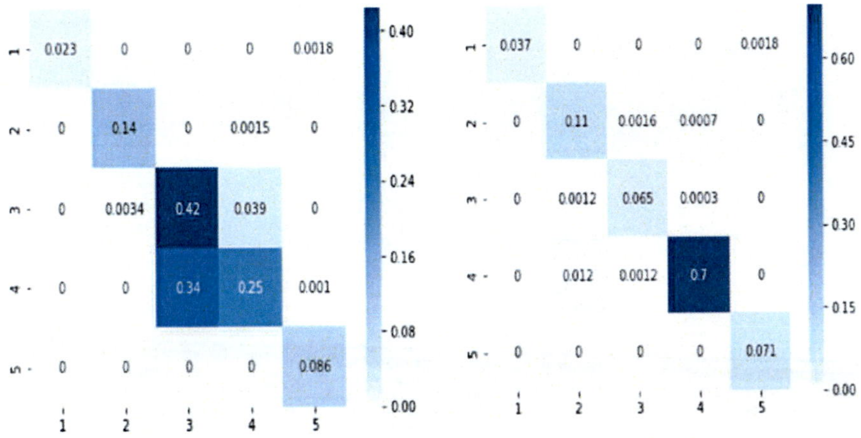


(a) (b)

**Fig. 5** Confusion matrix based on accuracy assessment **a** 2014, **b** 2022

**Table 3** Performance measures for the years 2014 and 2022

| Accuracy measures | 2014 | 2022 |
|---|---|---|
| Overall accuracy (%) | 91.9055 | 98.1448 |
| Overall kappa hat | 0.8804 | 0.9614 |

indicates the unchanged vegetation area, waterbody to vegetation, agricultural land to vegetation, built-up area to vegetation, and mining to vegetation. The yellow-coloured legend indicates agricultural lands which are not changed, waterbody to agricultural land, vegetation to agricultural land, built-up area to agricultural land, and mining to agricultural land. The red-coloured legend indicates the unchanged built-up areas, waterbody to built-up area, vegetation to built-up area, agricultural land to built-up area, and mining to built-up area. The last one pink-coloured legend indicates mining areas which remain unchanged, waterbody to mining, vegetation to mining, agricultural land to mining, and built-up area to mining, respectively. It is also visualized that the built-up area cover is maximum during the period 2014 to 2022.

The total area coverage of LULC change for each class from 2014 to 2022 is shown in Table 4. Here, it is observed that there are no changes in 5605.2, 4822.2, 4135.5, 65492.1 and 13032.9 $km^2$ area for waterbody, vegetation, agricultural land, built-up area and mining, respectively. It is also observed that 57.6, 1058.4, 277.2 and 399.6 $km^2$ are converted to vegetation, agricultural land, built-up area and mining respectively from the waterbody and so on.

For better interpretation, other changes in different classes from 2014 to 2022 are shown in Fig. 7. Here, W, V, A, B and M indicate waterbody, vegetation area, agricultural land, built-up area and mining, respectively which remains unchanged. Here, V-W, A-B, V-M, etc. indicate the changes of vegetation area to waterbody,

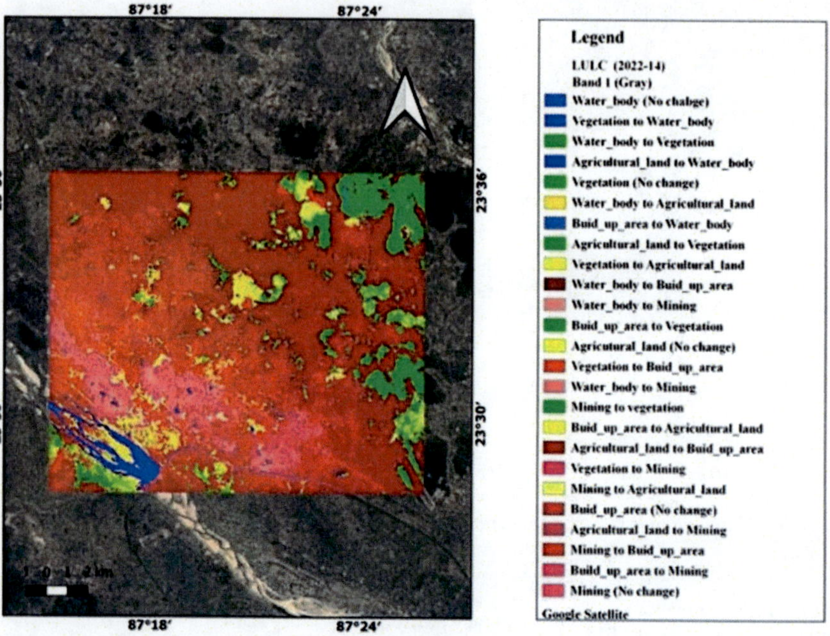


**Fig. 6** LULC changes in Durgapur during (2014–2022)

**Table 4** Overall area-based LULC change detection

| | Waterbody | Vegetation | Agricultural land | Built-up area | Mining |
|---|---|---|---|---|---|
| Waterbody | 5605.2 | 57.6 | 1058.4 | 277.2 | 399.6 |
| Vegetation | 459.9 | 4822.2 | 10,720.8 | 21,884.4 | 4018.5 |
| Agricultural land | 129.6 | 20,541.6 | 4135.5 | 104,276.7 | 202.5 |
| Built-up area | 225 | 5399.1 | 1029.6 | 65,492.1 | 2296.8 |
| Mining | 4518 | 243.9 | 1575.9 | 6396.3 | 13,032.9 |

agricultural land to built-up area, vegetation to mining, etc. Hence, it is observed in Fig. 7 that built-up area is constituted from agricultural land progressively which is gradually increased from 2014 to 2022. The change in vegetation to mining is also gradually increased in the mentioned years.

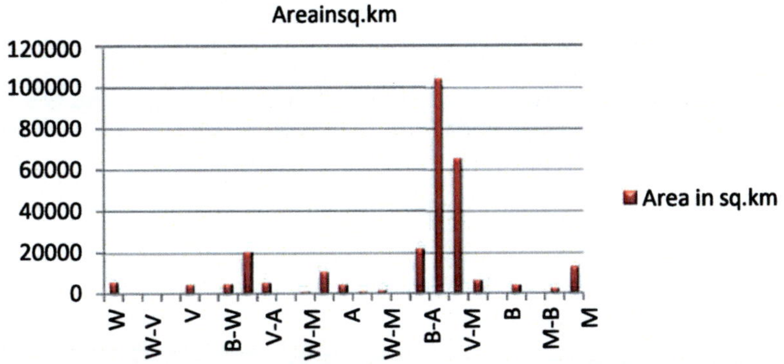


**Fig. 7** Bar plot for LULC change detection analysis from 2014 to 2022

## 4 Conclusion

A change of the LULC detection method is proposed here. The proposed GIS-based framework is focused to highlight the long-term change in LULC for different classes such as waterbody, vegetation, agricultural land, built-up area and mining along with urbanization. The areas under built-up and mining sectors are observed as increasing excessively from 2014 to 2022 in Durgapur. Alternatively, the areas under vegetation and agricultural land are decreased for the same period in Durgapur. The steady increase in waterbody during the study period is noticed. Here, more built-up area converted from agricultural land provides people with suitable accommodation. However, the farmers are affected by excessive decrease in the agricultural land. Therefore, the information on LULC variation between two specific years can indicate the essentials required for the selection and planning of land use schemes to meet the growing burden of fundamental human wishes and their associated benefits. However, a balance in the changes of LULC for each class can be obtained in future.

## References

1. Mawenda J, Watanabe T, Avtar R (2020) An analysis of urban land use/land cover changes in Blantyre City, Southern Malawi (1994–2018). Sustainability 12:2377. https://doi.org/10.3390/su12062377
2. Singh RK, Singha M, Singh SK, Debjeet PAL, Tripathi N, Singh RS (2018) Land use/land cover change detection analysis using remote sensing and gis of dhanbad district, India. Euras J For Sci https://doi.org/10.31195/ejejfs.428381
3. Roy P, Roy A (2010) Land use and land cover change: a remote sensing & GIS perspective. J Indian Inst Sci 90:489–502
4. Kaul HA, Ingle S (2012) Land use land cover classification and change detection using high resolution temporal satellite data. J Environ 1:146–152

5. Mohinuddin SK, Basak S, Roy P, Roy B, Ghosh MT (2020) Changing nature of land-use land cover and its impact on water quality using GIS and remote sensing techniques in Sagar Island West Bengal India
6. Ramanamurthy B, Victorbabu N (2021) Land use land cover (LULC) classification with waste-land demarcation using remote sensing and GIS techniques. IOP Conf Ser Mater Sci Eng 1025:012035. https://doi.org/10.1088/1757-899X/1025/1/012035
7. Ahmed I, Rihan M, Shahfahad, Rahman A (2021) Land use and land cover (LULC) change dynamics in the urban watershed of Guwahati
8. Kafy AA, Faisal AA, Sikdar S, Hasan M , Rahman M, Khan MH, Islam R (2020) Impact of LULC changes on LST in Rajshahi District of Bangladesh: a remote sensing approach. J Geograph Stud 3:11–23. https://doi.org/10.21523/gcj5.19030102
9. Li H, Wang C, Zhong C, Zhang Z, Liu Q (2017) Mapping typical urban LULC from landsat imagery without training samples or self-defined parameters. Remote Sensing 9:700. https://doi.org/10.3390/rs9070700

# Morphological Operation-Based Background Subtraction Method for Shadow Removal in Outdoor Video Sequences

**Tannistha Pal**

**Abstract** The difficult tasks of shadow identification and removal span a wide range of contexts, including computer vision, indoor and outdoor scenarios, and reconnaissance framework. An activity observation framework's performance may suffer if a shadow is incorrectly identified as a foreground object. Although many algorithms are available for removing shadow from images there are some limitations. The goal of this article is to provide a brief overview of the various algorithms used in shadow detection, shadow removal, and ghost elimination, along with their benefits and drawbacks. The article also offered a novel background subtraction-based technique using morphological operation for eliminating shadows from video frames that successfully identifies the shadow region in a frame and eliminates it from the frame. Furthermore, a comparison is made between the proposed method and the existing approaches, and the experimental findings demonstrate its robustness. The experimental outcomes exhibit that the proposed approach shows its effectiveness in terms of an accuracy of 81.37%, precision of 0.81, recall of 0.84, and f1 score of 0.83 compared to other existing approaches.

**Keywords** Background subtraction · Median filter · Morphological operator · Object detection · Shadow removal

## 1 Introduction

Moving object detection is a trending topic and it has gained considerable popularity in the last decades due to the enormous demand for computer vision applications, especially in video surveillance mainly consisting of robots, license plate reader, tracking human motion, anomaly detection, traffic monitoring, and security, etc. [1–7]. Shadow detection and removal is a very challenging task in moving object

T. Pal (✉)
Department of Computer Science and Engineering, National Institute of Technology, Agartala, Barjala, Jirania 799046, India
e-mail: tannisthapaul@gmail.com



S. Borah et al. (eds.), *Advanced Computational and Communication Paradigms*, Lecture Notes in Networks and Systems 535,
https://doi.org/10.1007/978-981-99-4284-8_32

detection. The shadow area can be misinterpreted as an entity when dealing with moving object detection from surveillance systems [8]. The shadows thus need to be eliminated in the pre-processing stage. Shadows can be commonly categorized as shadows of oneself [9]. When a medium's light intensity cannot pass through the object field, it emits a shadow on the item. The reflections that an object makes on another material or surface are, in fact, called projected shadows [8, 10–12]. Such shadows provide two major regions when numerous light rays enhance the scene. When cast darkness, the dark internal region is depicted as the umbra, and the bright external portion is the penumbra.

Numerous researchers have been carried out in recent times to remove shadows from videos and photos. Along with this, shadows have two distinctive properties which help to detect them in the initial stages and remove them later on. These properties are:

(i) The luminance of a shadow area significantly differs from its neighboring area but the impact on color is negligible.
(ii) In a narrow area, the shadow intensity is nearly constant.

Although many researchers have evolved with many algorithms and got a considerable amount of success in differentiating the foreground and background for detecting the moving object, many researchers encounter the challenge of shadows which also gets detected as foreground and this issue needs to be addressed. So, this article mainly concentrates on eliminating the shadows for moving object detection followed by detecting the object accurately. In this article, firstly, the whole video is segmented into consecutive frames, and secondly, all the frames have been processed through the shadow removal algorithm, which is based on Otsu's method [8]. After the shadows have been completely removed, the background subtraction algorithm is carried out to successfully determine the foreground object(s) which is/are moving. Then, on all the output frames of background subtraction, the post-processing method is accomplished for removing noises and distortions. Finally, desired frames devoid of any shadow, noise, and distortion have been achieved. This in terms has enhanced the accuracy in detecting any moving object from a video.

## 2 Literature Review

The shadow removal techniques may generally be divided into five groups. The brief analysis of the above mentioned methods are as follows.

### *2.1 Reintegration Methods*

The shadow removal technique for reintegration is based on the premise that a shadow-free image would be generated by negating the image declivity with the shade boundaries and reintegrating the changed declivity. This method ensures that the alteration of the ambient light of the picture is at a reduced amount than the reflection.

**Advantage:** The shadow removal algorithms based on reintegration generally increase the performance of the overall input image.

**Disadvantage:** The overall downside of such approaches is the use of image features of shadow in the detection efficiency.

## *2.2 Relighting Methods*

The goal of shadow removal relighting methods is to consider a component that plays an important role to improve the dark screen's lightness.

**Advantage:** Typically the method is easy and convenient.

**Disadvantage**: The key problem is to identify different sample indicators for shades of the umbra and penumbra, and the adjustment of the shadow cap.

## *2.3 Patch-Based Methods*

The patch-based shadow extraction methods prefer to function on packets instead of intensity values. Both techniques operate under the premise that the emission and perception inside a shadow region are more or less non-variable.

**Advantage:** This approach takes much less time to eliminate the shadows.

**Disadvantage:** Often, for a shadow patch, identifying the corresponding matching image patch accurately is hard.

## *2.4 Color Transfer Methods*

The strategies of color transition aim to convey color images from light areas to background.

**Advantage:** To boost the results, the assessment of optimal requirements and the selection of suitable non-shadow areas for calculating the metrics were obtained.

**Disadvantage:** This strategy can lead to an adverse effect against surfaces with heavy shades and strongly colored, inhomogeneous surfaces.

### 2.5 Interactive Methods

Methodologies for eliminating the shadow are either dynamic or manual. Some writers take short community feedback that will allow for better recognition of shadows. In addition, the provision of a forum for the individual to communicate with the saliency detection framework will result in better performance.

**Advantage**: These techniques are typically more efficient than the auto-assembled approaches because the client can give sequential decisions to identify the shadow in a frame.

**Disadvantage:** The method doesn't modify the entire image and has light or camera dependency.

## 3 Research Gap and Contribution

### 3.1 Research Gap

After investigating the related available papers on object detection with shadow removal as well as an improved version of the traditional methods, it has been found that background subtraction is the most commonly used technique, and ViBe (Visual Background extractor) algorithm is a very less time-consuming background subtraction technique [10] with some limitations. Inspired by [10], the article proposed a technique to detect moving objects with shadow removal.

Noise, distortion, and shadow in the final images are only a few of the common drawbacks associated with the background removal method of object recognition. The causes of this noise phenomenon are many and include both slow and abrupt changes in lighting, viewpoint motions, shifts in ambient topology, etc.

Similarly, occurrences of shadow to have many underlying causes. Shadows are generally classifiable to self-shadow as well as cast shadows. Self-shadow occurs around an entity while light intensity from some kind of origin cannot penetrate a certain entity's location. Cast shadows include two major zones, where numerous reflective surfaces enhance the image. In extended darkness, the deep internal area is labeled as the umbra, whereas the bright exterior zone is labeled its penumbra. The shadows are being treated as part of the real moving foreground object by the background subtraction techniques. Thus, it leads to false classification of real moving objects and also produces low accuracy rates.

### 3.2 Contribution

The problem of shadow in identifying a moving object is also addressed in this article. The proposed algorithm has efficiently removed the shadows from the video frames and helped to produce more accurate results in moving object detection. As the shadows are removed from the output images, the accuracy rate of the proposed algorithm also increases.

## 4 Proposed Method

### 4.1 Preprocessing on Input Images

The proposed technique aims in identifying the shadows by applying the Normalized Difference metric and Otsu Method is premised on consequent thresholding. In this article, at first, the HSV color model localizes the shadow area and then a buffer area is used to eliminate the shadow region. The obtained data is either a static image or a collection of image sequences. This ambient image(s) is transformed from RGB to HSV space, as the feature vector of Hue Saturation Value is symmetric toward light, i.e., it expresses the perceptual measures of feature representation, irrespective of changes in the unconstrained situations of the image. Otsu's method of thresholding is used in Shadow segmentation in HSV color space. Within a blend of light waves, a feature called Hue is related to the extent of wavelength supremacy. Hue represents the color that corresponds to the specific wavelength viewed by an observer. Saturation makes reference to the perceived worth or indicator of hue-mixed white light. As a result, shadow areas in the spectral domain have the highest assessment of saturation and the core value component.

(1) Low value, since elevated particles block the imminent light from the Sun.
(2) High absorption of short blue-violet spectrum is the outcome of the Rayleigh scattering effect in the atmosphere.
(3) High hue esteems since shadow regions are dark. So as to identify shadow regions, a positive threshold T fragment the shadowed pictures. The S and V components of I are utilized to remove shadows utilizing NDI as

$$NDI = \frac{S - V}{S + V} \tag{1}$$

A non-negative threshold T can be employed to segment the NDI images, in order to identify shadow regions. The Otsu thresholding algorithm is used for automated threshold evaluation. This method seeks an optimal T level, which maximizes-

$$NDI(T) = \frac{\mu.\omega(T) - \mu(T)}{\omega(T).\mu(T)} \tag{2}$$

$$\omega(T) = \sum_{i=0}^{T} P_i,\, \mu(T) = \sum_{i=T+1}^{255} P_i, \mu(T) = \sum_{i=0}^{255} i.P_i \tag{3}$$

$P_i$ is the possibility of gray-level pixels in the image. Dark pixels are those that have an image pixel's NDI value greater than the limit (T).

The final result will be a binary image, designated by the letters $I_{shadow}$, with all shadow pixels assigned to 1 and all non-shadow pixels to 0.

When the shadows are observed, they are graded according to the definition of the components attached. First, morphological dilation action is operated to increase the effectiveness of the shadow boundaries and the average neighborhood of the region is used to accommodate the darkness.

## 4.2 Mechanism of Pixel Classification

A pixel is categorized in relation to nearby neighbors in a selected color space in order to mitigate the impact of any outliers. For this reason, a series of samples rather than an explicit pixel model is used to describe each backdrop pixel. Since the background pixel's pdf cannot be estimated, the closest samples in the collection are used to compare the background pixel's current value to them. In contrast to current algorithms, particularly those of consensus-based approaches, this represents a fundamental distinction. The background samples are used to compare new values, with the adjacent pixels. The threshold value in the ViBe**(Visual Background extractor)** approach is the minimal number of matches, and it is indicated by #min. In this case, #min's value has been adjusted to 2. The key concept behind this is that comparing an intensity with neighboring intensity, rather than a broad sample size, is more sensible. Formally, let v(x) define the input in a defined Euclidean spatial domain at x and vi is a reference image. Each pixel x is defined by N no. of values in the antecedent images.

$$M(x) = \{v_1, v_2 \ldots . v_N\} \tag{4}$$

The set of samples is compared to v(x). To count the records that fall within it, a sphere termed SR(v(x)) is developed. R represents the sphere's center radius in this scenario (x). The collection of samples and the values in the set that traverse this sphere are determined by the following:

$$S_R(v(x)) \cap \{v_1, v_2, \ldots\ldots, v_N\} \tag{5}$$

The Euclidean Distance is used to compare each pixel in the present frame to that in the related model. To assess how similar an intensity is, we use the term Euclidean distance. The minimum distance is taken as 20, which is the standard. Unless it is noted as a foreground pixel, the current intensity will be a backdrop pixel if the

Euclidean distance between any two pixels is less than or equal to 20, and they are located in the appropriate model.

## *4.3 Morphological Operation*

The findings in foreground detection include fractures as well as noises in pictures. This article incorporates the principle of template matching intervention upon its observed foreground to eliminate certain cracks and disruptions. The main premise is to handle square shaped structural components to transform the picture's matching structure. This operation is helpful in removing tiny openings and covering cracks, incorporating erosion and dilation. Dilation processing helps the coordinates to detect the edge of the objects in an image. The structuring element decides the dilation's appropriate impact on the pixel values. If X is the source image, let Y be the structural element. The X by Y dilatation is specified by the following:

$$X \oplus Y = \bigcup_{y \in Y} Xy \tag{6}$$

where X is translated into $X_y$ by y. With the Erosion process, pixels on object borders are removed. The coordinates of the structural components from the picture are subtracted in this procedure. The erosion of X by Y is also given by the following expression:

$$X = \bigcap_{y \in Y} Yy \tag{7}$$

Close operation incorporates structural element Y first to deform image X and also to erode it. The expression also provides the closing of X by Y

$$X \bullet Y = (X \oplus Y)\Theta Y \tag{8}$$

There are still some noises in the image after the morphological close technique. A modified median filter is employed to eliminate such noise. This filter keeps the edge features of the foreground entities while also removing noise from them. Its basic goal is to traverse through the signal entry by entry, supplanting each entry with the median of neighboring entries. It is a stochastic image analysis filter that magnifies the vague information issues that linear filters can generate. By using these above mentioned steps, a noise-free output image is generated. The architecture of the proposed method is depicted in Fig. 1.

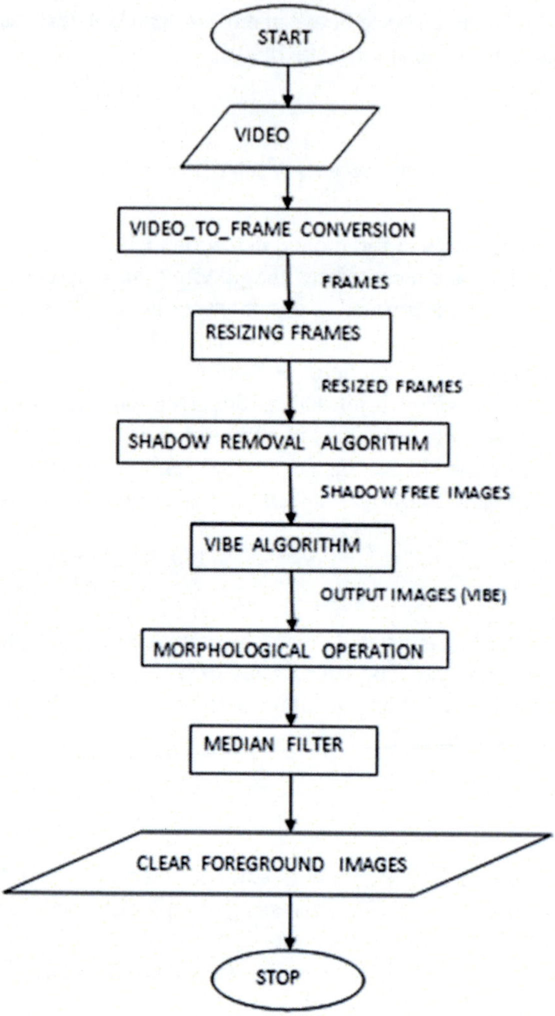


**Fig.1.** Workflow of the proposed method

## 5 Result and Discussion

In this article, the proposed method with some existing approaches has been tested in different Benchmark datasets [2, 10]. From experimental results, it is found that the proposed algorithm achieves more remarkable result than some previous strategies which is depicted in Table 1. As depicted in Figs. 2 and 3, the output of the proposed algorithm has less amount of noise, and unwanted shadow areas are efficiently removed. Some existing algorithms fail in detecting and removing the shadows cast by the moving objects, and hence have given a much lower accuracy rate. Along with it, these algorithms are also incapable of removing noises from the output images such as salt and pepper noises, and thereby low-quality output image is obtained with less accuracy rate. The proposed algorithm gives better output result for all the benchmark dataset that has been implemented.

Based on the predicted samples, six performance metrics are used: accuracy, precision, recall, specificity, F1 score, and Matthews correlation coefficient (MCC). F1 score and MCC both measure the accuracy of the model. Equation (9) shows a combined index and overall accuracy (OA) to obtain the complete and final model performance.

$$OA = Accuracy + F1 + MCC \tag{9}$$

Kim et al. [12] and Pal [7] used a background subtraction method using a hue and texture model to detect moving objects in the presence of noise, and they attained an accuracy of almost 68 and 74%, respectively. Sajid et al. [4] used a multimodal background subtraction technique in detecting moving objects with shadow removal and achieved an accuracy of 78%. Yun et al. [5] analyzed the scene conditions and formulated a background updation model for detecting objects with an accuracy of 80.45%. Although they achieve good accuracy, their methods incur substantial additional costs. Minaeian et al. [9] achieved 77% accuracy in detecting objects and eliminating shadows by using collaborative learning and correlation filter. Finally, the proposed method shows its efficiency in terms of an accuracy of 81.37%, precision of 0.81, recall of 0.84, and f1 score of 0.83 compared to other existing approaches. Also, this method is very less expensive as well as very efficient in terms of accuracy in detecting the object by eliminating shadows. The comparative analysis of the proposed method with other existing approaches [4, 5, 7, 9, 12] in terms of accuracy, recall, precision, and F1 score is illustrated in Table 1.

**Table 1** Comparative analysis of the proposed method with some existing approaches in respect of evaluation metrics

| Methods used | Accuracy(%) | Recall | Precision | F1 score | Specificity | MCC |
|---|---|---|---|---|---|---|
| Hue and texture based [12] | 68.02 | 0.68 | 0.64 | 0.66 | 0.65 | 0.49 |
| Multimodal based[4] | 78.01 | 0.80 | 0.75 | 0.76 | 0.75 | 0.72 |
| Scene conditional [5] | 80.45 | 0.81 | 0.81 | 0.81 | 0.80 | 0.78 |
| Background subtraction[7] | 74.87 | 0.74 | 0.76 | 0.73 | 0.72 | 0.71 |
| Correlation-based [9] | 77.71 | 0.78 | 0.79 | 0.72 | 0.70 | 0.67 |
| Proposed method | 81.37 | 0.84 | 0.81 | 0.83 | 0.84 | 0.80 |

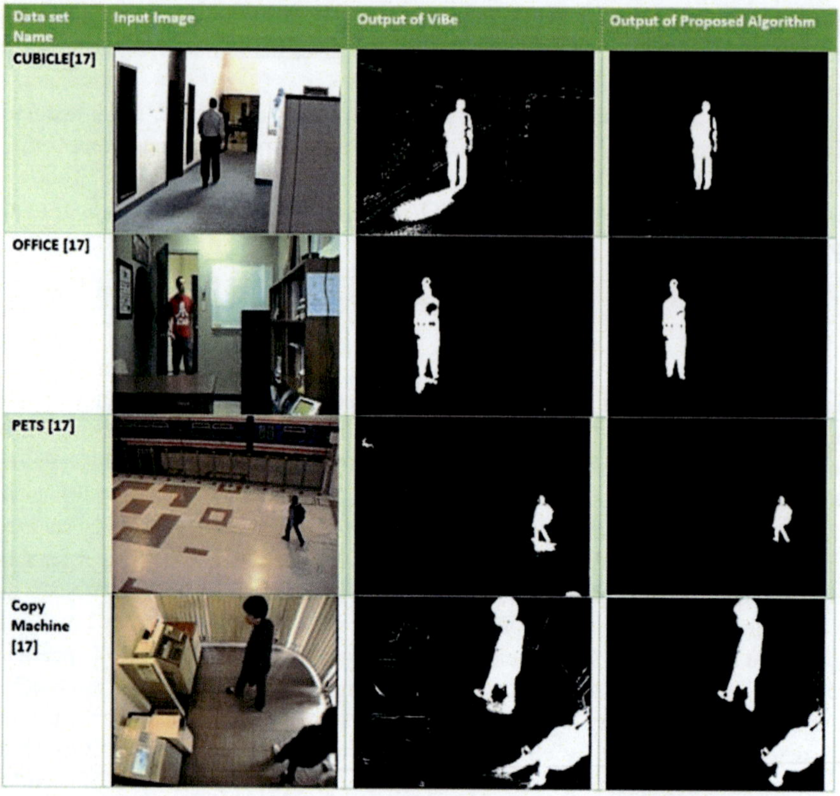


**Fig. 2** Experimental results on Benchmark dataset [10, 11]

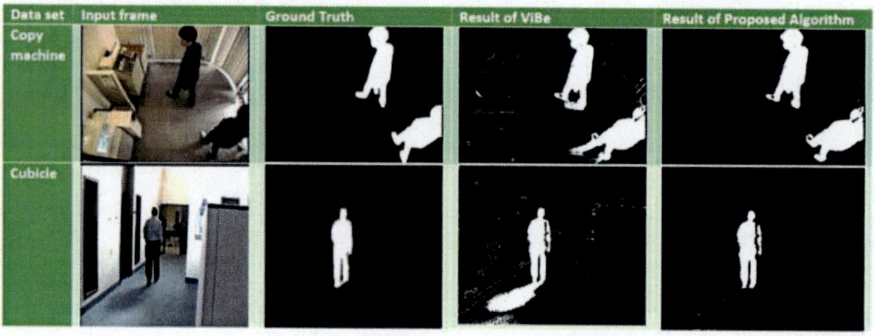


**Fig. 3** Experimental results on Benchmark datasets with ground truth [10, 11]

# 6 Conclusion

In this article, a technique has been put forth that may reliably detect the object by removing noise and shadow from the foreground images. The suggested technique is more accurate and also identifies the shadow zone. With this method, accurate experimental results are provided that aid in efficient object recognition. However, the output frames sometimes still contain noise. To overcome this limitation, foreground detected images are subjected to improvised versions of morphological and median filtering in order to get clear foreground images at a little additional cost. Furthermore, a comparison is made between the proposed method and the existing approaches, and the experimental findings show its effectiveness. The experimental outcomes exhibit that the proposed method shows its supremacy in terms of an accuracy of 81.37%, precision of 0.81, recall of 0.84, and f1 score of 0.83 compared to other existing approaches.

# References


1. Tom AJ, George SN (2020) Simultaneous reconstruction and moving object detection from compressive sampled surveillance videos. IEEE Trans Image Process 29:7590–7602
2. Chen Y, Wang J, Zhu B, Tang M, Lu H (2019) Pixelwise deep sequence learning for moving object detection. IEEE Trans Circuits Syst Video Technol 29(9):2567–2579
3. Eltantawy A, Shehata MS (2019) An accelerated sequential PCP-based method for ground-moving objects detection from aerial videos. IEEE Trans Image Process 28(12):5991–6006
4. Sajid H, Cheung S-CS (2017) Universal multimode background subtraction. IEEE Trans Image Process 26(7):3249–3260
5. Yun K, Lim J, Choi JY (2017) Scene conditional background update for moving object detection in a moving camera. Pattern Recogn Lett 88(1):57–63
6. Zheng A, Zhang L, Zhang W, Li C, Tang J, Luo B (2017) Local-to-global background modeling for moving object detection from non-static cameras. Multimedia Tools Appl 76(8):11003–11019
7. Pal T (2021) Improved background subtraction technique for detecting moving objects. Rec Adv Comp Sci Commun 14(9)
8. Minaeian S, Liu J, Son YJ (2018) Effective and efficient detection of moving targets from a UAV's camera. IEEE Trans Intell Transp Syst 19(2):497–506
9. Wibowo SA, Lee H, Kim EK, Kim S (2018) Collaborative learning based on convolutional features and correlation filter for visual tracking. Int J Control Autom Syst 16(1):335–349
10. Barnich O, Droogenbroeck MV (2011) Vibe: a universal background subtraction algorithm for video sequences. IEEE Trans Image Process 20(6):1709–1724
11. Zivkovic Z (2004) Improved adaptive gaussian mixture model for background subtraction. In: Proceedings of the 17th international conference on pattern recognition (ICPR), vol 2
12. Kim Y, Bae T, Ahn S (2020) Background subtraction with shadow removal using hue and texture model for moving object detection. In: International conference on electronics, information, and communication (ICEIC). Barcelona, Spain, pp 1–2

# Pre-processing Pipelines for Effective Segmentation of Lymphatic Filariasis Parasite Images

**Priyanka Kumar and Kanojia Sindhuben Babulal**

**Abstract** Lymphatic Filariasis commonly known as 'Elephantiasis' is a vector congenital disease that creates disfiguring as well as disabling in the forbearer. This gruesome disease is acquired in the childhood phase, but its clinical manifestation arises subsequently in life. It may cause short-term or perpetual disability. The significance of developing robust and efficient techniques to diagnose the Lymphatic Filariasis parasites within the peripheral blood smear has been accentuated by W.H.O. as well as the government counterpart. So, automated screening of the peripheral blood smear to detect the presence of microfilariae (mf) has to be initiated. The acquired images may contain various noises due to different illumination environments as well as the restricted image-acquiring resolution capabilities. Therefore, this research aims to eradicate the flaws of noise with the pipeline of various pre-processing techniques as follows: (i) color space transformation, (ii) image filtration using median filter, (iii) morphological operators of dilation and erosion, (iv) binary segmentation, and (v) smoothing and discretization. The use of various pre-processing techniques has produced resultant images which can be visually and numerically verified by 11 morphological image quality indexes. After observing these diversified image quality indexes, it may be concluded that the aforementioned pre-processing pipeline embraces image enhancement for better-segmented results.

**Keywords** Color space transformation · Dilation · Erosion · Segmentation · Smoothing · Discretization

## 1 Introduction

Lymphatic Filariasis (LF) is a deadly archetype of vector-borne diseases, and one of the major causes of disfiguring or disabling in the forbearer. The physical out-turn of this gruesome disease is very painful with primarily swollen limbs and other body

P. Kumar · K. S. Babulal (✉)
Department of Computer Science and Engineering, Central University of Jharkhand, Ranchi 835205, India
e-mail: sindhukanojia@gmail.com

 
S. Borah et al. (eds.), *Advanced Computational and Communication Paradigms*, Lecture Notes in Networks and Systems 535,
https://doi.org/10.1007/978-981-99-4284-8_33

parts. The patient finds it arduous to accomplish everyday routine work because the affected area becomes prone to frequent infections. In the early stages, this disease is asymptomatic and continues to damage the body's lymphatic system. The causative agent of this disease is thread-like nematode parasites. In India, almost 98% of the total cases are due o the filarial parasite species—Wuchereria bancrofti covering a total of 250 cities. The early detection of LF could certainly be recommended to cease further damage to the lymph nodes. The conventional approach used for the pathological confirmation of the LF is through the presence of the microfilariae (mf) in the peripheral night blood smear. Once the presence of mf is confirmed by the diagnostic center, the patient undergoes treatment.

The digital images acquired to automate this conventional approach are having different varieties of noises which could be only eliminated by different pre-processing techniques. The resultant of the pre-processed methodology will provide balanced segmented images resulting in good data input for the classifier's prediction. The pre-processing techniques are generally proved to be an effective way of resulting in better-segmented regions of the image, but, certainly, it comprises the processing speed. The principal goal of this research study is to investigate the different effects of pre-processing techniques to eliminate the noise on the acquired microscopic images along with the evaluation of quality indexes to confirm the caliber of the proposed work.

The core contribution of this research work unfolds three different objectives:

1. The standard methodology of image acquisition for the creation of LF database.
2. The application of different pre-processing techniques to enhance the segmented outcome of mf within the peripheral blood smear digital images.
3. The efficacy of the pre-processing techniques used on the acquired images is verified by the standard morphological image quality performance evaluation metrics.

AI-augmented systems can certainly enhance the efficiency of automated digital pathology by overwhelming the drawbacks of manual pathology. Since LF is pervasive in nature, therefore, the diagnosis of its parasite within the peripheral blood smear through the image processing tools plays a very vital role.

Below are some literature relevant to the field determining the relevance of pre-processing for better segmentation results.

The Electrocardiogram (ECG) data in [1] are having artifacts by distinct factors like interference caused by the electrooculogram, and electro-cardio meter are resolved by the techniques of cascade three adaptive filters and least mean square to produce less noisy data. In the research paper [2], the medical images of mammography are used to detect breast cancer, but it consists of unwanted parts which act as obstacles to correct prediction. So, binarization based on Otsu's thresholding the image quality is enhanced. Bandyopadhyay [3] deals with the transition of the image enhancement of the mammogram images by using the CLAHE mechanism which resulted in better feature extraction of the extracted region of interest (ROI). The research work of [4] aims for pre-processing tools such as color format conversion and color normalization for the removal of noise and artifacts from the digital microscopic images for better segmentation and characterization of erythrocytes. These noises were due to the uneven illumination of the blood smear. He et al. [5] showed an improved appearance of the images across the periphery of breast images by the use of different filters like the mean filter, and also the use of smoothing has increased the quality of the images. The research paper [6] deals with the magnetic resonance imaging dataset to assess brain tumor. It has used 3D spherical coordinate transform during the pre-processing of the images hypothesized to increase the DCNN model accuracy.

## 2 Materials and Methods

This section details the steps followed to determine the method of sample collection of the peripheral blood smear, acquisition of microscopic images of the mf parasite, pre-processing of the acquired images, segmentation, smoothing, and discretization defined in the below sections.

### *2.1 Proposed Framework for Segmentation of mf*

Figure 1 illustrates the workflow of this research idea, with emphasis on the utility of the pre-processing steps for better image enhancement. The first four steps of the workflow determine the color conversion, normalization, filtration, segmentation, and smoothing with discretization which proceeds the binary classification further.

### *2.2 Sample Collection*

The various types of mf samples are gathered from National Centre for Disease Control (NCDC), Varanasi, and Bhubaneswar, India. These samples consist of peripheral blood samples of patients suffering from Filaria. Subsequently, the collected samples are stained using JSB-1 for clear visualization of the parasite and also used to remove the debris of other blood components like erythrocyte, and W.B.C. Figure 2 represents some of the samples collected from the aforementioned center.

### *2.3 Image Acquisition*

The JSB-1 stained slide with smear upwards on the microscopic stage and focus on the smear with a 10x objective. When the organism or any suspicious material is seen, it is switched to an oil immersion objective with a magnification of 100x,

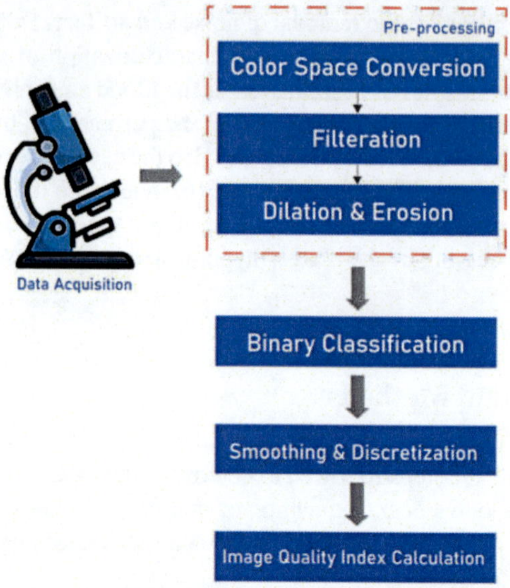


**Fig. 1** Flow diagram of proposed framework

**Fig. 2** Collected samples of Microfilariae parasite (mf)

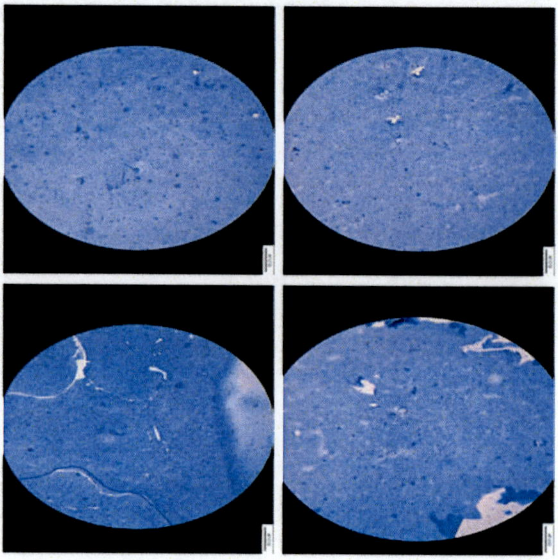

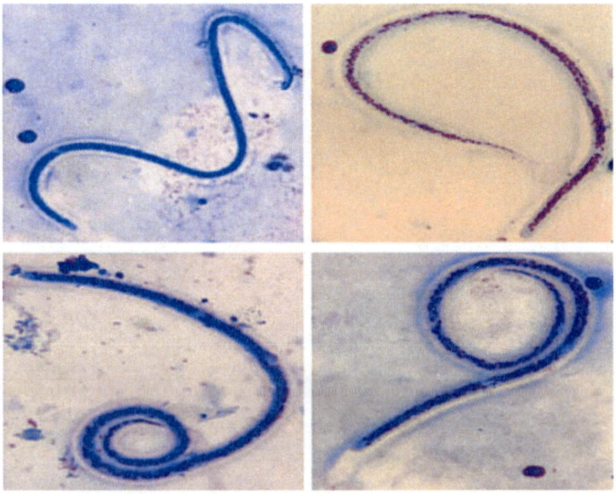

**Fig. 3** Acquired images of Microfilariae parasite

and the image is captured, Fig. 3. The dataset consists of two distinct species of LF—Wuchereria bancrofti and Brugia malayi.

## 3 Pre-processing

The main vision of pre-processing is to improve the image dataset that has unwilling distortions. It enhances the image features vital for further processing and results in more accurate predictions by the machine learning tools. In general, it aims to improve the quality of the acquired image so that we can eliminate undesirable distortion and also enhance the quality of the features which are necessary for our application. Pre-processing also helps to eliminate the unwanted pixel called noise. Further, in our experiment, we found that the edge of the parasite has to be preserved for better predictive accuracy of the machine. So, different filtering techniques are implemented in the below sections to preserve the edges of the parasite.

### *3.1 Color Space Transformation*

Basically, the hematological images grabbed by the microscope are obtained in the RBG color space format. Since the correlation between all three color space components in RGB is cohesive in nature, it is not very suitable for the data collected in this research work. This difficulty was eliminated by the conversion of RGB color images to the L*a*b color model. This color model certainly reduces the color dimension and hefty correlation between the different channels. In the L*a*b color model, 'L'

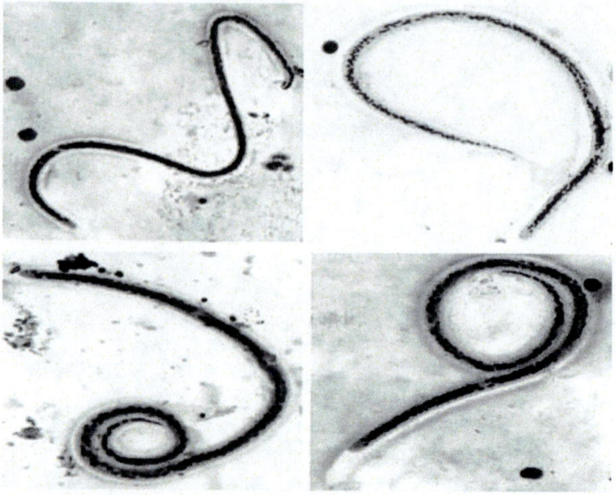

**Fig. 4** Images transformed into L*a*b color space

denotes luminosity, $a^*$ denotes the chromaticity, and $b^*$ denotes the color information [7, 8]. Consequently, the hematological images are transformed to CEILAB color space to reduce the complication of color dimension dependability. So, to reduce the discrepancies of color space dimension dependability with each other, color space conversion is used. Figure 4 represents the images of the parasite after the conversion into CEILAB.

### 3.2 *Image Filtration*

Although, necessary precautions were taken to reduce the uneven illumination of the blood smear exposed for image acquisition, some of the captured images were containing air bubbles and also the debris of other components of the blood. So, it accelerates the need for filtration within the acquired digital microscopic images. Heterogeneous filtering techniques could be deployed for the elimination of noise within the image but certainly depending upon the variety of the noise present. In this research methodology, median filter is used to furnish a better trade-off between noise cut out in addition to edge preservation [9]. Usually, a median filter is applied to reduce the random noise consisting of noise amplitude probability density having large tails along with periodic patterns. The output-filtered image is evaluated by placing the median of the values in the input window. When an edge is trespassed, one side or the other dominates the window, then the output switches sharply between the values. As a result, the edges are not blurred.

Figure 5 gives a glimpse of the enhanced images after the application of median filters.

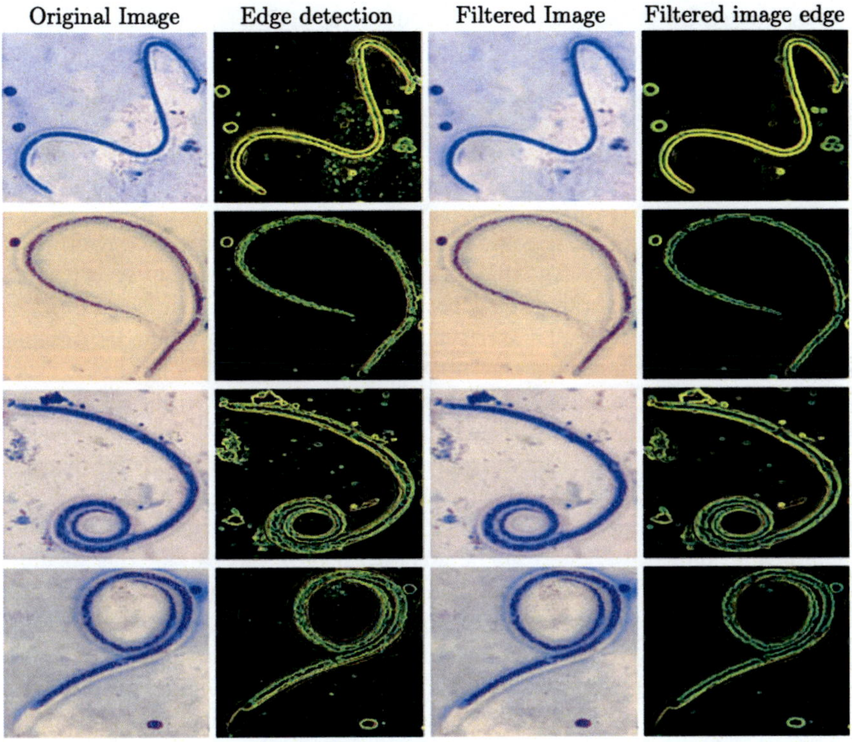


**Fig. 5** Enhanced images after application of median filter

## 3.3 Morphological Operator

In this research work, filters are employed for the application of two different morphological operators, namely dilation and erosion. Dilation is applied to fabricate the admissible features within the image more prominently by filling in the tiny holes of noise. It appends the pixels to the borderline of objects within an image. The dilation function fixes the output pixel to 1 because the structural element has defined the neighboring element as one.

Erosion is also a morphological operator that evaluates the output pixel considering the minimum among the neighborhood [10]. If any of the nearer pixels is 0, then the output pixel is zero. It eliminates the floating pixels and contour such that just the substantive elements remain [11]. The remaining lines appear to be thinner and smaller in shape. In particular, a filter used in this research work restores the value of every pixel with the median of its adjoining pixels for a window size given in Eq. 1:

$$(2R + 1) \times (2R + 1) pixels \tag{1}$$

where $R =$ radius. Although, in this work, the value of $R = 2$, it retains satisfactory outcomes in the experiment.

### *3.4 Binary Segmentation*

Automatic detection of the edge points involves a procedure that distinguishes the parasite pixels from the background. In this research work, a total of seventeen different segmentation algorithms have been implemented, although it was observed that the best-segmented results were evaluated by the Otsu algorithm. In particular, Otsu is an automatic threshold clustering method where the idea lies in the classification of the pixel based on the threshold value 'T' derived by the minimization of the intensity variance between the two different classes [12, 13]. This is defined as the weighted sum of the variance between two classes separated by the threshold value 'T' [14, 15]. It is noted that this algorithm demands bimodal distribution in the evaluated histogram of the required image for better performance. This requires a profound valley between the two different peaks of the histogram. This condition generally contains a clear object adjoining uniform background.

The outcome of the Otsu segmentation algorithm is represented in Fig. 6. It is rendered effectively by the size of the image object to be segmented because its performance degrades for very minute sizes. The apparent solution for the above-mentioned problem is to observe a close-up of the grabbed image to magnify the dimension of ROI. Furthermore, the algorithm can be distressed by the presence of various noises in the image. However, this step is solicited on the filtered images, hence such flaws do not appear in this research progress.

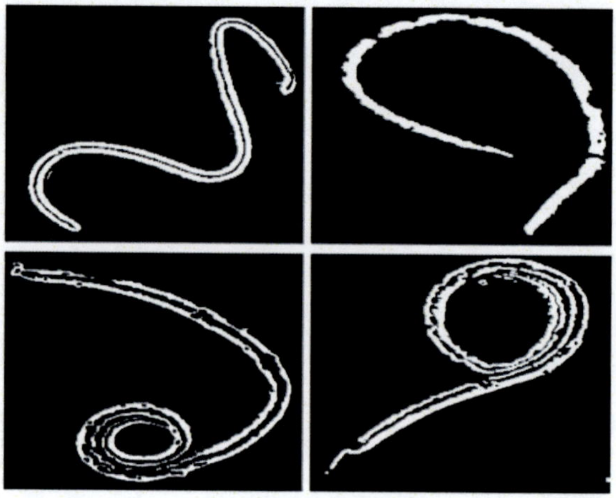

**Fig. 6** Binary segmented images using Otsu

### 3.5 Smoothening and Discretization

The result of the above step is a cluster of pixels that belongs to the mf parasite. Since the boundary distinguishing between ROI and the image background is very noisy and irregular, to abolish the uneven boundary curve, 'smoothing' is applied. It preserves the edges by pulling out the small islands and filling the holes [16]. Since smoothing of the desired region is achieved, discretization of the boundary points bifurcating mf and the background image is processed. This research aims to apply the 'Arc-length discretization' technique to select the equally spaced boundary points while advancing the boundary curves [17, 18]. The distance traced between the boundary points and their adjacent pixel is accounted for by the parameter 'h'. The value of this parameter could be evaluated either manually by the parasitologist or automatically prior to segmentation. The output generated after smoothening and discretization of segmented images is viewed in Fig. 7.

## 4 Evaluation of Image Quality Parameters

This research work aims to develop an indigenous framework of the pre-processing mechanism for the better segmentation of the mf parasite within the peripheral blood smear microscopic image. Table 1 presents the experimental setup environment required for this experiment. Table 2 refers to numerical morphological index parameters which stipulate the sheer quality of the segmented results evaluated after the application of techniques discussed in Sect. 2—(i) color space conversion; (ii) median filter; (iii) dilation and erosion; (iv) smoothening and discretization. The segmentation algorithm used to bifurcate the foreground from the background is the 'Otsu' segmentation algorithm.

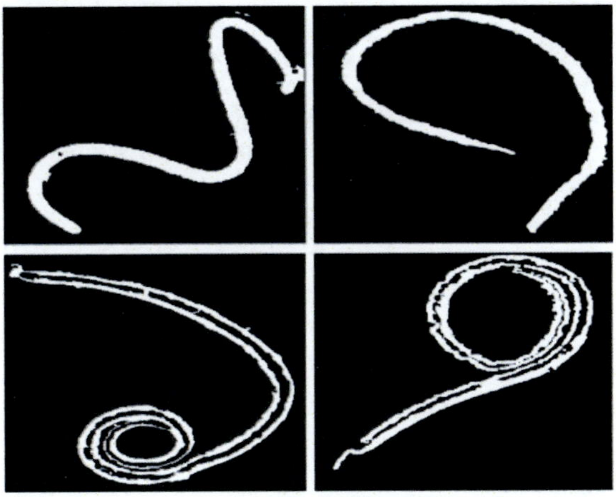

**Fig. 7** Smoothening and discretization of segmented images

**Table 1** Experimental setup environment

| Setup | Environment |
|---|---|
| Operating system | Windows 11 Home Single Language, 64-bit |
| Memory | 16 GB |
| CPU | I7-8650U CPU@ 1.90 GHz (8 cores), 2.1 GHz |
| Graphics | GeForce GTX 1650 with Max-Q design, 4 GB |
| Python | 3.9.7 |
| Anaconda | 4.9.2 |

**Table 2** Image quality assessment index

| Sl. no. | Image quality index | Value |
|---|---|---|
| 1 | Root mean squared error (RMSE) | 30.45 |
| 2 | Mean squared error (MSE) | 927.20 |
| 3 | Peak signal-to-noise ratio (PSNR) | 18.45 |
| 4 | Universal quality image index (UQI) | 0.98 |
| 5 | Multi-scale structural similarity index (MS-SSIM) | 0.98 |
| 6 | Erreur relative globale adimensionnelle de synthèse (ERGAS) | 9325.72 |
| 7 | Spectral angle mapper (SAM) | 0.13 |
| 8 | Visual information fidelity (VIF) | 0.38 |
| 9 | Block sensitive—peak signal-to-noise ratio (PSNR-B) | 18.45 |

Inferences of the observation points from this research activity:

- Structural Similarity Index (SSIM) is a parameter based on the attributes of luminance, contrast as well as structure. Its loft value signifies a consistent quality outcome.
- Peak Signal-to-Noise Ratio (PSNR) for all four examples are of higher values which surely indicate the good quality of the segmented image.
- Universal Quality Image Index (UQI) is also elevated which certainly briefs the quality work of the segmented result.

**Conflict of Interest** The authors declare no conflict of Interest.
**Declaration of competing Interest** The authors report no declaration of competing interest.

**Acknowledgements** The authors acknowledge sincere thanks to Dr. N. M. Somalkar (Regional Director, R.O.H.F, Bhubaneswar), Dr. S. S. Mishra (Senior Technician, R.O.H.F, Bhubaneswar), Dr. N. Velayudham (Filaria Consultant, R.O.H.F, Bhubaneswar), Dr. N. Kumar (Zonal Coordinator, W.H.O., Varanasi) for providing guidance and assistance in data collection of microfilariae parasite. The authors would also like to extend their heartiest thanks to Mr. C. Kumar (Research Scholar, N.I.T, Sikkim) for useful discussions.

## References

1. Correa AG, Laciar E, Patiño H, Valentinuzzi M (2007) Artifact removal from EEG signals using adaptive filters in cascade. J Phys Conf Ser (IOP Publishing) 90:012081
2. Fazilov SK, Yusupov O, Abdiyeva KS (2022) Mammography image segmentation in breast cancer identification using the Otsu method. Web Sci Int Sci Res J 3(8):196–205
3. Bandyopadhyay SK (2010) Pre-processing of mammogram images. Int J Eng Sci Technol 2(11):6753–6758
4. Kumar P, Babulal KS (2022) Hematological image analysis for segmentation and characterization of erythrocytes using FC-TriSDR. In: Multimedia tools and applications, pp 1–26
5. He W, Hogg P, Juette A, Denton ER, Zwiggelaar R (2015) Breast image pre-processing for mammographic tissue segmentation. Comput Biol Med 67:61–73
6. Russo C, Liu S, Di Ieva A (2022) Spherical coordinates transformation pre-processing in deep convolution neural networks for brain tumor segmentation in mri. Med Biol Eng Comput 60(1):121–134
7. Kumar P, Thakur RS (2021) An approach using fuzzy sets and boosting techniques to predict liver disease. CMC-Comput Mater Contin 68(3):3513–3529
8. Kumar P, Thakur RS (2019) Diagnosis of liver disorder using fuzzy adaptive and neighbor weighted K-NN method for LFT imbalanced data. In: 2019 international conference on smart structures and systems (ICSSS). IEEE, pp 1–5
9. Kumar P, Thakur RS (2021) Liver disorder detection using variable-neighbor weighted fuzzy k nearest neighbor approach. Multimed Tools Appl 80(11):16515–16535
10. Liu P, Zhang C, Qi H, Wang G, Zheng H (2022) Multi-attention densenet: a scattering medium imaging optimization framework for visual data pre-processing of autonomous driving systems. IEEE Trans Intell Transp Syst
11. Russo C, Liu S, Di Ieva A (2022) Spherical coordinates transformation pre-processing in deep convolution neural networks for brain tumor segmentation in mri. Med Biol Eng Comput 60(1):121–134
12. Goyal P, Duval Q, Seessel I, Caron M, Singh M, Misra I, Sagun L, Joulin A, Bojanowski P (2022) Vision models are more robust and fair when pretrained on uncurated images without supervision. arXiv:2202.08360
13. Babulal KS, Das AK (2022) Deep learning-based object detection: an investigation. In: Futuristic trends in networks and computing technologies: select proceedings of fourth international conference on FTNCT 2021. Springer, pp 697–711
14. Murugesan M, Kaliannan K, Balraj S, Singaram K, Kaliannan T, Albert JR (2022) A hybrid deep learning model for effective segmentation and classification of lung nodules from CT images. J Intell Fuzzy Syst 1–13
15. Reshma G, Al-Atroshi C, Nassa VK, Geetha B, Sunitha G, Galety MG, Neelakandan S (2022) Deep learning-based skin lesion diagnosis model using dermoscopic images. Intell Autom Soft Comput 31(1):621–634
16. Singla C, Sarangi PK, Sahoo AK, Singh PK (2022) Deep learning enhancement on mammogram images for breast cancer detection. Mater Today Proc 49:3098–3104

17. Raja NSM, Arunmozhi S, Lin H, Dey N, Rajinikanth V (2023) A study on segmentation of leukocyte image with Shannon's entropy. In: Research anthology on improving medical imaging techniques for analysis and intervention. IGI Global, pp 1131–1152
18. Başakın EE, Ekmekcioğlu Ö, Çıtakoğlu H, Özger M (2022) A new insight to the wind speed forecasting: robust multi-stage ensemble soft computing approach based on pre-processing uncertainty assessment. Neural Comput Appl 34(1):783–812

# Fish Classification Using CNN and Logistic Regression from Underwater Images

**Pratick Gupta, Pratima Sarkar, Bijoyeta Roy, and Shivam Kumar**

**Abstract** Nowadays curiosity of human being is expanded toward sea. It is difficult to explore the uses of robots by human being. So, people use robots to explore sea and that robot is popularly known as Autonomous Underwater Vehicle (AUV). The work aims toward classification of fish. Embedded systems are generally used to explore seabed. Embedded system is capable of operating heavy software and sensitive technology. In this work, CNN is used for the purpose of classification and Logistic Regression is used as a classifier. Fish4knowledge dataset is used for validating the work. Using evaluation criteria including accuracy, precision, recall, and F1-score, the suggested work is contrasted with several classifier types. In the result section, it is clearly shown that logistic regression is giving best result out of all the classifiers. In this proposed work 95% accuracy was achieved.

**Keywords** Fish classification · Machine learning · Logistic regression · Convolution neural network

## 1 Introduction

The ocean is filled with mysteries, and exploring them underwater has always been a fascinating subject. The state of the ocean varies from location to location [1, 2]. One of the most difficult problems is how to detect and distinguish fish from the hazy seawater. Fish categorization has received much attention in the areas of picture segmentation, pattern recognition, and information retrieval (FC). Based on their physical traits, fish species and families are identified and categorized. For industrial and environmental uses, such as fish farming and weather monitoring,

P. Gupta · P. Sarkar (✉) · S. Kumar
Computer Science and Engineering, Techno International New Town, New Town, India
e-mail: pratima.sarkar@tint.edu.in

B. Roy
Sikkim Manipal Institute of Technology, Pakyong, India
e-mail: bijoyeta.r@smit.smu.edu.in

 
S. Borah et al. (eds.), *Advanced Computational and Communication Paradigms*, Lecture Notes in Networks and Systems 535,
https://doi.org/10.1007/978-981-99-4284-8_34

they have been studied [3]. To identify and categorize the target fish into species, it makes advantage of similarities to the representative specimen image. FC is seen as useful for estimating and counting fish populations, monitoring ecosystems, and describing fish associations. FC should take into account a number of factors, including the fish's orientation and arbitrary size, unpredictability of features, environmental changes, low picture quality, segmentation errors, imaging circumstances, and physical shaping. With the enhancement in technology the level quality of image processing has also been enhanced, which helps in various fields like Under Water Image Detection [4, 5] of under deep water, there exists a variety of aquatic life which cannot be monitored because of blurry video and images. But with the help of CNN [6] (Convolutional Neural Networks) which comes under the domain of Machine Learning, it is quite possible to process and classify these images.

The classification of fish from unprocessed underwater photos is the major goal of this work. Fish4knowledge dataset images mostly consist of one type of fish in a single image. Logistic Regression is utilized as a classifier for fish species while CNN is used for classification. The proposed work utilized assessment parameters such as accuracy, precision, recall, and F1-score to compare various state-of-the-art classifiers like random forest Naive Bayes and K-Nearest neighbors.

## 2 State-of-the-Art Techniques

Since the K-NN Algorithm is non-parametric [7], it works on original data and does not work on imaginary data. This algorithm is often known as a slow learner, since it slowly learns from the training set and then it performs actions for categorization. K-NN is based on Euclidean formula between two points A (X1, Y1) and B (X2, Y2) which is

$$\text{Euclidean Distance} = \sqrt{(X_2 - X_1)^2 + (Y_2 - Y_1)^2} \quad (1)$$

A classification technique based on the Bayes' Theorem is known as Naive Bayes' classifiers [3]. It is a collection of algorithms based on conditional probability, and they are predicted using independent pair. Naïve Bayes Equation can be derived from Bayes' Theorem which is

$$P(A|B) = \frac{P(B|A)P(A)}{P(B)} \quad (2)$$

In order to boost the model accuracy using dataset, Random Forest [4] is a classifier that used many decision trees on various subsets of the supplied dataset and averages the outcomes. The random forest calculates predicted values from each of the values and collectively takes decision depending on majority.

## 3 Background Study

As quicker and more affordable methods to assess fish biomass and assemblage in oceans and freshwater bodies, computer vision and image processing techniques for autonomous underwater fish detection are piquing the interest of marine scientists. [1, 2]. Various strategies have been employed in the past for automatic fish recognition using various image and video processing methods. In [8], the authors have analyzed, the unwater aquatic training dataset was collected from Gulf of Mexico using Artificial neural network by incorporating the concepts of Dropout algorithm to address the issue of overfitting. It is suggested to use a hybrid approach that combines optical flow and Gaussian mixture models with a YOLO deep neural network to identify and classify fish in unconstrained underwater videos [9]. It was implemented on two data sets: The Life CLEF and a dataset provided by The University of Western Australia (UWA). For the purpose of recognizing and identifying fish in the wild, a maximum probability, partial ranking method based on sparse representation-based classification (SRC-MP) is put forth. The fish-category database is utilized to extract feature data using Eigenfaces and Fisher faces [10]. An approach uses fish features such as Efficient Match Kernels (EKM) and Kernel Descriptors (KDES) and trains a multi-class SVM classifier to identify underwater fish and achieved 84.4% classification accuracy on a dataset of 50,000 fish photos from 10 different species [11]. By capturing the scale texture pattern and fish outline data, an image processing and pattern recognition-based algorithms for unrestricted underwater fish categorization is proposed by Spampinato et al. in 2010 [12]. Region-Based Convolutional Neural Network, a cutting-edge machine learning technique is used to solve general object detection and localization challenges, by offering a unified strategy to recognize freely moving fish in unrestricted underwater environments. By using background subtraction and optical flow, a novel method is deployed to find fish motion information in videos, and then the results are integrated with the raw image to provide fish-dependent candidate regions [13]. By learning the background pixel distribution, Gaussian Mixture Modeling is a potent technique for separating foreground fish from background items. In [14], an approach for fish detection in complex background settings that combines pixel-wise posteriors and Gaussian Mixture Models is explored on Fish4Knowledge repository. Based on the density of shrimps, the image data were divided into three categories: low density, medium density, and high density. Using the parameter calibration technique, an enhanced Mask Regional Convolutional Neural Network (Mask R-CNN) model was introduced in [15]. Using ResNet-101, either by default or with parameter modification, the accuracy and effectiveness of shrimp recognition was confirmed.

# 4 Proposed Methodology for Fish Classification

Fish classification is an important issue to maintain ecosystem of underwater environment. The work has used CNN for feature extraction and logistic regression as classifier. The work follows the following phases as shown in Fig. 1.

## 4.1 Image Processing

Each image in train and test folder are loaded using load_img () function. Due to the lack of a powerful GPU on the machine on which the algorithm was executed, the entire image size was reduced to 150 × 150 pixels and converted into Numpy array and it is stored as X_train and X_test respectively. From.csv file the true values of the images are stored as Y_train and Y_test. A DataGenerator function was created and then train and test was transformed in batches so that it can easily be trained and tested.

## 4.2 The Model for Convolutional Neural Network (CNN)

Convolutional neural network works very well when we give input as image, voice, or audio. Convolutional, pooling, and fully connected (FC) layers are their three major varieties of layer as shown in Fig. 2. Convolutional layer is the first layer of CNN in which images convolve with some filters to extract features. Second layer is pooling layer which is used to reduce feature size and computation overhead. The last layer is fully connected layer which is responsible for classification of images. While working with large size images, it consists of huge amount of feature and so different layers are used for aggregation of features.

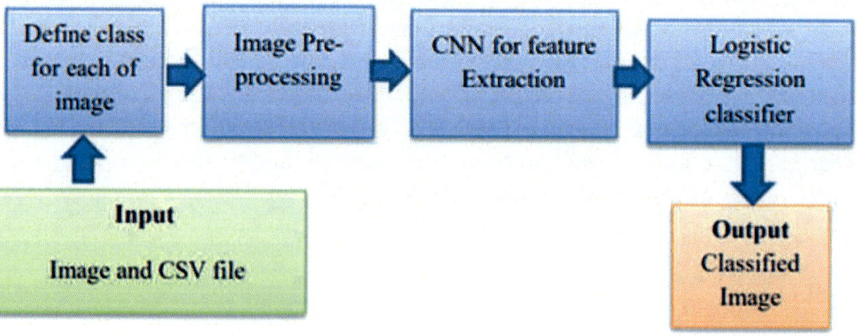


**Fig. 1** Block diagram for flow of work

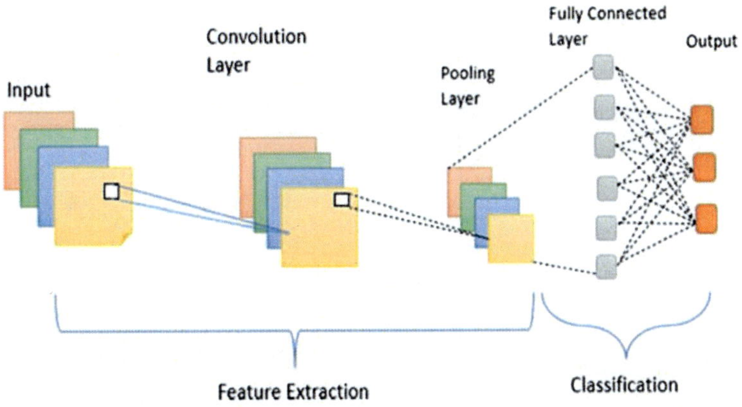


**Fig. 2** Convolution Neural Network Architecture

Here we are using different classifiers along with CNN and compare the result among them to get the most use out of it. The different classifiers included are K-NN, Logistic Regression, Naïve Bayes, and Random Forest.

The CNN model was created using Sequential and Convolution2D, imported from Keras. Two convolutional layers and two pooling layers are created. In the first layer of convolution: filter = 32, kernel_ size = (3, 3), input size = (150, 150, 3), activation = ReLU, then pooling layer is added: pool_size = (2, 2). In second layer of convolutional layer: filter = 64, kernel_size = (3, 3), input_size = (150, 150, 3), activation = ReLU, then finally pooling layer is added: pool_size = (2, 2). After creating layers successfully its features are flattened.

### 4.3 *The Model for Logistic Regression*

Logistic regression is a probability-based classification model. It is a supervised learning technique so learning from dataset and no pre-requisites are available. Logistic regression follows sigmoid curve for classification as shown in Fig. 3. In linear regression, it sometimes crosses the probability value 1 which is meaningless. To resolve the issue sigmoid function is used for classification in logistic regression formula as follows.

Using the CNN model which was created previously train and test data was predicted and reshaped to single row. Now, pipeline was created. Pipeline is the combination of many procedures that can be cross-validated while using various parameters. To scale down the data StandardScaler technique is used. In pipeline list of tuple is given and in this tuple it will be having two values one is the type of operation that is done so here it is given StandardScaler and second is classification technique, i.e., Logistic Regression. Now the pipeline was fitted and score was calculated. Using the pipeline test data is predicted and classification report is generated.

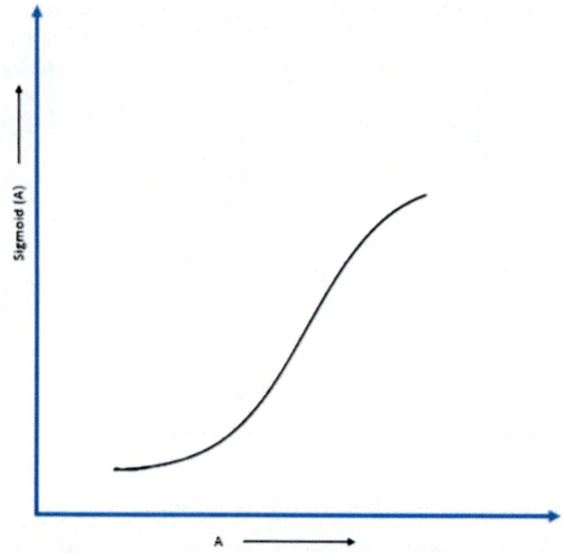


**Fig. 3** Logistic regression curve

# 5 Dataset Description

Fish4knowledge data set is used for fish detection of the work [16]. The dataset is created by Fish4knowledge project with the main aim to understand marine environment. The project does not work on synthetic data. Live videos are used for capturing data. The entire dataset is consisting of more than 27,370 data and 23 Species of fish. Table 1 represents the species names with corresponding number of fish. The Python programming language has been used for all of the work. The dataset was downloaded from [16] and using Python dataset was divided into Train, Test, and validation. In which Train folder contains 16,268 images and 1 csv file, Test folder contains 5440 images and 1 csv file, and Validation folder contain 5662 images and 1 csv file.

# 6 Evaluation Metric

In this work precision, recall F1-Score, and accuracy are used as evaluation parameter. Precision and recall is the primary parameter for evaluation.

1. C1 class is predicted by fish classifier, and it is a correct prediction—True Positive (TP).
2. C1 class is predicted by fish classifier, and it is a wrong prediction—False Positive (FP).
3. C1 class is not predicted by fish classifier, and it is a wrong prediction—False Negative (FN).

**Table 1** Dataset description

| Sl. no | Species | No. of fish | Sl. no | Species | No. of fish |
|---|---|---|---|---|---|
| 1 | Dascyllus reticulatus | 12,112 | 13 | Pomacentrus moluccensis | 181 |
| 2 | Plectroglyphidodon dickii | 2683 | 14 | Zebrasoma scopas | 90 |
| 3 | Chromis chrysura | 3593 | 15 | Hemigymnus melapterus | 42 |
| 4 | Amphiprion clarkii | 4049 | 16 | Lutjanus fulvus | 206 |
| 5 | Chaetodon lunulatus | 2534 | 17 | Scolopsis bilineata | 49 |
| 6 | Chaetodon trifascialis | 190 | 18 | Scaridae | 56 |
| 7 | Myripristis kuntee | 450 | 19 | Pempheris vanicolensis | 29 |
| 8 | Acanthurus nigrofuscus | 218 | 20 | Zanclus cornutus | 21 |
| 9 | Hemigymnus fasciatus | 241 | 21 | Neoglyphidodon nigroris | 16 |
| 10 | Neoniphon sammara | 299 | 22 | Balistapus undulatus | 41 |
| 11 | Abudefduf vaigiensis | 98 | 23 | Siganus fuscescens | 25 |
| 12 | Canthigaster valentini | 147 | | | |

4. C1 class is not predicted by fish classifier, and it is a correct prediction—True Negative (TN).

Parameters are calculated by the following formula:

$$\text{Precision} = \frac{TP}{TP + FP} \tag{3}$$

$$\text{Recall} = \frac{TP}{TP + FN} \tag{4}$$

$$\text{Accuracy} = \frac{TP + TN}{TP + TN + FN + FP} \tag{5}$$

$$F1 - score = \frac{2 \cdot \text{Recall} \cdot \text{precision}}{\text{Recall} + \text{precision}} \tag{6}$$

## 7 Result and Discussion

In this work, Logistic Regression for fish classification is used. 23 classes are used for fish classification from Fish4Knowledge dataset. Google Colab pro with Python is used for execution of the work. Table 2 presents the specification of the proposed work.

**Table 2** Specification of the developed model

| Parameters | Specification |
|---|---|
| Total images | 27,370 |
| Train images | 16,268 |
| Test images | 5440 |
| Validation images | 5662 |
| Batch size | 64 |
| Classes | 23 |
| RAM | 25 GB |
| Processor | GPU |

Table 3 depicts results for fish classification by logistic regression. It is observed from all the results that for all the classes the proposed work gives good precision recall and F1-score. From row number 11, 20, 22, 18 of Table 3 the results are not good as others due to unbalanced dataset.

**Table 3** Fish classification results for Logistic Regression

| SL No | Species | Precision | Recall | F1-score |
|---|---|---|---|---|
| 1 | Dascyllus reticulatus | 0.97 | 0.96 | 0.96 |
| 2 | Plectroglyphidodon dickii | 0.95 | 0.99 | 0.97 |
| 3 | Chromis chrysura | 0.93 | 0.94 | 0.93 |
| 4 | Amphiprion clarkii | 0.98 | 0.99 | 0.98 |
| 5 | Chaetodon lunulatus | 0.98 | 0.99 | 0.99 |
| 6 | Chaetodon trifascialis | 0.78 | 0.86 | 0.82 |
| 7 | Myripristis kuntee | 0.92 | 0.93 | 0.93 |
| 8 | Acanthurus nigrofuscus | 0.61 | 0.43 | 0.51 |
| 9 | Hemigymnus fasciatus | 0.89 | 0.86 | 0.88 |
| 10 | Neoniphon sammara | 0.94 | 1 | 0.97 |
| 11 | Abudefduf vaigiensis | 0.67 | 0.7 | 0.68 |
| 12 | Canthigaster valentini | 0.81 | 0.73 | 0.77 |
| 13 | Pomacentrus moluccensis | 0.92 | 0.92 | 0.92 |
| 14 | Zebrasoma scopas | 0.73 | 0.61 | 0.67 |
| 15 | Hemigymnus melapterus | 0.83 | 0.56 | 0.67 |
| 16 | Lutjanus fulvus | 0.87 | 0.93 | 0.9 |
| 17 | Scolopsis bilineata | 0.73 | 0.8 | 0.76 |
| 18 | Scaridae | 0.57 | 0.67 | 0.62 |

(continued)

**Table 3** (continued)

| SL No | Species | Precision | Recall | F1-score |
|---|---|---|---|---|
| 19 | Pempheris vanicolensis | 0.83 | 0.83 | 0.83 |
| 20 | Zanclus cornutus | 0.6 | 0.6 | 0.6 |
| 21 | Neoglyphidodon nigroris | 1 | 0.5 | 0.67 |
| 22 | Balistapus undulatus | 0.62 | 0.56 | 0.59 |
| 23 | Siganus fuscescens | 1 | 0.8 | 0.89 |

**Table 4** Comparative analysis

| Technique | Precision | Recall | F1-score | Accuracy (%) |
|---|---|---|---|---|
| CNN and Naïve Bayes classifier | 0.75 | 0.55 | 0.61 | 58 |
| CNN and Random Forest Classifier | 0.79 | 0.78 | 0.76 | 78 |
| CNN and K-Nearest Neighbor Classifier | 0.88 | 0.88 | 0.87 | 87 |
| CNN and Logistic Regression (Proposed) | 0.95 | 0.95 | 0.95 | 95 |

The proposed work is also compared with Naïve Bayes, Random Forest, K-Nearest Neighbor classifier and it is observed from the experiment that the logistic regression classifier with CNN is working better than other classifiers. Even though the work is being compared with Ensemble Learning, it resulted in 19% better accuracy. Table 4 represents the results of different classifiers and the proposed one.

## 8 Conclusion

Fish classification is an important task for monitoring of marine environment. Years by years' population of particular species needs to be identified for different scientific analyses. In this work, fish classification was done on Fish4knowledge data that was collected from raw videos. Fish classification is done by using CNN with Logistic Regression classifier. From experiments, it is clearly observed that CNN with Logistic Regression produces 8%, 17%, and 37% improved accuracy than K-NN, Random forest, and Naïve Bayes classifier respectively. In this classification, most of the images consist of single type of fish and therefore extracted entire image feature by CNN. However, if an image consists of multiple kinds of fish, it is important to extract feature for each particular region.

## References


1. Donahue J, Jia Y, Vinyals O, Hoffman J, Zhang N, Tzeng E, Darrell T (2013) DeCAF: a deep convolutional activation feature for generic visual recognition. arXiv preprint 32
2. Chatfield K, Simonyan K, Vedaldi A, Zisserman A (2014) Return of the devil in the details: delving deep into convolutional nets. In: BMVC 2014-proceedings of the british machine vision conference. https://doi.org/10.5244/C.28.6
3. Boiman O, Shechtman E, Irani M (2008) In defense of nearest-neighbor based image classification. In: IEEE conference on computer vision and pattern recognition, pp 1–8. https://doi.org/10.1109/CVPR.2008.4587598
4. Breiman L (2001) Random forests. Mach Learn 45:5–32. https://doi.org/10.1023/A:1010950718922
5. Fan Z, Xie J-K, Wang Z-y, Liu P-C, Qu S-J, Huo L (2021) Image classification method based on improved KNN algorithm. J Phys Conf Ser 1930:012009. https://doi.org/10.1088/1742-6596/1930/1/012009
6. Sarkar P, De S, Gurung S (2022) Fish detection from underwater images using YOLO and its challenges. Doctoral Symposium on intelligence enabled research
7. Smith A (2002) Principles of data mining. In: Hand D, Mannila H, Smyth P (eds) MIT Press, Cambridge, MA, 2001, 546 p. +xxxii, ISBN: 0-262-08290-X. Artificial Intelligence in Medicine, vol 26, pp 175–178. https://doi.org/10.1016/S0933-3657(02)00058-1
8. Cui S, Zhou Y, Wang Y, Zhai L (2020) Fish detection using deep learning. Appl Comput Intell Soft Comput
9. Jalal A, Salman A, Mian A, Shortis M, Shafait F (2020) Fish detection and species classification in underwater environments using deep learning with temporal information. Eco Inform 57:101088
10. Hsiao YH, Chen CC, Lin SI, Lin FP (2014) Real-world underwater fish recognition and identification, using sparse representation. Eco Inform 23:13–21
11. Palazzo S, Murabito F (2014) Fish species identification in real-life underwater images. In Proceedings of the 3rd ACM international workshop on multimedia analysis for ecological data, pp 13–18
12. Spampinato C, Giordano D, Di Salvo R, Chen-Burger YHJ, Fisher RB, Nadarajan G (2010) Automatic fish classification for underwater species behavior understanding. In: Proceedings of the first ACM international workshop on Analysis and retrieval of tracked events and motion in imagery streams, pp 45–50
13. Salman A, Siddiqui SA, Shafait F, Mian A, Shortis MR, Khurshid K, Ulges A, Schwanecke U (2020) Automatic fish detection in underwater videos by a deep neural network-based hybrid motion learning system. ICES J Marine Sci 77(4):1295–1307
14. Salman A, Maqbool S, Khan AH, Jalal A, Shafait F (2019) Real-time fish detection in complex backgrounds using probabilistic background modelling. Eco Inform 51:44–51
15. Hong Khai T, Abdullah SNHS, Hasan MK, Tarmizi A (2022) Underwater fish detection and counting using mask regional convolutional neural network. Water 14(2):222
16. Fish4Knowledge ground-truth datasets. Fish4Knowledge Ground Truth (nd) https://homepages.inf.ed.ac.uk/rbf/fish4knowledge/GROUNDTRUTH/. Accessed 9 Dec 2022

# A Short Survey of Elucidating the Emotion Recognition Methodologies Using Facial Images and EEG Signals

**Dilsheen Kaur, Anuradha Misra, and O. P. Vyas**

**Abstract** Since the high demand for Human–Computer Interaction (HCI), developing an automated model for recognizing facial gestures or emotions becomes challenging. Some experts have used facial images as a constructive part of recognizing different emotions of humans. Simultaneously, emotions can also be classified by using some electrical signals such as Electrocardiogram (ECG), Electroencephalogram (EEG), Electromyogram (EMG), and so on. Among all these signals, EEG captures the major information of brain activities. Recently, along with face images, EEG signal plays a pivotal role in emotion recognition. Though these methods have been estimated the promising results, the real-world implications and diverse feature identification of human states are still in place to gain more attention. Considering these factors, this survey aims to review the literature for analyzing emotion recognition performance using EEG signals and facial images. It also demonstrates the techniques that are utilized in machine learning and deep learning as well. The drawback of such implemented models is discussed, leading to future development. Subsequently, the survey part is followed by exploring the chronological review of the emotion recognition work, dataset utilization, methodologies employed, and experimental analysis with divergent metrics and features, and challenges. Finally, the research challenging factors provoke to raise the novel effective system for emotion recognition using facial images and EEG signals.

**Keywords** Emotion recognition · EEG signals · Facial images · Dataset utilized · Performance metrics · Merits and demerits · Challenging problems and future trends

D. Kaur (✉) · A. Misra
Computer Science and Engineering, Amity School of Engineering and Technology, Amity University Lucknow Campus, India
e-mail: dilsheen.kaur1@s.amity.edu

A. Misra
e-mail: amisra@lko.amity.edu

O. P. Vyas
Information Technology, Indian Institute of Information Technology Allahabad, Prayagraj, India
e-mail: opvyas@iiita.ac.in



S. Borah et al. (eds.), *Advanced Computational and Communication Paradigms*, Lecture Notes in Networks and Systems 535,
https://doi.org/10.1007/978-981-99-4284-8_35

# 1 Introduction

Today, human–machine interaction emerges with evolutionary growth that perceives human activities like their expressions, poses, and gestures via machines [1]. As humans contain various emotions, face emotion recognition aims to forecast the expressions of individuals. In general, emotions like anger, sadness, happiness, disgust, surprise, fear, etc. have existed universally for everyone [2]. These kinds of emotions are exhibited via faces, thus facial images play a pivotal input source for the recognition systems [3]. Humans can showcase their emotions because of inner feelings; hence, the consideration of internal activities is also required [4]. Therefore, physiological signals like EEG are highly demandable to recognize human expression as they possess brain-related activities [5]. Emotion recognition is widely used in different industries like health care, the robotics world, HCI, the animation field, education, alert systems for patients, etc. [6]. Enormous studies regarding facial expression identification are explored with the help of face images and EEG signals.

To develop the system, it exists with multiple challenges to consider. While using the face image as input, the system faces the issues like occlusion, illumination, head poses, and pixel variation based on human states. Another challenging factor is feature representation as it lacks intra- and inter-class variability and also dimension issue rectification [7]. Similarly, when considering the EEG signal, the major issue is the occurrence of noise present in the signal that degrades the recognition performance. Also, the feature attainment of the signal becomes a significant challenging problem due to the signal being non-stationary and non-linear [8]. Therefore, designing an effective emotion recognition model with maximal accuracy and minimal complexities is challenging.

Learning models are recently involved in recognition and classification models. Firstly, machine learning algorithms are immensely utilized to identify human emotional states [9]. Certain models like support vector machines (SVM), neural networks, fuzzy logic systems, etc. are adopted to improve accuracy [10]. Still, it faces some issues in achieving higher results. To overcome this, an advanced model of deep learning is deployed. A common deep learning method is Convolutional Neural Networks (CNN) [11], which has been widely used in the emotion recognition system. Nevertheless, the above systems are still paving the way for further enhancement. The prime intention of the survey paper is pointed out as follows: (a) to review the existing face emotion recognition models, (b) to analyze the chronological view and dataset description of several systems, (c) to present the performance metrics, advantages, and disadvantages of recognition systems, and (d) to provide the information regarding challenging factors for further development.

The introduction part is as follows: Part II demonstrates the related works taken from 20 recent papers. The algorithms and dataset utilized are given in Part III. Part IV discusses the implementation tools and performance measures. Part V illustrates the research gaps and future scopes. Finally, the conclusion is shown in Part VI.

# 2 Literature Review on Traditional Emotion Recognition Methodologies with Chronological Review

## 2.1 Related Works

In 2019, Zhang et al. [12] have utilized CNN to recognize the face emotions using images. Initially, the input images were normalized, and the edge of every layer in the image was fetched using the convolution operation. The obtained edge information was integrated with feature-represented images to consume the textural features of face images. Further, the feature dimension was mitigated by the max pooling approach, and classification was carried out by the softmax classifier function. Finally, the experimentation was performed, and its results were computed using multiple measures. Thus, the model's outcome has delivered that it has attained a maximum recognition rate.

In 2017, Ryu et al. [13] have framed the novel method as Local Directional Ternary Pattern (LDTP) for face emotion identification. The major objective of this model has been focused on retrieving the features from regions like the upper nose, eyes, mouth, and eyebrows. With the admittance of these features, two-level descriptors have been raised to determine the coarse features of expressions. Hence, the suggested method has overwhelmed the recognition performance in acquiring higher results.

In 2021, Mohan et al. [14] have described the deep learning models for facial expression recognition. The proposed model has contained two major parts. On the one hand, the "local gravitational force descriptor" has been utilized for extracting the significant features. On the other hand, these features were given as input to the "Deep Convolution Neural Network (DCNN)", which was processed with two criteria. The first criterion has explored the geometrical features such as curves, edges as well as lines. Then, the second has been considered the holistic-level features. Finally, these feature scores have been used to estimate the final classified value. The performance was validated and measured across divergent metrics. Thus, the results have declared that it has enhanced the recognition performance for finding facial expressions.

In 2022, Ullah et al. [15] have explored the improved DCNN learning model for face emotion recognition. It has comprised of three stages: super-resolution, recognition, and classification. Firstly, the Viola–Jones face detection approach was used to detect the required objects. It was followed by the feature extraction process, where the important features were retrieved by "Texton, Bag of Words (BOW), and Gray-Level Co-occurrence Matrix (GLCM), and LXGP". Finally, these acquired features were subjected to the DCNN model to perform the recognition. The model's performance has ensured to increase in the system efficiency by estimating the desired values.

In 2021, Devi and Satyanarayana [16] have presented a novel deep learning model for facial expression recognition. Initially, the input was pre-processed with the Gamma-HE method, and facial points were obtained through a "Pyramid Histogram of Oriented Gradients (PHOG)-based Supervised Descent (SMD)". Further, the facial parts were identified by the Viola–Jones approach, and their relevant features were extracted. From this, the optimal features were selected by "Modified Monarch Butterfly Optimization (MMBO)". Finally, these features were given into the classifier model, where the recognition occurred. The experimentation was carried out, and the results have proved that it has improved efficiency.

In 2021, Kumar et al. [17] have developed a novel approach to Fuzzy Support Vector Machine (FSVM) for emotion recognition. The input images were collected from standard data sources, in which the spectral and spatial-related features were obtained by wavelet transform. To reduce the dimensional problem, the "Pearson–kernel principal component analysis" has been deployed. Finally, these reduced features were fed into the FSVM classifier models, where the different emotions were recognized. Thus, the extensive results have illustrated that it has increased recognition performance.

In 2018, Mert and Akan [18] have designed the "Multivariate Empirical Mode Decomposition (MEMD)" for face emotion identification using EEG signals. The proposed MEMD model extracted the Intrinsic Mode Functions (IMFs) features regarding signal frequency and time traits. The validation was done with the standard data sources and compared against classical approaches. Therefore, the impressive results have demonstrated that it has led to significantly recognizing the face emotions.

In 2019, Taran and Bajaj [19] have developed a two-way filtering process using EEG signals for emotion recognition. Firstly, the correlation process ignored the noisy IMFs in the signal. This denoised signal was decomposed into several bands using "Variational Mode Decomposition (VMD)". Secondly, the frequency-based features were retrieved. Finally, the "Multi-Class Least Squares SVM (MC-LS-SVM)" has been deployed to provide the classified results by giving the feature as input. After the implementation, the proposed model has acquired outstanding results compared with other baseline approaches.

In 2021, Salankar et al. [20] have raised the effective classifier model for facial expression recognition. Initially, the pre-processing was done by EMD, where the decomposed signal was represented with the oscillatory signal of IMFs. Further, relevant features were determined through "second order difference plots (SODP)". Finally, the SVM with 2 layers of multilayer perceptron has been considered for exploiting the recognized results. The performance was measured with various metrics to ensure the effectiveness of the model. Hence, the findings have proved that it has the potential to differentiate emotions.

In 2021, Maheshwari et al. [21] have experimented with multi-channel CNN for facial expression recognition with the help of multiple channels of EEG signals. The channels were selected by influencing the bandpass filters. It was then given into DCNN, which was structured with average pooling, drop-outs, global average, and batch normalization. Using standard data sources, the experimentation was carried out with an accuracy measure. Thus, impressive results were obtained in contrast with conventional methodologies.

In 2020, Gao et al. [22] have proposed the feature fusion methodology for face emotion recognition. Here, the two types of features were retrieved by wavelet and spectrum-based characteristics of EEG signals. Further, the Principal Component

Analysis (PCA) model was utilized to deduce the dimension of features. Finally, the Relevance Vector Machines (RVM) and SVM have been considered for classification. Thus, the simulation was done and its results were obtained. The more accurate value has tended to increase the system's robustness.

In 2021, Patel et al. [23] aimed to deploy distinct entropy-based measures for recognizing emotions. It has resolved the past insights by employing a standard feature extraction process. Hence, the findings of the model have elucidated that it has exploited better recognition performance.

In 2022, Iyer et al. [24] have implemented the hybrid model of CNN-LSTM for face emotion recognition. The raw EEG signals were initially gathered and given to determine the significant features. Subsequently, it was fed into the hybrid CNN-LSTM model, where emotions were recognized. Finally, the proposed model was computed with an accuracy measure. It has elucidated that the model has yielded higher performance.

In 2020, Zhang et al. [25] have framed the "Improved Radial Basis Function Neural Network (I-RBF-NN)" for emotion identification. Firstly, the EEG signals were garnered and subjected to feature extraction. Further, the resultant noteworthy features were given as input to the I-RBF-NN for evaluating the classification result. After the implementation had been done, the proposed model overwhelmed with the recognition performance rather than other existing models.

In 2020, Zhang [26] have considered the Deep AutoEncoder (DAE) model for multimodality-based emotion recognition. Initially, the decision tree was employed to find the objective features of facial expression signals. Further, the Bimodal DAE (BDAE) has been taken for fusing the EEG signals with facial feature signals. The third layer of BDAE has retrieved the features, which were given into Library SVM (LIBSVM) classifier model to do the recognition job. The validating results were achieved that have helped to improve the performance.

In 2021, Chao and Dong [27] have proposed the CNN deep learning model for facial expression recognition. Initially, the original EEG signals were collected from benchmark datasets. The features were obtained from all the signal channels and represented in 3D matrix form. Finally, the CNN model was utilized to carry out the uni- and multivariate convolutions to determine the human emotion states. Thus, the performance was measured, and its individual results were acquired. The results have demonstrated that it has increased recognition performance.

In 2020, Wu et al. [28] have utilized the LSTM model for emotion recognition. Initially, the signal features were obtained by a "fully connected neural network" from the input signal. These acquired features were given to LSTM to predict the higher recognition result. In addition, the attention mechanism was employed to estimate the key signal features for improving the result. Finally, the outstanding results have elucidated that the model has improved the accuracy rate.

In 2019, Pane et al. [29] have taken the Random Forest (RF) classifier for recognition systems using parameter optimization. Initially, the raw signals were congregated, and the relevant features were extracted. In order to further attain the optimal parameters, grid search optimization has been employed to tune the parameters. On the contrary to SVM and Linear Discriminant Analysis (LDA), the RF model has enhanced the recognition rate to easily find out the different emotions.

In 2021, Pandey and Seeja [30] have explored the CNN model for emotion recognition using EEG signals. After collecting the signals, the features were extracted. It was then given to the CNN model to provide the classified results. Finally, the experimentation was carried out with two datasets. Hence, the results have ensured that the recommended model has attained higher results.

In 2021, Meshach et al. [31] have presented the "Multi-Dimensional SVM (MDSVM)" model for emotion recognition. The input images were gathered and used for subsequent sections. This model was developed with two levels of SVM. In the first level, the input has classified on valence, whereas the second level was to differentiate the data based on arousal. Finally, the recognition was performed with the proposed concept, and corresponding results were exploited after implementation. Thus, the suggested method has outperformed with more accuracy value rather than traditional approaches.

## 2.2 *Chronological Review*

The chronological analysis of emotion recognition methodologies is represented in Fig. 1. The recently developed and implemented research works for recognition using both face images, and EEG signals have been shown. Hence, the literature work is taken from the year as follows: 10% for 2018, 15% for 2019, 20% for 2020, 45% for 2021, and 10% for 2022, accordingly.

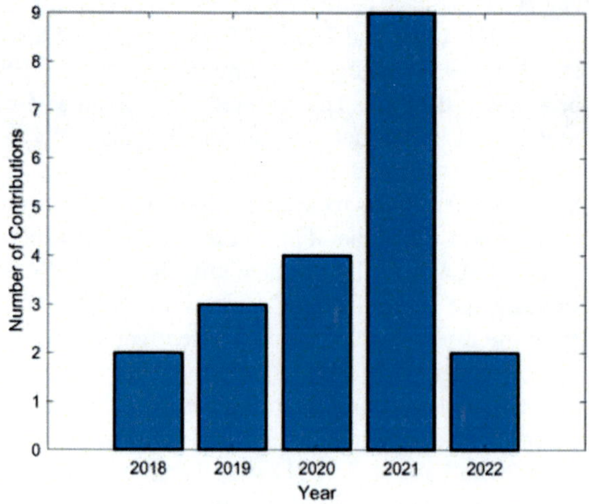


**Fig. 1** Chronological analysis of existing emotion recognition model using face images and EEG signals

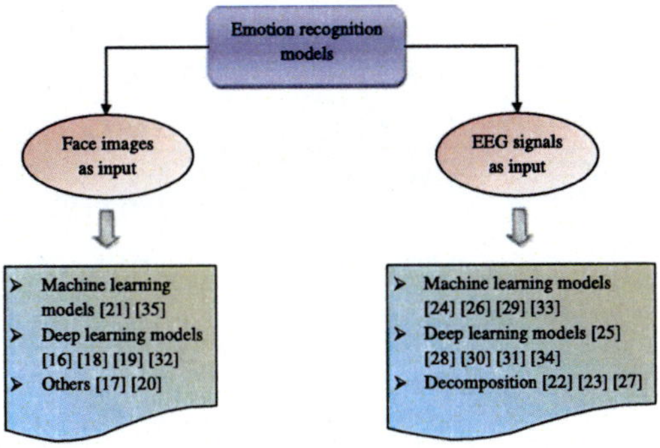


**Fig. 2** Algorithmic categorization of emotion recognition using face images and EEG signals

# 3 Methodologies Suggested and Dataset Utilized on Traditional Emotion Recognition Models

## *3.1 Algorithmic Categorization*

Over several research papers, different algorithms have been adopted for building effective emotion recognition. Distinct approaches of feature extraction, deep learning, and machine learning model are introduced for facial expression identification systems. It is mainly focused on the process of feature extraction of images and signals and recognition by employing different classifiers. Figure 2 shows the algorithms used for emotion recognition models.

## *3.2 Dataset Description*

To implement the recognition systems, multiple benchmark datasets have been deployed to perform emotion recognition. Here, the dataset is collected for both face images and EEG signals. Table 1 explores the dataset description and types of input sources used.

**Table 1** Dataset description and types of data taken for emotion recognition systems

| Citations | Author | Dataset utilized | Types of input source |
|---|---|---|---|
| [12] | Zhang et al. [12] | Fer-2013 | Face images |
| [13] | Ryu et al. [13] | CK+, JAFFE, MMI, CMU-PIE, GEMEPFERA and BU-3DFE | Face images |
| [14] | Mohan et al. [14] | FER2013 [32], JAFFE [33], CK+ [34], KDEF [35], and RAF | Face images |
| [15] | Ullah et al. [15] | JAFFE and CK+ | Face images |
| [16] | Devi and Satyanarayana [16] | JAFFE and CK+ | Face images |
| [17] | Kumar et al. [17] | JAFFE, FG Net, and CK+ | Face images |
| [18] | Mert and Akan [18] | DEAP | EEG signals |
| [19] | Taran and Bajaj [19] | Audio–video stimulus-based EEG-dataset | EEG signals |
| [20] | Salankar et al. [20] | DEAP | EEG signals |
| [21] | Maheshwari et al. [21] | DEAP, DREAMER, and DASPS | EEG signals |
| [22] | Gao et al. [22] | – | EEG signals |
| [23] | Patel et al. [23] | DEAP and SEED | EEG signals |
| [24] | Iyer et al. [24] | DEAP and SEED | EEG signals |
| [25] | Zhang et al. [25] | – | EEG signals |
| [26] | Zhang [26] | – | EEG signals |
| [27] | Chao and Dong [27] | DEAP | EEG signals |
| [28] | Wu et al. [28] | DEAP | Facial video frame and EEG |
| [29] | Pane et al. [29] | – | EEG signals |
| [30] | Pandey and Seeja [30] | DEAP and SEED | EEG signals |
| [31] | Meshach et al. [31] | International Affective Picture System (IAPS) | Face images |

# 4 Analysis of Performance Analysis, and Merits and Demerits of Traditional Emotion Recognition Models

## *4.1 Performance Measures*

The efficacy of the implemented model is validated with diverse metrics, as given in Table 2. It mostly concentrated on classification accuracy, recognition rate, etc. It helps ensure the system's robustness and efficiency in appropriately forecasting human emotions.

**Table 2** Different measures are taken for evaluating the emotion recognition models using face images and EEG signals

| Citations | Accuracy | Recognition rate | Classification accuracy | Sensitivity | F1-Score | Precision | Miscellaneous measures |
|---|---|---|---|---|---|---|---|
| [12] | ✓ | ✓ | – | – | – | – | – |
| [13] | ✓ | ✓ | – | – | – | – | – |
| [13] | ✓ | – | – | – | ✓ | ✓ | Recall |
| [15] | ✓ | – | – | ✓ | ✓ | ✓ | Specificity, FPR, FNR, PSNR, and SSIM |
| [16] | ✓ | – | – | ✓ | ✓ | ✓ | Specificity, FPR, and FNR |
| [17] | – | ✓ | ✓ | ✓ | – | – | – |
| [18] | – | ✓ | – | – | – | – | – |
| [19] | ✓ | – | – | – | ✓ | – | Kappa coefficient |
| [20] | ✓ | – | – | ✓ | ✓ | ✓ | Specificity |
| [21] | ✓ | – | – | ✓ | ✓ | – | Specificity and Kappa |
| [22] | ✓ | – | – | – | – | – | – |
| [23] | – | – | ✓ | – | – | – | – |
| [24] | ✓ | – | – | – | – | – | Standard deviation |
| [25] | – | – | ✓ | – | – | – | – |
| [26] | ✓ | – | – | – | ✓ | – | – |
| [27] | – | ✓ | – | – | – | – | ROC curve |
| [28] | – | – | ✓ | – | – | – | – |
| [29] | ✓ | – | – | – | – | – | – |
| [30] | ✓ | – | – | – | – | – | – |
| [31] | – | – | ✓ | – | – | – | Confusion matrix |

**Table 3** Merits and demerits of traditional emotion recognition methods using face images and EEG signals

| Author name | Algorithm | Merits | Demerits |
|---|---|---|---|
| Zhang et al. [12] | CNN | • It increases more convergence speed and recognition rate | • It may degrade the performance when the image contains occlusion, burring, and face posture |
| Ryu et al. [13] | LDTP | • The higher results make the system more robust and discriminable | • Time complexity occurs as it handles a massive amount of images |
| Mohan et al. [14] | DCNN | • It increases the recognition rate in terms of positive measures | • It requires more training time, but it makes the computational burden |
| Ullah et al. [15] | Recurrent Neural Network | • It offers more positive values and less negative values. It also avoids the overfitting issue | • It does not consider the face descriptor and parameter tuning process |
| Devi and Satyanarayana [16] | Gamma-HE and MMBO | • It increases the more accuracy, F-measure, and sensitivity | • It needs more training time |
| Kumar et al. [17] | FSVM | • It lessens the dimension issue and also improves the recognition rate | • It does not support to use of real-time images and becomes fragile when occlusion occurs |
| Mert and Akan [18] | MEMD | • It maximizes the accuracy rate while using both KNN and ANN | • It does not consider other error measures for validating efficiency |
| Taran and Bajaj [19] | EMD and VMD | • It effectively reduces the time complexity problem | • It suggests including some optimization algorithms for tuning the parameters |
| Salankar et al. [20] | SVM | • It attains high arousal features of signal to improve recognition performance | • It further stimulates to consider of other brain activities like depression and epilepsy |
| Maheshwari et al. [21] | DCNN | • It achieves the desired value of recognition accuracy | • As it contains multiple channels, it causes computation complexity |
| Gao et al. [22] | RVM and SVM | • It effectively distinguishes the three kinds of emotions, which is guaranteed by more accuracy value | • It suggests using ensemble-based learning model for further improvement |
| Patel et al. [23] | Entropy measures | • It retrieves more discriminative signal features for accuracy enhancement | • It may cause an overlapping issue that evades achieving reliability |

(continued)

**Table 3** (continued)

| Author name | Algorithm | Merits | Demerits |
|---|---|---|---|
| Iyer et al. [24] | CNN-LSTM | • This hybrid model exploits more accuracy value to recognize emotions effectively | • It creates an impact of overfitting issues |
| Zhang et al. [25] | RBFNN | • It surpasses the existing complexities and increases the accuracy value | • For further improvement, ensemble-based feature extraction will be used |
| Zhang [26] | DAE | • It improves the recognition ability in terms of the desired value of the F1-score | • It cannot apply to real-time implications |
| Chao and Dong [27] | CNN | • It enhances the system's feasibility by considering 3-dimensional way of computation | • It fails to implement the psychological related brain signals |
| Wu et al. [28] | LSTM | • It attains more spatial information related to three-channel bands. It helps to achieve the desired value | • It cannot use real-time emotions as in natural situations |
| Pane et al. [29] | RF | • It obtains the more optimum value as it tunes the parameter significantly | • It further develops with an ensemble feature selection model to focus on a particular part of the brain |
| Pandey and Seeja [30] | CNN | • It improves the generalization capability | • The subjective features are not obtained in this model to further improve the performance |
| Meshach et al. [31] | Multilayered SVM | • It efficiently recognizes distinct human emotions using eightfold cross-validation | • As the image contains some external factors, misclassification occurs |

## 4.2 *Merits and Demerits*

Table 3 elucidates the merits and demerits of existing facial expression recognition models. While implementing distinct methodologies, it has shown the benefits and challenging factors that lead the way for future recognition systems.

## 5 Research Direction and Future Perspective Toward Emotion Recognition Methodologies Using Facial Images and EEG Signals

This section explains the future research direction of emotion recognition systems. Over the past few years, emotion recognition becomes an effective model that has been extensively used for various applications [32]. Face images and EEG signals are considered input sources to construct an effective model to forecast emotions. The major research gap of this model is finding the salient features of both images and signals. Since the image contains some unwanted background representation, it reduces the system's efficiency and performance. Henceforth, the appropriate selection of facial features is challenging. Similarly, the time or frequency-related traits are critical to extracting from EEG signals [33]. Some learning models have utilized multiple kernels or channels to perform the recognition, yet it still sticks with computation burden and overfitting issues [34]. It will be directed to future development by resolving these issues. Furthermore, the different model contains certain key points to increase accuracy. In the future, the hybrid-based or ensemble-based model is suggested for emotion recognition to further achieve the desired recognition rate [35]. Though it estimates the outstanding value, it sometimes achieves the non-optimal value due to inadequate parameters. It will be resolved in the future by influencing heuristic algorithms to find the optimal parameters to perform a better recognition process.

## 6 Conclusion

This survey work has the objective of analyzing emotion recognition models using face images and EEG signals. It started with an introduction regarding the emotion recognition system and the contributions given. The chronological view has explored the recently developed papers followed by algorithm and dataset categorization using face images and EEG signals. Subsequently, the distinct metrics, advantages, and disadvantages have been noted for conventional methodologies. Finally, challenging issues have been listed that helped future research direction of emotion recognition.

## References

1. Yang B, Cao J, Ni R, Zhang Y (2018) Facial expression recognition using weighted mixture deep neural network based on double-channel facial images. IEEE Access 6:4630–4640
2. Said Y, Barr M (2021) Human emotion recognition based on facial expressions via deep learning on high-resolution images. Multimedia Tools Appl 80:25241–25253

3. Liliana DY, Basaruddin T, Widyanto MR, Oriza IID (2019) Fuzzy emotion: a natural approach to automatic facial expression recognition from psychological perspective using fuzzy system. Cogn Process 20:391–403
4. Saurav S, Saini R, Singh S (2021) EmNet: a deep integrated convolutional neural network for facial emotion recognition in the wild. Appl Intell 51:5543–5570
5. An Y, Xu N, Qu Z (2021) Leveraging spatial-temporal convolutional features for EEG-based emotion recognition. Biomed Signal Process Control 69
6. Liu Y, Fu G (2021) Emotion recognition by deeply learned multi-channel textual and EEG features. Futur Gener Comput Syst 119:1–6
7. Alhussein M (2016) Automatic facial emotion recognition using weber local descriptor for e-Healthcare system. Clust Comput 17:99–108
8. Topic A, Russo M (2021) Emotion recognition based on EEG feature maps through deep learning network". Eng Sci Technol Int J 24(6):1442–1454
9. Hu M, Wang H, Wang X, Yang J, Wang R (2019) Video facial emotion recognition based on local enhanced motion history image and CNN-CTSLSTM networks. J Vis Commun Image Represent 59:176–185
10. Xiaohua W, Muzi P, Lijuan P, Min H, Chunhua J, Fuji R (2019) Two-level attention with two-stage multi-task learning for facial emotion recognition. J Vis Commun Image Represent 62:217–225
11. Gu H, Chen Q, Xing X, Zhao J, Li X (2019) Facial emotion recognition in deaf children: evidence from event-related potentials and event-related spectral perturbation analysis. Neurosci Lett 703:198–204
12. Zhang H, Jolfaei A, Alazab M (2019) A face emotion recognition method using convolutional neural network and image edge computing. IEEE Access 7:159081–159089
13. Ryu B, Rivera AR, Kim J, Chae O (2017) Local directional ternary pattern for facial expression recognition. IEEE Trans Image Process 26(12):6006–6018
14. Mohan K, Seal A, Krejcar O, Yazidi A (2021) Facial expression recognition using local gravitational force descriptor-based deep convolution neural networks. IEEE Trans Instrum Measure 70, 1–12 (5003512)
15. Ullah Z, Qi L, Hasan A, Asim M (2022) Improved deep CNN-based two stream super resolution and hybrid deep model-based facial emotion recognition. Eng Appl Artif Intell 116
16. Devi DAS, Satyanarayana CH (2021) An efficient facial emotion recognition system using novel deep learning neural network-regression activation classifier. Multimedia Tools Appl 80:17543–17568
17. Kumar RJR, Sundaram M, Arumugam N (2021) Facial emotion recognition using subband selective multilevel stationary wavelet gradient transform and fuzzy support vector machine. Vis Comput 37:2315–2329
18. Mert A, Akan A (2018) Emotion recognition from EEG signals by using multivariate empirical mode decomposition. Pattern Anal Appl 21:81–89
19. Taran S, Bajaj V (2019) Emotion recognition from single-channel EEG signals using a two-stage correlation and instantaneous frequency-based filtering method. Comput Methods Programs Biomed 173:157–165
20. Salankar N, Mishra P, Garg L (2021) Emotion recognition from EEG signals using empirical mode decomposition and second-order difference plot. Biomed Signal Process Control 65
21. Maheshwari D, Ghosh SK, Tripathy RK, Sharma M, Acharya UR (2021) Automated accurate emotion recognition system using rhythm-specific deep convolutional neural network technique with multi-channel EEG signals. Comput Biol Med 134
22. Gao Q, Wang C-H, Wang Z, Song X-L, Dong E-Z, Song Y (2020) EEG based emotion recognition using fusion feature extraction method. Multimedia Tools Appl 79:27057–27074
23. Patel P, Raghunandan R, Annavarapu RN (2021) EEG-based human emotion recognition using entropy as a feature extraction measure. Brain Inf 8(20)
24. Iyer A, Das SS, Teotia R, Maheshwari S, Sharma RR (2022) CNN and LSTM based ensemble learning for human emotion recognition using EEG recordings. Multimedia Tools Appl

25. Zhang J, Zhou Y, Liu Y (2020). EEG-based emotion recognition using an improved radial basis function neural network. J Ambient Intell Humanized Comput
26. Zhang H (2020) Expression-EEG based collaborative multimodal emotion recognition using deep autoencoder. IEEE Access 8:164130–164143
27. Chao H, Dong L (2021) Emotion recognition using three-dimensional feature and convolutional neural network from multichannel EEG signals. IEEE Sens J 21(2):2024–2034
28. Wu D, Zhang J, Zhao Q (2020) Multimodal fused emotion recognition about expression-EEG interaction and collaboration using deep learning. IEEE Access 8:133180–133189
29. Pane ES, Wibawa AD, Purnomo MH (2019) Improving the accuracy of EEG emotion recognition by combining valence lateralization and ensemble learning with tuning parameters. Cogn Process 20:405–417
30. Pandey P, Seeja KR (2021) Subject independent emotion recognition system for people with facial deformity: an EEG based approach. J Ambient Intell Humaniz Comput 12:2311–2320
31. Meshach WT, Hemajothi S, Anita EAM (2021) Real-time facial expression recognition for affect identification using multi-dimensional SVM. J Ambient Intell Humaniz Comput 12:6355–6365
32. Alphonse AS, Dharma D (2018) Novel directional patterns and a generalized supervised dimension reduction system (GSDRS) for facial emotion recognition. Multimedia Tools Appl 77:9455–9488
33. Momennezhad A (2018) EEG-based emotion recognition utilizing wavelet coefficients. Multimedia Tools Appl 77:27089–27106
34. Zhang T, Zheng W, Cui Z, Zong Y, Li Y (2019) Spatial-temporal recurrent neural network for emotion recognition. IEEE Trans Cybern 49(3):839–847
35. Wang M, Huang Z, Li Y, Dong L, Pan H (2021) Maximum weight multi-modal information fusion algorithm of electroencephalographs and face images for emotion recognition. Comput Electr Eng 94

# Novel Image Cryptography Method to Improve Security in Cloud Computing

**Prathamesh Vijay Lahande and Parag Ravikant Kaveri**

**Abstract** Cloud is a practice of storing information and accessing requests using internet-supporting devices from anywhere around the globe. Due to the simplicity and integrity of the cloud, the user trusts and decides to reside his personal and confidential data on the cloud instead of storing the same on his local computer or local server. But several times, the cloud safety is hampered when this personal information is compromised often due to the indulgence of uninvited anonymities, leading to the security breach of the cloud. To safeguard this private and confidential data, several encryption–decryption methodologies have been used for a long time. In cryptography, typically, the user's private information is transformed into a cipher text using a specific key and then transformed back into its original form using the same key. The major disadvantage of this technique is that, if the key is in the open and known by any particular untrusted entity, then the personal and confidential data will likely be exposed. The main purpose of the study in this paper is to propose a novel encryption-decryption methodology that generates a more robust and challenging to break cipher image from the private text instead of a simple plain cipher text. This proposed encryption–decryption methodology will provide higher security to the cloud, thereby maintaining its security and integrity.

**Keywords** Cloud · Image cryptography · Integrity · PCICA · Security

P. V. Lahande (✉) · P. R. Kaveri
Symbiosis International (Deemed University), Symbiosis Institute of Computer Studies and Research, Pune, India
e-mail: prathamesh.lahande@sicsr.ac.in

P. R. Kaveri
e-mail: parag.kaveri@sicsr.ac.in



S. Borah et al. (eds.), *Advanced Computational and Communication Paradigms*, Lecture Notes in Networks and Systems 535,
https://doi.org/10.1007/978-981-99-4284-8_36

## 1 Introduction

In a public environment with many illegal data hijackers present, communication between the sender and the recipient can be secured using an age-old technique called cryptography [1]. This technique of cryptography can be seen as a practice of safeguarding the personal and confidential data of the user using its encryption and decryption methodologies [2, 3]. Today, with the simplicity and ease of using the cloud, the user decides to reside this personalized and secret information on the cloud by trusting its security and integrity measures. However, due to several uninvited data hijackers, this personalized and secret information residing on the cloud is often hijacked, thereby compromising cloud security [4–6]. The study of this research paper improves the security as well as the integrity of the cloud by proposing a novel encryption and decryption method called Prathamesh's Cloud Image Cryptography Algorithm (PCICA). The proposed algorithm PCICA also includes two phases: PCICA-Encryption and PCICA-Decryption. Image cryptography [3] is implemented for the PCICA-Encryption method, where the personalized and secret information of the end-user is converted into a cipher image instead of the traditional cipher text. This cipher picture is restored to plain text using the suggested PCICA-Decryption technique. A key is also employed, as in the usual approach, as an additional security measure both before and after the suggested encryption and decryption method is put into use. The main upper hand of this proposed method is the disguising factor of the cipher image is more than the traditional cipher text since the cipher image is entirely in an unintelligible format. The unintelligibility of the generated cipher image has been implemented as well as validated to enhance personal confidentiality in cryptographic communication.

The remaining paper is organized as follows: Sect. 2 provides the study of the relevant literature. The suggested methodology is provided in Sect. 3. Section 4 provides the results. The conclusions and recommendations are presented in Sect. 5.

Figure 1 shows the flow diagram of the suggested methodology.

## 2 Relevant Literature

Cloud environment is a challenging platform where securing the data is a crucial area for many researchers to focus upon. Many researchers have studied and addressed the security measures of the cloud. A novel technique that provides dual encryption and data fragmentation techniques has been proposed called HBDaSeC [7]. This technique not only secures sharing of the data but also withstands various security attacks. Cloud-based cryptographic solutions for sharing and data assurance are provided in this research paper [8]. The study in this research paper also offers an analysis of the numerous cryptographic techniques created to protect sensitive data in the

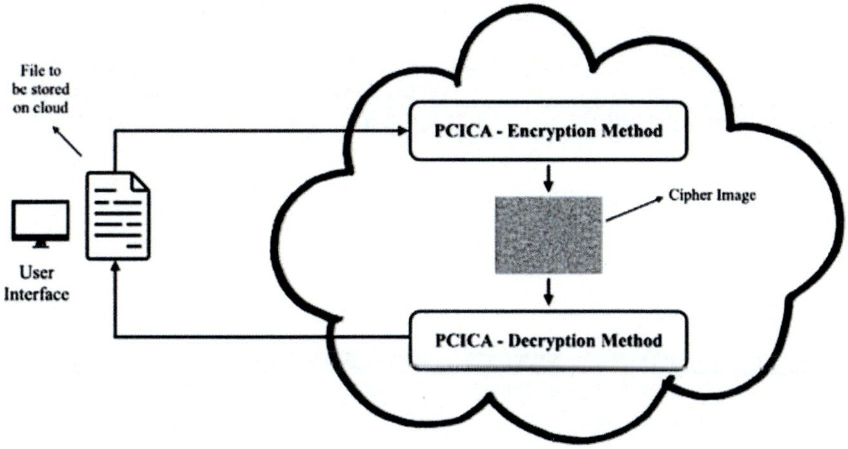


**Fig. 1** Flow diagram of the proposed system

cloud. A method called CP-WABE is proposed, which has the potential to withstand chosen-plaintext assaults in both the random oracle and generic group models [9]. The researchers have suggested an encoding method which is block based including the most-significant-bit encoding process for hiding and reversing information in encrypted images [10]. This encoding technique's benefits include increased security, higher embedding rates, and great visual quality. To maintain the integrity and protect privacy from unwanted parties, a novel model based on a genetic method is created [11]. A hybrid cryptography method-based end-to-end encryption technique has also been proposed in this work to safeguard secrecy in a cloud computing environment supporting multi-medias [12]. A secure hybrid cryptography approach has been proposed in this study, and an end-to-end encryption scenario enables the transmission and storage of media files online in a secure manner [13]. The study in this research provides a model combining the Blowfish and AES algorithms. The text files uploaded to the cloud under this strategy are more secure. Researchers have surveyed data security in the cloud, which covers cloud security aspects [14]. Cryptography mechanism of elliptic curviness is used to create an authentication and key agreement mechanism that is both secure and effective [15]. The proposed model can satisfy secrecy and provides robustness and resistance against attacks. The study suggested by the researchers includes a cloud-based encryption method based on chaotic sequences and a single-round dictionary [16]. The proposed study not only provides effectiveness but also provides a higher security level. For the objective of picture encryption and decryption, researchers examined and contrasted the concepts and methods now in use for vintage cryptography and quantum cryptography [17]. A

**Table 1** Literature study of the proposed method and tool used

| Ref. no | Method proposed | Technology/tool used |
|---|---|---|
| [7] | HBDaSeC | Oracle virtual machine and fog environment |
| [10] | BAME-RDHEI | MATLAB R2017a |
| [11] | CryptoGA | Microsoft's private cloud, OpenStack, and VMWare's vCloud Suite |
| [12] | A hybrid cryptography algorithm | Eclipse and java |
| [13] | Blowfish and AES combined Algorithm | Simulated environment |
| [15] | Method using ECC | Scyther tool |
| [18] | Client-side encryption method | Java |
| [19] | RC4 & RGB pixel shuffling method with HLSB | MATLAB |

cryptographic method is introduced that encrypts information on the user side before it is uploaded to the cloud [18]. This proposed model provides better performance regarding various performance parameters. A hybrid method is suggested using a streaming cipher method with steganography which is easy to encrypt and provides high security as well [19]. A lossless image cryptography algorithm is proposed, which increases the reliability since the actual and deciphered images are similar and difficult to break [6]. On the basis of discrete cosine transformation, a lossless image cryptography algorithm is proposed [3].

Table 1 depicts the proposed methods and tools used to design them.

## 3 Proposed Methodology

### *3.1 PCICA-Encryption Method*

The pseudo-code of the proposed PCICA-encryption methodology is as follows.

**Pseudo code of Encryption Algorithm**

```
Allocate memory for arrays red_array, green_array and blue_array to hold 255 values.
For i = 0 to 255 do
   red_array[i] = i // Initialise value of 'i' to value of red_array at ith index
   green_array[i] = I // Initialise value of 'i' to value of green_array at ith index
   blue_array[i] = I // Initialise value of 'i' to value of blue_array at ith index
End For
For i = 0 to length_of_file:
   character = Fetch the ith character of the inputted file
   character_ascii = ASCII (character) // Convert character to corresponding ASCII value
   red_value = red_array[character_ascii]       // Initialise value at character_ascii index of
                                                   red_array to r_value
   green_value = green_array[character_ascii] // Initialise value at character_ascii index of
                                                   green_array to g_value
   blue_value = blue_array[character_ascii]     // Initialise value at character_ascii index of
                                                   blue_array to b_value
   pixel = Create_Pixel (red_value, green_value, blue_value)      // Create a Pixel

End For
```

Figure 2 depicts the proposed PCICA-encryption method.

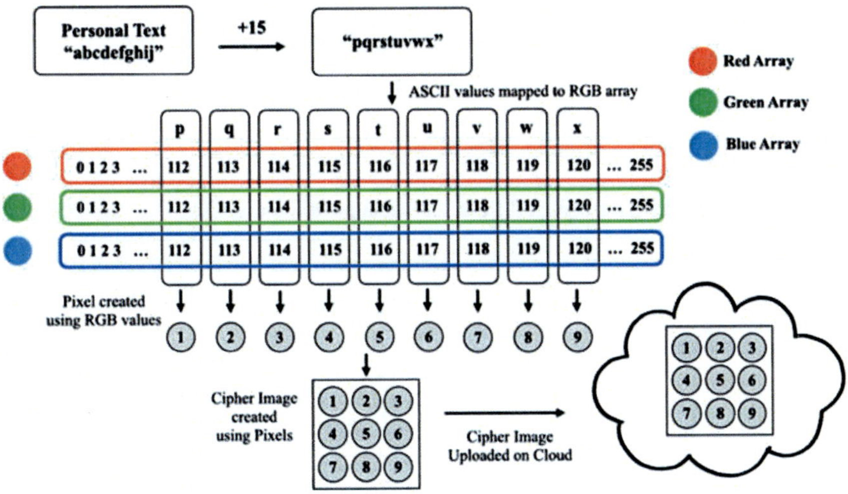


**Fig. 2** Proposed PCICA-encryption method

### 3.2 PCICA-Decryption Method

The pseudo-code of the proposed PCICA-decryption methodology is as follows.

```
Pseudo code of Decryption Algorithm
Allocate memory for arrays red_array, green_array and blue_array to hold 255 values.
For i = 0 to 255 do
    red_array[i] = i // Initialise value of 'i' to value of red_array at ith index
    green_array[i] = I // Initialise value of 'i' to value of green_array at ith index
    blue_array[i] = I // Initialise value of 'i' to value of blue_array at ith index
End For
For i = 0 to Number_of_Pixels in the Cipher Image:
   pixel = Fetch pixel at ith index of the cipher image
   red_pixel_value = red (pixel) // Fetch red value from the pixel
   green_ pixel_value = green (pixel) // Fetch green value from the pixel
   blue_ pixel_value = blue (pixel) // Fetch blue value from the pixel
   red_value = red_array[red_pixel_value] // Initialise value at red_pixel_value index of
                                                                  red_array to red_value
   green_value = green_array[green_pixel_value] // Initialise value at red_pixel_value
                                                           index of green_array to green_value
   blue_value = blue_array[blue_pixel_value] // Initialise value at blue_pixel_value index
                                                                of blue_array to blue_value
   character = Convert value of red_value, green_value & blue_value to character format
   character = character – key // Subtract the key to get the original character
   Append the character to the file
End For
```

Figure 3 depicts the proposed PCICA-Decryption method.

## 4 Results

To test and validate the proposed PCICA method, four scenarios are considered. The first scenario includes inputting a text file of size of 1 kilobyte (KB), the second scenario of 10 KB, the third scenario of 100 KB, and the last scenario of 1,000 KB.

Figure 4 represents the results obtained in the form of cipher images.

Table 2 represents the results obtained for the proposed PCICA method when a particular file is inputted.

From the above table, we can observe that the size of the cipher image is lower than the inputted text file size. Figure 5 represents the comparative graph of the existing system with the PCICA method.

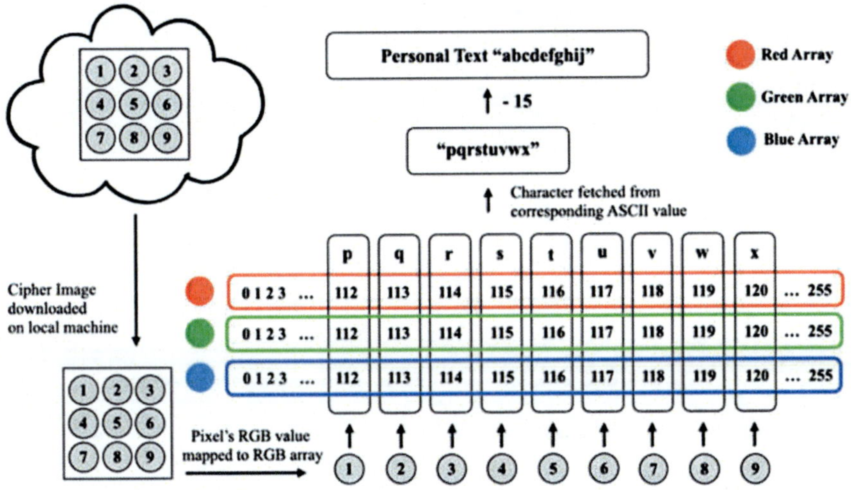


**Fig. 3** Proposed PCICA-decryption method

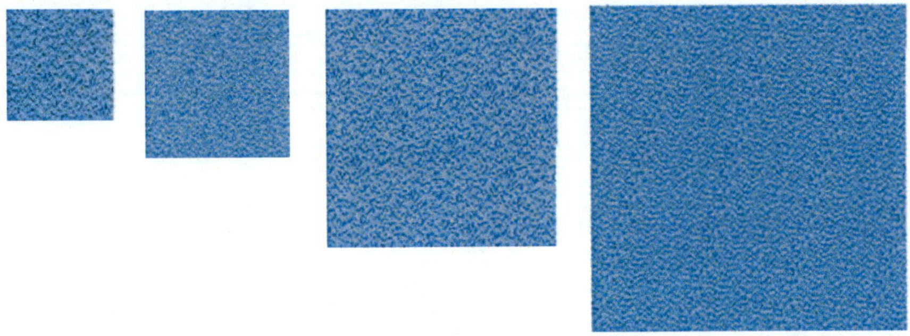

**Fig. 4** Cipher images obtained from Scenarios I, II, III, and IV, respectively

**Table 2** Results of the proposed PCICA method

| Scenario no | Inputted text file size (Kilo-Bytes) | PCICA method cipher image size ( Kilo-Bytes) |
|---|---|---|
| Scenario 1 | 1 | 0.15 |
| Scenario 2 | 10 | 1.45 |
| Scenario 3 | 100 | 10.89 |
| Scenario 4 | 1000 | 100.23 |

Table 3 represents the $R^2$ analysis performed for the PCICA method.

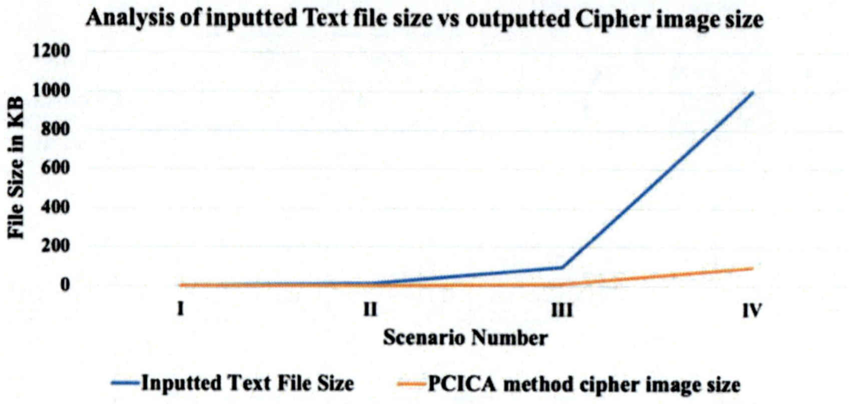


**Fig. 5** Analysis of input text versus output cipher image

**Table 3** $R^2$ analysis for the PCICA method

| Parameter | PCICA method |
|---|---|
| Linear regression equation | y = 30.968x–49.24 |
| Regression line slope | 30.968 |
| Slope sign | Positive |
| Regression line Y-Intercept | –49.24 |
| $R^2$ value | 0.686 |

## 5 Conclusions and Future Work

This research paper includes a novel encryption–decryption method called PCICA that can be used to improve and strengthen the security of the personalized and secret information of the end-user. An additional key can be used as an extra security element where just before encryption, the personalized and secret information of the user would be converted into cipher text as the first-level encryption using this key. The second level would be converting this cipher text into a cipher image using the proposed encryption method. The proposed decryption method would be the reverse of this encryption method. Table 4 represents parameter-wise conclusions drawn.

The proposed PCICA approach would increase the data secrecy of the cloud, allowing users to save any customized and private information in the form of cipher pictures without worrying about the security of their data. This method of encryption and decryption can also be implemented for audios and videos in the future.

**Table 4** Conclusions with respect to various parameters

| Parameter | Existing method | PCICA method |
|---|---|---|
| Cipher format | Cipher format is simple text | Cipher format is the image |
| Comprehension | Simple to comprehend | Difficult to understand |
| Complexity | Fundamentally simple | Complex technique |
| Privacy | Privacy of data is minimum | Privacy is higher than the existing system |
| Security | Security of data is minimum | The proposed system has higher security |

## References


1. Choo KKR, Domingo-Ferrer J, Zhang L (2016) Cloud cryptography: theory, practice and future research directions. Futur Gener Comput Syst 62:51–53
2. Subramanian N, Jeyaraj A (2018) Recent security challenges in cloud computing. Comput Electr Eng 71, 28–42
3. Tedmori S, Al-Najdawi N (2012) Lossless image cryptography algorithm based on discrete cosine transform. Int Arab J Inf Technol 9:471–478
4. Singh A, Chatterjee K (2017) Cloud security issues and challenges: a survey. J Netw Comput Appl 79:88–115
5. Tari Z, Yi X, Premarathne US, Bertok P, Khalil I (2015) Security and privacy in cloud computing: vision, trends, and challenges. IEEE Cloud Comput 2(2):30–38
6. Potey MM, Dhote CA, Sharma DH (2013) Cloud computing understanding risk, threats, vulnerability and controls: a survey. Int J Comput Appl 67(3), 9–14
7. Zhang B, Zeng Z, Shi X, Yang J, Veeravalli B, Li K (2022) A novel cooperative resource provisioning strategy for Multi-Cloud load balancing. J Parallel Distrib Comput 152:98–107
8. Zhang L, Xiong H, Huang Q, Li J, Choo KKR, Li J (2022) Cryptographic solutions for cloud storage: challenges and research opportunities. IEEE Trans Serv Comput 15(1):567–587
9. Li H, Yu K, Liu B, Feng C, Qin Z, Srivastava G (2022) An efficient ciphertext-policy weighted attribute-based encryption for the internet of health things. IEEE J Biomed Health Inform 26(5):1949–1960
10. Wang X, Chang CC, Lin CC (2021) Reversible data hiding in encrypted images with block-based adaptive MSB encoding. Inf Sci 567:375–394
11. Tahir M, Sardaraz M, Mehmood Z, Muhammad S (2020) CryptoGA: a cryptosystem based on genetic algorithm for cloud data security. Clust Comput 24(2):739–752
12. Harnal ES, Chauhan DR (2019) Hybrid cryptography based E2EE for integrity & confidentiality in multimedia cloud computing. Int J Innovative Technol Exploring Eng 8(10):918–924
13. Srivalli BSS, SwarupMedikonda SVN (2019) Development of a cloud-based secure text file application using hybrid cryptography and steganography. Int J Recent Technol Eng
14. Gunavathy S, Meena (2019) A survey: data security in cloud using cryptography and steganography. Int J Recent Technol Eng
15. Amintoosi H, Nikooghadam M (2019) A novel provably-secure ECC-based authentication and key management protocol for telecare medical information systems. In: International conference on computer and knowledge engineering
16. Li Y, Yu H, Song B, Chen J (2018) Image encryption based on a single-round dictionary and chaotic sequences in cloud computing. Concurrency Comput Pract Exp 33(7):1–1

17. Pawar HR, Harkut DG (2018) Classical and quantum cryptography for image encryption & decryption. In: 2018 International conference on research in intelligent and computing in engineering (RICE)
18. Sohal M, Sharma S (2018) BDNA-A DNA inspired symmetric key cryptographic technique to secure cloud computing. J King Saud Univ Comput Inf Sci 34(1):1417–1425
19. Abood MH (2017) An efficient image cryptography using hash-LSB steganography with RC4 and pixel shuffling encryption algorithms. In: Annual conference on new trends in information & communications technology applications (NTICT)
20. Bhowmik S, Acharyya S (2011) Image cryptography: the genetic algorithm approach. In: IEEE international conference on computer science and automation engineering

# CT Image Super-Resolution: A Perspective on Tissue-Specific Performance Analysis

**Bakul Gohel, Nisarg Doshi, and Krishna Veer Singh**

**Abstract** The potential of various image super-resolution (SR) techniques has been widely discussed and analyzed for medical imaging. Different body tissue segments in an image have different characteristics and levels of importance in evaluating the image for diagnosis. Therefore, evaluating the performance of the SR methods for different tissues is essential to assess the usability of the approach. In the present work, we evaluated SR methods performance for various tissues or organs, e.g., liver, bladder, lung, kidney, and bone, in CT images. We used bicubic interpolation and modified-ESPCN deep learning model as SR methods and peak signal-to-noise ratio (PSNR) and similarity index measure (SSIM) as performance measures. We observed significantly ($p < 0.01$, paired t-test) higher performance ratings for performance metrics computed over a full CT image than computed over body tissue region. PSNR and SSIM (primarily structural component) values were significantly ($p < 0.01$, paired t-test) lower for lung and bone tissues compared to liver, bladder, and kidney. Therefore, tissue-specific performance measure computation is essential for evaluating the SR methods for a medical imaging application. It would be good practice for researchers to report tissue-specific performance analysis as opposed to performance metrics computed over the entire images, to avoid overestimating the performance capabilities of new methods.

**Keywords** SSIM · Image super-resolution · CT scan · ESPCN · Image enhancement

B. Gohel (✉) · N. Doshi · K. V. Singh
Dhirubhai Ambani Institute of Information and Communication Technology (DA-IICT), Gandhinagar, India
e-mail: bakul_gohel@daiict.ac.in



S. Borah et al. (eds.), *Advanced Computational and Communication Paradigms*, Lecture Notes in Networks and Systems 535,
https://doi.org/10.1007/978-981-99-4284-8_37

## 1 Introduction

Super-resolution (SR) methods generate a higher-resolution (HR) image from a lower-resolution (LR) image. Recently, many studies have shown the usefulness of SR techniques in enhancing the resolution for several imaging modalities and applications [1, 2]. In recent years, the field of image SR has witnessed remarkable progress using deep learning (DL) techniques. DL-based SR models have been actively explored and often achieve state-of-the-art performance based on various benchmarks [1–3]. Various DL architectures have been used for image SR, ranging from the very first simple convolutional networks-based approach to the recent promising generative adversarial network (GAN) [1–4]. The peak signal-to-noise ratio (PSNR) and structural similarity index (SSIM) are widely used performance metrics for evaluating the various SR approaches [1–3, 5].

SR methods have also been discussed and explored for various medical imaging modalities [3, 6–8]. HR images provide incredible structural information about tissue and aid clinicians in quantitatively analyzing pathologies and diagnoses. However, HR images increase storage space requirement, image acquisition time, and cost (for example, MRI) or require higher radiation exposure (for example, CT scan) [3, 6–8]. An approximation of HR image from LR image using SR methods could overcome these challenges. Recently, convolutional neural network (CNN)-based approaches have been increasingly used for medical image SR applications in different imaging modalities, e.g., MRI, CT, X-ray, fundoscopy, histopathological images, etc., and showed impressive performance [3, 6, 7, 9–11].

Different medical radio images differ in characteristics in contrast to routing camera-based RGB images of objects and natural scenes. Additionally, medical radio images contain background regions (non-body tissue), which is outside the region of interest. Therefore, this region should not be considered during evaluation. Moreover, different tissue or organ or pathologies have different characteristics; therefore, image enhancement performance is likely to differ among them. Most of the literature on medical radio imaging has evaluated SR methods by measuring the similarity metrics over entire images [9, 12–16]. In contrast, in the present work, our aim was to look at tissue- or region-specific analysis of the performance measure, e.g., PSNR and SSIM for SR approaches, e.g., bicubic interpolation and modified-ESPCN deep learning model. For this study, we have used CT scans containing different tissues, e.g., liver, bladder, lung, kidney, and bone.

# 2 Methods

## 2.1 Dataset and Data Processing

We used the publicly available CT-ORG dataset for the evaluation [17]. It contained 3D CT scans from 140 subjects with six different organ classes, e.g., liver, bladder, lung, kidney, bone, and brain. Dataset also provides organ or tissue-specific segmentation masks. Because of the scarcity of brain CT volume, we do not perform the tissue-specific analysis for it.

Different CTs had different spatial resolutions. Most CTs have a spatial resolution higher than 1 mm in an axial plane (e.g., X–Y plane). Therefore, each CT was resized using bicubic interpolation to make 1 mm spatial resolution in the axial plane. Most studies evaluated the SR methods for down-sample factors $\times$ 2, $\times$3, or $\times$ 4 [3, 6–8]. In the present work, we evaluated the SR methods for the down-sample factor of $\times$ 4. HR CT slices were down-sampled by a factor of 4 to get LR images.

## 2.2 Model Architecture and Training

Many CNN-based approaches have been proposed for image SR [3, 7, 8, 15]. We modified the Efficient Sub-Pixel Convolutional Neural Network (ESPCN) architecture for the present study and used it for our analysis. ESPCN is a lightweight, relatively less resource-demanding CNN architecture [4]. We added two additional convolutional layers and skip connections to the classical ESPCN model (Fig. 1). Image intensity was rescaled between zero and one before training and testing. We trained the model for 100 epochs, and there were five steps per epoch and a batch size of 16 CT slices per step. In each training step, 16 CT slices were randomly chosen from a CT volume, which was randomly chosen from the training dataset. We used "adam" optimizer for training the network with a learning rate of 0.001. The computation resource consists of a workstation (Intel(R) Xeon(R) Silver 4214R CPU @ 2.40 GHz) with GPU NVIDIA GeForce RTX 2080ti having 11 GB of RAM. We trained two different models with two different loss functions, e.g., mean square error (MSE) and structural similarity index metric (SSIM) loss [18]. Half of the total CT volumes, e.g., 70, were used for the training, and the remaining CT volumes were used for the testing.

## 2.3 Performance Measurement

PSNR and SSIM metrics are widely used performance measures for image SR studies [2, 3]. SSIM is a perceptual metric that quantifies the similarity between two images and is widely used for evaluating the SR methods [5, 19]. SSIM is a perceptual quality

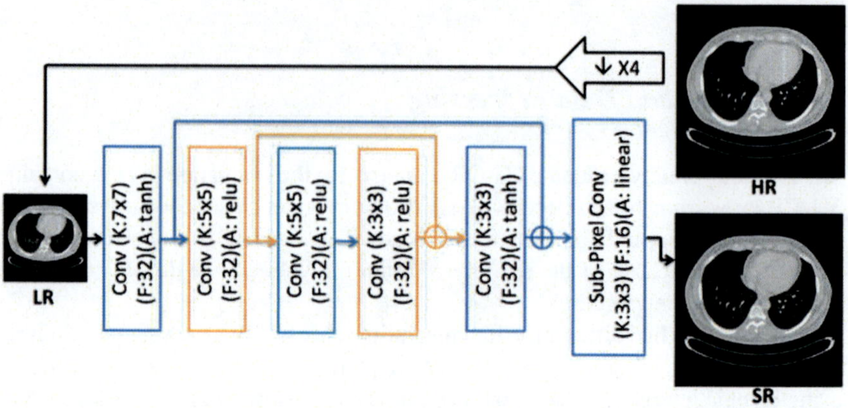


**Fig. 1** Modified-ESPCN architecture for image super-resolution

metric that loosely mimics the human visual system. SSIM models image degradation by combining three components—luminance (*L*), contrast (*C*), and structure (*S*).

$$SSIM(x, y) = f(L(x, y), C(x, y), S(x, y)) \tag{1}$$

$$\mu_x = \frac{1}{N}\sum_{i=1}^{N} x_i,\ \sigma_x = \left(\frac{1}{N-1}\sum_{i=1}^{N}(x_i - \mu_x)^2\right)^{1/2} \tag{2}$$

where $N$ is the number of pixels in a given image window. $\mu$ and $\sigma$ are the mean and standard deviation of the pixels in the image window, respectively.

Different components of the SSIM metric, e.g., luminance (*L*), contrast (*C*), and structure (*S*) are computed as follows:

$$L(x, y) = \frac{2\mu_x\mu_y + C_1}{\mu_x^2 + \mu_y^2 + C_1},\ C(x, y) = \frac{2\sigma_x\sigma_y + C_2}{\sigma_x^2 + \sigma_y^2 + C_2},\ S(x, y) = \frac{\sigma_{xy} + C_3}{\sigma_x\sigma_y + C_3} \tag{3}$$

where $\sigma_{xy}$ is the covariance matrix. In above equations, $C_1$, $C_2$, and $C_3$ are the constant to avid instability when denominator reaches zero.

PSNR for an individual pixel is defined as follows:

$$PSNR = 10 \cdot log_{10}\left(\frac{R^2}{MSE}\right) \tag{4}$$

MSE represents the mean squared error and R represents the maximum possible intensity magnitude of the image. We took R = 4096, considering CT pixel intensity (12 bits of quantization), e.g., Hounsfield Unit (HU) range of −1024 to 3071 for CT scan in the dataset.

For SSIM, we used the structural similarity index function available with the scikit-image Python library. We customized it to access the individual SSIM component, e.g., luminance ($L$), contrast ($C$), and structure ($S$) values. We used an $11 \times 11$ window size. We first computed a local pixel-wise SSIM map between the SR and HR images. PSNR and SSIM scores were computed for the entire CT slice and individual tissues, e.g., liver, bladder, lung, kidney, and bone. Tissue-specific PSNR and SSIM scores were computed by averaging them at pixels belonging to a particular tissue (e.g., tissue-specific mask).

## 3 Results

Mean PSNR and SSIM scores for a different CT image region for bicubic interpolation and modified-ESPCN methods are provided in Table 1. Modified-ESPCN approach with SSIM loss approach showed a significantly ($p < 0.01$, paired t-test) higher PSNR and SSIM score for all the regions compared to bicubic interpolation and modified-ESPCN approach with MSE loss. PSNR and SSIM scores were significantly ($p < 0.01$, paired t-test) higher for the entire image than the tissue-only region. PSNR and SSIM scores were significantly ($p < 0.01$, one-way ANOVA) different among tissues, e.g., liver, bladder, lung, kidney, and bone. The lung and bone tissues showed significantly lower PSNR and SSIM scores than the liver, kidney, and bladder. The SSIM score for structural component (S) was significantly ($p < 0.01$, paired t-test) lower than the luminance (L) and contrast (C) components across the regions for all SR methods (Table 2).

**Table 1** Region or tissue-specific comparison of performance metrics for CT images

| Performance metrics → | PSNR | | | SSIM | | |
|---|---|---|---|---|---|---|
| SR methods → regions ↓ | Bicubic | Modified ESPCN (MSE loss) | Modified ESPCN (SSIM loss) | Bicubic | Modified ESPCN (MSE loss) | Modified ESPCN (SSIM loss) |
| Full image | 49.10 | 46.63 | 50.80 | 0.80 | 0.84 | 0.88 |
| Tissue-only | 40.23 | 40.77 | 43.20 | 0.67 | 0.75 | 0.79 |
| Liver | 48.56 | 47.21 | 49.71 | 0.84 | 0.86 | 0.88 |
| Bladder | 48.12 | 46.36 | 49.51 | 0.83 | 0.85 | 0.87 |
| Lung | 39.82 | 40.30 | 43.39 | 0.63 | 0.72 | 0.77 |
| Kidney | 44.81 | 45.92 | 47.32 | 0.80 | 0.86 | 0.88 |
| Bone | 34.83 | 36.65 | 38.35 | 0.61 | 0.72 | 0.76 |

**Table 2** Region or tissue-specific comparison of luminance (*L*), contrast (*C*) and structure (*S*) of SSIM metric for CT images

| SR methods → | Bicubic | | | Modified ESPCN (MSE loss) | | | Modified ESPCN (SSIM loss) | | |
|---|---|---|---|---|---|---|---|---|---|
| Performance metrics → regions ↓ | SSIM (*L*) | SSIM (*C*) | SSIM (*S*) | SSIM (*L*) | SSIM (*C*) | SSIM (*S*) | SSIM (*L*) | SSIM (*C*) | SSIM (*S*) |
| Full image | 0.98 | 0.96 | 0.84 | 0.99 | 0.97 | 0.87 | 0.99 | 0.97 | 0.90 |
| Tissue-only | 0.99 | 0.96 | 0.71 | 1.00 | 0.96 | 0.78 | 1.00 | 0.96 | 0.82 |
| Liver | 1.00 | 0.98 | 0.85 | 1.00 | 0.99 | 0.87 | 1.00 | 0.99 | 0.89 |
| Bladder | 1.00 | 0.99 | 0.85 | 1.00 | 0.99 | 0.86 | 1.00 | 0.99 | 0.89 |
| Lung | 0.98 | 0.93 | 0.68 | 0.99 | 0.96 | 0.75 | 1.00 | 0.94 | 0.81 |
| Kidney | 1.00 | 0.98 | 0.81 | 1.00 | 0.99 | 0.87 | 1.00 | 0.98 | 0.89 |
| Bone | 1.00 | 0.96 | 0.63 | 1.00 | 0.95 | 0.76 | 1.00 | 0.96 | 0.79 |

## 4 Discussion and Conclusion

In the present work, we aimed to evaluate the CT image SR approach for different tissues, separately. We observed significantly lower SSIM scores when computed locally, e.g., specific to a particular tissue or region (Table 1), than computed over the entire image. It suggests that background regions (non-body tissue regions) give spuriously higher performance ratings. Therefore, the background region should be excluded during the computation of performance metrics while reporting performance outcomes. Note that most prior studies on medical radio-imaging SR have evaluated SR methods by measuring the similarity metrics over entire images [9, 12–16]. Different body tissues have different image characteristics or patterns. SR methods are sensitive to the information content, e.g., higher spatial frequencies or variance quantities. In the present work, different tissues in CT images showed significantly different performances for SR methods (Tables 1 and 2). Performance was significantly lower for lung and bone tissues. This is important as lung tissue is a region of interest in chest CT for radiologists and physicians, especially in the case of the subject with respiratory symptoms and COVID-19 infection. Therefore, SR quality should be evaluated for the region of interest. In line with our outcome, a recent study also reported lower SSIM scores for lung tissue patches in an SR CT image constructed with different SR methods [20].

The score for the structural component (*S*) of the SSIM measure is highly penalized by SR methods than contrast (*C*) and luminance (*L*) components (Table 2). It is expected as the structural component represents high spatial frequency or variance information, and approximating them is the difficult task. A CNN-based approach, e.g., modified-ESPCN with SSIM loss model, will likely learn that high spatial frequency pattern to some extent compared to bicubic interpolation.

Although we demonstrated our analysis on a CT scan of different organs, the outcome of this study can be generalized to other medical imaging modalities with different tissue types. In conclusion, we should exclude the non-tissue region during the computation of SR methods performance measures. It is essential to show the performance measure for specific tissue or region of interest for the medical image enhancement application using a SR approach.

## References

1. Chen H, He X, Qing L, Wu Y, Ren C, Sheriff RE, Zhu C (2022) Real-world single image super-resolution: a brief review. Inf Fusion 79:124–145. https://doi.org/10.1016/j.inffus.2021.09.005
2. Bashir SMA, Wang Y, Khan M, Niu Y (2021) A comprehensive review of deep learning based single image super-resolution. PeerJ Comput Sci 7:1–56. https://doi.org/10.7717/PEERJ-CS.621
3. Li Y, Sixou B, Peyrin F (2021) A review of the deep learning methods for medical images super resolution problems. https://doi.org/10.1016/j.irbm.2020.08.004
4. Shi W, Caballero J, Huszár F, Totz J, Aitken A, Bishop R, Rueckert D, Wang Z (2016) Sub-pixel convolutional neural network. Comput Vis Patter Recognit
5. Bakurov I, Buzzelli M, Schettini R, Castelli M, Vanneschi L (2022) Structural similarity index (SSIM) revisited: a data-driven approach. Expert Syst Appl 189:116087. https://doi.org/10.1016/j.eswa.2021.116087
6. Prince JL, Carass A, Zhao C, Dewey BE, Roy S, Pham DL (2019) Image synthesis and superresolution in medical imaging. In: Handbook of medical image computing and computer assisted intervention. https://doi.org/10.1016/B978-0-12-816176-0.00006-5
7. Yamashita K, Markov K (2020) Medical image enhancement using super resolution methods. Springer International Publishing. https://doi.org/10.1007/978-3-030-50426-7_37
8. Isaac JS, Kulkarni R (2015) Super resolution techniques for medical image processing. In: Proceedings of the international conference on technologies for sustainable development (ICTSD 2015). https://doi.org/10.1109/ICTSD.2015.7095900
9. Yu H, Liu D, Shi H, Yu H, Wang Z, Wang X, Cross B, Bramler M, Huang TS (2018) Computed tomography super-resolution using convolutional neural networks. In: Proceedings of the international conference image process (ICIP 2017), pp 3944–3948. https://doi.org/10.1109/ICIP.2017.8297022
10. Kolarik M, Burget R, Uher V, Povoda L (2019) Superresolution of MRI brain images using unbalanced 3D Dense-U-Net network. In: 2019 42nd International conference on telecommunications and signal processing, TSP 2019. https://doi.org/10.1109/TSP.2019.8768829
11. Rajeshwari P, Shyamala K (2021) Chest CT image super resolution using deep learning network models. In: 2021 International conference on smart generation computing, communication and networking, SMART GENCON 2021. https://doi.org/10.1109/SMARTGENCON51891.2021.9645873
12. Umehara K, Ota J, Ishida T (2018) Application of super-resolution convolutional neural network for enhancing image resolution in chest CT. J Digit Imaging 31:441–450. https://doi.org/10.1007/s10278-017-0033-z
13. Zhu J, Tan C, Yang J, Yang G, Lio P (2021) Arbitrary scale super-resolution for medical images. Int J Neural Syst 31. https://doi.org/10.1142/S0129065721500374
14. Zhang Z, Yu S, Qin W, Liang X, Xie Y, Cao G (2021) Self-supervised CT super-resolution with hybrid model. Comput Biol Med 138:104775. https://doi.org/10.1016/j.compbiomed.2021.104775
15. Zhang X, Feng C, Wang A, Yang L, Hao Y (2021) CT super-resolution using multiple dense residual block based GAN. Signal Image Video Process 15:725–733. https://doi.org/10.1007/s11760-020-01790-5

16. Jiang C, Zhang Q, Fan R, Hu Z (2018) Super-resolution CT image reconstruction based on dictionary learning and sparse representation. Sci. Rep. https://doi.org/10.1038/s41598-018-27261-z
17. Rister B, Yi D, Shivakumar K, Nobashi T, Rubin DL (2020) CT-ORG, a new dataset for multiple organ segmentation in computed tomography. Sci Data 7. https://doi.org/10.1038/s41597-020-00715-8
18. Zhao H, Gallo O, Frosio I, Kautz J (2016) Loss functions for image restoration with neural networks. IEEE Trans Comput Imaging. https://doi.org/10.1109/tci.2016.2644865
19. Wang Z, Bovik AC, Sheikh HR, Simoncelli EP (2004) Image quality assessment: from error visibility to structural similarity. IEEE Trans Image Process. https://doi.org/10.1109/TIP.2003.819861
20. de Farias EC, di Noia C, Han C, Sala E, Castelli M, Rundo L (2021) Impact of GAN-based lesion-focused medical image super-resolution on the robustness of radiomic features. Sci Rep 11. https://doi.org/10.1038/s41598-021-00898-z

# Co-registration of the Digitised Sensor Information and the Brain MRI Scan for Electrical Source Imaging: Fiducial-Based Alignment Versus ICP-Based Approach

**Bakul Gohel**

**Abstract** Source signal reconstruction from MEG/EEG sensor recordings requires co-registration of the 3D digitised head and sensor information to the brain MRI scan. Fiducial-based alignment and surface-fitting approaches are widely used for this co-registration procedure. The fiducial-based co-registration approach is simple and faster. In contrast, the surface-fitting approach may give poor co-registration in the case of spherical or quasi-spherical surfaces like a head surface. However, the surface-fitting approach, e.g. interactive closest point (ICP) algorithm, is extensively used in the current practice. We evaluated the fiducial- and ICP-based co-registration approaches' performance through simulations. In the present work, we also used one more fiducial point, a nasolabial angle, in addition to commonly used fiducial points such as nasion, left and right pre-auricular points for fiducial-based registration. Fiducial-based co-registration showed comparable or better performance than ICP-based co-registration while considering fiducial points digitisation error in magnitude of up to 5 mm. The addition of the fourth fiducial point, e.g. the nasolabial angle, improved the performance of the fiducial-based co-registration. Moreover, objective registration error, a cost function for the ICP-based approach, showed a weak correlation (correlation coefficient $< 0.25$) to the co-registration error. Fiducial-based approach with four fiducial points is simple, faster and better than the commonly used ICP-based approach for co-registering digitised head information to the brain MRI scan.

**Keywords** MEG · EEG · Inverse solution · Sensors-MRI co-registration · ICP

---

B. Gohel (✉)
Dhirubhai Ambani Institute of Information and Communication Technology (DA-IICT), Gandhinagar, India
e-mail: bakul_gohel@daiict.ac.in



S. Borah et al. (eds.), *Advanced Computational and Communication Paradigms*, Lecture Notes in Networks and Systems 535,
https://doi.org/10.1007/978-981-99-4284-8_38

# 1 Introduction

Electrical source imaging (ESI) of the brain using EEG or MEG needs the computation of the inverse model [1–3]. It requires bringing the sensor's locations and brain anatomical information from the MRI volume into a common cartesian three-dimensional (3D) space. Therefore, we need to digitise fiducial landmarks, the scalp surface, the sensor positions (in the case of EEG) or the locations of the head positioning coils (in the case of MEG) and need them to co-register with a subject's MRI scan [2–7]. There are two commonly used approaches for the co-registration of digitised head information (DHI) and MRI head information (MHI). The first method is based on the alignment of fiducial points, and another is based on the surface fitting approach. In a fiducial-based alignment, three fiducial points from the MHI, namely the nasion (NAS), left pre-auricular (LPA) and right pre-auricular (RPA) points, are linearly co-registered to the corresponding fiducial points in DHI. This fiducial-based alignment is simple, fast and does not require information about the entire scalp surface. However, any ambiguity or mismatch in defining the fiducial points between the two modalities of MHI and DHI, known as the fiducial placement error, causes co-registration error [4, 8, 9]. Head surface fitting provides an alternative or auxiliary approach to a fiducial-based alignment[4, 8, 9]. In this approach, the initial fiducial-based alignment is followed by the co-registration of the digitised head surface points and MRI head surface points such that the mean distance between the two surfaces(point cloud) is minimised. The surface fitting approach, e.g. the iterative closest point (ICP) algorithm, is commonly used in various studies, toolboxes and libraries for the ESI [4, 8, 9].

Although fiducial-based alignment and surface-fitting approaches are widely used in M/EEG-based studies, less effort has been made towards comparative evaluations of these co-registration approaches. Multiple prior studies have evaluated surface fitting-based approaches for the co-registration of DHI and MHI [10–14]. Whalen et al. showed that surface fitting based on the Levenberg Marquardt algorithm (ICP-LMA) performs better than fiducial-based or moment-based alignment [8]. Further analysis by Charribeli et al. showed that ICP-P2P error minimisation performs better or is comparable to the ICP-LMA approach [9]. A shared dataset was used in both of these studies [8, 9].

The surface-fitting algorithm performs better when two surfaces or point clouds are non-linear. In other words, co-registration performance is likely to be degraded if the surface is smoother and symmetrical in shape. There were exits many optimum local minima with poor solutions. As the head surface is spherical or quasi-spherical, surface fitting approaches, including the ICP algorithm, may fail to provide optimum solutions, especially in estimating rotation parameters. Moreover, inhomogeneity in the magnetic field may cause an inevitable structural distortion to MRI images [8, 15]. If not adequately addressed, it further leads to co-registration error [8]. A rigid-body ICP estimates rotation and translation parameters and is widely used in various toolboxes for ESI imaging; therefore, widely used in practice [4, 5, 16, 17]. On the other hand, as per our experience, identification and localisation of fiducial markers, e.g. NAS, LPA and RPA, are straightforward, and localisation error is less likely to exceed 5 mm. Therefore, we performed the comparative analysis of fiducial-based alignment and surface-fitting, e.g. ICP co-registration approaches. We used a simulation-based approach for our evaluation, in which digitised fiducial landmarks and head surface were simulated from the MRI scan. A nasolabial angle (NLA) landmark can be easily digitised on the face and identified in the head MRI scan; therefore, we also used it in addition to NAS, LPA and RPA for a fiducial-based co-registration. It is expected that the addition of one more landmark will improve the performance of the fiducial-based co-registration.

## 2 Materials and Methods

### 2.1 *Datasets*

We used two publicly available datasets for the performance evaluation. In these datasets, the face is not defaced, so all fiducial and surface information required for evaluation is available for analysis. Dataset I has 36 MRI scans (out of it, we used 35), and the mean age of the subjects was $20.17 \pm 1.73$ years [18, 19]. Dataset II has 36 MRI scans (out of it, we used 35), and the age of the subjects ranged from 18 to 24 years [20].

Personalising online interactions improves customer relationships and increases desirable behaviours, such as positive word-of-mouth and increased purchase intent [9]. Users develop more cognitive and emotional trust with a significant increase in personalisation [3, 4, 16]. However, other research suggests that the use of personal information stimulates privacy concerns in the user and affects the decision to continue using websites. The research examines constructs affecting users' behavioural intentions in the form of continue to revisit SNS and also studies inter-relations of users' control, privacy concerns and trust towards websites during the interaction. Results show that increasing perceived information control reduces the negative effect of privacy concerns on behavioural intentions.

### 2.2 *MRI Head Information*

In a brain MRI scan, we manually localised the fiducial landmarks that are NAS, LPA, RPA and NLA. Head surfaces mesh was extracted from head volume after segmenting and MRI scan. ESI imaging requires the computation of sourcemodel and headmodel modelled from a brain MRI scan [2, 3, 16, 21]. Sourcemodel is the 3D grid of point cloud representing the location information inside the brain volume. We computed the sourcemodel from the brain MRI scan and used it for the performance evaluation of the co-registration approaches.

## 2.3 Simulated Digitised Head Information

We added a random localisation error of up to 5 mm to each fiducial landmark identified in a brain MRI scan to derive the digitised fiducial landmarks. We derived the simulated digitised scalp and face surfaces from the head surface derived from a brain MRI scan. As per standard practice, surface points over the scalp, around the eye orbit and over the nose were considered (Fig. 1, Scalp + Face). We sampled 64, 128, and 256 points (vertices) among them. To achieve uniform sampling, we first performed k-mean clustering on the MRI head surface vertices, where k is the number of digitised head points sampled. In each cluster, the point nearest to the cluster centroid was chosen. We added random error sampled from Gaussian distribution with mean zero and one mm standard deviation. As mentioned earlier, magnetic field inhomogeneity induces scaling distortion in MRI [8, 15]. Therefore, we introduced up to 1% and 2% anisotropic scaling error to the simulated digitised head information. Recently, various researchers have explored the utility of face and forehead surfaces digitised with various optical scanners for performing co-registration [12, 22, 23]. Optical scanner-based methods are faster and easier than pivot probe-based scalp surface digitisation. In order to simulate optical scanner-based forehead and face surface digitisation, we also extract another set of points from the forehead, nose, and surrounding eye orbit, excluding the scalp (Fig. 1, Face only). For each MRI scan, we generated ten sets of simulation DHI.

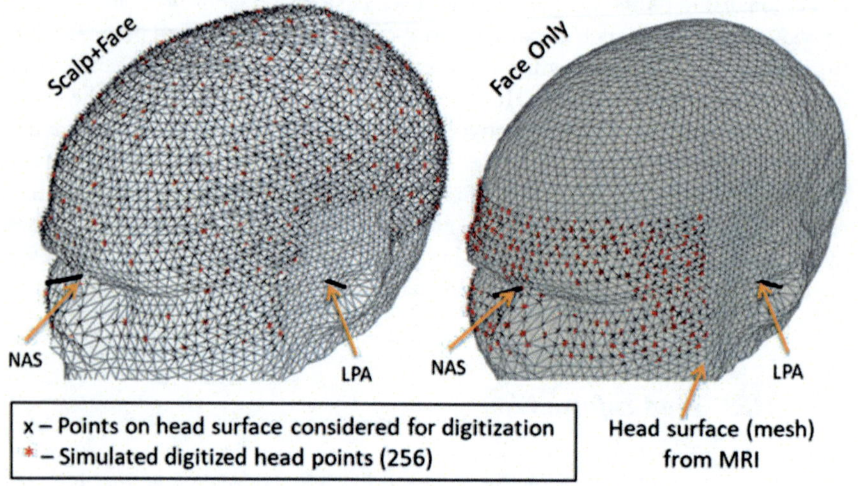


**Fig. 1** Head surface from brain MRI scan and simulation of digitised head points

### 2.4 *Fiducial-Based Alignment*

The co-registration of DHI and MHI can be performed exclusively using fiducial landmark information. We performed free rigid-body alignment for the Fiducial-based alignment [24]. Here, rigid-body transformation parameters with six degrees of freedom (DOF) were computed using a least-squares approach with a minimum of three fiducial points. This approach finds the parameters that minimise the error between the corresponding fiducial points. We performed fiducial-based alignment using three fiducial points, e.g. NAS, RPA and LPA. We will call it the FID3 approach in the subsequent text. We also performed fiducial-based alignment using four points, e.g. NAS, RPA, LPA and NLA. We will call it the FID4 approach in the subsequent text.

### 2.5 *ICP-Based Alignment*

Plenty of surface-fitting (registration) approaches are available in the literature [25–27]. However, we will use only the ICP algorithm for our evaluation as it is widely used across different studies, various ESI toolboxes and software libraries for ESI. The majority of the EEG/MEG data processing toolboxes or libraries use the ICP-P2P (points to a plane) algorithm for MHI and DHI co-registration [4, 5, 17, 28]. We have MRI surface and digitised head points. Therefore, registration of the points to the plane is the more intuitively appropriate approach. The ICP algorithm requires an initial solution to be adequately close. Therefore, as commonly used in standard practice, we performed the fiducial-based alignment before performing ICP-based refinement.

The ICP-based surface-fitting approach minimises the objective registration error (ORE), which is the averaged Euclidian distance between two surfaces (e.g. between MRI scalp surface and digitised head points) [8, 28]. The present study used the ICP-P2P registration algorithm provided in the FieldTrip toolbox library for surface fitting [5, 27]. The ICP-P2P algorithm considers the normal to the destination surface (i.e. the curvature of the MRI scalp mesh) in addition to the Euclidian distance between two surfaces [9, 27]. ICP-P2P estimates a rigid-body transformation. We performed the ICP-based alignment using digitised head points from scalp & face and face alone (Fig. 1). We will refer to these approaches as ICP-SF (e.g. scalp + face) and ICP-F( face only), respectively, in the subsequent text. Moreover, we performed co-registration using the ICP-SF approach with different numbers of digitised head points, e.g. 64,128, and 256.

### 2.6 Co-registration Error Measures

Any error in the co-registration process also deviates the grid point in the source-model from its original position, causing source localisation error in ESI. Hence, we used the source grid point position error (SGPPE) for performance evaluation [29]. It represents the source grid point with maximum error, e.g. Euclidean distance with respect to ground truth. We computed SGPPE for both fiducial- and ICP-based alignment approaches. We also computed ORE for the ICP-based approach. Note that ORE is the objective function for the ICP algorithm.

## 3 Results

As aforesaid, we generated ten sets of the simulated DHI per MRI scan, and we used 35 MRI scans from each dataset. Therefore, we had 10X35 evaluations for each dataset. Outcomes for the fiducial-based approach and ICP-based approach using 256 digitised head points are depicted in Fig. 2. Median SGPPE lies between 3 and 5 mm for both approaches, for different scaling factors, and for both datasets. A FID4 approach showed significantly lower or comparable SGPPE than FID3 (Table 1, 'FID3 > FID4'). A FID4 approach showed significantly lower SGPPE than the ICP-SF method for scaling factor 2% (Table 1, 'FID4 < ICP-SF'). However, it showed significantly lower SGPPE than the ICP-F method across all comparisons (Table 1, 'FID4 < ICP-F'). An ICP-SF approach showed significantly lower SGPPE compared to the ICP-F approach for scaling factors 0.5% and 1% (Table 1, 'ICP-SF < ICP-F').

ICP-SF performed with a lesser number of digitised head points, e.g. ICP-SF (128) and ICP-SF (64) showed significantly ($p < 0.01$, Wilcoxon rank-sum test, $n = 350$) higher SGGPE than ICP-SF performed with comparatively higher digitised head points, e.g. ICP-SF (256) (Fig. 3). A weaker correlation (correlation coefficient < 0.25) was observed between ORE and SGPPE for ICP-SF and ICP-F approaches (Fig. 4).

## 4 Discussion and Conclusion

We compared the performance of the fiducial-based alignment and surface-fitting approach, e.g. ICP-P2P-based methods for the co-registration of DHI and MHI. Overall comparable or better performance was observed for the fiducial-based alignment than the ICP-P2P approach.

There could not be an exact match (e.g. mainly scaling error) between DHI and MRI due to magnetic field inhomogeneity in the MRI scanner [8, 15] or error from the digitisation process. Moreover, the head surface is spherical to an elliptical shape.

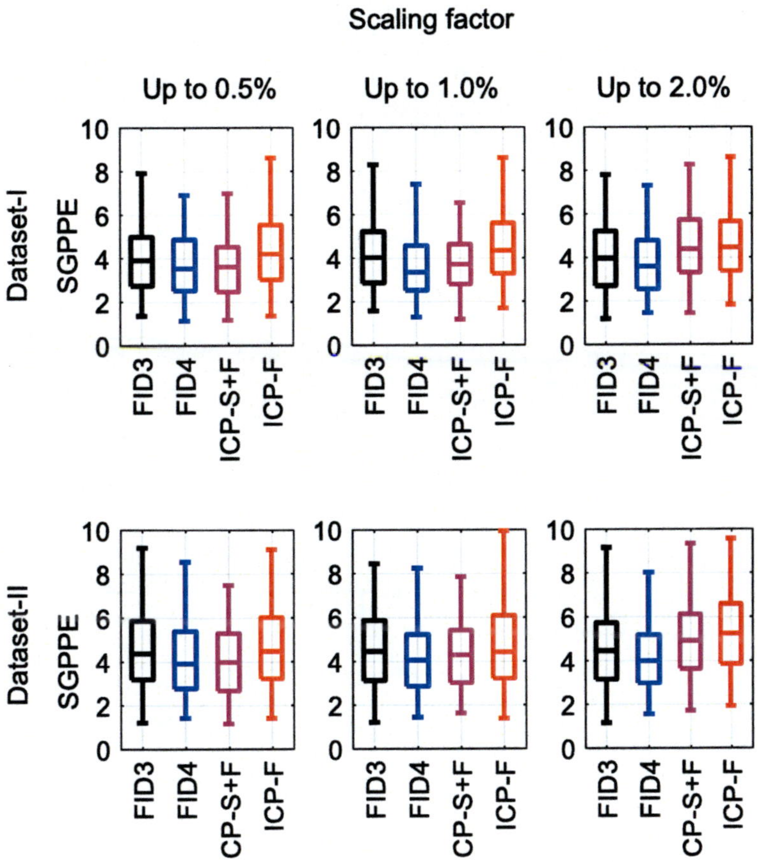


**Fig. 2** SGPPE for different approaches, at different scaling factors, and for different datasets

**Table 1** Comparison of SGPPE among different co-registration approaches

| Scale factor | FID4 < FID3 | FID4 < ICP-SF | FID4 < ICP-F | ICP-SF < ICP-F |
|---|---|---|---|---|
| DATASET-I | | | | |
| 0.5% | S | NS | S | S |
| 1.0% | S | NS | S | S |
| 2.0% | NS | S | S | NS |
| DATASET-II | | | | |
| 0.5% | S | NS | S | S |
| 1.0% | NS | NS | S | S |
| 2.0% | S | S | S | NS |

*Note* 'S' means significant ($p < 0.01$, Wilcoxon rank-sum test, $n = 350$). 'NS' means not-significant

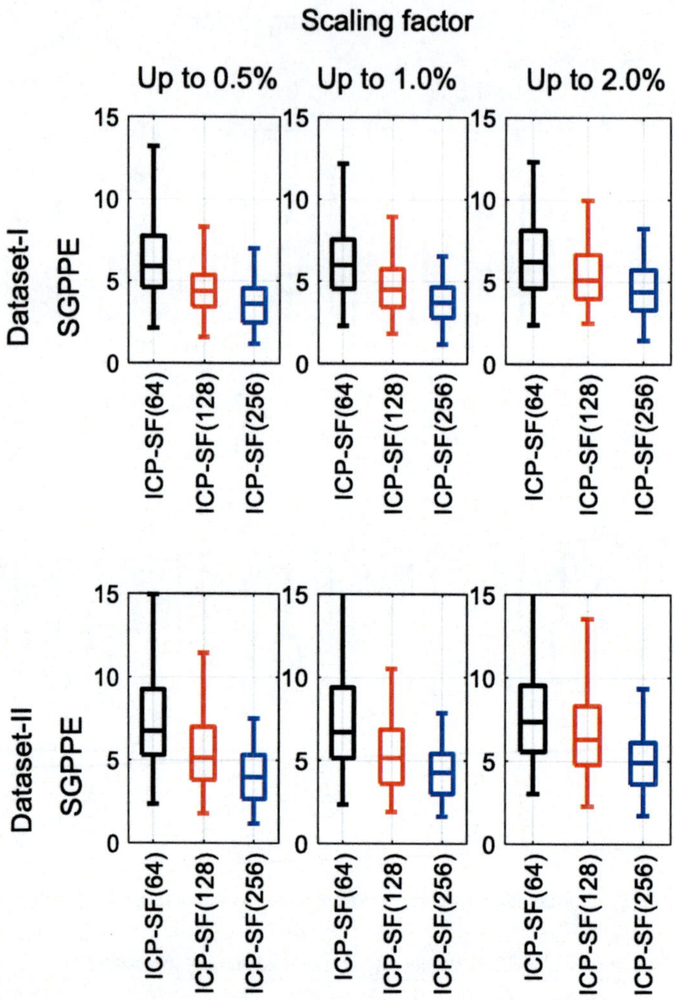


**Fig. 3** SGPPE for ICP-SF approach using different numbers of digitised head points. ICP-SF(N), where N represents the number of digitised head points

There are, therefore, exists multiple solutions with multiple local minima. Additionally, the loss function for the ICP-based algorithm is ORE. However, minimum ORE alone does not necessarily represent a minimum co-registration error. Our outcome also observed that ORE does not correlate well with co-registration error, e.g. SGPPE (Fig. 4). Therefore, using the ICP approach for co-registration, especially the refinement of co-registration between DHI and MRI, is conceptually less intuitive. It is worth mentioning here because the ICP approach is frequently used to refine co-registration following the fiducial-based alignment. In contrast to our

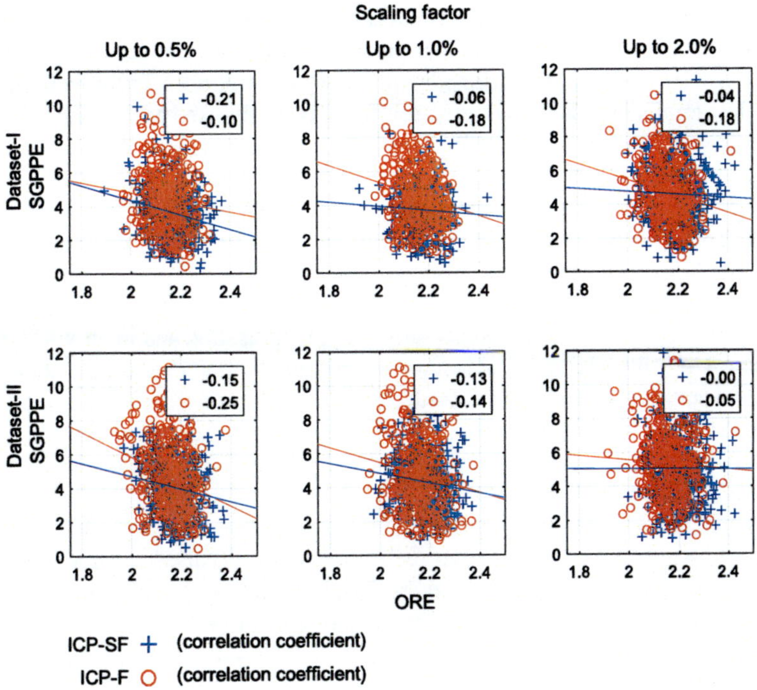


**Fig. 4** Association between ORE and SGPPE for ICP-SF and ICP-F approaches. The lines in the scatter plot represent the fitted regression lines. Values in the legend box are the correlation coefficients

results, previous studies [8, 9] have shown a lower co-registration error for the surface-fitting (ICP-based) approach than for fiducial-based alignment. In these studies, 32 ring-shaped reference MRI contrast markers were placed on a scalp, and the head surface between them was digitised. Since these markers are placed directly on the skin, it produces marker-size shallow bumps on the extracted MRI scalp surface after MRI volume segmentation. Consequently, the digitised head surface and the bumps on the MRI scalp surface produce a complementary non-linear geometry. Therefore, an ICP-based approach can be biased to produce a lower co-registration error in their evaluation.

Recently, the optical scanner- and photogrammetry-based approach used only facial surface (excluding the scalp) to perform a DHI and MRI co-registration [12, 22, 23]. It allows faster and easier digitisation. Prior studies showed promising outcomes; however, their evaluation is limited to ORE performance measures only. As stated earlier, achieving lower ORE does not necessarily represent a lower co-registration error (Fig. 4). In the present simulation-based study, co-registration based on only facial surfaces showed comparable or poor performance compared to combining the use of facial and scalp surfaces (Fig. 2 and Table 1). An optical scanner also helps in digitising the electrode position [30–32]. We can determine the digitised points on the scalp surface from these electrode locations. Therefore, we need not digitise the scalp surface separately and minimise the subject preparation time. The high-density EEG base ESI is usually performed with 64 or 128 electrodes. In this case, we may have less number of electrodes and, therefore, fewer digitised head points. However, the co-registration error, e.g. SGPPE for ICP-SF, increases as available scalp surface points decrease (Fig. 3).

The co-registration error from fiducial-based alignment mainly results from discrepancies in the placement of fiducial points between the two modalities (i.e. DHI and MHI). Therefore, the co-registration error depends on how we choose the fiducial landmarks. The determination of the NAS fiducial point is less ambiguous than that of the pre-auricular points during head digitisation and in an MRI volume. Like the NAS, the NLA fiducial landmark can be easily identified using both imaging modalities. We, therefore, also recorded the NLA fiducial information and observed through simulation that registration could be improved by additional consideration of NLA over three standard landmarks, e.g. NAS, LPA and RPA, for the fiducial-based approach (Fig. 2 and Table 1).

In conclusion, a fiducial-based alignment is simple, fast and efficient compared to an ICP-based approach for the co-registration of the digitised electrodes and brain MRI scan. Moreover, adding one more fiducial point, e.g. NLA, can improve the co-registration performance of the fiducial-based alignment.

## References


1. Wendel K, Väisänen O, Malmivuo J, Gencer NG, Vanrumste B, Durka P, Magjarević R, Supek S, Pascu ML, Fontenelle H, De Peralta G, Menendez R (2009) EEG/MEG source imaging: methods, challenges, and open issues. Comput Intell Neurosci 2009:1. https://doi.org/10.1155/2009/656092
2. Hansen P, Kringelbach M, Salmelin R (2010) MEG: an introduction to methods. https://doi.org/10.1093/acprof:oso/9780195307238.001.0001
3. Michel CM, Brunet D (2019) EEG source imaging: a practical review of the analysis steps. Front Neurol 10:1. https://doi.org/10.3389/fneur.2019.00325
4. Tadel F, Baillet S, Mosher JC, Pantazis D, Leahy RM (2011) Brainstorm: a user-friendly application for MEG/EEG analysis. Comput Intell Neurosci 2011:1. https://doi.org/10.1155/2011/879716
5. Oostenveld R, Fries P, Maris E, Schoffelen JM (2011) FieldTrip: Open source software for advanced analysis of MEG, EEG, and invasive electrophysiological data. Comput Intell Neurosci 2011:156869. https://doi.org/10.1155/2011/156869
6. Zorzos I, Kakkos I, Ventouras EM, Matsopoulos GK (2021) Advances in electrical source imaging: a review of the current approaches, applications and challenges. Signals 2:378–391. https://doi.org/10.3390/signals2030024
7. Lin SK, Lo RC, Lee RG (2020) Automatic co-registration of MEG and 3D digitization using 3d generalized hough transform. Biomed. Eng. Appl. Basis Commun. 32. https://doi.org/10.4015/S1016237220500192
8. Whalen C, Maclin EL, Fabiani M, Gratton G (2008) Validation of a method for coregistering scalp recording locations with 3D structural MR images. Hum Brain Mapp 29:1288–1301. https://doi.org/10.1002/hbm.20465

9. Chiarelli AM, Maclin EL, Low KA, Fabiani M, Gratton G (2015) Comparison of procedures for co-registering scalp-recording locations to anatomical magnetic resonance images. J Biomed Opt 20:16009. https://doi.org/10.1117/1.JBO.20.1.016009
10. Titto Mangalapallil A, Puthusserypady S (2004) Embedding EEG data on 3D head modeled using MRI data. In: 26th Annual international conference of the IEEE engineering in medicine and biology society, vol 3, pp 1833–1837. https://doi.org/10.1109/IEMBS.2004.1403546
11. Špiclin Ž, Hans A, Duffy FH, Warfield SK, Likar B, Pernuš F (2008) EEG to MRI registration based on global and local similarities of MRI intensity distributions. In: Lecture Notes in Computer Science (including subseries Lecture Notes in Artificial Intelligence and Lecture Notes in Bioinformatics), pp 762–770. https://doi.org/10.1007/978-3-540-85988-8_91
12. Murthy SVK, MacLellan M, Beyea S, Bardouille T (2014) Faster and improved 3-D head digitization in MEG using Kinect. Front Neurosci 8. https://doi.org/10.3389/fnins.2014.00326
13. Lamm C, Windischberger C, Leodolter U, Moser E, Bauer H (2001) Co-registration of EEG and MRI data using matching of spline interpolated and MRI-segmented reconstructions of the scalp surface. Brain Topogr 14:93–100. https://doi.org/10.1023/A:1012988728672
14. Kozinska D, Carducci F, Nowinski K (2001) Automatic alignment of EEG/MEG and MRI data sets. Clin Neurophysiol 112:1553–1561. https://doi.org/10.1016/S1388-2457(01)00556-9
15. Jafar M, Jafar YM, Dean C (2017) Assessment of geometric distortion in six clinical scanners using a 3D-printed grid phantom. J Imaging 3:28. https://doi.org/10.3390/jimaging3030028
16. Litvak V, Mattout J, Kiebel S, Phillips C, Henson R, Kilner J, Barnes G, Oostenveld R, Daunizeau J, Flandin G, Penny W, Friston K (2011) EEG and MEG data analysis in SPM8. Comput Intell Neurosci 2011. https://doi.org/10.1155/2011/852961
17. Gramfort A, Luessi M, Larson E, Engemann DA, Strohmeier D, Brodbeck C, Goj R, Jas M, Brooks T, Parkkonen L, Hämäläinen M (2013) MEG and EEG data analysis with MNE-Python. Front Neurosci. https://doi.org/10.3389/fnins.2013.00267
18. Boekel W, Wagenmakers EJ, Belay L, Verhagen J, Brown S, Forstmann BU (2015) A purely confirmatory replication study of structural brain-behavior correlations. Cortex 66:115–133. https://doi.org/10.1016/j.cortex.2014.11.019
19. Boekel W, Wagenmakers EJ, Belay L, Verhagen J, Brown S, Forstmann BU (2014) A 3T MRI data set on structural brain-behavior correlations. https://www.nitrc.org/projects/confrep2014/
20. Wimmer GE, Li J, Gorgolewski KJ, Poldrack RA (2021) rewardBeast. https://doi.org/10.18112/openneuro.ds002738.v1.0.2
21. Oostenveld R, Fries P, Maris E, Schoffelen JM (2011) FieldTrip: open source software for advanced analysis of MEG, EEG, and invasive electrophysiological data. Comput Intell Neurosci 2011. https://doi.org/10.1155/2011/156869
22. Taberna GA, Marino M, Ganzetti M, Mantini D (2019) Spatial localization of EEG electrodes using 3D scanning. J Neural Eng 16. 1https://doi.org/10.1088/1741-2552/aafdd1
23. Koessler L, Cecchin T, Caspary O, Benhadid A, Vespignani H, Maillard L (2011) EEG-MRI co-registration and sensor labeling using a 3D laser scanner. Ann Biomed Eng 39:983–995. https://doi.org/10.1007/s10439-010-0230-0
24. Eggert DW, Lorusso A, Fisher RB (1997) Estimating 3-D rigid body transformations: a comparison of four major algorithms. Mach Vis Appl 9:272–290. https://doi.org/10.1007/s001380050048
25. Billings SD, Boctor EM, Taylor RH (2015) Iterative most-likely point registration (IMLP): a robust algorithm for computing optimal shape alignment. PLoS ONE 10. https://doi.org/10.1371/journal.pone.0117688
26. Myronenko A, Song X (2010) Point set registration: coherent point drifts. IEEE Trans Pattern Anal Mach Intell 32:2262–2275. https://doi.org/10.1109/TPAMI.2010.46
27. Low K (2004) Linear least-squares optimization for point-to-plane ICP surface registration. Chapel Hill, Univ. North Carolina, pp 2–4
28. Park SY, Subbarao M (2003) An accurate and fast point-to-plane registration technique. Pattern Recognit Lett 24:2967–2976. https://doi.org/10.1016/S0167-8655(03)00157-0
29. Gohel B, Lim S, Kim M-Y, Kwon H, Kim K (2017) Approximate subject specific pseudo MRI from an available MRI dataset for MEG source imaging. Front Neuroinform 11:1–12. https://doi.org/10.3389/fninf.2017.00050

30. Taberna GA, Guarnieri R, Mantini D (2019) SPOT3D: spatial positioning toolbox for head markers using 3D scans. Sci Rep 91(9):1–9. https://doi.org/10.1038/s41598-019-49256-0.
31. Shirazi SY, Huang HJ (2019) More reliable EEG electrode digitizing methods can reduce source estimation uncertainty, but current methods already accurately identify brodmann areas. Front Neurosci. https://doi.org/10.3389/fnins.2019.01159
32. Homölle S, Oostenveld R (2019) Using a structured-light 3D scanner to improve EEG source modeling with more accurate electrode positions. J Neurosci Methods 326. https://doi.org/10.1016/j.jneumeth.2019.108378

# Communication Techniques

# Relative Analysis of OCDMA-FSO and FSO Under Normal, Haze and Fog Conditions

**Himali Sarangal, Butta Singh, Manjit Singh, Vinit Grewal, and Satveer Kour**

**Abstract** Free Space Optical communications offers a promising response for future broadband organizations offering high information data transmission when diverged from RF development. While optical code division multiple access (OCDMA) is forming into a promising opportunities for access associations as a result of its true capacity for updated information security, improved and decentralized organization control, and its organization similarly as capable utilization of bandwidth. In this paper, the comparison among the performance of FSO (free space optics) and OCDMA-FSO for five users is done under normal, haze and fog conditions at a bit rate of 10 Gb/s. The evaluation is completed at a distance of 5 km. The outcomes show that the OCDMA-FSO system performs better when contrasted with the straightforward FSO system. Attenuation increases, bit error also increases with increasing attenuation, and the proposed system, performs better than a simple FSO system. Under fog condition, the Quality factor of OCDMA-FSO is 8.26, whereas for FSO it is 3.60 at a distance of 5 km.

**Keywords** Optical code division multiple access · Free space optics · Bit error rate · Multi Access Interference · Spectral direct detection

H. Sarangal · B. Singh (✉) · M. Singh · V. Grewal
Department of Engineering and Technology, Guru Nanak Dev University, Regional Campus Jalandhar, India
e-mail: bsl.khanna@gmail.com

M. Singh
e-mail: manjit.ecejal@gndu.ac.in

V. Grewal
e-mail: vinit.ecejal@gndu.ac.in

S. Kour
Department of CET, Guru Nanak Dev University, Amritsar, India
e-mail: satveer.dcet@gndu.ac.in



S. Borah et al. (eds.), *Advanced Computational and Communication Paradigms*, Lecture Notes in Networks and Systems 535,
https://doi.org/10.1007/978-981-99-4284-8_39

## 1 Introduction

Without the cost of laying a fibre-optic link, FSO communication offers the possibility to send adequate information safely through free space. FSO correspondence link is unrealistic simply because of atmospheric obstructions that may prompt a high BER of the system [1]. However, a channel determines the system capabilities. The propagation channel in FSO is the atmosphere. Therefore, atmospheric conditions like rain, fog, wind, objects (buildings and birds), etc., decide the reliability and availability of an FSO link [2]. Moreover, the transmitted light also gets reflected or absorbed by the above-said conditions. In other words, it can be said that atmospheric conditions directly affect the FSO link and the quality of communication.

OCDMA is a keen innovation liked because of its extraordinary benefits of rapid optical handling and completely asynchronous transmission with low dormancy access [3]. OCDMA permits the number of transmitters to send the information to every transmitter concurrently by spreading the available bandwidth. A distinct chip pattern created by a PRBS (pseudo-random binary sequence) generator is assigned towards every transmitter. The data signal is propagated using this chip pattern. Multiplexing and sending the encoded signals through an optical fibre to the receiver from several transmitters is achieved. The decoder uses the exact chip pattern as the target emitter to decode the targeted data signal [4, 5].

The OCDMA system has certain drawbacks in addition to its major benefits. Its performance decreases by Multiple access interference (MAI) and (phase intensity induced noise (PIIN). MAI affects it, causing the intended receiver to be set just to detect its data stream while disregarding the other signals as noise. Therefore, it is desired to use OCDMA with SAC (Spectral amplitude coding) for the optical access network. In SAC-OCDMA, a source with broad-spectrum is used whose spectra are further sliced down into a number of wavelengths. These wavelengths are encoded by using a series of FBG. The signal from different segments is combined using a power combiner, and with the help of power splitter, the received signal is distributed among the intended receiver, which decodes the data using balanced detection [6–9]. The balanced detection consists of two arms, the upper arm consists of FBG as a decoder/optical filter having a transfer function H (f), and the lower arm consists of a filter that is complementary of the upper part filters. These two signals are detected using photo detectors and subtracted using an optical subtractor, which subsequently cancels MAI. At last, by using a low pass filter, an unwanted signal is removed. In SAC-OCDMA, the code design with zero cross-correlation (ZCC) is required since it removes the effect of MAI and suppresses the effect of PIIN [10, 11].

Even though OCDMA offers access networks a high bandwidth, low cost solution, installing optical fiber in remote locations that serve tens to a few hundred consumers still poses the biggest obstacle for fiber optics. Additionally, as bandwidth needs rise and resources stagnate, fiber optic cable deployment problems in the last mile only get more difficult. High speed access needs may be met by optical wireless, which is based on the most recent free space optics (FSO) technology.

A lot of work is being done to provide a broadband wireless solution for last-mile access. Conventional radio links and optical fibre can be replaced by a technology called FSO. The generating techniques are used to enhance the FSO link by making it more reliable and available. FSO uses a wireless channel, so its performance is mainly affected due to atmospheric conditions and the location of the system where it is installed [12–14].

OCDMA is a powerful SS technique that allows multiple users to access the whole bandwidth simultaneously. A number of researchers have designed a number of codes to increase the intake of users. As the number of mobile users increases, it increases the demand for high bandwidth and high speed. To fulfil these demands, researchers should develop new techniques that support more subscribers using the existing infrastructure. Therefore, there is a great demand to study the details of OCDMA and FSO systems. A hybrid communication system can be developed by integrating both OCDMA and FSO technologies for last-mile access networks. OCDMA, when integrated with FSO, provide high data transmission with more security to last-mile access [15–20].

This work is carried out to fulfil the demands of high data rate, and high bandwidth to remote users in the influence of atmospheric turbulence. Further, the OCDMA-FSO system is compared with the simple FSO system.

## 2 System Description

An FSO system is comprised of three fundamental parts: a transmitter, the engendering channel, and a recipient appeared in Fig. 1. The transmitter area comprises of the PRBS generator that creates an irregular succession of paired numbers. At that point, the NRZ pulse generator shapes the information signal to the electrical signal. The Continuous-wave laser creates an optical sign, and the Mach–Zehnder Modulator balances each signal. This optical signal is sent on channel through the atmosphere. At the receiver, an optical signal is identified by the photo detector (PIN) [16, 18, 20, 21]. A low pass channel is utilized to channel the signal from commotion. BER (bit error rate) analyzer consequently computes the Q factor, BER value.

The block diagram of OCDMA-FSO with the SDD (Spectral Direct Detection) method is given in Fig. 2. In this setup, the SDD technique is used because compared to complementary subtraction procedures, which require two sets of decoders and detectors, direct detection only needs one pair [22]. Therefore, the cost of designing the system is deceased. The information with the unipolar computerized signal is optically regulated utilizing the Mach–Zehnder modulator at the transmitter. At that point the adjusted arrangements are joined together and sent through an FSO link. At the receiver, an optical splitter is utilized to separate distinctive adjusted code arrangements. The decoded signal is distinguished by the PIN photo detector [2, 19]. To recuperate unique information, the approaching signal is separated utilizing LPF (low pass filter).

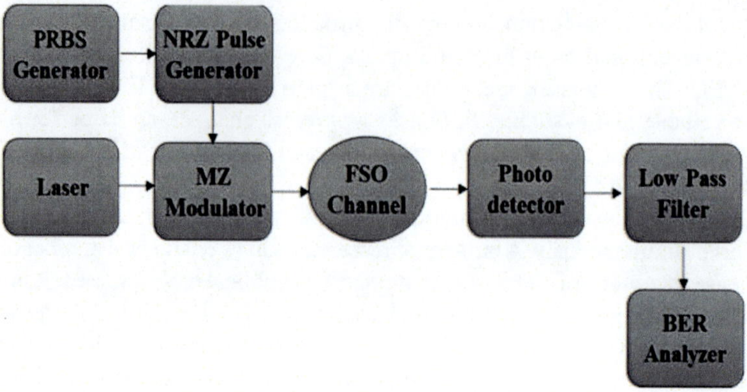


**Fig. 1** Block diagram of simple FSO system

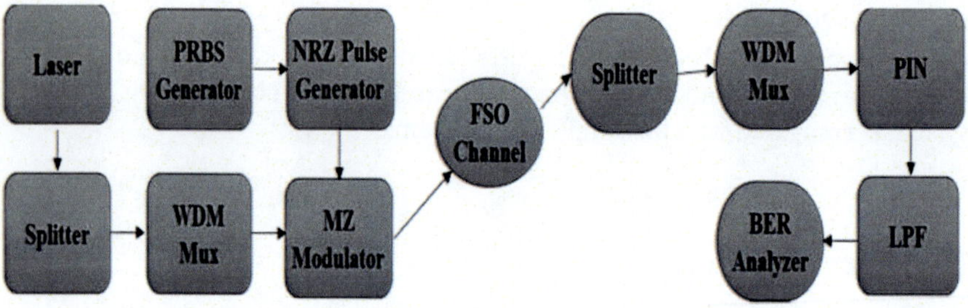


**Fig. 2** Block diagram of OCDMA-FSO system

## 3 Results and Discussion

To evaluate the performance of five users, theFSO and OCDMA-FSO system employing SDD technique and simulation software optiwave-13 is used. The direct technique supports large number of users due to the fact that the receiver has fewer filters than the traditional method [23, 24]. In addition, ZCC codes are used in OCDMA. ZCC code is designed by using the sequence of 0's (z's) and 1's (o's). These sequences are developed at a higher rate as compared to the baseband data. The cross-correlation ($\lambda c$) and the autocorrelation ($\lambda a$) are the two main code properties that should be satisfied by the code. Low cross correlation ($\lambda c = 0$ and $\lambda c \leq 1$) and high autocorrelation is required for the SAC-OCDMA system's implementation. The ZCC code for five users is constructed using [25, 26] as given below:

**Table 1** List of the parameters

| S. no | Parameters used | Values |
|---|---|---|
| 1 | Input power | 12 dBm |
| 2 | Transmission loss | 1db |
| 3 | Wavelength | 1550 nm |
| 4 | Receiver aperture diameter | 10 cm |
| 5 | Transmitter aperture diameter | 10 cm |
| 6 | Additional loss | 1 db |

$$\begin{bmatrix} O & Z & O & Z & Z & Z & O & Z & Z & Z & Z & Z & O & Z & Z & Z & Z & Z & Z & Z \\ Z & O & Z & Z & O & Z & Z & Z & O & Z & Z & Z & Z & Z & O & Z & Z & Z & Z & Z \\ Z & Z & Z & O & Z & O & Z & Z & Z & Z & O & Z & Z & Z & Z & Z & O & Z & Z & Z \\ Z & Z & Z & Z & Z & Z & Z & O & Z & O & Z & O & Z & Z & Z & Z & Z & Z & O & Z \\ Z & Z & Z & Z & Z & Z & Z & Z & Z & Z & Z & Z & Z & O & Z & O & Z & O & Z & O \end{bmatrix}$$

The wavelength used is 1550 nm and every chip has a spectral width of 0.8 nm. So, the wavelength used is from 1550 nm to 1565.2 nm. In simulation, attenuation for both systems are normal (0.43 dB/km), haze (4.3 dB/km) and fog (43 dB/km) respectively [2]. List of the parameters considered in the set up are shown in Table 1.

The performance of the FSO channel depends upon atmospheric conditions [27, 28]. The comparison of the FSO and OCDMA-FSO system under normal, haze and fog conditions in terms of Quality factor at a distance of 5 km is shown in Table 2. The results show that quality factor for the OCDMA-FSO system in all the environment conditions is better than the quality factor of the simple FSO system.

Figure 3 shows the range verses BER for two different FSO communication systems at different attenuation for five users. It is observed that with increase in attenuation, BER increases. The transmit power used was 12 dBm out of the broadband source. The results show that the BER value in case of normal OCDMA-FSO is 1E-65 at 1 km and 1E-20 at 5 km whereas in case of FSO, BER values are 1E-37 and 1E-12 at 1 km and 5 km respectively. Therefore, the performance of OCDMA-FSO is better than the simple FSO system.

In addition, OCDMA-FSO could perform sufficiently well up to 5 km for five users whereas a simple FSO system is confined to 3 km in case of fog. The noise generated is totally uncorrelated and random at the receiver.

**Table 2** Comparison of FSO and SAC-OCDMA-FSO

| Condition | Attenuation (dB/km) | FSO | OCDMA-FSO |
|---|---|---|---|
| | | Quality Factor | Quality Factor |
| Normal | 0.43 | 13.5 | 26.27 |
| Haze | 4.3 | 7.13 | 21.56 |
| Fog | 43 | 3.60 | 8.26 |

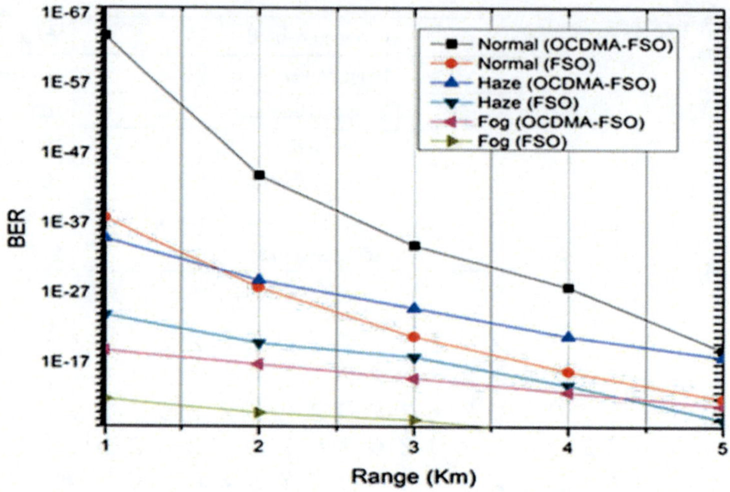


**Fig. 3** BER verses Range for five users

When the acquired findings are compared to reference [29], the comparison analysis shows that our system performs better when the input power is increased. The distance covered by the proposed system using the SDD approach is 5 km at 43 dB/Km attenuation, compared to 0.75 km at 8.68 dB/Km in [29]. Additionally, the proposed system's bit rate is 10 Gb/s. Further, the results can also be improved by incorporation of compensation techniques or by designing zero cross correlation code for OCDMA. In order words, it is noticed that OCDMA-FSO covers more distance as compared to simply FSO.

The eye diagrams of the FSO and hybrid OCDMA-FSO systems under Normal, Haze and Fog condition at a distance of 5 km are shown in Figs. 4, 5 and 6. It becomes clear that the proposed system performs better in comparison to FSO.

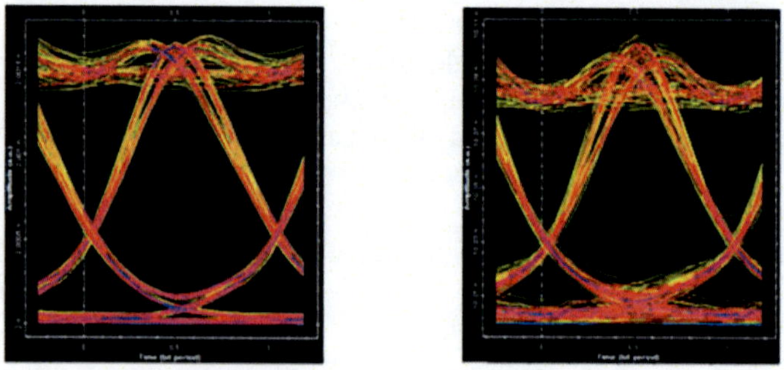

(a) OCDMA-FSO (b) FSO

**Fig. 4** Eye diagram under normal condition for five users

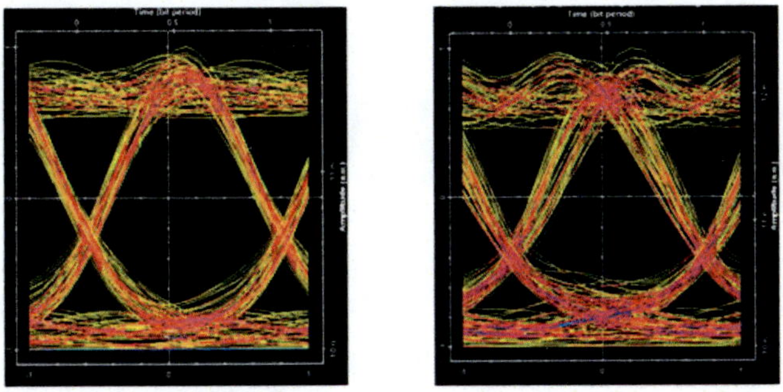

(a) OCDMA-FSO (b) FSO

**Fig. 5** Eye diagram under haze condition for five user

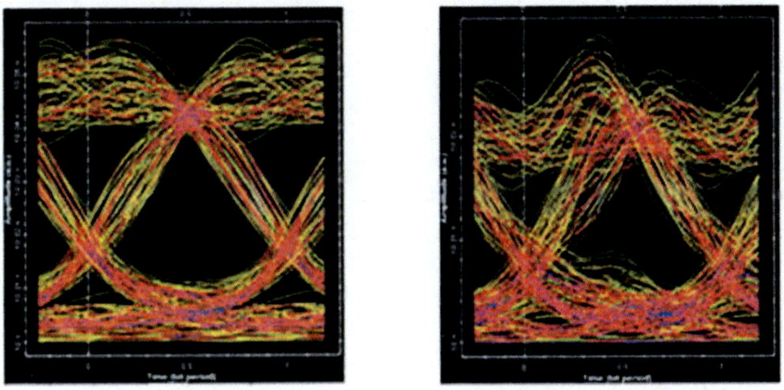

(a) OCDMA-FSO (b) FSO

**Fig. 6** Eye diagram under fog condition for five users

## 4 Conclusion

In this paper, the OCDMA-FSO system using spectral direct detection (SDD) with ZCC code is compared with the FSO System for five users. It has been shown through simulation that the quality factor and BER performance of OCDMA-FSO are better than FSO system at the same bit rate and distance. With the addition of OCDMA in the FSO system, the performance of system is improved. So, OCDMA, when integrated with FSO, provide high data transmission with more security to last-mile access. The Q-factor at a distance of 5 km under normal, haze, and fog conditions for OCDMA-FSO is 26.27, 21.56, and 8.26, respectively. Though, for FSO, Q-factor values are 13.5, 7.13, and 3.60, respectively.

## References


1. Sarangal H, Singh A, Malhotra J, Singh S (2018) Performance evaluation of hybrid FSO-SACO CDMA System under different weather conditions. J Opt Commun
2. Sarangal H, Singh A, Malhotra H, Chaudhary S (2017) A cost effective 100 Gbps hybrid MDM–OCDMA–FSO transmission system under atmospheric turbulence. Opt Quant Electron 49:184
3. Fyath RS, Ali HMM (2012) Transmission performance of optical code division multiple access network based on spectral amplitude coding. J Emerg Trends Comput Info Sci 3(3):444–455 (2012).
4. Nisar K, Sarangal H, Thapar SS, Qutubuddin M, Rahmath M (2018) Performance analysis of permutation matrix zero cross correlation code for SAC-OCDMA systems. Eur J Eng Res Sci 3(1):15–19
5. Singh M, Atieh A, Grover A, Barukab O (2022) Performance analysis of 40 Gb/s free space optics transmission based on orbital angular momentum multiplexed beams. Alex Eng J 61(7):5203–5212
6. Abd TH, Aljunid SA, Fadhil HA, Ahmad RB, Saad NM (2011) Development of a new code family based on SAC-OCDMA system with large cardinality for OCDMA network. Opt Fiber Technol 17:273–280
7. Abd TH, Aljunid SA, Fadhil HA (2013) A new code design for spectral amplitude coding optical CDMA systems using fiber bragg grating. J Optics 42. https://doi.org/10.1007/s12596-012-0110-0
8. Abdullah MK, Hasoon FN, Aljunid SA, Shaari S (2008) Performance of OCDMA systems with new spectral direct detection (SDD) technique using enhanced double weight (EDW) code. Opt Commun 281:4658–4662
9. Anuar MS, Aljunid SA, Saad NM, Hamzah SM (2009) New design of Spectral amplitude coding in OCDMA with zero cross-correlation. Opt Commun 282(14):2659–2664
10. Pendeza Martinez CA, Martinez ALM, Abrão T (2022) Linear programming-aided power allocation for large OCDMA networks. Opt Fiber Technol 74:103109. https://doi.org/10.1016/j.yofte.2022.103109
11. Abd El-Mottaleb SA, Singh M, Yousif Ahmed H, Zeghid M, Nisar KS, Alotaibi MF, Mahmoud M, Weera W (2023) Performance evaluation of a 160-Gbit/s OCDMA-FSO system via Laguerre-Gaussian beams under weather conditions. Alexandria Eng J 63:661–674. https://doi.org/10.1016/j.aej.2022.11.033
12. Moghaddasi M, Mamdoohi G, Shukri A, Noor M, Mahdi MA, Barrirah S, Anas A (2015) Development of SAC-OCDMA in FSO with multi-wavelength laser source. Opt Commun 356:282–289. https://doi.org/10.1016/j.optcom.2015.07.075
13. Mostafa S, Mohamad AENA, El-Samie FEA, Nabih A, Rashed Z (2017) Performance evaluation of SAC-OCDMA system in free space optics and optical fiber system based on different types of codes. Wireless Pers Commun 96:2843–61. https://doi.org/10.1007/s11277-017-4327-8
14. Kumar N, Rana AK (2013) Impact of various parameters on the performance of free space optics communication system. Optik 124:5774–5776
15. Nisar KS, Sarangal H, Thapar SS (2018) Performance evaluation of newly constructed NZCC for SAC-OCDMA using direct detection technique. Photonic Netw Commun
16. Chaudhary S, Tang X, Sharma A, Lin B, Wei X, Parmar A (2019) A cost-effective 100 Gbps SAC-OCDMA–PDM based inter-satellite communication link. Opt Quant Electron 51:148
17. Singh M, Sappal AS (2020) Radio over fiber (RoF) link modelling using cross term memory polynomial. J Opt Commun
18. Sarangal H, Singh A, Malhotra J (2017) Construction and analysis of a novel SAC-OCDMA system with EDW coding using direct detection technique. J Opt Commun
19. Singh M, Kriz J, Kamruzzaman MM, Dhasarathan V, Sharma A, El-Mottaleb S (2022) Design of a high-speed OFDM-SAC-OCDMA-based FSO system using EDW codes for supporting 5G data services and smart city applications. J Front Phys

20. Sarangal H, Nisar KS, Thapar SS, Singh A, Malhotra J (2021) Performance evaluation of 120 GB/s hybrid FSO-SACOCDMA-MDM system using newly designed ITM-Zero cross-correlation code. J Opt Quantum Electron 53(64)
21. Abd El-Mottaleb SA, Fayed HA et al (2018) Enhanced spectral amplitude coding OCDMA system utilizing a single photodiode detection. J Appl Sci 8(10):1861
22. Negi CM, Pandey A, Soni GG, Gupta SK, Kumar J (2011) Optical CDMA networks using different detection techniques and coding schemes. Int J Future Gener Commun Networking 4(3):25–34
23. Rashidi CBM, Anuar MS, Aljunid SA (2010) Study of direct detection technique for zero cross correlation code in OCDMA. In: International conference on computer and communication engineering (ICCCE'10). https://doi.org/10.1109/ICCCE.2010.5556830
24. Norazimah MZ, Aljunid SA, Fadhil HA, Md Zain AS (2011) Analytical comparison of various SAC-OCDMA detection techniques. In: 2nd International conference on photonics, Kota Kinabalu, Malaysia, pp 1–5. https://doi.org/10.1109/ICP.2011.6106864.
25. Panda S, Bhanja U (2015) Design and performance analysis of one dimensional zero cross correlation coding technique for a fixed wavelength hopping SAC-OCDMA. Int Sch Sci Res Innov 9(11):1330–1338
26. Hasoon FN, Samad MDA, Abdullah MK, Shabudin RKZ, Aljunid SA, Shaari S (2008) Spectral amplitude direct decoding OCDMA system using high power LED source with MDW codes at 2.5 Gbps. Int J Microwave Opt Technol 3(4):487–490
27. Vats A, Kaushal H (2014) Analysis of free space optical link in turbulent atmosphere. Optik 125:2776–2779. https://doi.org/10.1016/j.ijleo.2013.11.066
28. Pham AT, Luu TA, Dang NT (2010) Performance bound for turbo-coded 2-D FSO/CDMA systems over atmospheric turbulence channels. IEICE Trans Fundam Electron Commun Comput Sci 93(12):2696–2699
29. Sahbudin RKZ, Kamarulzaman M, Hitam S (2013) Mokhtar M, Anas SBA (2013) Performance of SAC-OCDMA-FSO communication systems. Int J Light Electron Optics 124(17):2868–2870

# An Analytical Model for a Proportional–Integral–Derivative Controlled Phase-Locked Loop for Communication Systems

**Geetamoni Konwar and Tulshi Bezboruah**

**Abstract** We propose here an analytical model for proportional–integral–derivative controlled phase-locked loop for communication systems. The transfer function of the system has been derived and simulated to analyze the various aspects, e.g. settling time, phase margin, damping factor, overshoot, bandwidth and stability of the system. From the analysis of simulation results it is observed that the settling time of the system decreases exponentially with increasing phase margin and it increases exponentially with increasing damping factor. It is also observed that the settling time and percentage of overshoot of the system is inversely proportional to each other. The settling time is also decreased exponentially with the increase in bandwidth. The Bode plot and Root Locus analysis show that the proposed system is highly stable. The system provides a fastest settling time of up to 238.57 ps. The system can also improve the settling time by 99.9% over the conventional system with a maximum frequency of 6.21802 GHz, which is the novelty of the proposed work.

**Keywords** Fast settling time · Stability of PLL · PID controller · Transfer function · High bandwidth

## 1 Introduction

In the design of the phase-locked loop (PLL), loop filter (LF) is crucial for the operation of the whole PLL as it has a major impact on the closed loop bandwidth (BW), stability as well as dynamic response of the loop [1]. The proportional–integral–derivative (PID) controller is one of the most widely used industrial feedback controller with its three individual control actions, such as, P stands for proportional control, I for Integral control and D for derivative control [2, 3]. The use of PID controller depends on proper application of combinations of these controllers for

G. Konwar (✉) · T. Bezboruah
Department of Electronics and Communication Technology, Gauhati University, Guwahati, Assam 781014, India
e-mail: geetak222@gmail.com



S. Borah et al. (eds.), *Advanced Computational and Communication Paradigms*, Lecture Notes in Networks and Systems 535,
https://doi.org/10.1007/978-981-99-4284-8_40

corrective action to minimize the error signals, which represents how far or near is the desired output from the actual output [4, 5] can be achieved. In today's process control scenario, more than 95% of the control loops are of PID type [6]. In spite of the straightforward structure, the popularity of PID controllers are also due to the simplicity of the design procedures, wide availability and effectiveness of the system performance. In addition, the PID controller provides further advance features, such as multi-loop or multivariable control solution, gain scheduling, continuous adaptation, and automatic tuning [7].

The PLL system have many aspects like settling time (ST), phase-margin (PM), bandwidth (BW), damping factor (DF) and overshoot (OS), which are contradictory to each other. Fast settling with minimum OS and suitable BW is considered one of the most important challenges in recent PLL systems. The ST is a very important challenge in applications which need fast frequency switching to protect the call by protecting data security and to avoid interference as well as multi-path fading in the communication system, e.g. latest mobile technology for the frequency synthesis to protect the call from interruption. Also, the frequency modulated transceiver needed a local oscillator with low ST for various receptions as well as carrier transmission [1, 8, 9]. The ST may be minimized by fixing the LF BW broader, but it may result in unwanted noise in voltage-controlled oscillator (VCO) output. So, the OS may go up significantly [10]. The PM is evaluated at the frequency that is equal to the loop BW and has a significant impact on the stability [11]. Another important aspect of PLL is the DF. The DF controls the magnitude of the frequency response. If DF is made significantly larger than one, then the dynamic response time of the system increases; while smaller values of DF gives an excessive OS and large oscillations before the system settles down. Large OS provides peaking in the transient response [3, 6]. In order to overcome these difficulties we proposed a PID controlled PLL which can improve the ST significantly.

## 2 The Proposed System

The fundamental block diagram as proposed for the PID controlled PLL is depicted in Fig. 1. It comprises of a PID controller, a phase detector (PD) and a VCO connected in a feedback loop through the frequency divider (FD). The PD produces an error signal $V_{\Phi}$, proportional to the phase difference between the incoming signal and the VCO output signal by comparing the phase at each input [12, 13]. The PID controller takes the output of PD and removes the high frequency noise by appropriate tuning of P, I and D gains to produce a dc voltage. The PID controller output voltage $V_0$ controls the frequency of the VCO. The voltage at the input of the VCO determines the frequency $f_{osc}$ of the periodic signal $V_{osc}$ at the output of the VCO. The FD divides the output frequency by $N$. So, loop gain (LG) is reduced by the factor of $N$.

A basic property of the PLL is that it attempts to maintain the frequency lock, that is, $f_{osc} = f_{in}$ between $V_{osc}$ and $V_{in}$ even if the frequency $f_{in}$ of the incoming signal varies with time. The phase difference between the VCO signal and the incoming

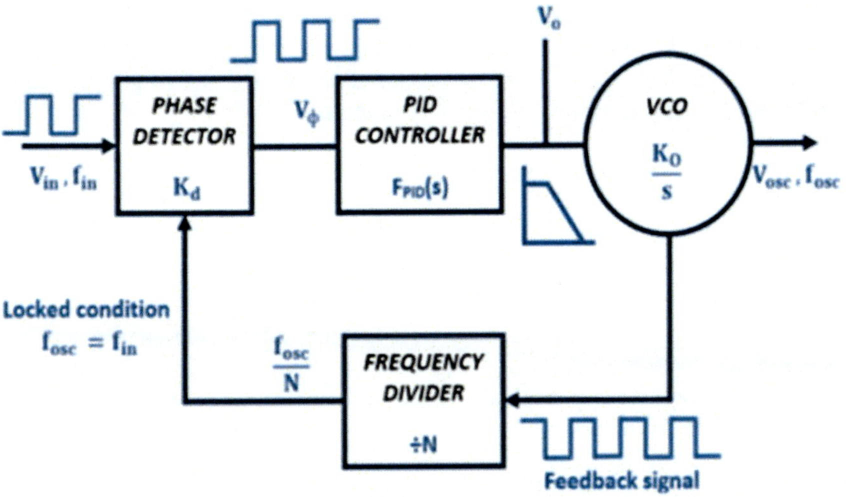


**Fig. 1** The PID controlled PLL system

signal will begin to increase in time. As a result, the PID controller output voltage $V_0$ increases as the VCO output frequency $f_{osc}$ increases until it matches fin, thus keeping the PLL in the locked condition [14].

By assuming the loop to be locked, the generalized closed loop TF of the PLL system can be derived as:

$$H_{system}(s) = \frac{K_d F_{PID}(s)\frac{K_0}{s}}{1 + K_d F_{PID}(s)\frac{K_0}{s}\frac{1}{N}} \tag{1}$$

where, $K_d$ is the TF of PD, $\frac{K_0}{s}$ is the TF of VCO, $F_{PID}(s)$ is the TF of PID controller and $\frac{1}{N}$ is the TF of FD.

TF of the PID controller can be derived as [4]:

$$F_{PID}(s) = K_P + \frac{K_I}{s} + sK_D \tag{2}$$

where, the proportional gain is denoted by $K_P$, the I-controller gain is denoted by $K_I$ and the D-controller gain is denoted by $K_D$. One of the greatest strength of the PID controller is that, a numerical change in any individual coefficients, $K_P$, $K_I$ and $K_D$ changes only the size of contribution in the path of the term [15].

The choice of $H_{system}(s)$ in Eq. (1) is very important. It determines the loop BW, and hence the performances of the whole PLL system.

Substituting Eq. (2) into (1), the overall TF of the system can be derived as:

$$H_{system}(s) = \frac{K_d K_0(s^2 K_D + sK_P + K_I)}{s^2(1 + K_D)K + sKK_P + KK_I} \tag{3}$$

Taking, $\frac{K_d K_0}{N} = K =$ loop gain constant.

The LG of the system should be large in order to minimize the phase error. Equation (3) is the standard TF of the proposed system that gives high flexibility in analyzing the various aspects and stability of the system.

## 3 The Simulation

The simulation of the system is carried out in MATLAB (R2018a) order to evaluate the performance and accuracy of the proposed system in terms of step response, bode plot and root-locus analysis.

Figure 2 shows the step response for few different test cases of the system. It can be seen that the frequency of oscillation increases with decreasing ST, that means, the ST for case 1 is faster than that of case 2, case 3, case 4 and case 5. It is also seen that the peaking increases as the ST decreases.

The bode magnitude and phase plots of the system is shown in Fig. 3. It is observed that the PM for all the five test cases is positive and lies on the left-hand side of the s-plane. Hence, the proposed system is stable. The loop BW is evaluated from Bode plot through use of "BANDWIDTH" function.

Figure 4 shows the root-locus analysis of the proposed system. As the LG of the proposed system increases, the real and imaginary parts increase and the locus becomes a circle centered on zero. For stability of the system, the poles must be in

**Fig. 2** Step responses

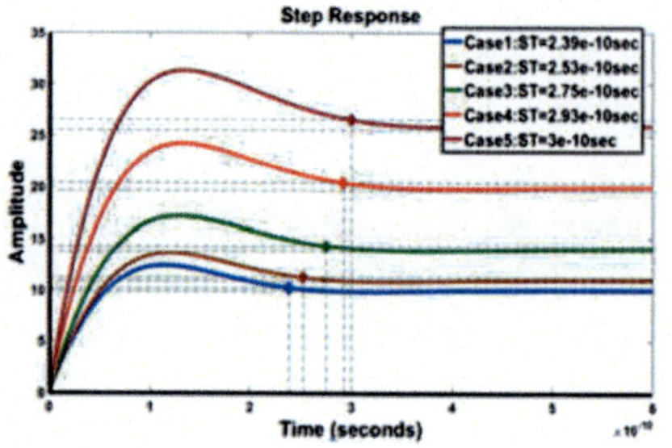


**Fig. 3** Bode plots

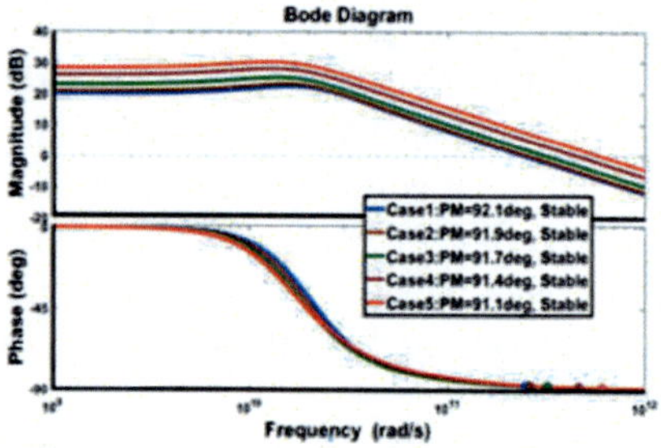

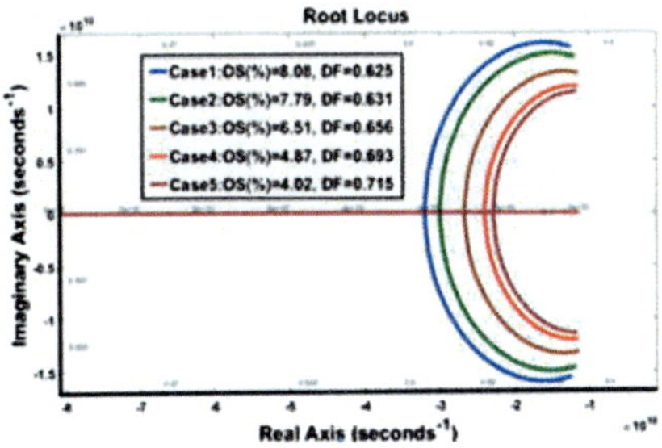


**Fig. 4** Root-locus response

the left side of the s-plane. The relationship between system poles and zeros then predicts the stability. The poles for different cases of the proposed system are within the unit circle, hence the system is stable. The DF and percentage of OS is measured from root-locus analysis.

Figure 5 shows the impact of PM, DF, OS and BW on ST of the system. It is observed from Fig. 5a that the ST decreases exponentially with increasing PM. The PM of 92.0781° provides the fastest ST of 238.57 ps. From Fig. 5b, it is observed that the ST increases exponentially with increasing DF. The DF of 0.625 provides the fastest ST of the system. From the analysis of Fig. 5c, it is seen that the ST and percentage of OS of the system is inversely proportional to each other. The fastest ST of the system is obtained at OS of 8.08%. Figure 5d shows the impact of BW on ST. The loop BW is directly related to the ST. Large BW provides a fast lock to the system but at the same time it degrades the ST of the system. From Fig. 5d, it is seen that the ST decreases exponentially with the increase in BW and the fastest ST is obtained at the maximum achievable loop BW of 6.21802 GHz.

Figure 6 shows the impact of loop BW on OS. The loop BW is directly proportional to the OS of the system. It is observed that by using the PID controller, less than 9% OS can be achievable.

Figure 7 shows the impact of DF on PM of the system. The PM decreases as the DF increases. The PM reduces from 92.0781° at DF of 0.625–91.0308° at 0.724.

Figure 8 shows the impact of percentage of OS on DF. It is seen that the OS and DF are inversely proportional to each other. The DF decreases as the OS increases.

## 4 Results and Discussion

The comparison of the results of the proposed system with the results of earlier works with different PLL approaches is summarized in Table 1. It is observed that the fastest ST obtained for the proposed system is 238.57 ps, while all other relevant works provide ST in the range of 0.03 µs–286 ns. Also, the PM of the system is

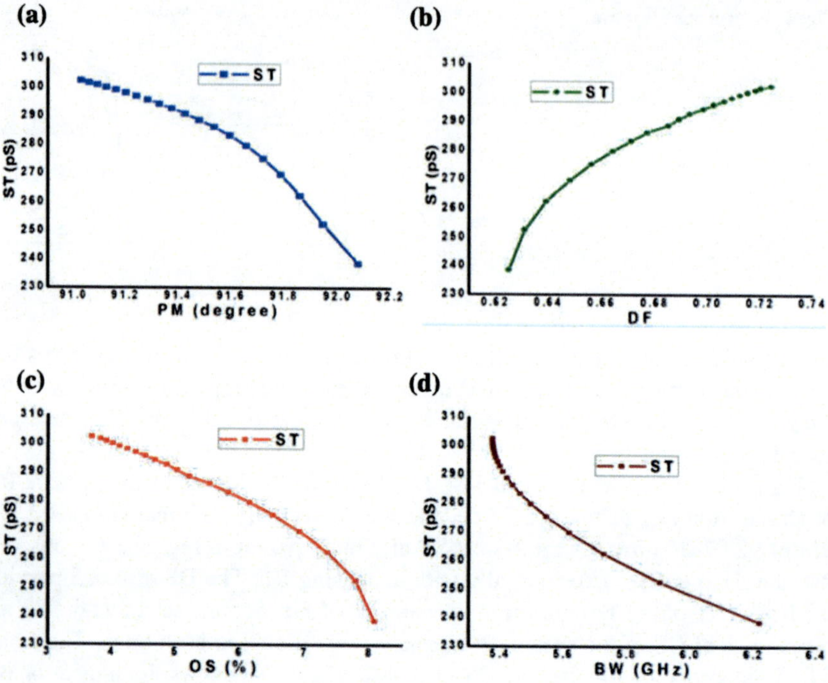


**Fig. 5** **a** Impact of PM on ST. **b** Impact of DF on ST. **c** Impact of OS on ST. **d** Impact of BW on ST

**Fig. 6** Impact of BW on OS

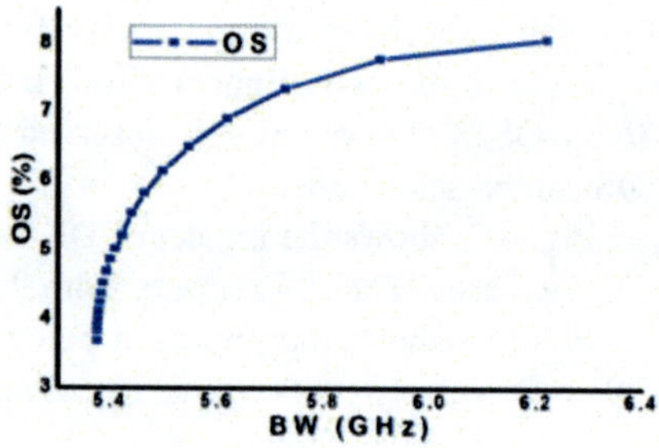


**Fig. 7** Impact of DF on PM

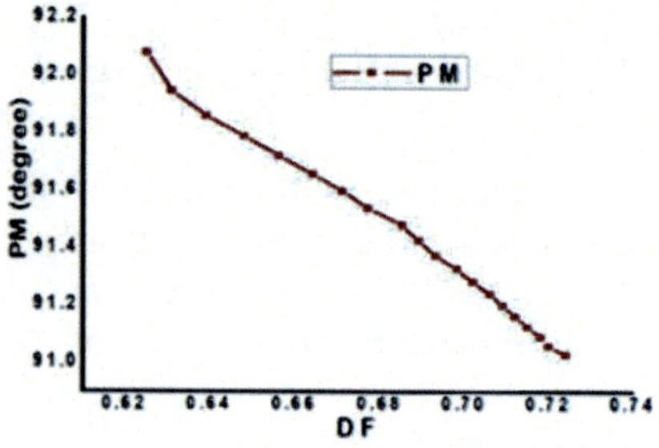

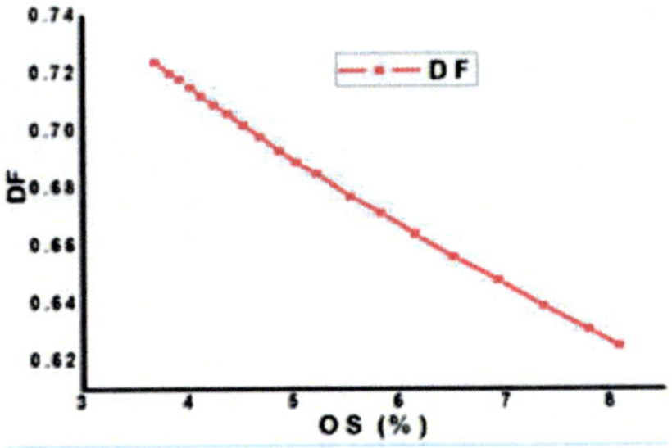


**Fig. 8** Impact of OS on DF

**Table 1** Performance summary and comparison

| Work | ST | PM | BW | OS (%) | DF | PLL architecture |
|---|---|---|---|---|---|---|
| *This work* | *238.57 ps* | *92°* | *6.21 GHz* | *8.08* | *0.625* | *PID controlled PLL* |
| Ref. [1] | 286 ns | NA | 210 MHz | NA | NA | All digital delay-locked loop |
| Ref. [2] | 3.1 μs | 60° | 1280 kHz | NA | NA | Conventional PLL |
| Ref. [4] | 0.03 s | 55.4° | NA | 32 | 0.707 | Multiple-complex coefficient filter (MCCF)-PLL |
| Ref. [5] | 0.11 μs | 45.1° | 93.1 MHz | 3.41 | 0.05 | PLL |
| Ref. [6] | 3.5 μs | NA | NA | NA | NA | All-digital PLL (ADPLL) |
| Ref. [7] | 22 μs | NA | 30 kHz | NA | NA | Charge-pump (CP)-PLL |
| Ref. [8] | 0.08 s | NA | NA | 16.31 | NA | PID controller |
| Ref. [9] | 4.03 s | NA | NA | 1.62 | NA | PID controller |
| Ref. [10] | 19.4 ps | 41.3° | 59.3 GHz | 4.43 | 0.704 | PIDC‖LF PLL |
| Ref. [11] | 10 μs | 53° | 66.67 kHz | NA | NA | PLL frequency-synthesizer (FS) |
| Ref. [12] | 31 ns | NA | NA | NA | 1.5 | PID controlled PLL |

92° at 6.21 GHz and system OS of 8.08% which satisfies the stability limits with the DF of 0.625. The percentage (%) of improvement in the ST with respect to other relevant works is summarized in Table 2. From Table 2, it is observed that the ST of the system improves up to 99.99% with PID controller in comparison to the earlier works.

Table 2 % of improvement in ST than other relevant works

| References | ST (ps) | | % of improvement |
|---|---|---|---|
| | Earlier work | This work | |
| [8] | $8e^{10}$ | 238.57 | 99.99 |
| [9] | $4.03e^{12}$ | | 99.99 |
| [10] | 19.4 | | 91.86 |
| [12] | 31,000 | | 99.23 |
| [13] | 225,000 | | 99.89 |
| [14] | $9e^{6}$ | | 99.99 |
| [15] | $381e^{6}$ | | 99.99 |

## 5 Conclusion

The various performance aspects of the proposed PID controlled PLL is illustrated in this work including the impact of one aspect as a function of other. It is observed that the ST of the proposed system improves significantly up to 99.9% as compared to the conventional system. The PM of the system lies in the range from 91.0308 to 92.0781°. The proposed system predicts a DF in between 0.625 to 0.724 and OS varying from 3.69 to 8.08% and is in good agreement with a standard criterion. The maximum frequency range of the system is 6.21802 GHz while the minimum achievable frequency range is 5.37418 GHz. From the analysis of results of various simulation results, it is found that the system is extremely stable. Since for the proposed PID controlled PLL system, all the parameters are well within the stability limit with very fast ST, hence, the system will be beneficial in research and industrial applications which need fast frequency switching as a trade-off.

## References


1. Rahimpour H, Gholami M, Ardeshir G, MiarNaimi H (2015) Low settling time all digital DLL for VHF application. Int J Eng (IJE), Trans C: Asp 28(3):419–425. https://doi.org/10.5829/idosi.ije.2015.28.03c.11
2. Ali MKM, Hashemipour O (2019) Fast locking technique for phase locked loop based on phase error cancellation. AEU-Int J Electron Commun 109:99–106. https://doi.org/10.1016/j.aeue.2019.06.025
3. Nguyen NH, Nguyen PD (2018) Overshoot and settling time assignment with PID for first-order and second-order systems. IET Control Theory Appl 12(17):2407–2416. https://doi.org/10.1049/iet-cta.2018.5076
4. Golestan S, Mohammad M, Francisco DF, Josep MG (2014) Performance improvement of a prefiltered synchronous-reference-frame PLL by using a PID-type loop filter. IEEE Trans Ind Electron 61(7):3469–3479. https://doi.org/10.1109/TIE.2013.2282607
5. Yang WB, Wang HH, Chang HI, Lo YL (2020) A fast-locking all-digital PLL with dynamic loop gain control and phase self-alignment mechanism for sub-GHz IoT applications. Jpn J Appl Phys 59(SGGL08). https://doi.org/10.35848/1347-4065/ab7276

6. Shubin L, Sun D, Ding R, Huang S, Zhu Z (2019) A multiplexing DAPD technique for fast-locking and charge pumps calibration in PLLs. IEEE Microwave Wirel Compon Lett 29(8):535–537. https://doi.org/10.1109/LMWC.2019.2923891
7. Babu RSR, Deepa S, Jothivel SA (2014) Closed loop control of quadratic boost converter using PID controller. Int J Eng (IJE), Trans B: Appl 27(11):1653–1662. https://doi.org/10.5829/idosi.ije.2014.27.11b.02
8. Uyanik HU, Tarim N (2007) PID-controlled PLL for fast frequency-hopped systems. In: 6th IEEE Dallas circuits and systems workshop on system-on-chip, Dallas, TX, USA, pp 1–3. https://doi.org/10.1109/DCAS.2007.4433198
9. Konwar G, Bezboruah T (2008) Some aspects of modelling and simulation of a high frequency and fast settling time proportional-integral-derivative controlled phase-locked loop. In: 2018 2nd IEEE international conference on power, energy and environment: towards smart technology, pp 1–9. https://doi.org/10.1109/EPETSG.2018.8658754
10. Kuangand N, Wu X (2006) A fast-settling PLL frequency synthesizer with direct frequency presetting. In: IEEE international solid-state circuits conference, pp 741–750. https://doi.org/10.1109/ISSCC.2006.1696113
11. Ahmet K, Songuler O, Bogosyan S, Gokasan M (2012) Fast locking of PLLs using fuzzy scheduled SMC. In: 2012 IV IEEE international congress on ultra modern telecommunications and control systems, pp 234–239. https://doi.org/10.1109/ICUMT.2012.6459671
12. Sulaiman DR (2020) Novel high-frequency PLL design for wireless applications. J Electr Syst 16(3):332–349. http://journal.esrgroups.org/jes/papers/16_3_5.pdf
13. Beiraghdar F, Ghobadi RA, Sheikhaei S, Tohidian MA (2021) Fast settling frequency synthesizer with switched-bandwidth loop filter. Int J Circuit Theory Appl 49:2021–2031. https://doi.org/10.1002/cta.2993
14. Shanan H, Dalton D, Chillara V, Dato P (2022) A 9–12 GHz coupled-RTWO FMCW ADPLL with 97 fs RMS Jitter, −120 dBc/Hz PN at 1 MHz offset, and with retrace time of 12.5 ns and 2 μs chirp settling time. In: 2022 IEEE international solid-state circuits conference (ISSCC), pp 146–148. https://doi.org/10.1109/ISSCC42614.2022.9731575
15. Bertulessi L, Cherniak D, Mercandelli M, Samori C, Lacaita AL, Levantino S (2022) Novel feed-forward technique for digital bang–bang PLL to achieve fast lock and low phase noise. IEEE Trans Circuits Syst I: Regul Pap 69(5):1858–1870. https://doi.org/10.1109/TCSI.2022.3146788

# Performance Study and Comparative Analysis of Routing Protocols in Wireless Sensor Networks

**Mandeep Kumar and Jahid Ali**

**Abstract** Wireless Sensor Network (WSN) is usually prolonged into huge geographical regions through inexpensive sensor devices that can tolerate themselves with very less power usage. The networking ability enables the sensor nodes for integrating, coordinating, and collaborating with each other, which is a basic alternation in network area and it distinguishes sensor network nodes from other networks, namely Ad-Hoc, Internet (IP) datagram and etc. At present, the routing protocol in a wireless sensor network experiences various issues, including coverage packet loss, video real time and real-time audio streaming, new scalability, energy restraints, severe weather environments and so on. Moreover, network routing is termed as the incorporation of a routing approach and routing protocol. The intent of the routing system is to afford consistent assessment of network nodes topology when the routing model offers intelligence with regards to optimal path computation. Thus, in this paper, a study is made with several routing systems in WSN and the performance is evaluated. Here, the performance evaluation is done by the existing routing methods along with the Taylor SailFish Optimizer (Taylor SFO) approach. Additionally, the performance of the routing technique is analyzed with throughput, distance, energy, end to end delay as well as Packet Delivery ratio (PDR). The Taylor SFO routing model achieved minimal distance and delay of 195.44 m, and 15.25 ms as well as maximum throughput, PDR, and residual energy of 119.98 kbps, 94.87%, and 221.65 J.

**Keywords** Packet delivery ratio · Taylor SailFish optimizer · Wireless senor network · Routing protocol

M. Kumar (✉)
Research Scholar, Computer Science and Engineering, I.K. Gujral Punjab Technical University, Jalandhar – Kapurthala Highway, VPO – Ibban, Kapurthala 144603, India
e-mail: mandeep_recj@yahoo.com

J. Ali
Professor, Computer Applications, Sri Sai Iqbal College of Management and Information Technology, VPO Badhani, Tehsil, Pathankot, Punjab 145001, India

 
S. Borah et al. (eds.), *Advanced Computational and Communication Paradigms*, Lecture Notes in Networks and Systems 535,
https://doi.org/10.1007/978-981-99-4284-8_41

# 1 Introduction

During the past few years, WSN area has attained high attention and it is more significant for military and civil applications. Moreover, various WSN applications are battlefields, manufacturing environments, emergency circumstances, and so on. The progression of sensor nodes is strengthened owing to the developments in the area of integrated electronics, and microelectronics [1]. The properties, which are integrated with the networks are wireless, sensors, ad hoc, and mobiles. Moreover, these properties are executed in real time for factory situation observation [2] and energy emergency response information. Additionally, these networks are susceptible to several types of security threats at the network layer from intruders [2, 3]. Usually, WSN is a circulated system that comprises of one or more base stations and various nodes. Usually, the nodes are inexpensive and small and also the base station has maximal processing proficiencies and reinforced electrical power source. Here, sensor nodes are prepared along with a mini Central Processing Unit (CPU) and application specific sensors. Furthermore, the radio transceiver is also combined to every sensor node, which affords the effective node for performing communication with the base station and other nodes. Usually, these nodes frequently operate on the batter power as well as utilize the undependable radio channels by fading and noise. Consequently, the increment of network lifetime and data throughput of nodes are the major issues during the operation and design of the network [4].

In addition, routing is an imperative process for choosing effectual routs in a network. The router offerings the traffic direction activated on the web [3]. Data packet routing from the source to the base station by a network is the most fascinating challenge for researchers in WSN. Since, energy is an imperative and basic element of the routing system in the discovery phase, the major restriction of energy resources is the most important factor [1]. Moreover, routing is functioned for different arrangements of sensor networks, which comprises of transportation networks, electronic information sensor networks, and also the public switched telephone network. In addition, the most important prerequisite of WSN is trust [2]. Besides, data packets are routed in a high routing network around expressive regions such that the entire network misfortune can be overloaded [3]. However, the traditional approaches for preventing and mitigating the attacks are not appropriate for WSN owing to the restricted abilities of WSN [1]. Moreover, the amount of attacks on WSN is usually enlarging speedily. Thus, slowly the attacks are increasing and also the difficulties of attack launch. Generally, attacks are creating huge damages and losses to businesses and industries [5]. WSN are commonly affected by both external and internal attacks. Accordingly, the common internal attack types are eavesdropping, sniffing, and Denial of Service (DoS) and so on, whereas internal attacks are namely grey hole, while black hole, etc., [6, 7]. In addition, various approaches are introduced for increasing the energy efficiency of WSN, which covers evert layers of the communication protocol extending from the physical to the application layer [4, 8, 9].

The main intention of this research paper is theoretical analysis and comparative study of various routing systems in WSN. The performance of the routing model is assessed by means of distance, throughput, PDR, end to end delay, and the energy is explained in detail. Moreover, the prevailing routing techniques are compared with the present Taylor SFO routing system to estimate the performance. In addition, analysis is performed with 50 and 100 nodes by altering the round quantity.

The outstanding sections of the paper are listed as follows, the review of the prevailing routing methods is presented in Sect. 2. Section 3 provides an explanation about the Taylor SFO routing algorithm and the analysis is deliberated in Sect. 4. Moreover, the results and analysis are exposed in Sect. 5 and Sect. 6 gives the conclusion.

## 2 Literature Survey

The literature survey of the existing attack recognition methods in WSN is explicated in this section. Tsiota et al. [10] designed multi-tier HWN for black hole attack detection in WSN. The network node positions were modelled based on a manageable multi-tier scheme of Poisson Point Processes (PPPs), and it calculated the network coverage in a closed structure. Moreover, the network coverage was designed for two setups in this model. The reliability was highly increased in this model, but the observation error was not reduced. Mehetre et al. [3] presented a trustable and secure routing model for attack detection in WSN. Here, security techniques, such as identification of attack node through detection packet, prevents the node from the attacker, and trust to identification was exploited for choosing trusted paths, which were depending on the minimal threshold rate. In addition, the dual assurance model was separated into two categories for transmitting secure data packets from the source to the destination. Energy loss in this model was highly reduced, although failed to further improve the packet to delivery ratio. Almazaideh et al. [4] modelled a novel energy-aware and reliable routing protocol for WSN. The multi-hop routing approaches were utilized for increasing the lifetime of WSN, wherein the source node transmits the packet to the base station over other nodes. Here, energy efficiency was obtained through increasing the minimal residual energy. Furthermore, entropy was utilized for computing the consistency of residual energy distribution. Besides, a look up table was exploited for maintaining the information regarding the network's present energy state. The complexity was less in this model, but failed to manage the performance through a striking optimal trade-off between the parameters. Aljumah et al. [7] developed a futuristic algorithm for attack detection in WSN. The clusters were formulated based on sensor nodes and it comprises the nodes, which were the available zone of other. Additionally, sensor nodes choose the cluster coordinator and it was selected as the cluster coordinator. This approach efficiently reduced delay duration, even though it failed to verify the stability.

# 3 Taylor SFO-Based Routing

The Taylor SFO routing technique is explicated in this section, which performs energy aware routing and attack detection in WSN. In SFO, the optimal route is selected by considering three factors, such as delay, energy, and distance. The network lifespan is increased by this method through exploiting energy aware routing and it ensures a secure communication. This approach mainly comprises of two phases, namely routing as well as black hole attack identification at the base station. Originally, when WSN is simulated consequently, energy aware routing and also attack detection is in a network. After the completion of the network simulated, routing is performed using the Taylor SFO algorithm. Usually, routing is more imperative in WSN because it helps to convey the data among the base station and sensor nodes. Moreover, routing is essential to transmit the data amongst the base station and nodes for generating the communication. In addition, fitness measures, like distance, energy and delay are included for the routing process. Meanwhile, black hole attack identification is performed at the base station in WSN. WSN is prone to several types of attacks in which black hole attack is more common, although it is difficult to identify and avoid the attack. For that, data is pre-processed to enhance the data quality. Then, feature extraction is carried out for effectual identification of black hole attack, wherein significant features are extracted. Finally, a deep stacked auto encoder [11, 12] is employed for performing attack detection at the base station. Besides, the deep stacked auto encoder is trained by the optimization model, named the Taylor SFO approach. Usually, black hole attack happens, while the attacker captures sensor nodes in a network to block packets and breaks them to reach the base station. Accordingly, the Taylor SFO model is the combination of the Taylor series [13] along with the SFO approach [14].

# 4 Performance Analysis Based on End to End Delay, Throughput, Energy, Distance, Packet Delivery Ratio

The performance of the routing algorithm is usually evaluated based on several metrics, namely distance, energy, throughput, PDR, end to end delay, and these are explained as follows.

## *4.1 Definitions*

This section provides the detail explanation about various metrics employed in the routing process.

***(a) End to end delay***: It is referred to as the time taken by a pack for transmitting data over a network from source to destination.

$$D = N \times \frac{L}{R} \tag{1}$$

$L$ is the packet length, $R$ is the transmission rate, and $N$ is the number of links.

***(b) Energy***: Usually, network lifetime is increased by means of energy nodes, because sensor nodes utilize the power from battery [15]. The energy structure in WSN operations comprises of transmitting, sleeping, idle and receiving mode. The maximal energy consuming node is selected in order to identify the route. Moreover, it utilizes the routing path with less energy to broadcast the data, and also the path is collected in the routing table. The energy rate in each node is calculated with respect to the time interval. The energy consumption is evaluated by,

$$S^r(x) = P_t^r * V_t^r + P_k^r * V_k^r \tag{2}$$

***(c) Distance***: Let us consider the sensor node as $c$ and $q$ with location at initial as $(v_1, w_1)$ and $(v_2, w_2)$. Here nodes shift with various velocities at a specific direction along with angle $\phi_1$ and $\phi_2$ in $x$-axis [16]. The node $c$ movse with distance $t_1$, whereas node $q$ moves with distance $t_2$ at time duration $n$. The distance among the nodes $c(v_1, w_1)$ and $q(v_2, w_2)$ is computed by,

$$\omega_{(c,q)} = \sqrt{|v_1 - v_2|^2 + |w_1 - w_2|^2} \tag{3}$$

where, $\omega_{(c,q)}$ implies distance among nodes $c(v_1, w_1)$ and $q(v_2, w_2)$.

***(d) PDR***: This metric is referred as the entire amount of data packets delivered to the total amount of packets transmitted from the source to the destination in a network, which is expressed as,

$$\beta = \frac{G}{B} \tag{4}$$

where, $G$ indicates the amount of delivered packets and $B$ is the packet quantity transferred from the source.

***(e) Throughput***: It is stated as the total sum of data efficiently transferred from one region to other region for a specific time duration.

$$\lambda = \frac{\mu}{H} \tag{5}$$

where, $\mu$ implies delivered nodes at simulation period $H$.

## 4.2 Theoretical Conclusion of All Performance Metrics

The metrics, like end to end delay, throughput, energy and PDR considered for assessing the performance of routing is explicated in this section. The performance of trustable and secure routing in [3] was evaluated by means of various measures, including PDR, throughput, and energy. Moreover, the method in [3] attained the rate with a throughput of 160, energy of 85 J and PDR of 18%. Additionally, the method in [7] utilized the metrics, such as throughput and end to end delay of 29.8 and 69,999.9.

## 4.3 Additional Parameters

Additionally, the other metrics, namely network size, latency are also considered for estimating the performance of routing algorithms. The routing technique in [3] attained latency of 20,000 and network size is 18. Furthermore, coverage probability is also considered in [10] for evaluating the routing performance.

## 4.4 Performance of SFO Routing Algorithm in WSN

In addition, the performance of the Taylor SFO routing algorithm is evaluated using the False negative Rate (FNR) and False Positive Rate (FPR), and it is deliberated in this section. The FNR of Energy efficient routing, Futuristic method routing, Trustable and secure routing, Multitier HWN, and Taylor SFO routing is 0.1054, 0.1007, 0.1266, 0.0852, and 0.0614 in 90% of train data. Moreover, FPR of prevailing routing approaches and Taylor SFO routing is 0.1325, 0.1751, 0.1501, 0.1359, and 0.1211. Figure 1 represents the analysis of Taylor SFO routing with FPR and FNR.

## 4.5 Analytical Study Based on Different Routing Techniques

This section illustrates the analytical study of several prevailing routing algorithms with the Taylor SFO routing method. The method in [10] calculated the performance by coverage probability with transmit power. Besides, the metrics, like latency, time, energy, and routing path length is analysed with network size in [3]. However, the approach in [4] utilized life span, residual energy and success probability as evaluation metrics by varying number of nodes, data loss percentage and number of hops, respectively. Additionally, the technique employed in [7] exploited throughput and end to end delay as performance metric, which are computed depending on plot quantity.

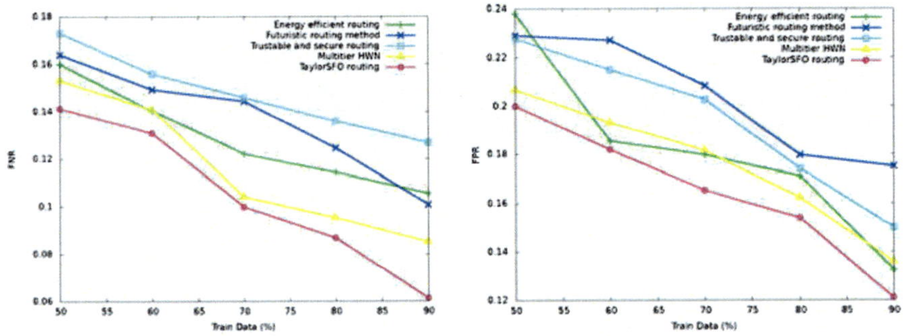


(a) (b)

**Fig. 1** Analysis using Taylor SFO routing technique **a** FNR, **b** FPR

### *4.6 Superiority of SFO Based Routing*

Table 1 illustrates comparative analysis of Taylor SFO routing model with 50 and 100 nodes using various performance metrics by shifting the number of rounds and 30th number of rounds is considered here for discussion. The delay attained by the existing method and Taylor SFO routing is 24.73, 26.48, 23.56, 21.92, and 15.25 ms. The distance of Energy efficient routing is 224.29 m, Futuristic method routing is 218.10 m, Trustable and secure routing is 213.54 m, Multitier HWN is 199.55 m, and Taylor SFO routing is 195.44 m. The residual energy of Taylor SFO routing is 221.65 J, whereas existing techniques are 258.03, 247.87, 240.66, and 232.95 J, and also the rate of PDR attained by the prevailing and Taylor SFO routing model is 90.73, 90.59, 91.10, 91.38, and 93.89%. The range of throughput attained by the existing routing models and Taylor SFO routing is 107.26, 106.97, 107.30, 108.58, and 109.40 kbps. Hence, the Taylor SFO routing technique attained better performance with delay of 15.25 ms, distance of 195.44 m, and residual energy of 221.65 J in 50 nodes. Additionally, the PDR and throughput of 94.87%, 119.98 kbps is obtained in 100 nodes.

## 5 Results

The performance analysis outcomes of the Taylor SFO routing model along with the experimental setup, simulation outcomes and dataset description are exposed in this section.

**Table 1** Comparative discussion of Taylor SFO model with other routing techniques

| Number of nodes | Metrics | WMC | Futuristic method | Trustable and secure routing method | Multitier HWN | Taylor SFO routing |
|---|---|---|---|---|---|---|
| 50 | Delay (ms) | 24.73 | 26.48 | 23.56 | 21.92 | **15.25** |
| | Distance (m) | 224.29 | 218.10 | 213.54 | 199.55 | **195.44** |
| | Residual energy (J) | 258.03 | 247.87 | 240.66 | 232.95 | **221.65** |
| | PDR (%) | 90.73 | 90.59 | 91.10 | 91.38 | 93.89 |
| | Throughput (kbps) | 107.26 | 106.97 | 107.30 | 108.58 | 109.40 |
| 100 | Delay (ms) | 30.16 | 28.17 | 30.66 | 26.63 | 19.91 |
| | Distance (m) | 258.03 | 247.87 | 240.66 | 232.95 | 221.6 |
| | Residual energy (J) | 20.97 | 18.72 | 19.58 | 17.90 | 21.65 |
| | PDR (%) | 94.53 | 94.09 | 93.06 | 94.85 | **94.87** |
| | Throughput (kbps) | 117.97 | 118.62 | 117.53 | 119.83 | **119.98** |

## 5.1 Experimental Setup and Dataset Description

The execution of the Taylor SFO routing scheme is implemented in MATLAB tool using PC with Intel i3 processor and 8 GB RAM.

## 5.2 Simulation Results

This section exposes the simulation consequences of the Taylor SFO routing method; here, Fig. 2a refers to the simulation of results for 50 nodes as well as the simulation result for 100 nodes is depicted in Fig. 2b.

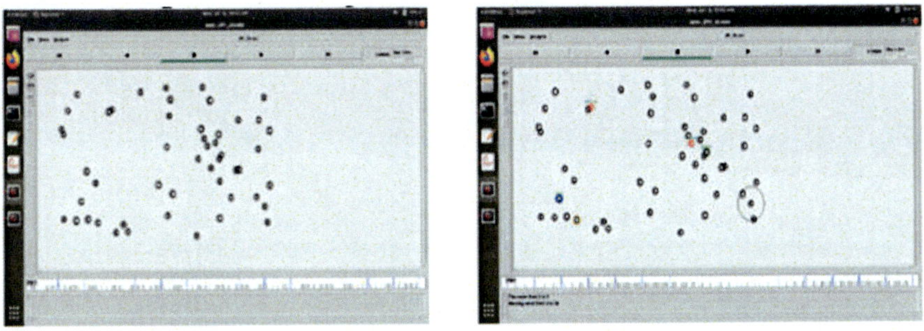

(a)

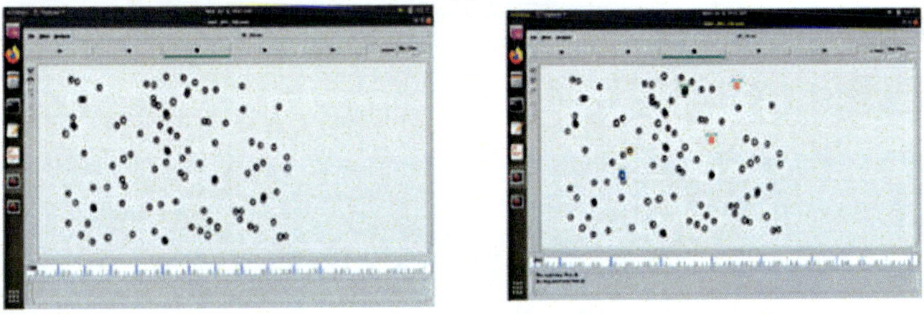

(b)

**Fig. 2** Simulation outcomes of Taylor SFO-based routing (**a**) 50 nodes (**b**) 100 nodes

## 5.3 *Comparative Techniques*

The performance of the existing routing methods, namely energy multitier HWN [10], trustable and secure routing model [3], energy efficient routing scheme [4], and futuristic routing model [7] are compared with the Taylor SFO routing technique.

## 5.4 *Comparative Analysis*

The comparative analysis of the Taylor SFO routing algorithm is deliberated in this section with regards to 50 and 100 nodes with respect to various performance metrics.

(1) ***Analysis with 50 nodes***

The comparative analysis of the Taylor SFO routing model with 50 nodes is based on various performance metrics with various number of rounds as displayed in Fig. 3. The analysis of the Taylor SFO routing technique for delay, distance, residual energy, PDR, and throughput is specified in Fig. 3a–e. The delay attained by the existing method and Taylor SFO routing is 33.41, 44.1, 41.1, 32.32, and 20.93 ms in 25 number of rounds. When the number of rounds is 25, the distance of Energy efficient

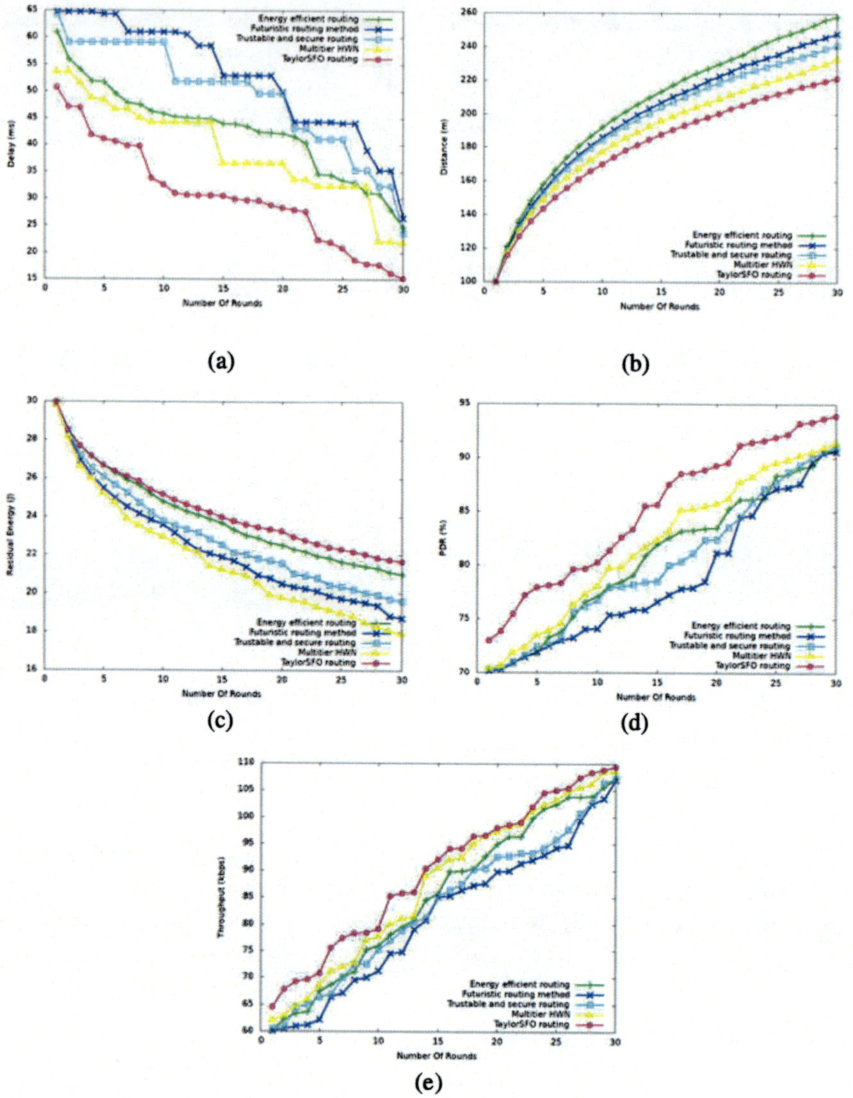


**Fig. 3** Comparative analysis of Taylor SFO routing model with 50 nodes **a** delay, **b** distance, **c** residual energy, **d** PDR, **e** throughput

routing, Futuristic routing model, Trustable and secure routing, Multitier HWN, and Taylor SFO routing is 212.63, 207.01, 202.78, 189.50, and 185.59 m. In 25th number of rounds, the residual energy of Taylor SFO routing is 22.64 J, whereas for existing techniques are 21.91, 19.98, 20.08, and 18.22 J. The PDR attained by the prevailing and Taylor SFO routing model is 88.34, 87.15, 87.71, 89.55, and 91.94%, while number of rounds is 25. While considering 25 as the number of rounds, the

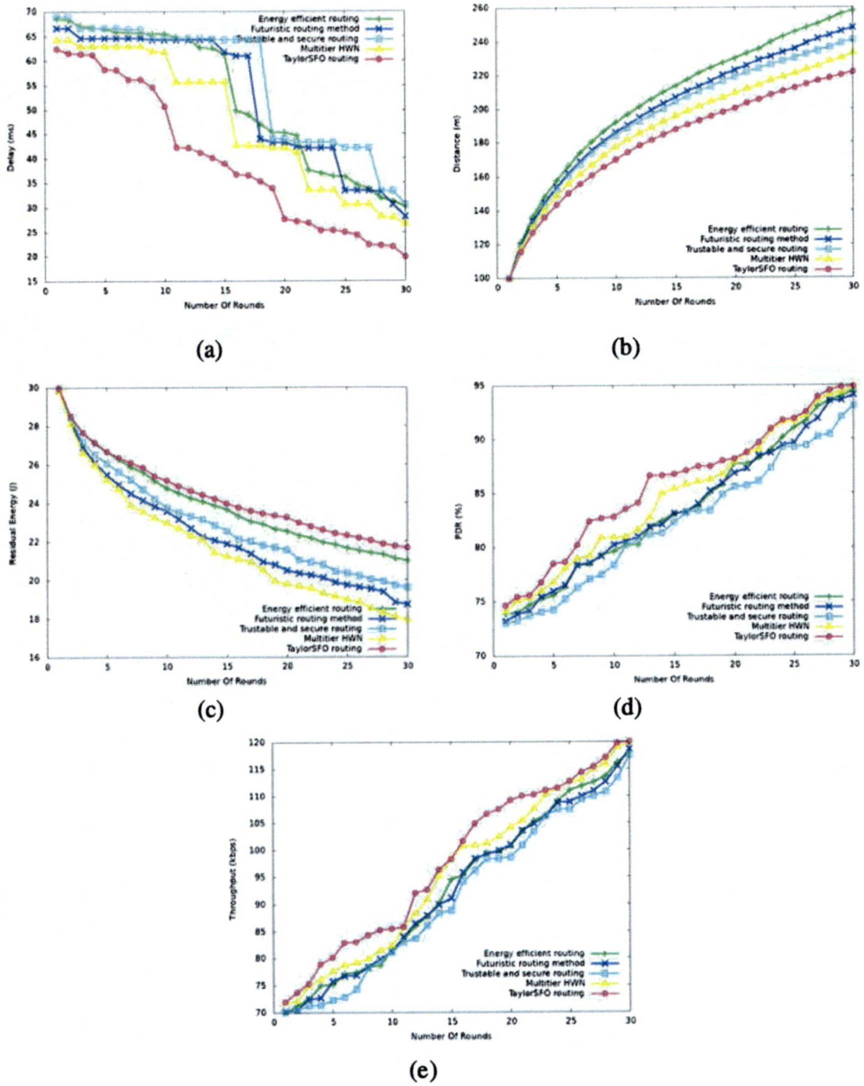


**Fig. 4** Comparative analysis of Taylor SFO routing technique using 100 nodes **a** delay, **b** distance, **c** residual energy, **d** PDR, **e** throughput

throughput of the existing routing models and Taylor SFO routing is 102.45, 94.26, 95.77, 103.28, and 105.14 kbps.

(2) ***Analysis with 100 nodes***

Figure 4 exposes the analysis of the Taylor SFO routing approach with 100 nodes using different performance metrics by varying the amount of rounds. The analysis of the Taylor SFO routing method for delay, distance, residual energy, PDR, and throughput is shown in Fig. 4a–e and 25th number of rounds is considered for analysis. The delay of Taylor SFO routing is 25.08 ms, whereas for the existing routing methods are 36.28, 33.42, 42.08, and 30.66 ms. The delay attained by the existing method and Taylor SFO routing is 245.65, 235.72, 230.33, 220.92, and 212.55 m, whereas the residual energy is 21.64, 19.72, 20.33, 18.96, and 22.32 J. The PDR of the existing routing models and Taylor SFO routing is 91.04, 89.62, 89.20, 91.62, and 91.88% and also the range of throughput is 111.05, 108.88, 107.47, 112.65, and 112.73 kbps.

## 6 Conclusion

This paper deliberates a performance study and comparative analysis of the Taylor SFO routing model in WSN. Here, the comparison is performed with various routing techniques to highlight the downsides by evaluating the performance. This paper presents a detailed study regarding the existing routing approaches along with the Taylor SFO routing model based on various metrics. Moreover, different figures as well as tables are summarized that afford the illustration routing performance. The metrics, such as residual energy, distance, PDR, delay, and throughput are considered. The metrics employed for evaluating the performance of the routing algorithms are analysed in this paper. In addition, the performance is evaluated with 50 and 100 nodes with the above metrics through altering quantity of rounds. Here, the prevailing routing algorithms, including the Trustable and secure routing method, Multitier HWN, energy efficient routing, and Futuristic method routing are analysed with the Taylor SFO routing technique. Hence, the Taylor SFO routing approach attained better performance with minimal delay of 15.25 ms, minimum distance of 195.44 m, high residual energy of 221.65 J, maximal PDR of 94.87%, and maximum throughput of 119.98 kbps.

## References

1. Kellner A, Alfandi O, Hogrefe D (2012) A survey on measures for secure routing in wireless sensor networks. Int J Sens Netw Data Commun 1(10):1–17
2. Alajmi NM, Elleithy K (2016) A new approach for detecting and monitoring of selective forwarding attack in wireless sensor networks. In: 2016 IEEE long island systems, applications and technology conference (LISAT). IEEE, pp 1–6
3. Mehetre DC, Roslin SE, Wagh SJ (2019) Detection and prevention of black hole and selective forwarding attack in clustered WSN with active trust. Clust Comput 22:1313–1328
4. Almazaideh M, Levendovszky J (2020) Novel reliable and energy-efficient routing protocols for wireless sensor networks. J Sens Actuator Netw 9(1):5
5. Islam MS, Rahman SA (2011) Anomaly intrusion detection system in wireless sensor networks: security threats and existing approaches. Int J Adv Sci Technol 36(1):1–8

6. Khan ZI, Afzal MM (2017) Security in wireless sensor networks: DoS perspective. Int J Eng Res Technol (IJERT) 6:311–316
7. Aljumah A, Ahanger TA (2017) Futuristic method to detect and prevent blackhole attack in wireless sensor networks. Int J Comput Sci Netw Secur (IJCSNS) 17(2):194
8. Rault T, Bouabdallah A, Challal Y (2014) Energy efficiency in wireless sensor networks: a top-down survey. Comput Netw 67:104–122
9. Khan JA, Qureshi HK, Iqbal A (2015) Energy management in wireless sensor networks: a survey. Comput Electr Eng 41:159–176
10. Tsiota A, Xenakis D, Passas N, Merakos L (2019) On jamming and black hole attacks in heterogeneous wireless networks. IEEE Trans Veh Technol 68(11):10761–10774
11. Liu G, Bao H, Han B (2018) A stacked autoencoder-based deep neural network for achieving gearbox fault diagnosis. Math Probl Eng 2018
12. Jayapriya K, Mary NAB (2019) Employing a novel 2-gram subgroup intra pattern (2GSIP) with stacked auto encoder for membrane protein classification. Mol Biol Rep 46:2259–2272
13. Sahin E, Hamzaoglu I (2007) An efficient intra prediction hardware architecture for H. 264 video decoding. In: 10th Euromicro conference on digital system design architectures, methods and tools (DSD 2007). IEEE, pp 448–454
14. Shadravan S, Naji HR, Bardsiri VK (2019) The Sailfish optimizer: a novel nature-inspired metaheuristic algorithm for solving constrained engineering optimization problems. Eng Appl Artif Intell 80:20–34
15. Balachandra M, Prema KV, Makkithaya K (2014) Multiconstrained and multipath QoS aware routing protocol for MANETs. Wirel Netw 20:2395–2408
16. Yadav AK, Tripathi S (2017) QMRPRNS: design of QoS multicast routing protocol using reliable node selection scheme for MANETs. Peer-to-Peer Netw Appl 10:897–909

# Design and Implementation of Third and Fourth Order Phase Lock Loop Using Simulink

Sneha Singh, Kashmiri Deka, and Geetamoni Konwar

**Abstract** A phase-locked loop is an electronic circuit with a voltage or voltage-driven oscillator that constantly adjusts itself to match the frequency of an input signal. It is used to generate, stabilize, modulate, demodulate, filter or recover a signal from a noisy communication channel where there is data interruption. It is widely used in wireless or radio frequency applications, including Wi-Fi routers, broadcast radios, walkie-talkie radios, televisions and mobile phones etc. In this work, the idea of design, analysis and implementation of 3rd and 4th order phase lock loop using Simulink is presented to reduce the settling time. The simulation results also provide performance overviews such as phase margin and loop bandwidth. The proposed model improves the settling time upto 90.82% than the other conventional PLL.

**Keywords** Simulink · Loop filter · Settling time · Transfer function · PLL

## 1 Introduction

Analysis of phase lock loop (PLL) using Simulink is an efficient method. Designing a PLL on a chip is a tedious process as it requires a lot of sums to fabricate it and consume months of time for development. Hence, we need another alternative to know whether the design will work or not and so Simulink is the preferred choice. Here, we discuss, analyze, design and discover various uses of PLL using Simulink. PLL is an important circuit used in the communication system usually for demodulation purpose. W. H. Eccles and J. H. Vincent discovered in 1919 that two electronic oscillators tuned to oscillate at slightly different frequencies but coupled to a resonant circuit would soon oscillate at the same frequency [1–3]. Although PLLs have been around since the 1920s [4, 5], their popularity and applications skyrocketed with the introduction of the monolithic PLL. One of the first PLL ICs was the RCA 4046 CMOS Micropower PLL, introduced in the 1970s [6–8]. The integrated circuits are

S. Singh (✉) · K. Deka · G. Konwar
Department of Electronics and Communication Technology, Gauhati University, Guwahati, India
e-mail: snehasinghdec45@gmail.com



S. Borah et al. (eds.), *Advanced Computational and Communication Paradigms*,
Lecture Notes in Networks and Systems 535,
https://doi.org/10.1007/978-981-99-4284-8_42

used in a wide range of applications, including frequency synthesis (FS), frequency modulation and demodulation, voltage-to-frequency conversion, and data synchronization in today's day and age [9, 10]. Also, PLL is most efficient in rejecting high frequency noise signals and generation of standard reference signals and implementing the limits of PLL enhances the stability of the voltage source converter [11]. PLLs are also used in mobile phones, televisions, broadcast radios, radar systems, communication systems and so on [12–14]. Also, an enhanced three-phase phase-locked loop can be obtained by combining an improved moving average filter and a linear active disturbance rejection controller (IMAF-LADRC-PLL) [15].

In this work, we are focusing on designing a PLL using foundation blocks such as phase detector (PD), low pass filter (LPF), frequency divider (FD) and voltage-controlled oscillator (VCO). Further we simulate and analyze certain criteria such as settling time (ST), loop bandwidth (BW), phase margin (PM) etc. using step response and bode plot analysis for checking the performance of the PLL. The ST is a very important for applications which need fast frequency switching such as Frequency Hopping Spread Spectrum (FHSS) in communication systems, Step Frequency Radars and Wireless Local Area Networks (WLAN) [16–20]. Recent digital cellular mobile phone's system technology requires high channel switching performance, including a reduced ST for the frequency synthesis. The ST is reduced by making the LF-BW wider but increasing the LF-BW results in an undesired increase in the noise BW at the VCO output of the system [11, 21, 22]. Also, in FM the transceiver which has multiple frequency channels needs local oscillation frequency for reception and carrier transmission with low ST [10, 21]. Thus, it is essential to improve the ST without affecting the noise performance and power consumption of the system. Therefore, in this paper we present a Simulink based behavioral model to improve the ST.

## 2 Mathematical Estimation in Frequency Domain

The circuit diagram of the proposed 3rd and 4th order PLL is given in Figs. 1 and 2. For the 3rd order PLL, a 2nd order passive LPF (PLPF) is used while a 3rd order PLPF is used for the 4th order PLL. The PD produces an error signal, proportional to the phase difference between the incoming signal and the VCO output signal by comparing the phase at each input. The PLPF takes the output from PD and removes the high frequency noise to produce a dc voltage. The PLPF output voltage controls the frequency of the VCO. The voltage at the input of the VCO determines the frequency of the periodic signal at the output of the VCO. The FD divides the output frequency by N. So, loop gain (LG) is reduced by the factor of N.

The TF of the PLL can be derived as:

$$H(s) = \frac{K_d F(s) \frac{K_0}{s}}{1 + \frac{K_d K_0 F(s)}{Ns}} \tag{1}$$

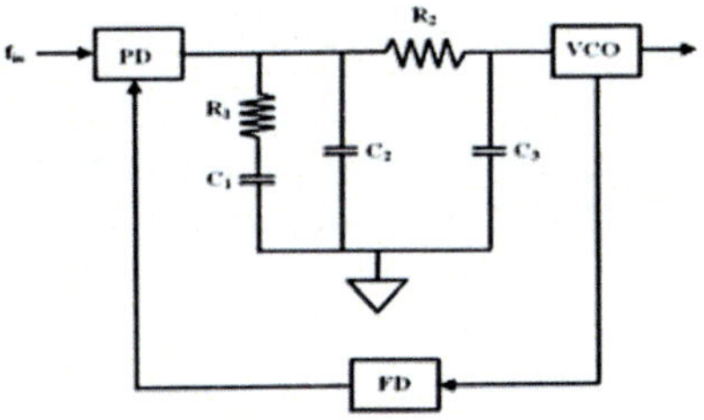


**Fig. 1** Circuit diagram of 3rd order PLL

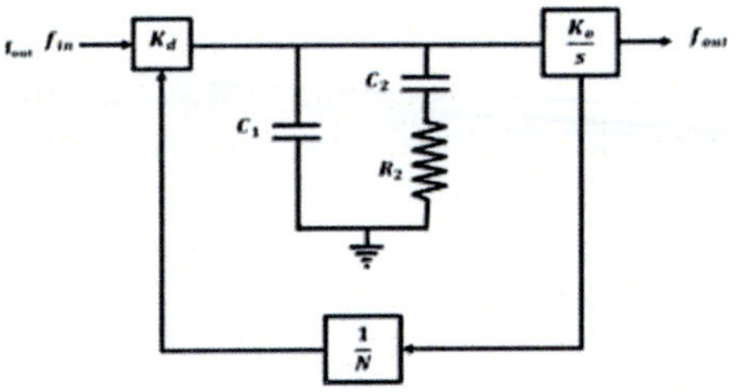


**Fig. 2** Circuit diagram of 4th order PLL

where, $K_d$ is PD gain in volt/radian, $K_0$ is the VCO gain, $N$ is the division ratio and $F(s)$ is the TF of LPF.

The TF of the 2nd order PLPF can be derived as:

$$F_{2\mathrm{nd}}(s) = \frac{1 + sC_2R_2}{s^2C_1C_2R_2 + s(C_1 + C_2)} \tag{2}$$

The TF of the 3rd order PLPF can be derived as:

$$F_{3\mathrm{rd}}(s) = \frac{1 + sC_2R_2}{s^3C_1C_2C_3R_2R_3 + s^2\left\{\begin{array}{c} C_1C_2R_2 + C_3C_3R_2 \\ + C_3R_3(C_1 + C_2) \end{array}\right\} + s(C_1 + C_2 + C_3)} \tag{3}$$

The 3rd order PLL TF can be derived by substituting Eq. (2) in (1) as:

$$H_{2\mathrm{nd}}(s) = \frac{K_dK_0(1 + sC_2R_2)}{s^3C_1C_2R_2 + s^2(C_1 + C_2) + sC_2R_2K + K} \tag{4}$$

where $K = \frac{K_dK_0}{N}$ = loop gain constant [17].

The TF of 4th order PLL can be derived by substituting Eq. (3) in (1) as:

$$H_{3\mathrm{rd}}(s) = \frac{K_dK_0(1 + sC_2R_2)}{\begin{array}{l} s^4C_1C_2C_3R_2R_3 + s^3\left\{\begin{array}{c} C_1C_2R_2 + C_3C_3R_3 \\ + C_3R_3(C_1 + C_2) \end{array}\right\} \\ + s^2(C_1 + C_2 + C_3) + sC_2R_2K + K \end{array}} \tag{5}$$

## 3 Simulink Model

In the linearized phase-domain analytical model, each component is represented by a MATLAB TF. As a result, we can analyze the phase of the signal at each point in the loop. Figure 3 shows the phase-domain model in Simulink, replacing the MATLAB equations with predefined blocks that implement the TF for each component. For 3rd and 4th order PLL design in Simulink, only the parameters of the TF of PLPF are changed. So, the overall design remains the same.

The Simulink response for both the 3rd and 4th order PLL is shown in Fig. 4 through Fig. 8. The Figs. 4 and 8 show the reference and output voltage for the VCO within the 3rd and 4th order PLL respectively. During simulation, the VCO is oscillating at its free-running oscillating frequency (1000 Hz) at the beginning, while the input signal is fixed at 10,000 Hz. Once the PLL achieves lock, the output signal is locked in frequency to the input signal (i.e. the VCO is oscillating at 10,000 Hz). The phase difference signal is taken at the output of the PLPF. This indicates that the output signal is locked in frequency to the input signal, but there is a phase error. It can be seen from Figs. 5 and 9 that there is small ripple in the phase difference signal. This leads to phase noise (PN) in the output signal. Figures 6 and 10 show the output of the 2nd and 3rd order PLPF while Figs. 7 and 11 show the VCO output response of the 3rd and 4th order PLL.

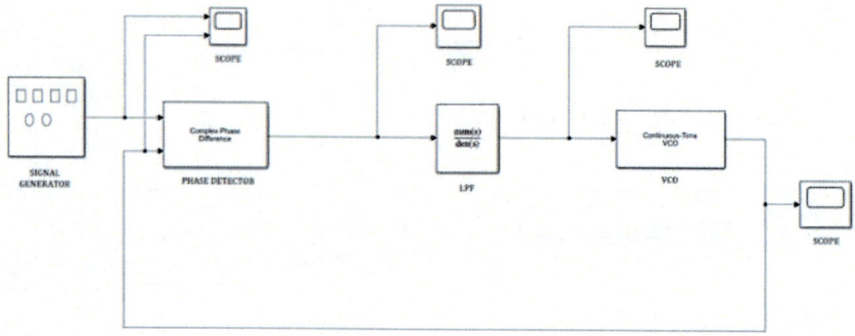


**Fig. 3** Simulink model of PLL

**Fig. 4** Phase difference input at PD for 3rd order PLL

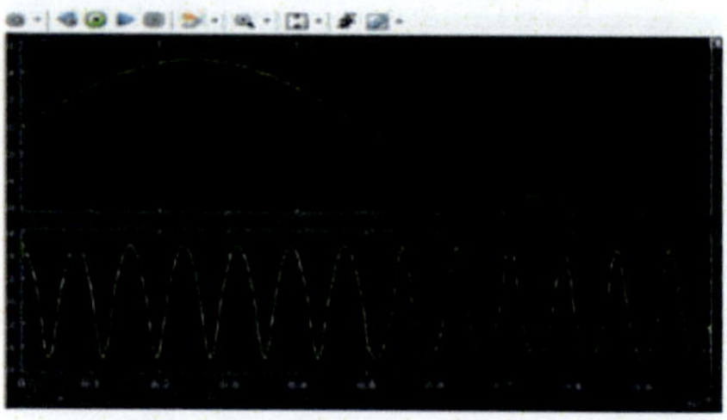

**Fig. 5** PD output for 3rd order PLL

**Fig. 6.** 2nd order PLPF output

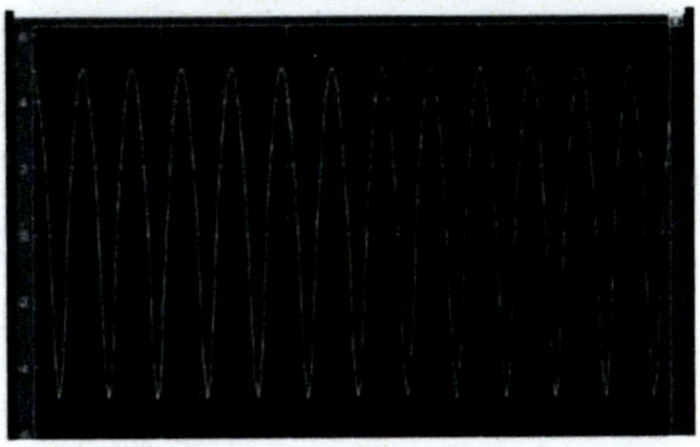

**Fig. 7** VCO output for 3rd order PLL

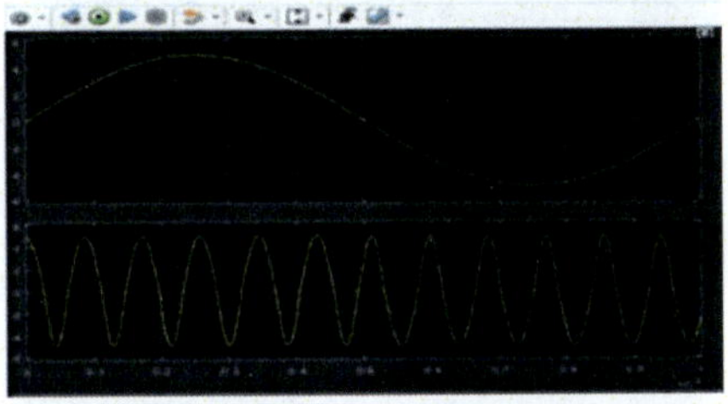

**Fig. 8** Phase difference input at PD for 4th order PLL

**Fig. 9** PD output for 4th order PLL

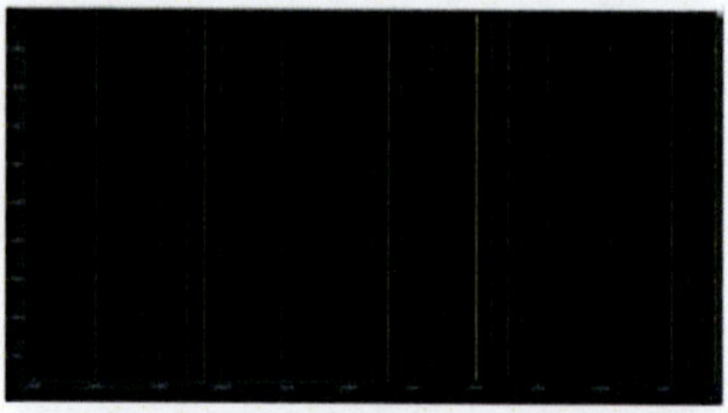

**Fig. 10.** 3rd order PLPF output

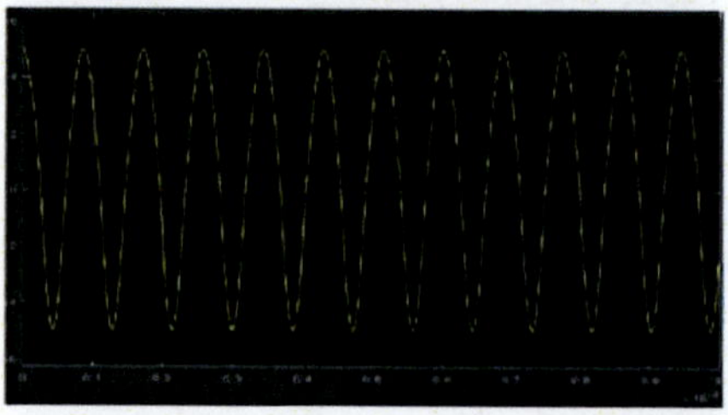

**Fig. 11** VCO output for 4th order PLL

# 4 Modelling Results

The aim of the proposed model is to simulate the PLL ST, PM and loop BW. Figure 12 through Fig. 15 shows the impact of PM and loop BW on ST for both the 3rd and 4th order PLL system. The ST is used to optimize the PM of the system. It is observed from Fig. 12 that the ST of the 3rd order PLL decreases exponentially with increasing PM while for the 4th order PLL it decreases linearly in a zigzag manner with increasing PM as shown in Fig. 13. The PM of 76° provides the fastest ST of 2.06 ns for the 3rd order PLL while the fastest ST of 1.39 ns for the 4th order PLL is observed at PM of 72.4°. Figures 14 and 15 show the impact of loop BW on ST. It is observed that the ST decreases exponentially with the increase in loop BW for both 3rd and 4th order PLL. The fastest ST is obtained at the maximum achievable loop BW of 614 MHz and 192 MHz for the 3rd and 4th order PLL respectively. In general, large BW results in slow ST. But, from the results of the simulation of the proposed model, it is observed that even at large BW, it is possible to obtain fast ST by simply increasing the order of the LF.

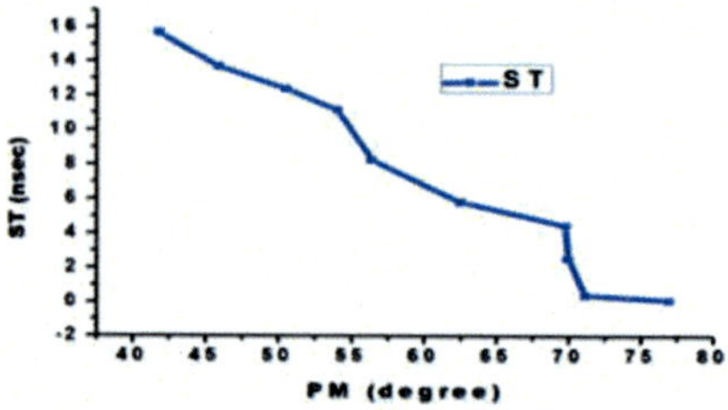


**Fig. 12** Impact of PM on ST of 3rd order PLL

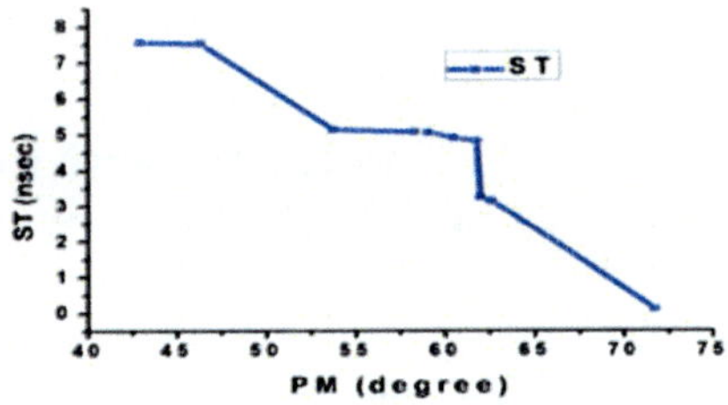


**Fig. 13** Impact of PM on ST of 4th order PLL

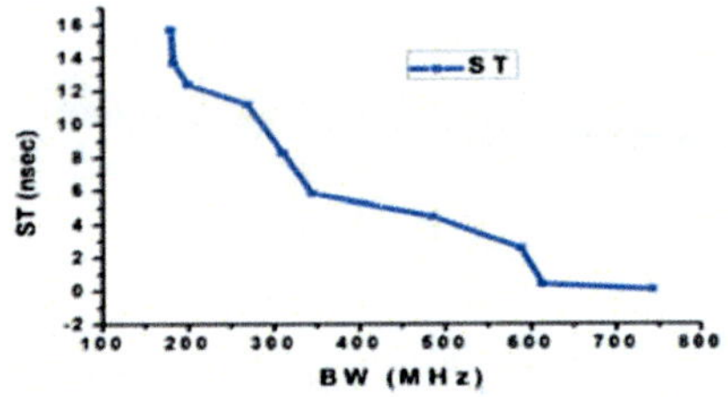


**Fig. 14** Impact of loop BW on ST of 3rd PLL

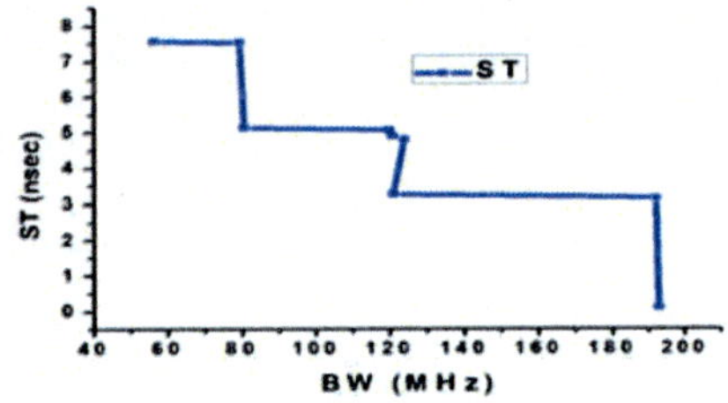


**Fig. 15** Impact of loop BW on ST of order 4th order PLL

The comparison of the results of the proposed system with the results of earlier work with different PLL approaches is summarized in Table 1. It is observed that the fastest ST obtained for the system is 1.39 ns for the 4th order PLL, while all other relevant works provides ST in the range of 5–98.6 ns. The percentage (%) of improvement in the ST with respect to other relevant works is also summarized in Table 1. It is observed that the ST of the system improves up to 99.21% in comparison to the earlier works.

**Table 1** Comparison of ST of our work with earlier relevant works

| ST (ns) | | Type of PLL | % of improvement (%) |
|---|---|---|---|
| This work | Earlier works | | |
| 2.06 (3rd order PLL) or 1.39 (4th order PLL) | 5 [2] | 3rd order PLL | 58.8 |
| | 176 [6] | 4th order PLL | 99.21 |
| | 98.76 [9] | 3rd order PLL | 97.91 |

## 5 Conclusion

A 3rd and 4th order PLL behavioral modelling in Simulink is presented. Each PLL component was modelled separately and was connected for an overall simulation. Results from the presented model were compared with each other and present fast switching speed of 2.06 ns and 1.39 ns respectively for the 3rd and 4th order PLL. Using this proposed Simulink model for both the 3rd and 4th order PLL, 32.52% improvement is achieved in ST for the 4th order PLL compared to the 3rd order PLL. Both the systems result in improved stability in terms of PM. The PM for both the systems is inversely proportional to the ST.

## References

1. Stephens DR (2002) The early history of phase-locked loops. In: Phase-locked loops for wireless communications: digital, analog and optical implementations. Springer, pp 1–9. https://doi.org/10.1007/0-306-47314-3_1
2. Mishra BK, Save S, Patil S (2011) Design and analysis of second and third order PLL at 450 MHz. Int J VLSI Des Commun Syst (VLSICS) 2(1):97–114. https://doi.org/10.5121/vlsic.2011.2109
3. Godave A, Choudhari P, Jadhav A (2018) Comparison and simulation of analog and digital phase locked loop. In: 2018 9th international conference on computing, communication and networking technologies (ICCCNT), pp 1–4. https://doi.org/10.1109/ICCCNT.2018.8494198
4. Patel GS, Das NN, Sinha SK (2020) Optimization of frequency settling time of PLL using 3rd MASH sigma delta modulator. Int J Innov Technol Explor Eng (IJITEE) 9(3):2853–2859. https://doi.org/10.35940/ijitee.B6222.019320
5. Majeed KKA, Kailath BJ (2017) PLL architecture with a composite PFD and variable loop filter. IET Circuits, Devices Syst 12(3):256–262. https://doi.org/10.1049/iet-cds.2017.0336
6. Al-Maaitaha GO, Al-Haraseesb AS (2018) Fast lock and settling time improvement for indirect frequency synthesizer phase locked loop. Jordan J Electr Eng 4(2):72–86. http://jjee.ttu.edu.jo/docs/vol4no2/JJEE%20vol.%204,%20no.%202,%20pp.%2072-86.pdf
7. Sulaiman DR (2020) A novel high-frequency PLL design for wireless applications. J Electr Syst 16(3):332–349. http://journal.esrgroups.org/jes/papers/16_3_5.pdf
8. Patel ND, Modi GR, Gandhi PP, Naik AP (2017) Design and analysis of phase-locked loop and performance parameters. Int J Microelectron Eng (IJME) 3(1/2/3):1–10. https://doi.org/10.5121/ijme.2017.3301
9. Konwar G, Borah M, Bezboruah T (2018) Modeling and behavioral simulation of PID-controlled phase-locked loop. In: Advanced computational and communication paradigms. Springer, Singapore, pp 55–66. https://doi.org/10.5121/ijme.2017.3301
10. Ali MK, Hashemipour O (2019) Fast locking technique for phase locked loop based on phase error cancellation. AEU-Int J Electron Commun 109:99–106. https://doi.org/10.1016/j.aeue.2019.06.025
11. Tang Y, Li Y (2023) Common Lyapunov function based stability analysis of VSC with limits of phase locked loop. https://doi.org/10.1109/TPWRS.2022.3233762
12. Xiang B, Fan Y, Ayers J, Shen J, Zhang D (2020) A 0.5 V-to-0.9 V 0.2 GHz-to-5 GHz ultra-low-power digitally-assisted analog ring PLL with less than 200 ns lock time in 22 nm FinFET CMOS technology. In: 2020 IEEE custom integrated circuits conference (CICC), pp 1–4. https://doi.org/10.1109/CICC48029.2020.9075897
13. Deng X, Li H, Zhu M (2021) A novel fast-locking ADPLL based on bisection method. Electronics 10(12):1–10. https://doi.org/10.3390/electronics10121382

14. Pan Z, Wang X, Hoang TGT, Tian L (2023) An enhanced phase-locked loop for non-ideal grids combining linear active disturbance controller with moving average filter. https://doi.org/10.1016/j.ijepes.2023.109021
15. Mouro J, Paoletti P, Sartore M, Tiribilli B (2023) Dynamical response and noise limit of a parametrically pumped microcantilever sensor in a phase-locked loop. https://doi.org/10.1038/s41598-023-29420-3
16. Beiraghdar F, Rad AG, Sheikhaei S, Tohidian M (2021) A fast settling frequency synthesizer with switched-bandwidth loop filter. Int J Circuit Theory Appl. https://doi.org/10.1002/cta.2993
17. Kamal N, Al-Sarawi S, Weste NHE, Abbott D (2010) A phase-locked loop reference spurs modelling using Simulink. In: 2010 international conference on electronic devices, systems and applications, pp 279–283. https://doi.org/10.1109/ICEDSA.2010.5503058
18. Kuznetsov NV, Kuznetsova OA, Leonov GA, Neittaanmäki P, Yuldashev MV, Yuldashev RV (2015) Limitations of the classical phase-locked loop analysis. In: 2015 IEEE international symposium on circuits and systems (ISCAS), pp 533–536. https://doi.org/10.1109/ISCAS.2015.7168688
19. Wang X, Choi YB, Je M, Yeoh WG (2007) A Simulink model for all-digital-phase-locked-loop. In: 2007 IEEE international workshop on radio-frequency integration technology, pp 70–73. https://doi.org/10.1109/RFIT.2007.4443922
20. Parkalian N, et al (2018) Modeling and simulation of digital phase-locked loop in Simulink. In: 2018 15th international conference on synthesis, modeling, analysis and simulation methods and applications to circuit design (SMACD), pp 121–129. https://doi.org/10.1109/SMACD.2018.8434865
21. Yu C, Sun Q, Shen L, Yan J, Yuan Q (2022) Phase-locked loop considering low-order harmonic and frequency fluctuation of power grid under high-penetration of renewable energy. In: 2022 IEEE international power electronics and application conference and exposition (PEAC), pp 1443–1448. https://doi.org/10.1109/PEAC56338.2022.9959528
22. Xu J, Qian H, Qian Q, Xie S (2023) Modeling, stability, and design of the single-phase SOGI-based phase-locked loop considering the frequency feedback loop effect. IEEE Trans Power Electron 38(1):987–1002. https://doi.org/10.1109/TPEL.2022.3201252

# Survey on IoT Based Smart Fisheries in Context with Himalayan Region of Sikkim

**Pritika Adhikari and Dhruba Ningombam**

**Abstract** Though the aquaculture in Sikkim has been growing steadily in the last few years, there is still a lot of under-exploited potential in the state's aquaculture sector. Smart fishery with IoT can be used to manage the ecosystem of the fish pond and make the process of fish production much more efficientas compared to the traditional techniques. It can help in optimizing the fish production by reducing diseases and fish fatality rate and in turn minimizing production cost. Be it for consumption or ornamental purposes, fishery and aquaculture sector could be expanded by combining it with user friendly IoT tools and attracting more hesitant fish-farmers to the trade. This will play a significant role in the development of the rural economy and the socio-economic condition of the fish farmers in Sikkim.

**Keywords** Water quality monitoring · Parameters · Sensors · Actuators · IoT · Automated fisheries · Smart fisheries

## 1 Introduction

Sustainable aquaculture is the key to meet the increasing food demand of our growing population. Fish is nature's super-food that has important role to play in nutrition and food security. Fisheries provide resources from ecology to economy. Fisheries are next to agriculture in terms of providing employment and food supply [1]. In addition to direct consumption, fisheries contribute to food and nutrition security through income generation, increasing the household's ability to purchase food and providing a source of employment for women who participate in fishing and postharvest activities [2]. From direct capture to other activities in the operational chain including transport, equipment supplies, etc., livelihood of millions of people around the world,

P. Adhikari (✉) · D. Ningombam
Sikkim Manipal Institute of Technology, Majhitar 737136, India
e-mail: pritikai1997@gmail.com

D. Ningombam
e-mail: dhruba.n@smit.smu.edu.in

 
S. Borah et al. (eds.), *Advanced Computational and Communication Paradigms*, Lecture Notes in Networks and Systems 535,
https://doi.org/10.1007/978-981-99-4284-8_43

is supported by the fishing sector. According to Indian Ministry of Fisheries, Animal Husbandry & Dairying, fisheries provide income and employment to more than 28 million people in India. In the Himalayas, aquaculture is undergoing an exponential development due to a progressive reduction in natural fish stocks and catch fishery, coupled with an increasing demand for food fish. Freshwater aquaculture represents approximately one-third of the total world aquaculture production [3]. Coldwater resources hold a considerable population of food, sport, and ornamental fish species [4]. The fishing industry supports rural development, expands export prospects, improves the management of natural resources, and protects biological diversity. However, there are many issues that need to be addressed like infrastructural development and culture practices for management of brood-stock, stock handling, feeding and health management. Other major identified constraints for the expansion of fish farming in the state are delay and insufficient supply of fish feed.

Sikkim is abundant with several water resources in the form of rivers, lakes and streams. This offers tremendous potential for developing freshwater inland fisheries for both culture and capture Trout (brown trout, rainbow trout) and carp (common carp, grass carp, silver carp) are the two primary fish breeds that are cultured for mass production in Sikkim. Among the two trout farming has interested and pulled in a lot of local farmers. Chettri [5] reports that among other Himalayan states, Sikkim shows an upward trend is fish production. The state had a growth of 245.44% between 2009–10 and 2013–14. With introduction of various programs to promote fisheries development in the state, the combined production of the two fishes has gone up from 120 to 400 tons annually in the last few years [6]. Trout, especially Rainbow Trout is found to be one of the most sensitive of the fishes, making monitoring of the water criterion extremely crucial. Along with several issues like poor transportation and marketing facilities in the rural areas, the absence of use of modern technology is identified as a key problem [5]. Training and providing fish farmers with resources for modern technologies like smart fisheries will allow to increase the level of fish production and to bring state's resources into proper utilization. And though there are several IoT based Smart Fishery techniques being applied in various countries and in other parts of India, there is still a lack of such smart systems in the Himalayan region including Sikkim. Given, that this hilly region has different environmental conditions than that of the tropical region in the country, the requirements and demands of pisciculture also varies. Hence the IoT system and its components required will be different in this region and that's where we will mainly focus on.

Aquaculture has greatly benefited from the use of modern technology, helping it to become more productive, eco-friendly, and labour-efficient. Fishes are sensitive to the physical and chemical parameters. The important water parameters that can drastically affect the marine life include physical ones like pH, temperature, dissolved oxygen and other chemical or biological contaminants like pathogens. Slight changes in these parameters can have drastic effects on the fish and thus need to be minutely monitored. The need for such smart monitoring and actuating systems is especially important in case of large fish production farms as small changes could mean big losses.

In this report existing work on IoT systems and different sensors and other tools are reviewed. Some of them are applied and tested and some of them may just be proposed systems. The systems that are used to remotely monitor and control fish farms have proven to be efficient in yielding maximum production with minimum losses. But there is no design and implementation of such smart systems in befitting the water conditions and types of fish cultured in Sikkim. There is also no proper research on the IoT devices needed to design smart fishery systems specific to the state's requirement. The goal of the research is to study the water parameters and the types of sensors and actuators needed to monitor and control those parameters. Existing system pipelines and architectures that could be customized and replicated to fit Sikkim's terrain is discussed in this paper. The objective of this survey paper is to thoroughly study the parameters, required hardware devices and existing systems that can be used as a base to curate a system that meets the need of fish farmers for fish production in Sikkim.

## 2 Existing Work

Many research studies have been conducted to show that aquaculture needs to be advanced to minimize loss and increase production to satisfy the food demand of the rapidly growing population. Lot of researchers have employed various sensors to monitor and manage fish production. And few have also designed mobile and/or web application for easier real time monitoring and managing various aspects of fish production.

The literature review is divided into 4 parts and each part will discuss the portion of the paper that addresses the heading under discussion.

### *2.1 Parameters*

Since a minute change in any of the following parameters of water qualities, temperature, pH, dissolved oxygen, turbidity, and conductivity has adverse effects on the aquatic animal and these parameters are common in almost all the research studies. There are also research studies that measures hydrogen peroxide [7] and ammonia [8, 9] as these have effects on the fish production. Additionally, there are research studies [7, 10, 11] that monitor the fish movement and study fish behavior. As the day shifts the requirements of the fish changes that affects their movement. Hence to understand and predict the pattern of fish movement and their needs throughout the day, these parameters prove to be effective.

Some research also shows that colour [12], oil layer [13] and light intensity [14] play a role in determining fish behavior and are directly linked to the quality of the water and the mortality rate of the fish. One research in particular [13] monitors these parameters and intensively studies the fish behavior to figure out the hunger level and makes predictions for the next day using regression.

## 2.2 Sensors

Industry standard temperature sensor, pH sensor, DO meter, turbidity, and water conductivity sensors are used to measure the temperature, pH, dissolved oxygen, turbidity, and conductivity respectively. Most research studies conducted so far have used electric sensors to measure the various factors affecting the water quality but few others [14] have opted for magnetic sensors to create a safer fish-pond environment. Colorimetric analyzers are used to measure the content of ammonia in the water [8]. Saha et al. [12] after thorough research have concluded that the chemical imbalance in some parameters cause the imbalance of some other parameters and from the values of some parameters, we can determine the values of others. Hence all parameters may not be monitored reducing the number of sensors needed.

Research focusing on application of regression analysis on aquaculture [15], measures temp, ORP and pH. In this system using the measured values of pH and ORP DO is calculated. DO meters are generally expensive and this eliminates the need for it without compromising the water quality [7, 10, 16]. Ultrasonic sensors and motion sensors are used in these research studies to detect the fish and study its movement and behavior for pattern prediction and future analysis.

## 2.3 Methods

After thorough review of the literature, it is clear that some technologies/hardware that are used in all the research studies are the same, varying only on the model used. They are as follows:

**Arduino Uno**: It is a microcontroller board used for sensor acquisition. A USB cable is used to connect the Arduino Uno to the computer so that it may be programmed using a USB to serial converter. It can exchange data with a computer via a serial connection.

**Raspberry Pi**: Raspberry Pi is a single board computer that provides full functionality of a computer. It includes built-in Bluetooth and Wi-Fi, and it can communicate serially with Arduino. You can use the Arduino to send data to and from the Raspberry Pi by connecting it to any sensors, actuators, and motors you may have.

**Smartphone with MTP**: Any android smartphone with MTP (Media Transfer Protocol) can act as the end point. MTP enables the automatic interchange of media files.

Older system uses c, c++ or java [10] as the programming language but most in recent years choose Python as it is simple and provides libraries that can easily be imported and used in any project. PHP, HTML, CSS, MySQL are popular among researchers for creating user-friendly applications. Technologies in the beginning [7, 16] collected data at regular intervals and stored them locally. Ya'acob et al. [16] collect data in the morning, mid-day and evenings to understand how the sensed parameters vary through-out the day and results are compared. Research studies in the recent years [10, 11, 15] have real-time monitoring systems that continuously monitor the changes in the parameters and send the data to be stored in a cloud facility. Ya'acob et al. [16] in particular use Blynk Server for data storage. Spandana and Seshagiri Rao [17] entire paper focuses on the use of WSN in aquaculture and compares the usage of WSN and wired networks for aquaculture. And the authors conclude that WSN in much cheaper and efficient, especially in remote areas.

The data collected can be used in one or more of the following ways:

- **Observation**: The results are displayed on the LCD connected with the Raspberry Pi processors only [18] or in most cases is transferred, stored and shown in the web or mobile application. Alerts can be sent and the farmers take the necessary steps manually. No actuators employed in these research studies [8, 12, 15–17].

- **Activation**: A system of actuators can be integrated to the system. And a threshold value for the range of parameters can be set such that when the sensor reads a value that is below or above the threshold value an action is automatically activated. Masum et al. [13], Hapsari et al. [18], Prasad et al. [10], Chatterjee and Pandey [14], Balakrishnan et al. [19] have various actuators as discussed in detail in the next part.

- **Prediction**: As the systems are getting smarter, the data collected can be studies in details to learn patterns of the fish movement, their behavior and requirements that changes throughout the day, or may depend on the weather conditions of a particular day. Masum et al. [13] have created a basic system to make predictions for the next day and hunger level and use regression to make predictions so that the farmers know what to expect and take precautionary step.

## 2.4 Actuators

Early researches were mainly focused only on gathering information about the various factors that affect the water quality and how the fish production is affected by these factors. However, with advancement in technology more and more researchers are including effective actuators in their systems. Water heaters are turned on and off to increase or decrease the temperature of the water [18]. Chatterjee and Pandey [14] incorporate a pH lowering and raising liquids module where the farmer can remotely release them into the water as per the requirement. Aeration or agitation can increase the DO. Karim et al. [11] uses air pumps for this purpose. Balakrishnan et al. [19] use an oxygen bank for the purpose and they also have used water pumps to increase

**Table 1** Different types of ultrasonic sensors

| Parameters | SHARP GP2Y0A21YKOF | (HC SR-04) |
|---|---|---|
| Range | 10–80 cm | 2–10 m |
| Beam-width | 75° | 30° |
| Frequency | 353 THz | 40 kHz |
| Unit cost | ~750 INR | ~130 INR |

the water level and water is withdrawn through the pipe in case of excessive rise in the water level [11, 18]. Some also use water pumps that are activated when needed. Chatterjee and Pandey [14] haves a system in place that controls light automatically depending on the time of the day and weather conditions. Ramya et al. [7] Prasad et al. [10] here the researchers have included an automated feeding system that feeds the fishes either on regular intervals and whenever needed from the web or mobile application remotely.

# 3 Hardware Components

## *3.1 Ultrasonic Sensors*

Ultrasonic sensor is used for observing the movement of different fishes in the pond, throughout the day and making advance decision based on the observations. The fish movement may vary depending on the time of the day and/or the weather. It can also be used to detect any unwanted particles or organism in the water. Depending on the size of the fish pond either of the two easily available ultrasonic sensor from Table 1 can be used. For smaller ponds HC SR-04 is cost effective and powerful enough.

## *3.2 Temperature Sensors*

DS18B2 is the best and most reliable of all the water sensors. It is highly accurate and drains very small amount of current. The measurement range of sensor is from −55 to 125 °C. In addition, the sensor can operate in normal or "parasite" power supplying mode. It means, that the DS18B20 can derive power directly from the data line ("parasite power"), eliminating the need for an external power supply. Hence it best temperature sensor that can be employed for accurate sensing in the cold water ponds in Sikkim (Table 2).

**Table 2** Temperature sensors

| Parameters | LM35DZ | DS18B2 | AM2320 |
|---|---|---|---|
| Measurement range (°C) | −55 to 150 | −55 to 125 | −55 to 125 |
| Supply voltage (V) | 4–30 | 3–5.5 | 3.1–5.5 |
| Accuracy (°C) | ±3/4 | ±0.5 | ±0.5 |
| Communication interface | 1—wire | 1—wire | $1^2$ C |
| Current drain (μA) | 60 | 10–12 | ~10 |

**Table 3** Turbidity sensors

| Manufacturer/ model | Current (mA) | Output | Size (cm) | Weight (g) | Range (NTU) |
|---|---|---|---|---|---|
| Global Waters WQ370 | 30 | 4–20 mA | 3.8 diam. × 21.6 length | 454 | 0–1000 |
| Green Span TS3000 | 35 | SDI12 | 4.7 diam. × 41.8 length | 500 | 0–1000 |
| Analite 9000 | 15 | −2.5 to 2.5 V | 3.2 diam. × 18.2 length | 380 | 0–3000 |

## 3.3 *Turbidity Sensors*

Turbidity is a significant parameter that affects the health of the marine animals. Since Sikkim gets heavy rainfall, the water is often muddy and turbid. So, choosing the turbidity sensor is paramount. Analite 9000, which is a high range, low response time and has lower current drain is best suited for the water conditions in Sikkim (Table 3).

## 3.4 *Microchip*

Several factors have to be considered while selecting the micro-controller. For starters the micro-controller should consume less energy. For this purpose, they usually come with low power mode where some functions are disabled and clock is reduced. Power active, power idle and power down are the most common power modes. Active mode allows all the hardware to be active and working. But it consumes power the most. It is useful for operations that require high processing speed such as calculation operations. ATmega328p with 32 kb flash memory which has low power consumption and greater CPU speed than others compared is the most feasible choice of micro-controller (Table 4).

**Table 4** Microchips

| Parameters | ATmega328p | PIC16F877A | ATmega16p |
|---|---|---|---|
| Program memory type | Flash | Flash | Flash |
| Program memory size (kb) | 32 | 14 | 16 |
| ADC channels | 8 | 8 | 8 |
| Temperature range | −40 to 85 | −40 to 125 | −40 to 85 |
| Pin count | 32 | 40 | 44 |
| Low power | Yes | No | No |
| CPU speed (MIPS/ DMIPS) | 20 | 5 | 16 |

# 4 Existing Architectures

With simple design with just few sensors monitoring to machine learning (ML) based models that can make smart decisions the system designs vary greatly. Architectures of the recent systems for smart fish farming is discussed in the section below. Some system designs are simple and but can be the basic model for advanced system designs. They are easy to implement and are cost effective. They may be simple but are efficient and can reduce the fish fatality rate by a great degree. Two of the architectures have a machine learning based prediction or decision-making capabilities [13, 20]. This newer systems with machine learning models and mobile application could be powerful tool to minimize the loss. These ML based smart models have high accuracy and have proven to be efficient in aiding the fish farmers make informed decision. Mobile applications make remote sensing and controlling convenient. However, these complex systems are not easy to design and can be too costly for the farmers. Some parameters of the existing design may not have to be monitored for the type of fish and water condition in Sikkim. While there are other conditions to be monitored in the existing systems. Understanding of the systems already in use will give us a head start to design a system specific to meet the requirements of the state.

## *4.1 Guppy Fish Farm Monitoring and Controlling [18]*

The architecture of this system is simple comprising of two distinct modules for monitoring and control. The flow chart of the control model and monitoring model is shown in Figs. 1 and 2 respectively. They monitor only pH and salinity and only activate water pumps when the sensor values are out of range. But their simple and clear design allows any easy addition of any sensor and activators.

The control unit consists of the pH and salinity sensors. The data from the sensors are processed by the microcontroller and is displayed on the LCD. The servo motor of

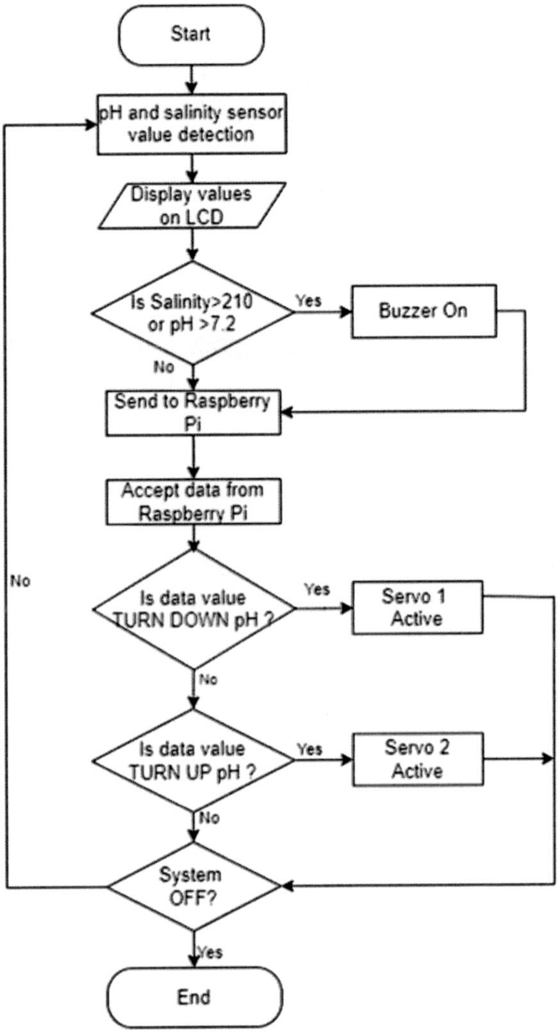


**Fig. 1** Controlling module [18]

this module is designed to automatically open the valves for pH-raising and lowering liquids. When the pH or salinity level of the water is out of range the buzzer is used to alert the farmers. Web based monitoring is enabled by the MM. The processor in this module is Raspi. The microcontroller processes all of the sensor's data and displays it on the LCD. A cloud database is used to store the data from the system. The pH neutralizing valve can be controlled virtually through the web application. The CM will read the sensor values and the buzzer will be turned on if the salinity is greater than 210 or pH is greater than 7.2. After receiving the sensor values, the MM sends a control signal to the virtual button. The system checks to see if pH raise or pH lower button was pressed. Depending on the button pressed the valve for respective valves

Fig. 2 Monitoring module [18]

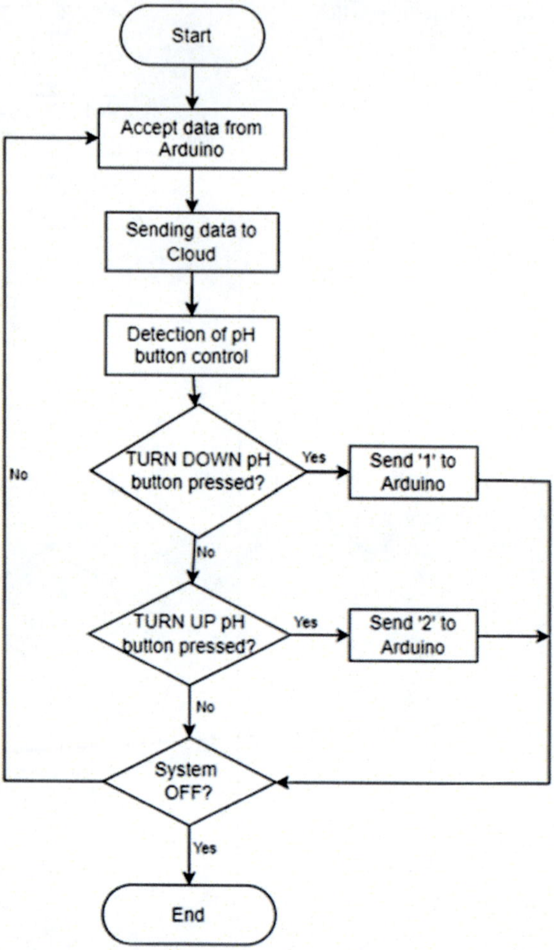


are opened. The MM sends the data received from the CM to the cloud database. Through the web application, the button to reduce pH transmits data '1' to the CM and data '2' is transmitted in case pH raise button is pressed.

## 4.2 *Design and Implementation of IoT Based Ideal Fish Farm [13]*

In this architecture, at the beginning all the sensors are initialized and their readings are taken. The servo meter first has to be rotated 30° to the left and right for the pH sensor to give any valid reading. 5 values from the pH sensor, temperature sensor and IR optical sensor, in the interval of 3 min is taken. The mean and median is calculated for each sensor and the best value is considered. In this case the optimum range for pH is between 7–8.5 and 22–27 °C is the accepted range of temperature. If the average of the mean and median calculated is out of the desired range for rapid growth of the fish, then the alarm is turned on or the farmer is alerted through the web interface or mobile application. In a minute, 6 subsequent readings, that is a reading every 10 s for a minute, is taken from the optical sensor to understand the movement of the fish. Mean, median and the best value for this are also calculated. For hunger prediction, 4.5 is set as the threshold value. RMS of the estimated average of all the sensor is calculated and compared with the threshold value. If it exceeds the threshold the fish farmer is notified through the web interface or mobile application. Integrated ESP8266 WIFI module with the Wemos D1 microcontroller transmits the data from the device to the mobile app and web interface. Based on these important features of the water, the system forecasts how much fish food is necessary in real time.

### 4.2.1 Model for Fish-Feed Consumption Prediction

To make the system unique, a prediction model for forecasting the amount of food for the next day is designed. Figure below demonstrates the machine learning process. For the machine learning algorithm, the dataset used is processed data collected from the local fish farmers in the spam of 8 months. The collected raw data is sent through a multi-layer pre-processing module which extracts important features and normalizes the range of independent variables. Oil layer, pH, temperature are the input parameters and the amount of food needed in the output. One hot encoding is used to convert the categorical data into numerical data. Water level 1 indicates that the water level is acceptable and 0 will set off the alarm. Hunger level 0 refers to extreme level of hunger, 1 indicates medium level and 2 indicated hunger is satisfied. Backward elimination is carried out to remove any insignificant feature that are features that have no effect on the dependent variable or the final outcome of the model.

The dataset is split into 80 and 20%, where 80% is the training data and 20% is the testing data that is used to evaluate the prediction of the trained model. As regression is a technique for predictive analysis that gives the relationship between features (independent variables) and the outcome (dependent variable), it is selected as the model for this design. The researchers have deployed various regression algorithms such as Polynomial Regression, Linear Regression, Support vector Regression and Decision Tree Regression. MAE, MSE, RMSE, cross validation of the outputs of various models are calculated. Using these, the performance and accuracy of the models are compared. It is concluded that SVR gives the best outcome.

## 4.3 Smart Fish Aquaculture Monitoring System [11]

This system focuses on continuous monitoring of the water parameters and storing and processing the collected data at a reduced cost. Different sensors used here are pH, water level, turbidity, motion detection and temperature sensors. The motion detector is used to detect dead fish in the pond and to monitor the fish movement. Analyzing the fish movement allows to optimize the fish production by reducing the fatality rate and predicting future patterns. Cloud based databases used in the existing models require high amount of internet usage making the systems expensive. To overcome this, the authors have proposed a new system where a local computer system acts as a server host to store the data from the sensors. The data can be retrieved by the user from anywhere at a low cost via the internet. This reduces the cost of transmitting data over the internet by reducing the number of regular computations needed. The system also has an Android and web application with interactive GUI (Graphic User Interface) for remote sensing. The GSM (Global System for Mobile communication) modem embedded system design allows the farmers to be alerted without internet connection. For remote places that has no access to the internet this system is best suited. It also is connected with a water filtration plant which can be automated to kick in when the turbidity level of the water exceeds the threshold level. This system is cost effective and well suited for fish farms in remote location.

## 4.4 Smart Water Quality Prediction for Biofloc Aquaculture [20]

A method of wastewater treatment known as the Biofloc system has become extremely important in aquaculture. The idea behind the procedure is to add a carbohydrate source to maintain a greater C–N (Carbon to Nitrogen) ratio while also improving the quality of the water produced. Biofloc technology allows fishes to be cultivated in recycled water and is ecofriendly and ensures high productivity. This system like the one discussed in Sect. 4.2 has a machine learning model to make smart decisions. The sensors used here are pH, temperature and Total Dissolved Solids (TDS). TDS is a sensor used to indicate the total solid dissolved level in liquids, in this case water ad is unit is in grams. Bio flocs are collections (flocs) of particulate organic waste, such as excrement and uneaten feed, as well as algae, bacteria, protozoans, and other microorganisms. TDS here is used to measure the level of flocs in the water body rearing fishes. This system monitors the water quality of the fish pond continuously and sends the data through a Wi-Fi module to the user via a mobile application. Data collected from samples of various fish farms that biofloc technology, a machine learning model was trained. Continuously collected sensor values are fed into the model and in if any of the sensor values is out of the desired range the model generates a decision. The information is sent to the user through an android application. The system checks the water quality and in

**Table 5** Pros and Cons of the discussed architectures

| System | Pros | Cons |
|---|---|---|
| 4.1 | • Simple design and implementation<br>• Clear and distinct monitoring and controlling system | • Design very specific to guppy fish<br>• Only monitors pH and salinity<br>• Can only control pH |
| 4.2 | • Machine learning model for following day feed consumption<br>• Mobile, web application and alarms to alert the farmers | • Oil layer prediction not needed for the freshwater fish farms<br>• Water level monitoring can also be excluded to reduce the cost |
| 4.3 | • Cost effective alternative method to cloud storage<br>• Motion detector to detect dead fish and fish movement pattern | • Lack of activation system<br>• Can remove some of the sensors and include some actuators |
| 4.4 | • Trained machine learning model for decision making<br>• Monitors bio floc level in water with TDS | • Comparison with other systems based solely on survey<br>• Accuracy of the model can be improved |

case of nay abnormal rise or fall in the sensor values, it predicts the best course of action. To validate the model various functionalities of the system has been tested and compared with the existing systems and that their system's accuracy 77.3% and the user satisfaction of the mobile application was found to be very high (Table 5).

# 5 Conclusion

Thus, through different studies conducted so far it can be observed that different fishes have different physical, chemical and biological requirements depending on their species and their location. This paper is a review of all the existing methods using which a system with the features that are relevant and effective can be designed based on the location and other factors of the fish farm. The main aim here is to make use of all the existing researches and systems designed so far to curate an IoT model that meet the requirements of the fishes that are reared and how different environmental and chemical factors can be monitored and manipulated in Sikkim. Interviews, surveys and case studies of the state's fish farmers will be carried out in the future to understand their needs and inputs.

Heavy Rainfall in Sikkim causes the water in the fish ponds to be extremely turbid and low in temperature. Thus, temperature and turbidity are important factors that need to be monitored and controlled for optimal fish production. Analite 9000, which is a high range, low response time and has lower current drain. Analite 9000 is best suited turbidity sensor for the water conditions in Sikkim. DS18B2 is the best and most reliable of all the water sensor and will work best with the State's environmental condition. Ultrasonic sensor helps us monitor the fish movement and behavior and detect any abnormality or presence of foreign matter in the fish pond. Furthermore,

Data collected from the ultrasonic sensor can be used to study the pattern of fish behavior/movement and for predictive analysis of the same. Though HC SR-07 is low width and frequency, it works good for a small fish farm. ATmega328p with 32 kb flash memory which has low power consumption and greater CPU speed than others compared is the most feasible choice of micro-controller.

## References


1. Ghosh A, Mohapatra BK, Roy AK (2017) Relationship between fish seed production and inland fish production of West Bengal, India. SM J Biom Biostat 2(2):1013
2. Contribution of Fisheries to Food and Nutrition Security in the SADC Region (n.d.) ReadkonG.com—create your own pages in a few simple steps and share with the friends easy and fast. https://www.readkong.com/page/contribution-of-fisheries-to-food-and-nutrition-security-in-4540159
3. Li D, Liu S (2018) Water quality monitoring and management: basis, technology and case studies. Academic Press
4. Chettri KB (2020) Ecology and economics of coldwater inland fisheries in Sikkim. Doctoral dissertation, Sikkim University
5. Chettri KB (2018) Present scenario of fisheries in the hills of Sikkim. Int J Creat Res Thoughts (IJCRT) 6(2):1784–19791
6. Animal Husbandry Livestock, Fisheries and Veterinary Services (n.d.) Animal husbandry livestock, fisheries and veterinary services
7. Ramya A, Rohini R, Ravi S (2019) Iot based smart monitoring system for fish farming. Int. J. Eng. Adv. Technol. 8(6):420–424
8. Khan M, Masud M, Akowuah B (2021) Development of IoT based fish monitoring system for aquaculture. Intell Autom Soft Comput 32. https://doi.org/10.32604/iasc.2022.021559
9. Tamim AT, et al (2022) Development of IoT based fish monitoring system for aquaculture. Intell Autom Soft Comput (Tech Science Press)
10. Prasad M, Majeed S, et al Cost effective IoT based automated fish farming system with flood prediction. In: IJATCSE 44913-I2020
11. Karim S, Hussain I, Hussain A, Hassan K, Iqbal S (2021) IoT based smart fish farming aquaculture monitoring system. Int J Emerg Technol 12(2):45–53
12. Saha S., Rajib RH, Kabir S (2018) IoT based automated fish farm aquaculture monitoring system. In: 2018 international conference on innovations in science, engineering and technology (ICISET). IEEE
13. Masum AKM, et al (2019) Design and implementation of IoT based ideal fish farm in the context of Bangladesh aquaculture system. In: 2019 1st international conference on advances in science, engineering and robotics technology (ICASERT). IEEE
14. Chatterjee R, Pandey AK (2020) IOT based smart water monitoring system for fish farming ponds. Int J Sci Eng Res. ISSN 2229-5518
15. Hsu W-C, et al (2020) Application of regression analysis to achieve a smart monitoring system for aquaculture. Information 11(8):387
16. Ya'acob N, et al (2021) Water quality monitoring system for fisheries using internet of things (IoT). IOP Conf Ser: Mater Sci Eng 1176(1). IOP Publishing
17. Spandana K, Seshagiri Rao VR (2018) Internet of things (Iot) based smart water quality monitoring system. Int J Eng Technol 7(3):259–262
18. Hapsari GI, Wakid Z, Mudopar S (2020) IoT-based guppy fish farming monitoring and controlling system. Telkomnika 18(3):1538–1545

19. Balakrishnan S, Sheeba Rani S, Ramya KC (2019) Design and development of IoT based smart aquaculture system in a cloud environment. Int J Ocean Ocean 13(1):121–127
20. Rashid MM, Nayan A, Simi SA, Saha J, Rahman MO, Kibria MG (2021) IoT based smart water quality prediction for Biofloc aquaculture. Int J Adv Comput Sci Appl 12(6)

# Optimal Frequency Control in a Microgrid Under Wind Power and Load Uncertainties

**Rachita R. Sarangi, Asit Mohanty, Prakash K. Ray, Ajit K. Barisal, and Suvendu M. Baral**

**Abstract** With the addition of electric vehicle (EV) to the power grid, the frequency stability of the system improves considerably. But it poses a problem in controlling the active power in a microgrid system since in the charging mode of the electric vehicle, it can act as a load and in charging mode, it acts as a source. This leads to a mismatch between the active power generation and load demand. Hence, in this paper, the impact of EV on the frequency control of microgrid is studied using PID controller. Further, the gains of PID controller are tuned using Particle Swarm Optimization (PSO) and Differential Evolution (DE) to suppress the uncertainties due to the uncertainties due to the wind power output and load demand. Again, the virtual inertia factor because of the converter-based distributed generation (DG) is also considered in the frequency control study. It is observed that DE-PID provides better frequency stability performance in comparison to that of the PSO-PID controller under the above operating scenarios.

**Keywords** Microgrid · Particle swarm optimization (PSO) · Differential evolution (DE) · Frequency control

## 1 Introduction

Electric vehicles (EVs) which are being incorporated into the present power system are growing at a totally fast pace. The additions of EVs together with the non-conventional have posed issues like power balance and frequency deviation. Due to the intermittent nature of the renewable electricity sources, the diesel generator and different electricity storage units like battery, flywheel, etc. assist in reliability of power supply. With the penetration of Electric Vehicles and Renewable Energy Sources manipulation of frequency deviation is a major challenge. This has been

R. R. Sarangi · A. Mohanty · P. K. Ray (✉) · A. K. Barisal · S. M. Baral
Department of Electrical Engineering, Odisha University of Technology and Research, Bhubaneswar, India
e-mail: pkray@outr.ac.in



S. Borah et al. (eds.), *Advanced Computational and Communication Paradigms*, Lecture Notes in Networks and Systems 535,
https://doi.org/10.1007/978-981-99-4284-8_44

a place of research in current times. Due to the bidirectional exchange of energy between the EV and the Grid a smart communication device between them can remedy many issues and additionally offer extra benefits. Microgrids offer an effective answer in far flung regions regarding the use of DGs for energy delivery [1]. Doolla and Bhatti defined load frequency control (LFC) for far flung small hydro plants in [2], and mentioned approximate load reduction with the assistance of on/off control valves. In [3], the authors defined a reconfigurable dual-mode device for small hydroelectric plant (AC) and Solar photovoltaic device (DC). which switches from grid connected to islanded mode of operation in case of any grid disturbance and reconnects to it whilst the source returns to normalcy. Kumari et al. [4] applied a non-minimal phase system, in hydro turbine control system, using a complimentary Sliding Mode Controller which is compared to the outcomes received from the usage of a PID controller To control the grid frequency, [5] has discussed about the virtual inertia aid capability of variable speed hydropower which improves the frequency and active power flow in HVDC links. Authors in [6, 7] considered distributed generation (DG) in microgrids with wind and solar structures for automated load frequency control using AGC (Automatic generation control) and AVR (Automatic voltage regulation).

In [8], three generating sources namely wind, thermal and hydro are considered and various kinds of tuned PID controllers are used to discover the satisfactory controller amongst them. In [9], the weighted frequency control is done with the usage of parameter adjustment in the Conventional Power System Stabilizer using improved Particle Swarm Optimization. It could be very critical to apply accurate controller together with desirable optimization strategies to get befitting advantage with the controller. Because the higher the controller, better could be frequency control for the source. The effect of electrical cars on LFC is the maximum vital region of research. The effect of electrical cars on multi-region structures changed into finished in [10–13]. This article introduces particle swarm optimization (PSO) and differential evolution (DE) strategies for optimizing PID controller parameters.

Concluding the thoughts, the main motivation for this paper is to

(a) Simulate a microgrid having diesel generator, Solar Power, Battery, Wind Generator along with the Electric Vehicle.
(b) Optimize the different gains associated with the PID controller using PSO and DE technique in presence and absence of EV.
(c) Obverse the response of the system with PID controller.

## 2 System Under Study

A microgrid is made up of many distributed sources like wind turbines, diesel generators, solar power sources or solar panels, and energy storage devices like flywheel energy storage and battery storage systems. DC power is generated by several renewable energy sources, necessitating the use of DC/AC converters. In a microgrid system, Diesel Engine Generator, Wind Turbine Generator, solar PV system, Electric Vehicles (EV) and Battery Energy Storage Systems (BESS) are utilized. The key focus of the Renewable Energy Sources (RES) is to meet the load demand. BESS is used as a backup source for the system's dynamic stability for a short period of time. The surplus energy produced by the RES is utilized to recharge the battery. Microgrid systems can also be scaled up as load demand grows. By varying the active power support, the output controllers that are connected to the Diesel Engine Generators and Battery Energy Storage Systems in each individual microgrid can be used to address frequency disturbances in both microgrid as well as variation of power in the tie-line, in order to achieve zero steady-state error in frequency response and flow of power in the tie line, as well as a faster steady-state response. The planned systems are highly stable because diesel generator will take over to balance the load requirement if erratic renewable energy sources are unable to supply the anticipated load demand. The microgrid structure and its linearized model are shown in Figs. 1 and 2 respectively. Also, the parameters used for different components in microgrid are presented in Table 1.

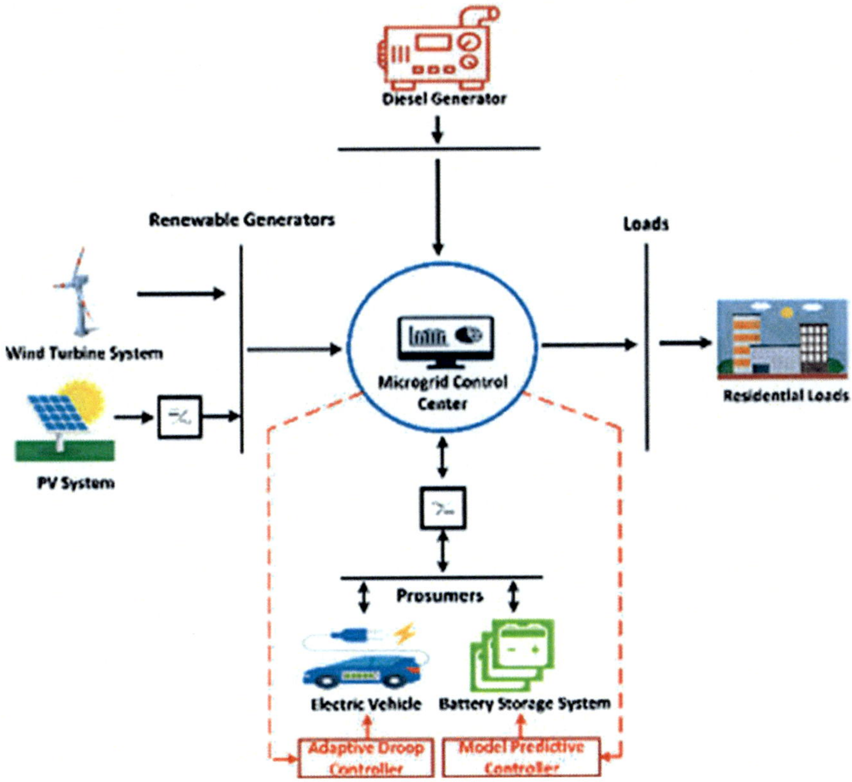


**Fig. 1** Block diagram of the microgrid configuration

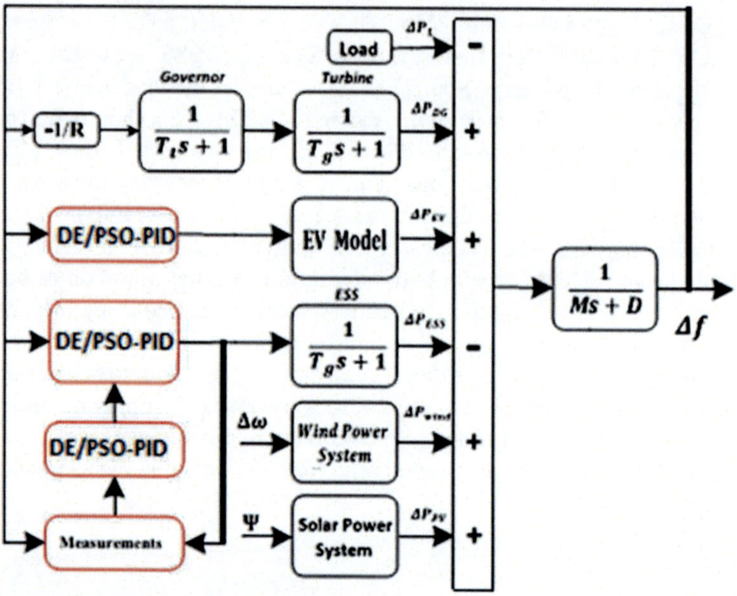


Fig. 2 Linearized model of the microgrid

**Table 1** System parameters

| Parameters | Value | Parameters | Value |
|---|---|---|---|
| DG rated power (MW) | 10 | WTS inertia constant | 5 |
| ESS rated power (MW) | 5 | WTS time constant | 1.5 |
| PV rated power (MW) | 3 | WTS $K$ parameter | 3.6 |
| WTS rated power (MW) | 5 | WTS $K$ parameter | 5.7 |
| Battery capacity of EVs (kWh) | 66.2 | EV gain $K_{EV}$ | 1 |
| Damping constant $D$ (pu/Hz) | 0.015 | EV battery time constant | 1 |
| Droop coefficient (Hz/pu) | 3 | $S^{min}$, $S^{max}$ | 0.2, 0.9 |
| DG generator time constant | 0.8 | DH | 0.1667 |
| DG generator time constant | 0.4 | $S_d$ | 0.6 |

The authors have considered an isolated microgrid in this article. To evaluate the dynamic behaviour of the system, a single PID controller is chosen. Particle Swarm Optimization (PSO) and Differential Evolution (DE) are used to calculate the different gains of the PID controller. The system is analysed in two different conditions: (a) with EV present and (b) without EV. The popularly known Integral Square Error (ISE) method is employed as cost function, $C$ as given in Eq. (1) with simulation time $T$ (s).

$$C = \int_0^T (\Delta f)^2 \mathrm{d}t \quad (1)$$

## 2.1 Diesel Engine Generator

Figure 3 gives the transfer function for DEG's first-order model. As a result of intermittency in solar and wind power, DEG's speed governor control balances the error between demand and generation in an islanded microgrid.

## 2.2 Wind Turbine Modelling

A wind turbine harnesses and converts wind energy's kinetic energy into mechanical energy. The mechanical energy is directly fed to the generator's rotor. Rotor rotation is converted into electrical energy in a wind turbine by the turbo generator shaft mechanism. Equation (2) represents the mechanical power obtained from the wind turbine.

$$P = \frac{1}{2}\rho A C_{\mathrm{p}}(\lambda, \beta) V_{\mathrm{w}}^3 \quad (2)$$

where

$\lambda$ is tip speed ratio, $\rho$ is air density factor (kg/cu m), $V_{\mathrm{w}}^3$ is wind speed (m/s), is power coefficient, $A$ is Swept area (m$^2$) and $\beta$ is blade pitch angle. The transfer function model of wind turbine is shown in Fig. 4.

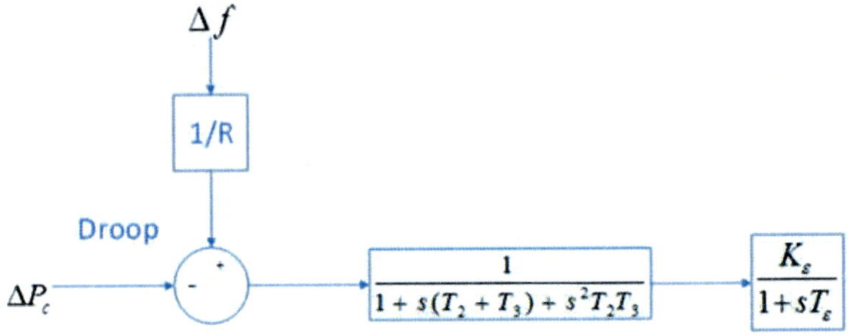


**Fig. 3** Transfer function modelling of diesel engine generator

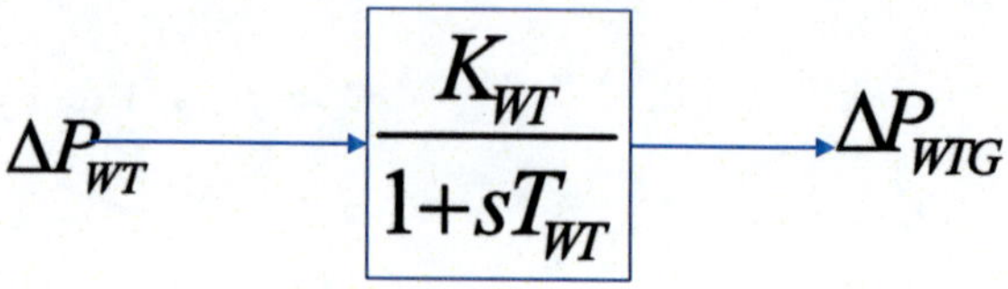


**Fig. 4** Transfer function model of a wind turbine

## 2.3 *Battery Energy Storage System (BESS)*

A microgrid has two types of storage systems. A battery energy storage system or flywheel energy storage systems are the two options. The output of renewable energy sources changes throughout the day. To achieve the load requirement, a storage system is required. When power generation falls short of load, the energy is released or absorbed by the storage system as needed. These energy storage systems must be small, efficient and devoid of emissions. Flywheel energy storage technology provides short-term storage and compensates for changes in wind output. The battery stores energy in the form of direct current (DC) for energy exchange, and a converter (DC/AC inverter) is required to convert DC to alternating current (AC). Traditional frequency regulating devices, such as DEG, are not suited for frequency regulation for dynamic load fluctuations due to the high inertia of spinning mass. As a result, a rapid and dynamic frequency regulating device, such as BESS, is necessary to account for abrupt load variations. BESS's first-order transfer function is as follows (Fig. 5).

## 2.4 *Solar Power*

Solar radiation is converted into electricity when it hits the surface of a solar cell. Solar panels based on semiconductors are routinely utilized to convert radiation energy into electrical energy. The solar cell's border resistance causes the voltage loss inside the cell. The transfer function model used for Solar PV power is (Fig. 6).

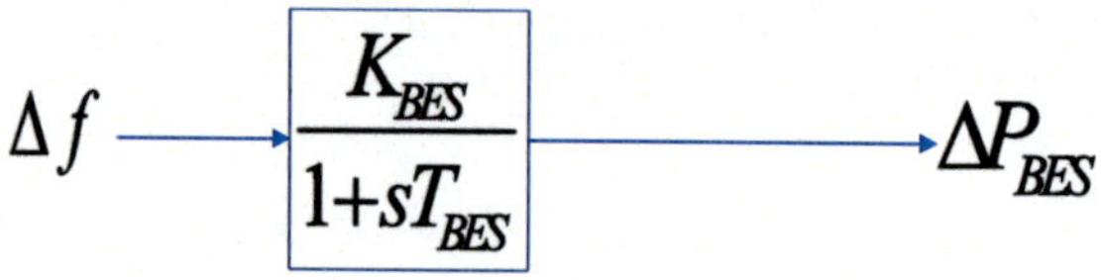


**Fig. 5** Model of the BESS first-order transfer function

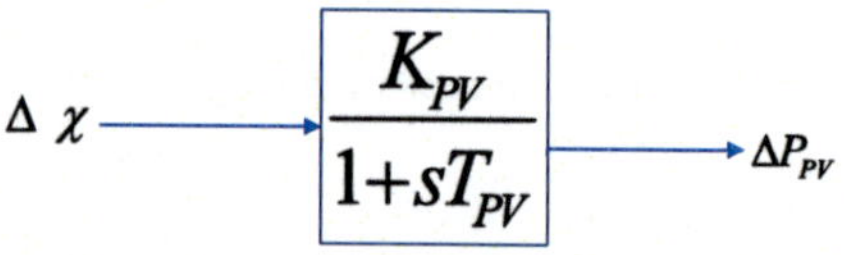


**Fig. 6** Model of solar power first-order transfer function

## 2.5 Model of Electric Vehicle

The EV is viewed as a form of energy storage (battery). Figure 7 depicts a comparable EV model for an LFC system. This model determines the charging or discharging rate in a controllable condition. Figure 7 depicts the electric vehicle (EV) model in charging mode, the frequency error signal ($\Delta f$) is an input to the EV controller when the vehicle is in charging mode and the resultant power required by the EV to get charged/discharged is obtained as an output. The total charge that can be stored in the battery is by ±Bkw, the EV charging time constant $T$, and the complex frequency term is $s$. $E$ is the battery energy at any instant of time $t$, and ($E_{\max}$) and ($E_{\min}$) are the upper and lower limits of $E$, that are 90% and 80%, respectively. The difference within the energies at any point of time $t$ and the $E_{\max}$ and $E_{\min}$ are denoted, respectively, as $K_1$ and $K_2$ and expressed as

$$K_1 = E - E_{\max}$$

$$K_2 = E - E_{\min}$$

The local control Centre, which acts as a channel for communication between the power grid and electric cars and further regulates the number of EVs, employs the plug-in EV energy storage model, depicted in Fig. 7, to measure the amount of energy stored in the batteries of the local control Centre.

At any point of time ($t$), the number of controllable EVs in the system can modelled mathematically as

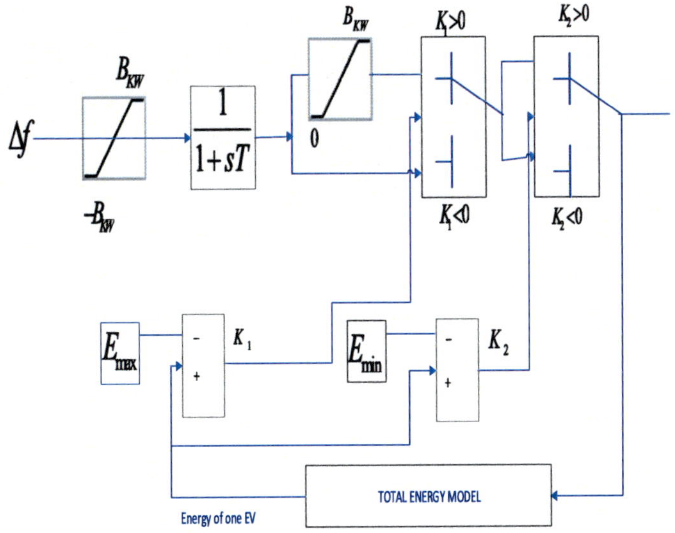


**Fig. 7** The modelling of the electric vehicle in MG load frequency control

$N_{\text{controllable}}(t)$ which can be calculated as shown in Eq. (3):

$$N_{\text{controllable}}(t) = N_{\text{initial}}(t) - N_{\text{plugout}}(t) + N_{\text{controlink}} \quad (3)$$

$N_{\text{initial}}(t)$—The number of controlled EVs at $t = 0$.

$N_{\text{plugout}}(t)$—EVs which are switching states from charging to driving mode.

$N_{\text{controlink}}(t)$—The number of EVs that have switched from charging to controllable mode over the time period $t$.

Energy equation can be given by Eq. (4) as

$$E_{\text{control}}(t) = E_{\text{initial}}(t) + E_{\text{controlink}}(t) - E_{\text{plugout}}(t) - \text{ELFC}(t) \quad (4)$$

$E_{\text{initial}}(t)$—energy at time $t = 0$.

$E_{\text{controlink}}(t)$—energy increase due to EVs.

$E_{\text{plugout}}(t)$—energy decrease due to plug out.

Energy at the connection is provided by the product of $N_{\text{controlink}}(t)$ and average charging energy as energy rises, and the (SOC) state of charge of the EV likewise changes from charging to controlled ($C$kWh).

$E_{\text{controlink}}(t)$ is then calculated by Eq. (5) as

$$E_{\text{controlink}}(t) = C\text{kWh} * N_{\text{controlink}}(t) = C\text{kWh} * N_{\text{controlink}}(t) \text{ [kWh]} \quad (5)$$

ELFC—energy equivalent of the LFC signal, can be calculated by integrating the PLFC (the power measurement at the local control centre), can be expressed through Eq. (6) as

$$\text{ELFC}(t) = \int \text{PLFC}t_0(t) \cdot \text{d}t \quad (6)$$

PLFC—power at local control centre.

The EV integration makes the overall system complex however it also delivers the benefit of frequency stabilization.

# 3 Optimization Techniques

## 3.1 *Particle Swarm Optimization*

Particle swarm optimization (PSO) is basically a bio-computational technique which optimizes a solution in the solution space. It does not require the optimization problem to be in a differential as in other classical methods, e.g. gradient descent or Quasi-Newton methods.

The ultimate aim of an effective optimization technique is to minimize or maximize a variable represented by the vector depending on the formulated problem $f(X)$. The vector $X$ is known as the position vector and $n$ gives the no. of variables which may be determined in the problem. $f(X)$ which is the optimization problem is also known as the objective function. Let us consider a swarm of $P$ particles which have a position vector and a velocity vector at $t$ iteration for each particle $i$ that composes it. These vectors are updated through a dimension $j$ according to the equations given below

$$V_{ij}^{t+1} = wV_{ij}^{t} + c_1 r_1^t \left(pbest_{ij} - X_{ij}^t\right) + c_2 r_2^t \left(gbest_j - X_{ij}^t\right)$$

and

$$X_{ij}^{t+1} = X_{ij}^{t} + V_{ij}^{t+1}$$

## 3.2 *Differential Evolution*

Differential Evolution algorithm is a metaheuristic method which makes minimal assumptions about the problem which is to be optimized and explores big spaces of solutions. It optimizes a problem through maintenance of a fixed population of solutions and creating new solutions through mutation and crossover in the existing ones and then keeps the solution which has the best fitness for the given optimization problem. This method can also be used $n$ problems which are non-continuous, noisy and change rapidly overtime. The three major steps involved are Initialization, Mutation and Crossover.

We can describe the objective function which needs to be minimized as

$$f\colon R^{\mathrm{D}} \rightarrow R$$

where D is the dimensionality of the problem to be optimized

$$x_{i,0}^{j} = x_{\min}^{j} + \mathrm{rand}(0, 1) \cdot \left(x_{\max}^{j} - x_{\min}^{j}\right)$$

whereas $x \in R^D$ gives the pool of the candidate solutions. The below algorithm describes the steps involved in this optimization process:

***Initialization***

The problem is initialized as described below

$$x_{i,0}^{j} = x_{\min}^{j} + \text{rand}(0, 1) \cdot \left(x_{\max}^{j} - x_{\min}^{j}\right)$$

where rand is a uniformly distributed random value varying within [0,1].

***Mutation***

A new mutant vector $V_{i,G}$ is generated by using one of the five methods as described below

$$V_{i,G} = X_{r_1^i,G} + F \cdot \left(X_{r_2^i,G} - X_{r_3^i,G}\right)$$

$$V_{i,G} = X_{best,G} + F \cdot \left(X_{r_1^i,G} - X_{r_2^i,G}\right)$$

$$V_{i,G} = X_{i,G} + F.(X_{r_{bestt},G} - X_{i,G}) + F.(X_{r_1^i,G} - X_{r_2^i,G})$$

$$V_{i,G} = X_{best,G} + F \cdot \left(X_{r_1^i,G} - X_{r_2^i,G}\right) + F \cdot \left(X_{r_3^i,G} - X_{r_4^i,G}\right)$$

$$V_{i,G} = X_{r_1^i,G} + F \cdot \left(X_{r_2^i,G} - X_{r_3^i,G}\right) + F \cdot \left(X_{r_4^i,G} - X_{r_5^i,G}\right)$$

where $r_1^i \ldots$ mutually exclusive integers varying from [1, NP] These indices are randomly generated once for each mutant vector. The scaling factor $F$ is a positive control parameter for scaling the difference vector.

**Cross Over**

After the mutation phase, crossover operation applied is applied to each $X_i$, $G$ and its corresponding mutant vector $V_{i,G}$ to generate $U_{i,G}^{j}$

$$U_{i,G}^{j} = \left\{V_{i,G}^{j}, \text{ if } \left(\text{rand}_j[0, 1) \le \text{CR}\right) \text{ or } (j = j_{\text{rand}})\right.$$

Otherwise

$$U_{i.G}^{j} = x_{i,G}^{j}$$

where CR denotes user defined cross over rate.

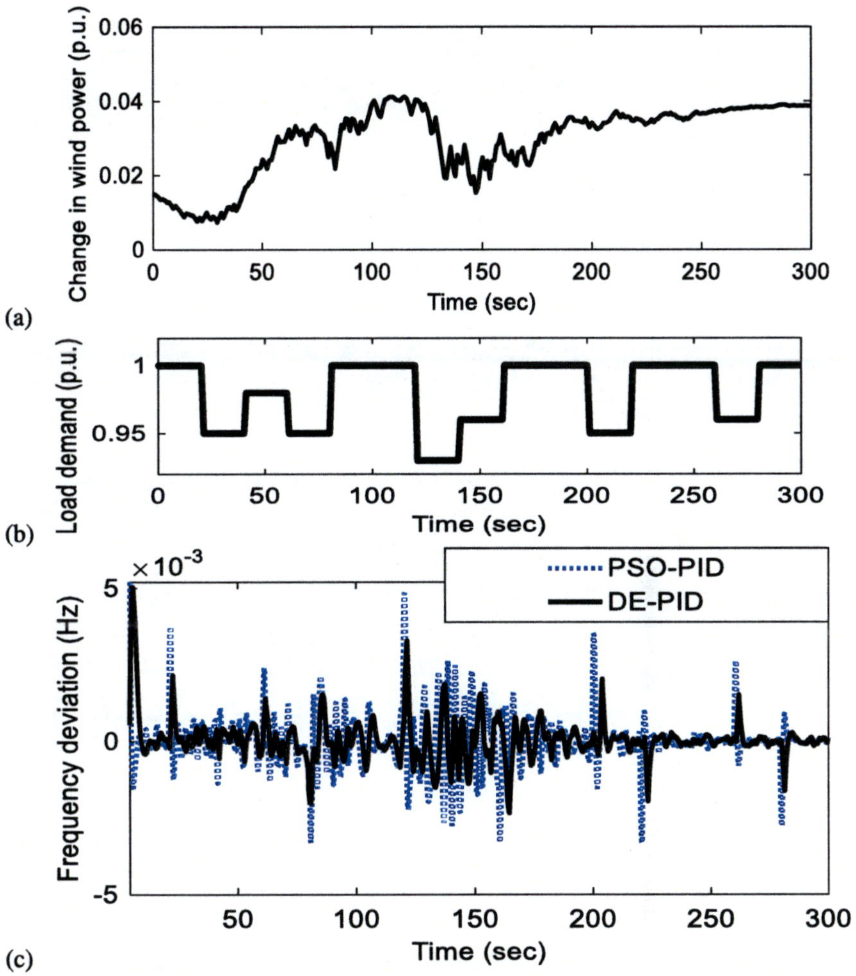


**Fig. 8** **a** Wind power **b** load demand and **c** frequency deviation for random change in wind power and continuous step change in load demand

## 4 Simulation Results and Analysis

### Case Study I: Random change in wind power and continuous step change in load demand

This case study presents the variation in the frequency deviation under the random change in wind power and continuous step change in load demand. The change in wind power and load are shown in Fig. 8a, b respectively. The frequency deviation under this operating condition is shown in Fig. 8c using PSO-PID and DE-PID controllers. It is observed that PSO-PID provides lesser frequency deviation as

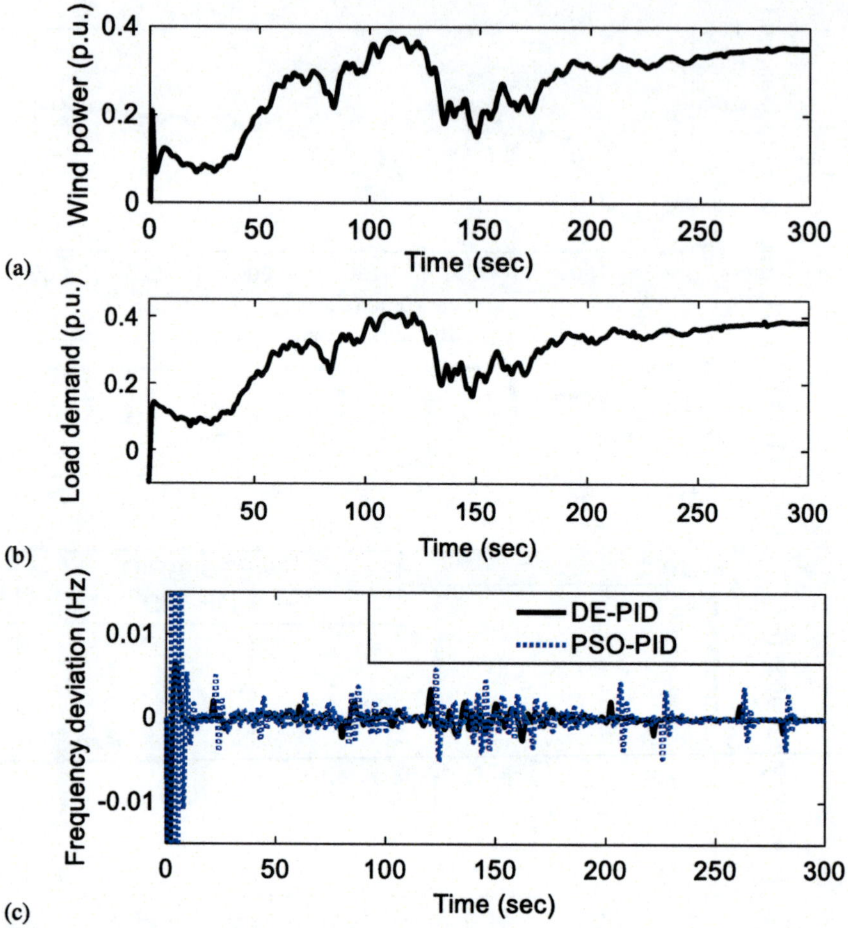


**Fig. 9** **a** Wind power **b** load demand and **c** frequency deviation for random change in wind power and load demand

compared to that of DE-PID controller. It is also observed that the frequency is controlled effectively even under the continuous change in load demand and nonlinear variation in the wind power output.

**Case Study II: Random change in wind power as well as load demand**

This case study presents the variation in the frequency deviation under the random change in wind power and load demand. The changes in wind power and load are shown in Fig. 9a, b respectively. The frequency deviation under this operating condition is shown in Fig. 9c using PSO-PID and DE-PID controllers. It is observed that the frequency is controlled effectively even under the random change in load demand and nonlinear variation in the wind power output. It is noticed that PSO-PID provides

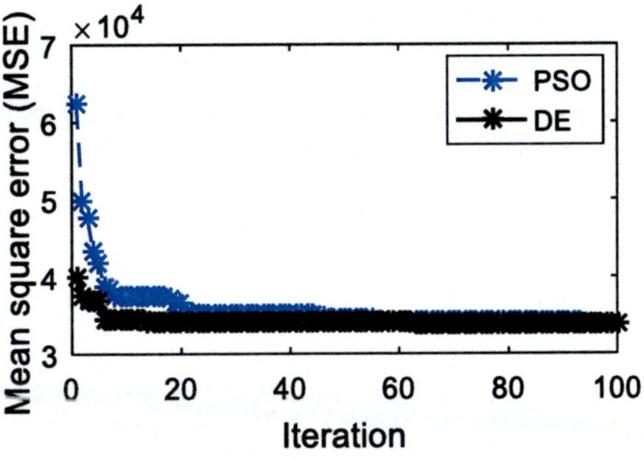


**Fig. 10** Convergence plot

minimum frequency deviation as compared to that of DE-PID controller under this operating condition.

The convergence curve representing the mean square error (MSE) versus number of iteration is shown in Fig. 10. It is being noticed that DE provides minimum MSE and hence better convergence characteristic in comparison to that of the PSO. This is the reason why the frequency control using DE is analysed to be better than that of the PSO under different operating scenarios as shown in the case studies.

## 5 Conclusions

This paper has presented the study of frequency control in microgrid with presence and absence of electric vehicle and virtual inertia under different operating conditions. The effect of presence of EV and virtual inertia on the frequency control of microgrid is analysed using PSO optimized PID and DE optimized PID controllers. The non-linearities and uncertainties are being suppressed using the above two optimal controllers. It is observed that the frequency deviation becomes comparatively less when using DE optimized PID than that of the PSO optimized PID and conventional PID controllers under the uncertain variations in the wind power and the load demand, i.e. under the above operating scenarios.

## References

1. Farrokhabadi M, Cañizares CA, Bhattacharya K (2017) Frequency control in isolated/islanded microgrids through voltage regulation. IEEE Trans Smart Grid 8(3):1185–1194. https://doi.org/10.1109/TSG.2015.2479576
2. Doolla S, Bhatti TS (2006) Load frequency control of an isolated small-hydro power plant with reduced dump load. IEEE Trans Power Syst 21(4):1912–1919. https://doi.org/10.1109/TPWRS.2006.881157

3. Reddy KM, Singh B (2018) Multi-objective control algorithm for small hydro and SPV generation-based dual mode reconfigurable system. IEEE Trans Smart Grid 9(5):4942–4952. https://doi.org/10.1109/TSG.2017.2676013
4. Kumari R, Prabhakaran KK, Desingu K, Chelliah TR, Sarma SVA (2021) Improved hydro turbine control and future prospects of variable speed hydropower plant. IEEE Trans Ind Appl 57(1):941–952. https://doi.org/10.1109/TIA.2020.3028798
5. Reigstad TI, Uhlen K (2021) Variable speed hydro power for provision of fast frequency reserves in the Nordic grid. IEEE Trans Power Syst. https://doi.org/10.1109/TPWRS.2021.3078098
6. Pan I, Das S (2016) Fractional order AGC for distributed energy resources using robust optimization. IEEE Trans Smart Grid 7(5):2175–2186. https://doi.org/10.1109/TSG.2015.2459766
7. Ray PK, Puhan PS, Priyadarshini P (2021) Automatic load frequency control of distributed generations in a microgrid. In: 2020 3rd international conference on energy, power and environment: towards clean energy technologies, pp 1–6. https://doi.org/10.1109/ICEPE50861.2021.9404423
8. Shaker HK, Zoghby HE, Bahgat ME, Abdel-Ghany AM (2019) Advanced control techniques for an interconnected multi area power system for load frequency control. In: 2019 21st international Middle East power systems conference (MEPCON), pp 710–715. https://doi.org/10.1109/MEPCON47431.2019.9008158
9. Morsali J, Esmaeili Z (2020) Proposing a new hybrid model for LFC and AVR loops to improve effectively frequency stability using coordinative CPSS. In: 2020 28th Iranian conference on electrical engineering (ICEE), pp 1–7. https://doi.org/10.1109/ICEE50131.2020.9260695
10. Naveed A, Sönmez Ş, Ayasun S (2021) Impact of electric vehicle aggregator with communication time delay on stability regions and stability delay margins in load frequency control system. J Mod Power Syst Clean Energy 9(3):595–601. https://doi.org/10.35833/MPCE.2019.000244
11. Khooban M (2018) Secondary load frequency control of time-delay stand-alone microgrids with electric vehicles. IEEE Trans Ind Electron 65(9):7416–7422. https://doi.org/10.1109/TIE.2017.2784385
12. Nguyen HNT, Zhang C, Zhang J (2016) Dynamic demand control of electric vehicles to support power grid with high penetration level of renewable energy. IEEE Trans Transp Electrification 2(1):66–75. https://doi.org/10.1109/TTE.2016.2519821
13. Debbarma S, Dutta A (2017) Utilizing electric vehicles for LFC in restructured power systems using fractional order controller. IEEE Trans Smart Grid 8(6):2554–2564. https://doi.org/10.1109/TSG.2016.2527821

# Author Index


**A**
Abhijit Adhikari, 337
Abhishek Bagepalli, 121
Ahamed Shafeeq, B. M., 3
Ajit K. Barisal, 531
Akshata Kumble, 147
Amish Kumar, 251
Amod Kumar Tiwari, 259
Aniket Inamdar, 237
Anirban Sarkar, 177
Anunay Maheshwari, 237
Anup Basumatary, 323
Anuradha Misra, 259, 425
Anu Sasidharan, 281
Aravapalli Rama Satish, 301
Arushi Tiwari, 281
Ashwini Bhatt, 147
Asit Mohanty, 531
Atif Mahmood, 259

**B**
Baisakhi Chakraborty, 133, 187
Bakul Gohel, 449, 457
Balajee Maram, 301
Bhabesh Nath, 113
Bhagavan, V. S., 337
Bhaswati Baishya, 113
Bijoyeta Roy, 415
Bipal Khanal, 99
Bishal Gupta, 363
Biswaraj Sen, 157
Boby Clinton Urikhimbam, 43
Butta Singh, 471

**C**
Chandan Kumar, 85, 99
Chandrima Chakraborty, 133

**D**
D'Angelo, Egidio, 281
Debashis Nandi, 201, 251
Devarani Devi Ningombam, 75
Devi Krishna, 281
Dhruba Ningombam, 515
Dilsheen Kaur, 425
Dipendra Gurung, 55
Divya M. Sasidharan, 281
D'Souza, Noel, 309

**G**
Gaidinlung Kamei, 15
Geetamoni Konwar, 481, 505
Giovanni Naldi, 281

**H**
Harshit Choudhary, 237
Hemalatha Sasidharakurup, 281
Himali Sarangal, 471
Himanshu Kumar Diwedi, 259

**I**
Indrajit Banerjee, 15

**J**
Jahid Ali, 491




S. Borah et al. (eds.), *Advanced Computational and Communication Paradigms*, Lecture Notes in Networks and Systems 535,
https://doi.org/10.1007/978-981-99-4284-8

**K**
Kailash Pati Mandal, 133, 187
Kalpana Sharma, 157, 363
Kamlesh Dutta, 225
Kanojia Sindhuben Babulal, 403
Karunakar A. Kotegar, 211
Kashmiri Deka, 505
Kasi Viswanadh, 281
Kazi Amrin Kabir, 381
Krishna Veer Singh, 449

**L**
Linthoingambi Takhellambam, 43

**M**
Mahesh Chandra Govil, 85
Mandeep Kumar, 491
Manish Godse, 187
Manjit Singh, 471
Margaret Kathing, 273
Md. Sarfaraj Alam Ansari, 85, 99
Moutusi Mondal, 251

**N**
Nandan Banerji, 133, 187
Narayan Murmu, 201
Nazrul Hoque, 43
Nisarg Doshi, 449
Nitya Jitani, 323

**P**
Palash Ghosal, 201, 251
Panigrahi, B. K., 291
Parag Kumar Guha Thakurta, 381
Parag Ravikant Kaveri, 439
Prakash K. Ray, 531
Prasenjit Mukherjee, 133, 187
Prathamesh Vijay Lahande, 439
Pratick Gupta, 415
Pratima Sarkar, 415
Preeti Kale, 237
Pritika Adhikari, 515
Priyanka Chakraborty, 177
Priyanka Kumar, 403

**R**
Rachita R. Sarangi, 531
Rahul Kumar Gupta, 225
Rajesh Gopakumar, 211
Ranjan Yambem, 43
Razia Loyola Furtado e Sardinha de, 351
Rebecca Shruti Soren,, 55
Rishang Kumar Brahma,, 273
Robindro Singh Khumukcham, 43
Rohan Raj Maram, 301
Rohit Agarwal, 201
Rosy Sarmah, 323

**S**
Sakshi Ahuja, 291
Sandeep Gurung, 55
Sanjai Mohan Verma, 31
Sanjay Singh, 121
Santosh Kumar, 31
Sarat Saharia, 273
Sasibhushana Rao Pappu, 301
Satveer Kour, 471
Shafiul Alom Ahmed, 113
Shivam Kumar, 415
Shyam Diwakar, 281
Siddalingaswamy, P. C., 309
Sital Sharma, 157, 363
Sneha Singh, 505
Soham Biswas, 85
Sriveda Medatati, 147
Subba Reddy, N. V., 309
Suman Som, 85
Sumathi, 337
Suvendu M. Baral, 531
Swapnil Kharade, 237

**T**
Tannistha Pal, 391
Tapan K. Gandhi, 291
Tawal Koirala, 363
Tulshi Bezboruah, 481

**V**
Varanasi L. V. S. K. B. Kasyap, 337
Vinit Grewal, 471
Vishal Anand, M., 211
Vivek Noel Soren, 291
Vyas, O. P., 425

**Z**
Zahid Ahmed Ansari, 3